I0131945

H. Bluhm, T. Brückel, M. Morgenstern, G. von Plessen, C. Stampfer

Electrons in Solids

Graduate Texts in Condensed Matter

Series Editor
Prof. Dr. Christian Enss
Heidelberg University
Kirchhoff-Institute for Physics
Im Neuenheimer Feld 227
69120 Heidelberg
Germany

Hendrik Bluhm, Thomas Brückel,
Markus Morgenstern, Gero von Plessen,
Christoph Stampfer

Electrons in Solids

Mesoscopics, Photonics, Quantum Computing,
Correlations, Topology

DE GRUYTER

Authors

Prof. Dr. Hendrik Bluhm
RWTH Aachen University
Institute of Physics (2C)
Otto-Blumenthal-Str. 28
52056 Aachen, Germany
bluhm@physik.rwth-aachen.de

Prof. Dr. Thomas Brückel
Forschungszentrum Jülich GmbH/
RWTH Aachen University
Faculty of Mathematics, Computer
Science and Natural Sciences
JCNS-2 & PGI-4
52425 Jülich, Germany
T.Brueckel@fz-juelich.de

Prof. Dr. Markus Morgenstern
RWTH Aachen University
Institute of Physics (2B)
Otto-Blumenthal-Str. 28
52056 Aachen, Germany
mmorgens@physik.rwth-aachen.de

Prof. Dr. Gero von Plessen
RWTH Aachen University
Institute of Physics (1A)
Otto-Blumenthal-Str. 28
52056 Aachen, Germany
gero.vonplessen@physik.rwth-aachen.de

Prof. Dr. Christoph Stampfer
RWTH Aachen University
Institute of Physics (2A)
Otto-Blumenthal-Str. 28
52056 Aachen, Germany
stampfer@physik.rwth-aachen.de

ISBN 978-3-11-043831-4
e-ISBN (PDF) 978-3-11-043832-1
e-ISBN (EPUB) 978-3-11-042929-9

Library of Congress Control Number: 2018955522

Bibliographic information published by the Deutsche Nationalbibliothek
The Deutsche Nationalbibliothek lists this publication in the Deutsche Nationalbibliografie; detailed bibliographic data are available on the Internet at http://dnb.dnb.de.

Cover image: Marco Pratzer
Typesetting: le-tex publishing services GmbH, Leipzig
Printing and binding: CPI books GmbH, Leck

www.degruyter.com

Preface

Cutting edge research in modern solid state physics is characterized on the one hand by highly complex systems with various interactions between quasi-particles. In this context, also a novel characterization scheme, that of topology, has entered solid state physics and allows new insights into the description of quantum materials. On the other hand, there is a strong tendency to trace back solid state phenomena and systems to their quantum physical basis, where phase coherence, wave function interference, superposition and entanglement play an essential role. This is particularly evident in research on quantum transport in nanostructures or in quantum engineering, where solid state nanosystems are studied with respect to quantum information or sensor applications. These topics usually go beyond the content of standard courses and textbooks in solid state physics. My physics colleagues at the RWTH Aachen University, therefore, have established an advanced course on top of the basic solid state physics lectures, where master and PhD students are exposed to these important fields of modern research. The present textbook based on this advanced course provides some topics which have exemplary character and are paradigmatic for this kind of latest current research in solid state physics. From a deeper understanding of these examples other fields of modern research are also easily accessible.

I therefore recommend the book to all those who want to dive deeper into this fascinating field of modern solid state physics, to master and PhD students, but also to scientists starting a new research field and who want to learn some essentials in a reasonably short time. I wish the book much success since it promotes our beautiful science.

Hans Lüth
Jülich Aachen Research Alliance

https://doi.org/10.1515/9783110438321-201

Overview

This textbook is the result of a compulsory, advanced lecture course in experimental solid state physics given within the masters course at RWTH Aachen University. It goes beyond the classical topics of ferromagnetism and superconductivity. Thereby, it bridges the corresponding basic knowledge in solid state physics to more recent research topics such as topology or quantum computation. In particular, it aims to guide students through the different levels of complexity inherently given by the partially classical, partially semiclassical and partially quantum mechanical description of solid state properties, where the latter often employs simplified Hamiltonians as so-called toy models. Switching between such different degrees of complexity is inherent to solid state physics, since a full quantum mechanical solution of the many-particle problem is not available. Thus, one has to choose adequate approximations, strongly guided by experimental results.

The concept of the book is to provide the conceptional skills needed to select an adequate model and, thus, an adequate level of complexity. At the same time, we aim to foster an intuition for the key properties of different types of electron systems under different conditions. Since we address experimentalists, we often build on intuitive, heuristic arguments instead of rigorous calculations and, moreover, always combine our arguments with experimental results. This way of reasoning provides a key insight into the required interplay between experiments and theory, which allows one to correctly understand the favorable properties of the unsolvable, basic Schrödinger or Dirac equations.

The selection of topics is not meant to be exhaustive in terms of modern solid state physics, but provides a few deliberately chosen examples, which represent a line of increasing complexity. We start with mesoscopics, where quantum mechanics enters mainly by the phase of the electronic wave functions rendering the phase coherence the decisive term. We continue with the description of optical properties of solid state electrons, where pure classical electrodynamics is partly sufficient to describe the phenomena. However, the description of, e.g., transition rates already requires the use of superpositions of stationary electronic states in a coherent quantum mechanical fashion. Chapter 3 of the book deals with solid-state based quantum computing, arguably the most important emerging application of the quantum mechanical dynamics of solid state electrons. Here, the coherent quantum mechanical description of the exploited degrees of freedom is key, which will eventually be used for a more efficient information processing via superposition states, i.e., multiple stationary states in parallel. The high degree of control of the quantum mechanical dynamics also enables an experimental insight into the transition between the quantum mechanical behavior on the microscopic scale and the phase-free classical behavior mostly guiding our macroscopic scale experience.

https://doi.org/10.1515/9783110438321-202

In Chapter 4 of the book, we tackle the notoriously difficult electron-electron interaction quantum mechanically using the well-studied example of transition metal oxides. We show how simplified toy models of the Schrödinger equation capture essential properties as, e.g., the bandgap opening, which appears for a nominally half filled single-particle electron band due to the electron-electron interaction (Mott–Hubbard transition). The toy models start with a tight binding description, which reduces the single-particle electron wave functions to single atoms, before introducing the complex additional terms. Hence, we deal with point-like starting electrons, such that the phase of the wave functions has a minor influence on the final properties of the system. This approach changes in the last chapter of the book, where we describe itinerant electrons being prone to different types of interactions, namely the electron-phonon, the electron-disorder and the electron-electron interaction. We begin with a diverging perturbative example, the Peierls transition. Afterwards, we will guide students through the complex description of band structures via topology. Finally, we will present emergent phenomena of the many particle character driven by the electron-electron interaction, such as the fractional charge of electrons appearing in fractional quantum Hall phases. The latter description stresses a key method in solid state physics, namely the description of the system by quasi particles with a very abstract meaning.

We believe that this selection of topics sets the students into a position to deal with the different levels of complexity, which are typical for current solid state research. It is, moreover, meant to provide a base to tackle other modern research topics such as, e.g., unconventional superconductivity, magnetoelectronics or advanced photonics employing metamaterials. Hence, we present a dedicated upgrade course to the well-established curriculum in solid state physics. We hope that this novel approach relying on a partly heuristic argumentation will be successful at other universities too. Being aware of the fact that a different selection of topics might be suited for the task as well, we are interested in any type of feedback.

We wish to thank Fabian Hassler, Volker Meden, and Marcus Liebmann for insightful discussions, Hans Lüth for very helpful comments after a complete cross-reading of the manuscript, Tjorven Johnsen for cross-reading of some parts of the manuscript, Gereon Graff and Federica Haupt for taking care of a multitude of formalities, Marco Pratzer for designing and making the cover image, as well as Thomas Graff, Tobias Cronert, Paul Doege, and Jonathan Conrad for help with the optimization of all other images.

Markus Morgenstern (on behalf of all the authors)
RWTH Aachen University and JARA-FIT

Contents

Christoph Stampfer

1 Electrons on mesoscopic length scales: the role of the electron phase

1.1 Introduction

The first chapter of this book deals with mesoscopic physics,[1] a field which describes the properties of electrons at length scales, where the phase of the electron wave function is crucial for the observed properties.

Generally, electrons in metals and semiconductors experience an irregular lattice potential. It arises from defects, lattice imperfections, grain boundaries, vacancies, doped impurities as well as from thermally induced lattice vibrations (phonons). An important length scale for the transport properties of such disordered systems is the temperature dependent phase coherence length l_φ, the scale over which the resulting electron dynamics is phase coherent. Exceeding this length scale, the transport properties can be described within the framework of quasi classical techniques such as the kinetic or Boltzmann theory (Drude model). However, at length scales smaller than l_φ, the quantum degrees of freedom influence the dynamics. Here, at the so-called mesoscopic scale, other physical principles become relevant.

- **What is mesoscopic physics?** As a subject, mesoscopic physics involves the domain of length scales in between the atomistic and the macroscopic one. Here, the influence of quantum phase coherence finds manifestations in the observed physical properties. In the macroscopic world, one often considers the so-called thermodynamic limit with $n = N/V =$ const. and $N, V \to \infty$, where N is the number of particles and V is the volume. Usually, an electron system approaches the macroscopic limit, once its size $L \sim V^{1/3}$ is much larger than the characteristic correlation length ξ_c of the charge distribution. In most cases, ξ_c is on the order of an atomistic length such as the interatomic distance $n^{-1/3}$, which is in the sub-nm range. However, for conducting systems at low temperature, it turns out that the length scale dividing microscopic from macroscopic behavior is more crucially l_φ, which can be as large as a few µm. This scale establishes the field of mesoscopic physics. Studies in the mesoscopic size range can be motivated by an interest in understands how the macroscopic limit of solids develops from the atomistic limit. But more importantly, many novel phenomena exist due to the phase coherence within such mesoscopic systems.

1 The word mesoscopics was coined by van Kampen in 1981.

https://doi.org/10.1515/9783110438321-001

- **Where can one find mesoscopic structures?** The development of microelectronics and modern fabrication techniques combined with the routine availability of millikelvin temperatures in laboratories have allowed the study of novel classes of devices with dimensions smaller than the typical scales over which phase coherence is established. The market that drives the semiconductor industry has established nanostructure technology at a level of unprecedented sophistication. These days, artificial semiconductor devices are manufactured with a size well below 100 nm. In addition, externally bottom-up fabricated nanostructures, such as carbon nanotubes with diameters down to 1 nm, can be incorporated into such devices, acting, e.g., as small conducting bridges.
- **Why study mesoscopic physics?** Leaving aside the technological benefits of the continual miniaturization, the thereby enabled quantum devices present a unique opportunity to study new physical phenomena. Where phase coherence is established, the manifestation of quantum mechanics is often substantial. The fabrication of mesoscopic structures such as, e.g., quantum dots, provides a laboratory, in which one can explore the fundamental properties of many-particle systems down to the few electron regime, while tuning the influence of disorder or of strong electron-electron interactions (Section 5.5.1.2) deliberately. This unprecedented control provides many new insights into the fundamentals of quantum mechanics.

1.2 Basics of electron transport

In this section, we review some of the basic transport properties of metals and semiconductors. We begin with a discussion of Ohm's law and its local version allowing the introduction of a number of important physical quantities. This is followed by a discussion of the Drude model with particular focus on its limitations. We then start, focusing on quasi-two-dimensional systems, to discuss the connection between electron drift and diffusion currents. Finally, we discuss the important characteristic length scales and summarize the different transport regimes providing us a guide for the following Sections 1.3–1.5.

1.2.1 Classical concepts of electronic transport

We will depart from the well-known Ohm's[2] law, which is probably the most important relation of the early quantitative description of the physics of electricity. Ohm's law,

2 Georg Simon Ohm, 1789–1854, German physicist. Ohm's law, which actually had been discovered by Henry Cavendish (1731–1810) in 1781, was first published in 1827 in Ohm's famous book *'Die galvanische Kette, mathematisch bearbeitet'*.

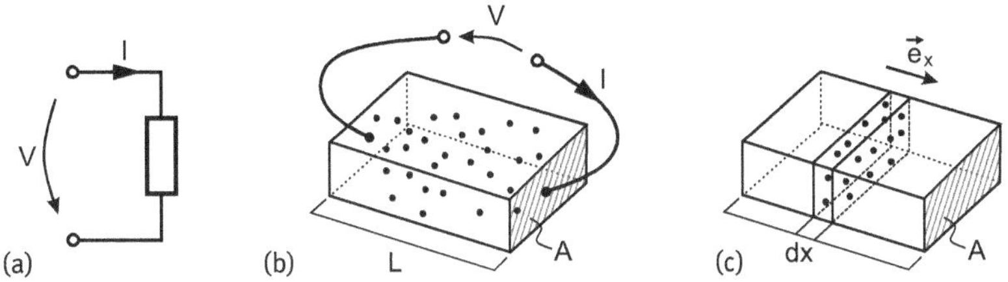

Fig. 1.1: (a) Circuit diagram of an electrical resistor and the orientation of voltage V and current I. (b, c) Electrical current through a metal piece with length L and cross-sectional area A, visualizing the terms introduced in the text.

$V = RI$, describes the relationship between voltage V, current I and resistance R of a macroscopic sample (Fig. 1.1(a), (b)). For a homogeneous (cuboid-shaped) solid we can express the resistance R by the resistivity ρ: $R = \rho L/A$, where L is the length of the resistor and A its cross-sectional area. For many of our discussions, it will turn out that the conductance G, which is the inverse of the resistance, $G = R^{-1}$, as well as the conductivity σ, which is the inverse of the resistivity, $\sigma = \rho^{-1}$, are more convenient physical quantities. For example by making use of the conductance, Ohm's law can be rewritten as $I = GV$. By expressing the conductance by the conductivity as $G = \sigma A/L$, this directly leads to $I/A = \sigma\ V/L$. By making use of the current density $|\vec{j}| = I/A$ and the electric field strength $|\vec{E}| = V/L$ and including their directions, which in our example for both quantities is $\vec{e}_x$ (Fig. 1.1(c)), we obtain the so-called local Ohm's law:

$$\vec{j} = \sigma\vec{E}\ . \tag{1.1}$$

Next, we revisit a microscopic picture for describing the current density. Remembering that we can express the current through an (oriented) area A by the total number of charge Q per time t passing this area, this leads to $I = \mathrm{d}Q/\mathrm{d}t$. By considering a small slab of our conductor with thickness $\mathrm{d}x$ (Fig. 1.1(c)), we can express $\mathrm{d}Q$ by $\mathrm{d}Q = -en_{e,x}A\mathrm{d}x$, where $e > 0$ is the elementary charge and $n_{e,x}$ is the density of the carriers (i.e., the number of carriers $N_{e,x}$ divided by the volume) moving in x-direction. This leads, by introducing the electron velocity in x-direction $\vec{v}_x = (\mathrm{d}x/\mathrm{d}t)\vec{e}_x$, to the following expression for the current density,

$$\vec{j} = -en_{e,x}\vec{v}_x = -en_\mathrm{e}\vec{v}_\mathrm{d}\ . \tag{1.2}$$

In the last step, we introduced the so-called drift velocity[3], which we defined such that $\vec{v}_\mathrm{d} = n_x\vec{v}_x/n_\mathrm{e}$, where n_e is the total carrier (electron) density, i.e., the total number of carriers N_e divided by the volume. For making the connection to the local Ohm's law (eq. (1.1)), we define the carrier mobility μ as the ratio of the drift velocity $\vec{v}_\mathrm{d}$ to the

3 The reason why we call this drift velocity can only be understood after the next two sections.

applied electric field $\vec{E}$:

$$\mu \equiv \frac{|\vec{v}_\mathrm{d}|}{|\vec{E}|} . \tag{1.3}$$

This finally allows the expression of the conductivity as

$$\sigma = en_\mathrm{e}\mu . \tag{1.4}$$

In the next section, we will review the Drude model, which provides a microscopic understanding of the drift velocity $\vec{v}_\mathrm{d}$ and thus of the carrier mobility μ.

1.2.2 The Drude model

A remarkably successful theory of metallic conduction was introduced by Drude[4] in 1900. Although the Drude model has some severe shortcomings, the successes are worth considering. It is still used today as a quick and practical way to form simple pictures and to obtain rough estimates of properties whose more precise comprehension requires analysis of considerable complexity.

It all started with the discovery of the electron in 1897 by J. J. Thomson[5], which had a vast and immediate impact on theories of the structure of matter, and also suggested a mechanism for describing electron motion in metals. Drude constructed his model of electrical and thermal conduction by applying the highly successful kinetic theory of gases, i.e., the Boltzmann[6] theory, to a metal, which he considered as the host of an ideal gas of electrons. The compensating positive charge to keep the solid charge neutral was assumed to be attached to much heavier particles, the ion cores, which Drude considered to be immobile. Importantly, the Drude model treats the dense electron gas with the methods given by the kinetic theory for a neutral dilute gas, with only slight modifications.

In the framework of the pre-quantum mechanical Drude model, the kinetic energy of the electrons is obtained from the theorem of equipartition of energy resulting, for three degrees of freedom, in

$$\frac{1}{2}m_\mathrm{e}v_\mathrm{th}^2 = \frac{3}{2}k_\mathrm{B}T .$$

Here, k_B is the Boltzmann constant, m_e is the the mass of the electron, and v_th is the absolute value of the average thermal velocity at temperature T. In the frame of this model, the electrons in a metal are therefore moving rapidly in all directions with

4 Paul Drude, 1863–1906, German physicist. It is worth having a look at his original publications: *Zur Elektronentheorie der Metalle*, Annalen der Physik 1, p. 566 and 3, p. 369, 1900.

5 Joseph J. Thomson, 1856–1940, British physicist.

6 Ludwig E. Boltzmann, 1844–1906 was an Austrian physicist and philosopher. Sadly both, Boltzmann and Drude, committed – in unrelated tragic circumstances – suicide in 1906.

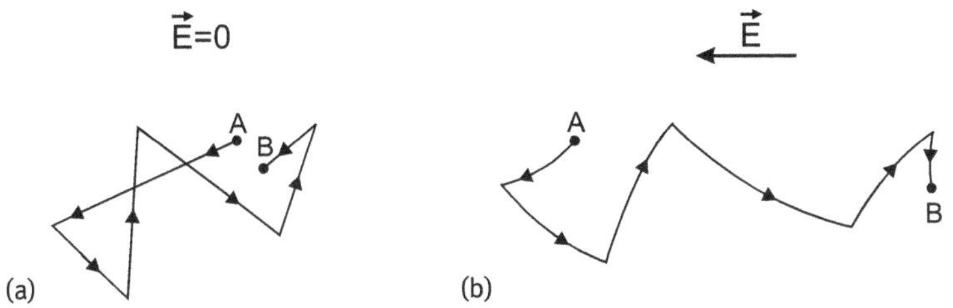

Fig. 1.2: Schematic illustration of the random scattering path of an electron in a conductor. (a) Random thermal motion at electric field $\vec{E} = \vec{0}\,\mathrm{V/m}$. (b) Random thermal motion combined with the motion due to an externally applied electric field $\vec{E} \neq \vec{0}\,\mathrm{V/m}$.

speed v_{th} (e.g., $v_{\mathrm{th}} \approx 10^5\,\mathrm{m/s}$ at $T = 300\,\mathrm{K}$). The thermal motion of an electron may be thought of as piecewise straight and interrupted by a succession of random scattering from collisions with the ion cores, impurity atoms and other scattering centers, as shown in Fig. 1.2 (a).[7] Importantly, in the absence of an electric field, the random motion leads to a zero net displacement of an electron over a sufficiently long period of time. Thus the drift velocity averages to zero, $\vec{v}_{\mathrm{d}} = \vec{0}$, and there is no net electric current density (eq. (1.2)).

In the presence of an electric field $\vec{E}$, however, there will be a mean electron velocity. It is directed opposite to the electric field, since the electron charge is negative (Fig. 1.2 (b)). The resulting mean velocity $\vec{v}$ can be computed by considering the total force $\vec{F}$ acting on the electron:

$$\vec{F} = -e\vec{E} = \dot{\vec{p}} = m_{\mathrm{e}}\dot{\vec{v}}\,,$$

$$\Rightarrow \quad \vec{v} = \vec{v}_0 - \frac{e\vec{E}(t - t_0)}{m_{\mathrm{e}}}\,.$$

Here, $\vec{p}$ is the momentum of the electron, $\vec{v}_0$ is its initial velocity at time t_0 and t is the time, during which $\vec{F}$ acts on the electron. Crucially, the later expression implies that $\vec{v}$ would grow infinitely with time t. Drude correctly concluded that there is a counteracting friction force, which takes into account that electrons change their momentum due to collisions (Fig. 1.2). Concerning these collisions, Drude made the following reasonable assumptions:

- Between collisions, the interaction of a given electron, both with other electrons and with ion cores, is neglected.
- Collisions in the Drude model, as in the kinetic theory of gases, are instantaneous events that abruptly alter the velocity of an electron.
- The electron experiences a collision, i.e., it suffers an abrupt change in its velocity, with a probability per unit time $1/\tau_{\mathrm{sc}}$. This means that the probability of an elec-

7 Quantum mechanically, the ion cores are only relevant for scattering, if displaced from their regular lattice position such as during their thermal motion as described by phonon excitations.

tron undergoing a collision in any infinitesimal time interval of length $\mathrm{d}t < \tau_{\mathrm{sc}}$ is just $\mathrm{d}t/\tau_{\mathrm{sc}}$. The time τ_{sc} is known as the collision or scattering time.
- Electrons are assumed to achieve thermal equilibrium with their surroundings only through collisions.

In the equilibrium case, all these assumptions lead to a force balance, i.e., to an average net zero force acting on the moving electrons, given by

$$\vec{F} = \dot{\vec{p}} = -e\vec{E} - m_{\mathrm{e}}\frac{\vec{v}_{\mathrm{d}}}{\tau_{\mathrm{sc}}} \stackrel{!}{=} \vec{0}\,, \tag{1.5}$$

where we have now included the friction force $m_{\mathrm{e}}\vec{v}_{\mathrm{d}}/\tau_{\mathrm{sc}}$ pointing in the opposite direction of the drift velocity. The friction force is chosen to be proportional to the velocity as usual for viscous friction forces. This expression will allow us to connect the drift velocity $\vec{v}_{\mathrm{d}}$ with the electric field $\vec{E}$. However, before doing so, we will make use of our knowledge about quantum mechanics to adapt the Drude model in a two-fold way:

I) First, we will consider that electrons in a crystal are represented as wavepackets of Bloch[8] states propagating as quasiparticles with charge $-e$ and energy E. For a parabolic band, the energy dispersion of such a crystal electron is given by

$$E(\vec{k}) = E_{\mathrm{s}} + \frac{\hbar^2|\vec{k}|^2}{2m^*}\,, \tag{1.6}$$

where $\hbar = 1.05 \cdot 10^{-34}$ Js is the reduced Planck's constant, m^* is the effective mass of the charged quasiparticle in the solid, i.e., the crystal electron, $\vec{k}$ is the wavevector, and E_{s} is the bottom of the band (Fig. 1.3 (a)). Importantly, the influence of the crystal lattice is incorporated in the effective mass m^*, which differs from the mass of the free electron, $m^* \neq m_{\mathrm{e}}$. For a parabolic band, one can extract the effective mass directly from the inverse curvature of the band dispersion $E(k)$, using $k := |\vec{k}|$, which leads to the well-known expression

$$m^* \equiv \left(\frac{1}{\hbar^2}\frac{\mathrm{d}^2E(k)}{\mathrm{d}k^2}\right)^{-1}\,. \tag{1.7}$$

For example, the effective mass of crystal electrons of copper is $m^*_{\mathrm{Cu}} = 1.01 \cdot m_{\mathrm{e}}$, of Pt, it is $m^*_{\mathrm{Pt}} \simeq 13 \cdot m_{\mathrm{e}}$ and in the conduction band of GaAs, it is $m^*_{\mathrm{GaAs}} = 0.063 \cdot m_{\mathrm{e}}$. Thus, as a first important consequence, we have to substitute m_{e} by m^* in eq. (1.5).

8 Felix Bloch, 1905–1983, was a Swiss physicist mainly working in the US. He was the first graduate student of Werner Heisenberg (1901–1976) and published his doctoral thesis in 1928. Therein, he established the quantum theory of solids, using waves (now known as Bloch waves) to describe electrons in periodic lattices.

II) Second, we will follow Sommerfeld[9], who generalized the Drude model in 1933 in order to incorporate the so-called Fermi[10]-Dirac[11] statistics, which is important for describing many electron systems by taking into account that two electrons are never in the exactly identical single-particle state (Pauli[12] exclusion principle). An important consequence for metals and semiconductors is that, for $T = 0\,\mathrm{K}$, the single-particle states are filled up to the Fermi energy E_F (Fig. 1.3 (a)). With the Fermi energy, which directly depends on the carrier density in the system, a new important energy scale enters allowing the definition of a Fermi wavevector $\vec{k}_\mathrm{F}$ by $E_\mathrm{F} = E(\vec{k}_\mathrm{F})$ and correspondingly a Fermi velocity by

$$\vec{v}_\mathrm{F} \equiv \frac{1}{\hbar}\vec{\nabla}_{\vec{k}} E(\vec{k})\big|_{\vec{k}=\vec{k}_\mathrm{F}} . \tag{1.8}$$

For example, a typical Fermi velocity in Cu is $|\vec{v}_\mathrm{F}| = 1.57 \cdot 10^6\,\mathrm{m/s}$. One can use this expression in connection with the momentum $\vec{p}_\mathrm{F} = \hbar\vec{k}_\mathrm{F} = m^*\vec{v}_\mathrm{F}$ for an alternative definition of the effective mass. For isotropic bands, it is given by

$$m^* \equiv \hbar^2 k \left(\frac{dE(k)}{dk}\right)^{-1} . \tag{1.9}$$

As this expression is only well-behaving for $k \neq 0$ it is not widely used. It can be helpful for systems with linear dispersion $E \propto k$, where eq. (1.7) would return $m^* = \infty$.

The Fermi velocity is given by $\vec{v}_\mathrm{F} = \hbar\vec{k}_\mathrm{F}/m^*$, where m^* of a parabolic band is a constant. As $\vec{k}_\mathrm{F}$ depends on the carrier density, $\vec{v}_\mathrm{F}$ also depends on it. Crucially, it turns out that the Fermi velocity, and not the thermal velocity, is responsible for the random motion depicted in Fig. 1.2. This is due to the fact that only electrons at the Fermi level are relevant for the electronic transport (Fig. 1.3). These electrons intrinsically move with the Fermi velocity $\vec{v}_\mathrm{F}$ already at $T = 0\,\mathrm{K}$. Usually, $\vec{v}_\mathrm{F}$ is also much larger than any additional thermal velocity. Thus; in the following discussion, we have to replace v_th by v_F. This is not only important for the conceptual understanding of the success of the Drude model, but gets particularly crucial when computing for example the distance an electron travels between two elastic collisions separated by the time τ_sc. This distance will be an important characteristic length scale and is known as the mean free path $l_\mathrm{m} = v_\mathrm{F}\tau_\mathrm{sc}$.

---

9 Arnold Sommerfeld, 1868–1951, was a German theoretical physicist and mentored a large number of students which significantly contributed to the new era of theoretical physics by developing quantum mechanics. Many of them won the Nobel Prize.

10 Enrico Fermi, 1901–1954, was an Italian physicist and his short name helped to get it attached to many different formulas and physical effects.

11 Paul Dirac, 1902–1984, was an English theoretical physicist. The biography "The strangest man: The Hidden Life of Paul Dirac" (from G. Farmelo) is worth reading.

12 Wolfgang Pauli, 1900–1958, was an Austrian-born Swiss (and American) theoretical physicist and one of the PhD students of Arnold Sommerfeld. He won the Nobel price for his contributions to quantum physics.

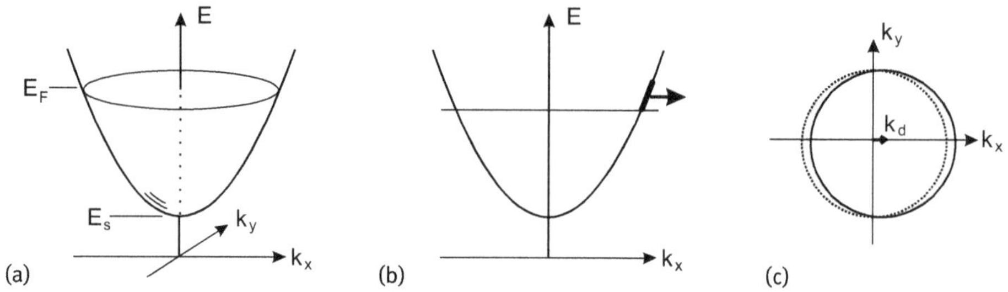

Fig. 1.3: (a) Parabolic band structure highlighting the band edge E_s and the Fermi energy E_F. (b) Cross-section through the parabolic band structure showing right moving excess carriers (thick line with arrow). (c) The equilibrium distribution (dotted line) and the drifted distribution (solid line) of electrons at low temperature. At equilibrium, all electrons are in states within a circle of radius k_F. Due to the electric field $\vec{E}$, this Fermi circle is shifted by the vector $\vec{k}_d$ in the direction opposite to $\vec{E}$.

By making use of these corrections, we firstly find from eq. (1.5) the following expression for the drift velocity,

$$\vec{v}_d = -\frac{e\tau_{sc}}{m^*}\vec{E}\,,$$

and from eq. (1.2) the current density,

$$\vec{j} = \frac{e^2 n_e \tau_{sc}}{m^*}\vec{E}\,.$$

In other words, the applied electric field breaks the equilibrium between right and left moving electrons (Fig. 1.3(b)) and shifts the Fermi distribution function[13] in the k_x–k_y-plane by the so-called drift vector (Fig. 1.3(c))

$$\vec{k}_d = -\frac{e\tau_{sc}}{\hbar}\vec{E}\,.$$

Here, we used $\hbar\vec{k}_d = m^*\vec{v}_d$. Finally, we find the mobility according to eq. (1.3) as

$$\mu = \frac{e\tau_{sc}}{m^*}\,, \tag{1.10}$$

and the conductivity according to eq. (1.4) as given by the so-called Drude–Sommerfeld formula

$$\sigma = \frac{e^2 n_e \tau_{sc}}{m^*}\,. \tag{1.11}$$

As the effective mass m^* is a parameter given by the band structure, all the remaining physics is captured by the time τ_{sc}. For small scattering rates, one can assume that the

13 For more details, see Section 1.4.3.2. For the moment, remember that the equilibrium Fermi distribution function can be written as $f(E(\vec{k}), T) = 1/[\exp((E(\vec{k}) - E_F)/(k_B T)) + 1]$.

rates for different scattering mechanisms are independent and, hence, can be added by the Matthiessen's[14] rule,

$$\frac{1}{\tau_{\text{sc}}} = \frac{1}{\tau_{\text{el-def}}} + \frac{1}{\tau_{\text{el-ph}}} + \frac{1}{\tau_{\text{el-el}}},$$

where $\tau_{\text{el-def}}$ is the relaxation time due to the scattering of electrons (el) at defects (def), $\tau_{\text{el-ph}}$ due to the scattering of electrons with phonons (ph) and $\tau_{\text{el-el}}$ due to the electron-electron scattering. This sums up the most important scattering processes. Section 1.2.6 describes that $\tau_{\text{el-def}}$ is related to an elastic scattering, which is phase preserving for the electron wave function, while $\tau_{\text{el-ph}}$ and $\tau_{\text{el-el}}$ relate to inelastic scattering processes, which are combined with a loss of phase coherence. In most cases, electron-electron scattering can be neglected, which is mainly due to the small number of electrons being able to contribute to relevant scattering mechanisms changing the average $\vec{v}_{\text{d}}$.

Using next the relation for the effective mass $m^* = \hbar k_{\text{F}}/v_{\text{F}} = h/(v_{\text{F}}\lambda_{\text{F}})$, where λ_{F} is the Fermi wave length, and the relation for the mean free path $l_{\text{m}} = v_{\text{F}}\tau_{\text{sc}}$, we can rewrite the conductivity of eq. (1.11) as

$$\sigma = \frac{e^2}{h} n_{\text{e}} l_{\text{m}} \lambda_{\text{F}} \,. \tag{1.12}$$

This formula reveals the importance of the prefactor e^2/h for conductance properties as discussed in detail in Sections 1.3 and 1.5.

1.2.3 Density of states in low dimensional systems

Before discussing the limitations of the Drude model, we will revisit the connection of carrier density n_{e} and density of states (DOS) for low-dimensional systems. Mainly because of simplicity, we will focus on 2D and 1D systems. The electronic density of states $\mathcal{D}(E)$ measures the number of electronic states per energy and per volume. It depends crucially on the dimensionality of the system. We consider a system with length L, width W and thickness t and use the following conventions:

- We will talk about three-dimensional (3D) systems when $L, W, t \gg \lambda_{\text{F}}$.
- We will talk about two-dimensional (2D) systems when $L, W \gg \lambda_{\text{F}}$ and $t < \lambda_{\text{F}}$. The terminology quasi-two-dimensional electron gas (system) is used in the case when $t \geq \lambda_{\text{F}}$, i.e., in the case where more than one subband is occupied with electrons.
- We will talk about one-dimensional (1D) systems when $L \gg \lambda_{\text{F}}$ and $W, t < \lambda_{\text{F}}$. The terminology quasi-one-dimensional electron gas (system) is used in the case when t, $W \geq \lambda_{\text{F}}$, again having more than one subband at the Fermi energy.

---

14 Augustus Matthiessen, 1831–1870, was a British chemist and physicist. He obtained his PhD in Germany at the University of Gießen at age 21.

For a parabolic band and $E > 0$, the density of states per volume, area, or length is given by

$$\mathcal{D}_d(E) = \frac{2}{\Gamma(d/2)} \left(\frac{m^*}{2\pi\hbar^2} \right)^{d/2} E^{(d-2)/2} \,, \tag{1.13}$$

where $\Gamma(x)$ is the gamma function[15], the factor 2 is due to a twofold spin degeneracy, and $d = 1, 2, 3$ is the dimensionality of the system. Moreover, we have set the energy of the band edge $E_s = 0$. For a detailed derivation of the density of states, see, for example, the book of T. Ihn [1][16].

The carrier density at zero temperature is then given by

$$n_e = \int_0^{E_F} \mathcal{D}_d(E) \mathrm{d}E \,. \tag{1.14}$$

For two-dimensional systems ($d = 2$), this directly results in $n_e = m^* E_F/(\pi\hbar^2)$, which leads to the important relation $n_e = k_F^2/(2\pi)$. Similar expressions can also be derived for three-dimensional ($n_e = k_F^3/(3\pi^2)$) and one-dimensional systems ($n_e = 2k_F/\pi$). The later expressions, which can also be derived by considering the state distribution exclusively in $\vec{k}$ space, are valid for each isotropic band independent of its dispersion relation $E(\vec{k})$.

1.2.4 Limitations of the Drude model

So far, we have discussed the semiclassical description of electron transport in the framework of the Drude model without discussing the range of validity of this description. For a 2D system, i.e., using $n_e = k_F^2/(2\pi) = k_F/\lambda_F$, we can express the conductivity of eq. (1.12) by

$$\sigma = \frac{e^2 n_e \tau_{sc}}{m^*} = \frac{e^2}{h} k_F l_m \,. \tag{1.15}$$

However, this expression is only valid when:

1. $L, W \gg l_m$: The condition makes sure that we are in the diffusive regime and are allowed to average over many scattering events, i.e., it makes sure that the situation illustrated in Fig. 1.2. holds.
2. $\lambda_F \ll l_m$: The condition makes sure that we are in the limit of geometrical optics, i.e., we can treat electrons like particles and do not have to care much about the wave nature of the electron.

---

15 $\Gamma(1/2) = \sqrt{\pi}$, $\Gamma(1) = 1$, $\Gamma(3/2) = \sqrt{\pi}/2$.

16 The derivation of the 3D density of states can be found there on page 37, the 2D one on p. 70 and the 1D version on p. 178.

This second condition can also be rewritten as

$$k_F l_m \gg 1 ,$$

which is known as the Ioffe[17]-Regel[18] criterion (1960).

1.2.5 Einstein relation

We will now introduce a different way to express the conductivity, which particularly highlights the fact that at zero temperature only a few electrons near the Fermi energy are involved in transport (Fig. 1.3(b), (c)). We therefore consider a rectangular piece of a two-dimensional conductor of length L and width W. The conductor is stretched between two large conducting pads, which will be called the left and the right lead or contact (Fig. 1.4(a)). The contacts are treated as reservoirs[19]. A bias voltage V is applied across the two leads creating an electric field

$$\vec{E} = -\frac{V}{L}\vec{e}_x ,$$

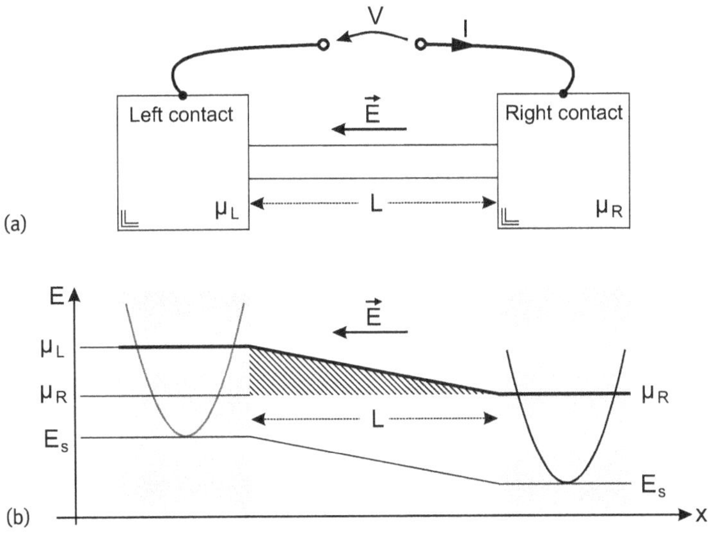

Fig. 1.4: (a) A two-dimensional conductor of length L and width W connected to two reservoirs (left and right contact) with electrochemical potentials μ_L and μ_R. (b) The band diagram under applied bias voltage V highlighting diffusive transport of the electrons in the hatched area.

17 Abram F. Ioffe, 1880–1960, Russian physicist.
18 Anatoli R. Regel, 1915–1989, Russian physicist.
19 A reservoir is a large system, which always remains in thermal and electrochemical equilibrium. In short, the system is large enough such that one can extract particles, e.g., electrons, without changing the electrochemical potential, the total charge, or the temperature T of the reservoir.

in the conductor. Consequently, the bottom of the band edge E_S follows the electrostatic potential energy and acquires a constant slope proportional to the electric field $\vec{E} = \vec{\nabla} E_S(\vec{x})/e$. It is crucial to realize that the total electron density is constant at all positions (along the x-axis) and thus there is no gradient in carrier density n_e, which would give rise to a diffusion current.

The situation changes, when we only look at the carriers with an energy larger than the electrochemical potential[20] in the right contact μ_R (hatched area in Fig. 1.4 (b)). All electrons below μ_R do not contribute to the current as all right moving states $+k$ are completely compensated by an equal number of left moving states $-k$ (Fig. 1.3 (b)). For the sake of simplicity, we concentrate on a two-dimensional system with parabolic dispersion $E(\vec{k})$, such that the density of states $\mathcal{D} := \mathcal{D}_2 = 4\pi m^*/h^2$ (eq. (1.13)) is independent of the energy. We can then express the excess carrier density above μ_R in the left contact by multiplying $\mathcal{D}$ with the energy $\mu_L - \mu_R$,

$$n_{L,e} = \mathcal{D}(\mu_L - \mu_R) .$$

Setting the excess carrier density in the right contact to $n_{R,e} = 0$, we end up with a concentration gradient of the excess carrier density from the left to the right contact (hatched area in Fig. 1.3 (b)) making the diffusion equation applicable. We therefore write

$$\vec{j} = -eD\vec{\nabla} n_e = e^2 D\mathcal{D}\frac{\mu_L - \mu_R}{|e|L}\vec{e}_x = e^2 D\mathcal{D}\vec{E} ,$$

where D is the diffusion constant. Comparing with the relation $\vec{j} = \sigma\vec{E}$, we obtain an expression for the conductivity:

$$\sigma = e^2 D\mathcal{D} . \tag{1.16}$$

This is the so-called Einstein relation. It has to be identical to the conductivity expressed by the carrier mobility $\sigma = en_e\mu = e^2 n_e \tau_{sc}/m^*$ (eq. (1.11)). Using $n_e = k_F^2/2\pi$ (text following eq. (1.14)), $\hbar k_F = m^* v_F$, and $\mathcal{D} = m^*/\pi\hbar^2$, we get

$$D = \frac{1}{2} v_F^2 \tau_{sc} . \tag{1.17}$$

1.2.6 Characteristic length scales

At the end of this introductory section, we will summarize the characteristic length scales that are playing a crucial role in quantum transport of mesoscopic samples:

Sample size L, W: It is typically in the range of micrometers down to nanometers.

---

20 The electrochemical potential is a thermodynamic measure of the chemical potential that includes the energy contribution of the electrostatics. The electrochemical potential of electrons can be expressed by $\mu = [\partial(U_{\mathrm{Int}} - TS)/\partial N]_{V,T} - e\Phi_{\mathrm{el}}$, where U_{Int} is the internal energy, S is the entropy and Φ_{el} is the electrostatic potential. The volume V and the temperature T are kept constant.

Interatomic spacing a: In solids, it is typically in the range of a few Angstroms (0.1 nm). Diamond has, for example, $a \approx 0.15$ nm.

Fermi wavelength λ_F: This is the wavelength of electrons at the Fermi energy, which dominates the electrical transport properties. For a two-dimensional electron gas (2DEG), we find:

$$\lambda_F = \frac{2\pi}{k_F} = \frac{2\pi}{\sqrt{2\pi n_e}} = \sqrt{\frac{2\pi}{n_e}} . \tag{1.18}$$

In semiconductors, such as two-dimensional systems made out of heterostructures (Section 1.3.1), the Fermi wavelength is in the range of 50 nm, whereas in typical metals, such as Cu or Ag, λ_F is in the range of a few Angstroms.

Elastic mean free path l_{el}: This is the average trajectory length covered by an electron before being elastically scattered into a different wave vector direction. It is often considered as a temperature-independent material property. A typical example is the scattering of the electron at a charged impurity, where effectively no energy is transferred. The elastic mean free path can be as large as 20 µm in a clean 2DEG or as short as a few Angstrom in a highly disordered alloy. At low temperatures, l_{el} dominates the scattering of electrons, i.e., $l_m \simeq l_{el}$.

Inelastic mean free path l_{in}: This is the typical distance that the electron travels before experiencing an inelastic scattering event. This length scale is strongly temperature dependent. It is typically limited by electron-phonon or electron-electron scattering processes.

Phase coherence length l_φ: At distances shorter than the phase coherence length l_φ (often also referred to as phase breaking length), quantum interference of electron paths must be considered, while over longer distances, the electrons lose their phase memory. The phase coherence length can be tens of µm in 2DEGs and up to several nm in regular metals. The phase coherence length is usually significantly temperature dependent. Neglecting spin-flip scattering[21] events, one finds $l_\varphi \simeq l_{in}$.

Thermal diffusion length l_T: Given a wave packet, which is built out of electrons with a spread in energy of $k_B T$, the thermal diffusion length is the distance that the wave packet travels before losing its phase memory due to the dephasing of its different contributions. It it is given by $l_T = \sqrt{(D\hbar)/(k_B T)}$, where D is the diffusion constant. It is often of the same order as l_φ.

1.2.7 Different transport regimes

Having introduced the different length scales, we can now classify different transport regimes. In particular, the sample size L, the Fermi wave length λ_F, the mean free path

21 For details, see, e.g., Section 9.9 of [2].

l_{m}, and the phase coherence length l_φ are employed. We distinguish the following regimes:
- $\lambda_{\mathrm{F}} \ll l_{\mathrm{m}}$: no localization to weak localization regime,
- $\lambda_{\mathrm{F}} \sim l_{\mathrm{m}}$: localization regime,
- $\lambda_{\mathrm{F}} > l_{\mathrm{m}}$: strong localization regime.

For $\lambda_{\mathrm{F}} \ll l_{\mathrm{m}}$, we can further classify the transport regimes as shown in Table 1.1.

Tab. 1.1: Different transport regimes.

	Classical	Quantum
Diffusive	$\lambda_{\mathrm{F}} \ll l_\varphi, l_{\mathrm{m}} \ll L$	$\lambda_{\mathrm{F}} \ll l_{\mathrm{m}} \ll l_\varphi, L$
Ballistic	$\lambda_{\mathrm{F}} \ll L < l_\varphi, l_{\mathrm{m}}$	$\lambda_{\mathrm{F}}, L < l_{\mathrm{m}}, l_\varphi$

1.3 Two-dimensional electron systems

In this section, we will shortly discuss different ways of obtaining two-dimensional electron systems. This is often the base for mesoscopic device structures or lower dimensional systems such as 1D or 0D structures. In particular, we will discuss two experimentally relevant 2D systems, namely gallium arsenide (GaAs) heterostructures and graphene. The epitaxially grown hetero-interface between GaAs and aluminum gallium arsenide (AlGaAs) provides an example of a mature 2D electron system, where the key concepts also apply to other III/V or silicon/silicon germanium (Si/SiGe) heterostructures. In contrast, graphene is a prominent example of the more recently emerging class of two-dimensional material systems.

1.3.1 GaAs based heterostructures

Many experiments in the field of mesoscopic transport have been based on GaAs/AlGaAs heterostructures, where a thin conducting layer is formed at the interface between GaAs and AlGaAs. The starting point is a heterostructure grown by molecular beam epitaxy as shown in Fig. 1.5. It typically exhibits an impurity concentration in the part-per-billion (10^{-9}) range and an extremely low defect density. The low defect density is partly based on the fact that AlAs and GaAs have nearly identical lattice constants, which avoids dislocations at the interface. Close to the surface, Si atoms are added as electron donors (Si^+) in order to populate the interface between GaAs and $\mathrm{Al_{0.3}Ga_{0.7}As}$ with electrons, hence allowing for the formation of a two-dimensional electron gas (2DEG). The remote positioning of the donors with respect to the 2DEG minimizes disorder in the 2DEG and thus improves the carrier mobility μ. It also im-

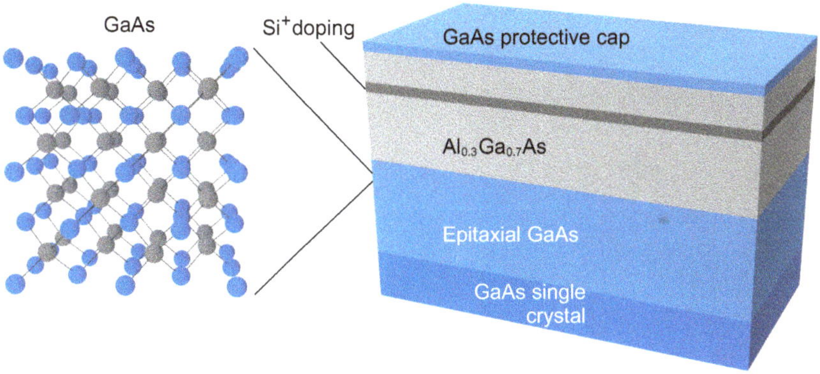

Fig. 1.5: Atomic lattice (left) and layer-structure (right) of a GaAs/AlGaAs heterostructure. Electrons will be confined to the interface between GaAs and $Al_{0.3}Ga_{0.7}As$ and are thus only able to move in two dimensions. This establishes a two-dimensional electron gas (2DEG).

proves the protection of quantum devices within the 2DEG, since the dopant layer can be a source of charge noise, which is detrimental for the device performance. The depth of the depicted GaAs–$Al_{0.3}Ga_{0.7}As$ interface is typically $\sim 100\,\text{nm}$ below the surface. Substantially larger depths would make it difficult to isolate individual electrons in nanostructures such as quantum dots (Section 1.3.1.1), while smaller distances would increase the detrimental influence of the dopant layer and of the disorder from the surface.

The heterostructure results in a band profile perpendicular to the surface, which is sketched on the left in Fig. 1.6 (black line: band profile, red line: confined wave function)[22]. The electrons of the 2DEG are confined in the roughly triangular potential well at the interface between GaAs and $Al_{0.3}Ga_{0.7}As$. For small enough electron densities, only one eigenstate of the confinement potential (red line) is below the Fermi energy (dashed, brown line), i.e., occupied with electrons. Hence, all the dynamics along the z-direction are frozen out and we obtain a 2DEG with one occupied subband.

1.3.1.1 Confinement in GaAs-based heterostructures

Lateral confinement within such a 2DEG is provided by gate electrodes on the surface (light grey areas on the top plane of Fig. 1.6). These electrodes can be negatively charged such that they deplete the electrons within the 2DEG areas below the gates. This way, one can induce, e.g., the lateral double-well potential sketched as a black line at the interface between GaAs and $Al_{0.3}Ga_{0.7}As$ (Fig. 1.6). Adequate gating can

---

22 The band profile is tuned by doping (marked Si^+) and the choice of the band offsets of the materials, which is tunable via chemical composition. The band profile is additionally influenced by the Fermi level pinning at the surface. The band profile is typically calculated numerically by employing commercial or freeware programs, which solve the one-dimensional Poisson–Schrödinger equation [3].

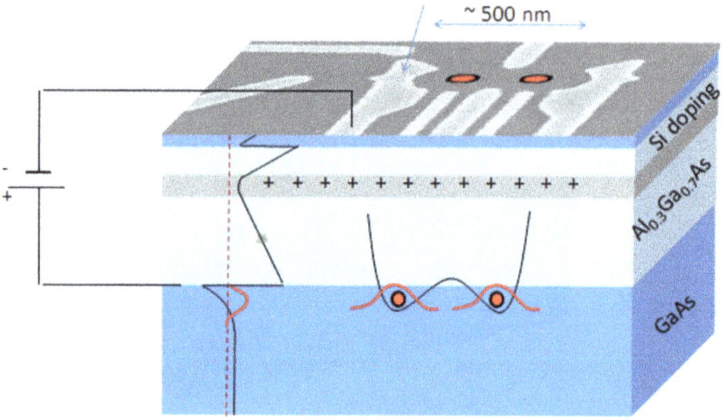

Fig. 1.6: Vertical heterostructure as in Fig. 1.5, where electrons (red dots) are confined vertically to the interface between GaAs and $Al_{0.3}Ga_{0.7}As$ as well as laterally into a double-well potential (black line at the GaAs–$Al_{0.7}Ga_{0.3}As$ interface). The double-well potential is defined using the electrostatic gates, visible as light gray structures on the top surface. Negative voltages applied to these gates push away electrons in the 2DEG, thus locally depleting it. Careful adjustment of the voltages creates the desired potential (black double well) with a controllable number of electrons in each well. On the left, the band profile perpendicular to the heterointerface is sketched as a black line with Fermi level as a dashed brown line. Red lines depict wave functions in vertical and lateral direction. The gate finger starting at the lower left of the top surface together with the two adjacent gates creates a quantum point contact (Section 1.4.2), which is used to readout the quantum dots used as a qubit (Section 3.3.1.6).

thus lead to two adjacent areas each confining a few electrons in lateral direction (red dots in Fig. 1.6). Such confinement areas are called zero-dimensional quantum dots, since the electrons can not move in any direction any more, i.e., they move in zero dimensions or on a dot. Quantum dots are also dubbed artificial atoms, since they provide discrete energy levels such as normal atoms. They are employed in multiple fundamental and applied experiments (Section 3.3.1, 3.3.2, 5.5.1) including, e.g., experiments aiming for quantum computation, where each quantum dot represents one quantum bit or qubit (Section 3.3.1).

The use of GaAs as the active layer for the quantum dots has the additional advantage that GaAs exhibits a parabolic conduction band with a small effective mass $m^*_{\mathrm{GaAs}} = 0.067 \cdot m_e$. This implies relatively large energy distances between adjacent confinement levels, because the confinement energies are proportional to $1/m^*$. At low enough temperatures ($T < 1$ K), one can, hence, typically neglect thermal excitations into higher orbital levels of the quantum dot, such that one obtains a high level of control on the occupation of different levels.

1.3.2 Graphene

Another possibility to get a 2DEG is by the use of materials that are only a few atoms in thickness. Here, graphene is the most famous example. It is an allotrope of carbon consisting of a single layer of carbon atoms arranged in a honeycomb crystal structure. As it consists only of surface atoms, the intimate environment has a huge influence on the material properties. For utilizing the intrinsic properties, state-of-the-art devices are based on heterostructures, where graphene is encapsulated in hexagonal boron nitride (Fig. 1.7). From an electronic point of view, graphene is at the borderline between a metal and a semiconductor, called a semi-metal or a zero bandgap semiconductor. The linearly dispersing valence and conduction band with $E(\vec{k}) \propto |\vec{k}|$ touch each other at the corners of the Brillouin zone, such that the Fermi surface consists only of two inequivalent points, if the graphene is charge neutral. For small energies, the band structure of graphene (Fig. 1.8(a)) can be approximated by [4]

$$E(\vec{k}) = \pm\hbar v_F |\vec{k}| \, . \tag{1.19}$$

The Fermi velocity $v_F = |\hbar^{-1}\nabla_{\vec{k}} E(\vec{k})| \simeq 10^6$ m/s remains constant, if the Fermi level is tuned away from charge neutrality, e.g., by doping the graphene via a gate electrode. The linear dispersion relation is distinct from the parabolic one (eq. 1.6) used to describe free electrons as well as the crystal electrons in many metals and semiconductors. The quasi-particles in graphene are rather mimicking massless relativistic particles, which becomes apparent when comparing eq. (1.19) with $E^2 = (c_0 p)^2 + (m_{rest} c_0^2)^2$. By substituting the speed of light in vacuum c_0 by v_F, writing $p = \hbar k$ and setting the rest mass $m_{rest} = 0$, we obtain the dispersion of graphene at low energies. As the quasi-particles in graphene behave like relativistic particles with linear dispersion, the concept of an effective mass determined via the curvature of the band is not very useful. Nevertheless, for quantities like the carrier mobility (eq. (1.10)), where the band struc-

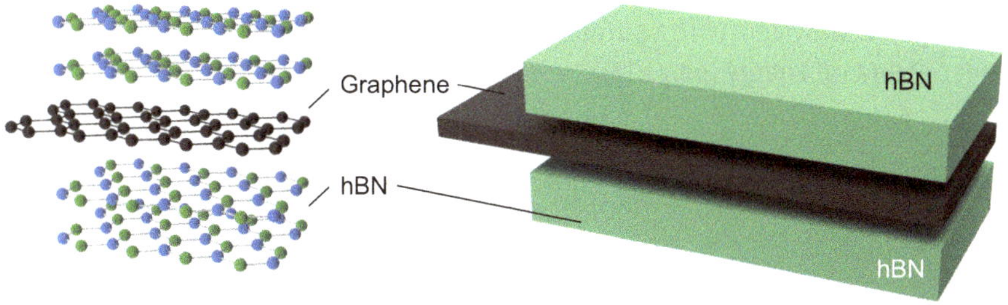

Fig. 1.7: Atomic lattice (left) and layer-structure (right) of a graphene based heterostructure. Van der Waals forces are keeping the graphene, which is encapsulated in hexagonal boron nitride (hBN), in place. As hBN is an insulator, the electrons will be confined to graphene and are only able to move in two dimensions, which establishes the 2DEG.

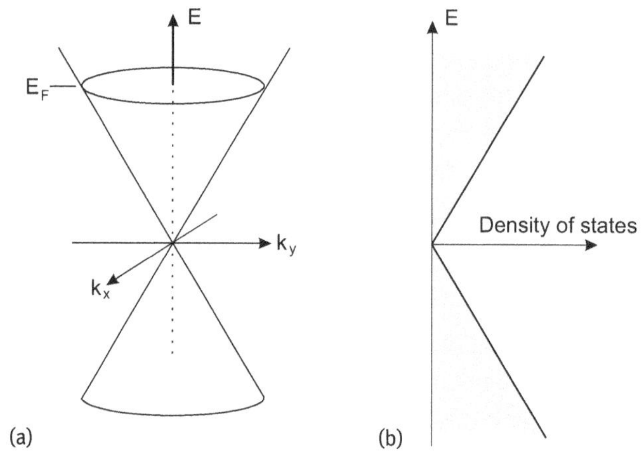

Fig. 1.8: (a) Bandstructure of graphene exhibiting a linear band dispersion.The Fermi level E_F is marked for the case of negatively charged graphene, e.g., via gating. (b) Energy dependent density of states of graphene.

ture does not enter, we can make use of the effective mass given by eq. (1.9) reading:

$$\mu = \frac{e\tau_{\mathrm{sc}}}{m^*} = \frac{e\tau_{\mathrm{sc}} v_{\mathrm{F}}^2}{E} .$$

In general, one has to depart from eq. (1.19) in order to obtain all the expressions as derived in the previous section. For example, the density of states gets (Fig. 1.8(b)),

$$\mathcal{D}(E) = \frac{2}{\pi\hbar^2 v_{\mathrm{F}}^2} E . \tag{1.20}$$

Here, we substituted the factor of two of eq. (1.13) by a factor of four due to an additional valley degeneracy in graphene, i.e., there are two Dirac cones in the Brillouin zone, which both contribute to the density of states. Notice that the result of eq. (1.20) is distinct from a 2D system with a parabolic band structure, where the density of states does not depend on energy (eq. (1.13)).

1.4 Ballistic transport

In Section 1.2.6 and 1.2.7, we have introduced different length scales, which are important for describing electron transport. In particular, an electronic system can be in the diffusive or ballistic regime depending on the ratio between the mean free path l_{m} and the characteristic system size L. In the ballistic regime, where $L \ll l_{\mathrm{m}}$, scattering inside the conductor loses relevance for the electron transport. Thus, the classical Drude theory breaks down. Transport under such conditions is governed by the ballistic motion of electrons inside the conductor and only the sample boundary plays the role of a scatterer. Since interference effects may be important in such systems, a quantum mechanical description of the conductance problem is required.

In this section, we will start with the quantum mechanical description of transport through an ideal one-dimensional quantum wire. In a next step, we will include a quantum cavity (or a region with moderate disorder), where scattering can occur allowing the connection between transmission and conductance to be made. Finally, by making use of a semiclassical approximation, we will gain a qualitative understanding of a number of mesoscopic phenomena.

1.4.1 Transport through a quantum wire

In this section, we will show that the conductance of a ballistic quantum wire with a truly one-dimensional band structure, i.e., with one open spin-degenerate conducting channel, is given by the conductance quantum $G_0 = 2e^2/h$. Strikingly, the conductance neither depends on any details of the band structure such as the effective mass or the Fermi velocity nor on any other material property such as the Fermi energy E_F. This is an important result and characteristic for many *universal* mesoscopic phenomena, that do not depend on material properties or system details. We will see more examples in subsequent sections of this book.

We will start by considering electron transport through a perfect wire with no scattering. Consider a wire with cross-sectional area A ($\sqrt{A} \sim \lambda_F/2$) and length L ($L \ll l_m$), which is stretched between two large conducting pads referred to as the left and right contacts (Fig. 1.9(b)).

The total charge in a small volume of length $\mathrm{d}x$ (Fig. 1.9(c)) is then again given by (Section 1.2.1)

$$\mathrm{d}Q = -en_e A\mathrm{d}x \,,$$

where n_e is the electron density (units: m^{-3}).

Now, we apply an external voltage $V = (\mu_L - \mu_R)/e$ (Fig. 1.9(b), (c)) such that the resulting electric field leads to the acceleration of electrons from the left contact (L) to the right contact (R). Here $\mu_L > \mu_R$ are the electrochemical potentials in the left and right contacts (Fig. 1.9(b)).

The current density is then again given by

$$\vec{j} = \frac{I}{A}(-\vec{e}_x) = \frac{-1}{A}\frac{\mathrm{d}Q_R}{\mathrm{d}t}(-\vec{e}_x) = -en_{R,e}\frac{\mathrm{d}x}{\mathrm{d}t}\vec{e}_x = -en_{R,e}v_x\vec{e}_x \,,$$

where Q_R is the total charge moving in $(-\vec{e}_x)$ direction and $n_{R,e}$ is the density of carriers (electrons) moving to the right. By introducing $N_{R,e} = n_{R,e}AL$ as the total number of electrons moving to the right and by using $I = \vec{j} \cdot (-\vec{e}_x)A$, we can rewrite the above expression as

$$I = \frac{e}{L}N_{R,e}v_x \,. \tag{1.21}$$

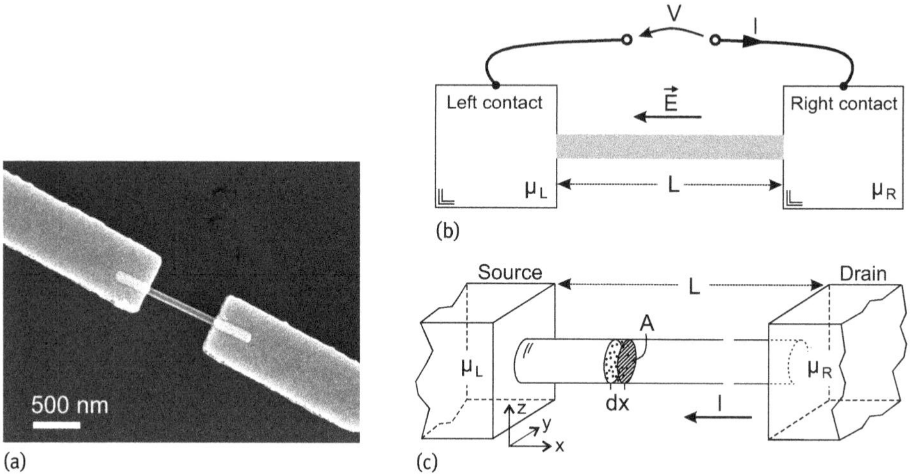

Fig. 1.9: (a) Scanning electron microscopy (SEM) image of an indium arsenid (InAs) nanowire with a diameter of around 100 nm (Courtesy of T. Schäpers, Forschungszentrum Jülich). (b) Schematic illustration highlighting the applied voltage and electric field direction. (c) More detailed illustration highlighting the geometrical dimensions. For more details on labeling see text.

Next, we make use of the fact that the wire is one-dimensional, meaning that normal to the wire axis (i.e., in the y–z plane) the wave functions are quantized. In the most simple case, we only have one quantized level, which is occupied with electrons. Along the wire axis, the electrons can propagate freely occupying states $\psi_{\mathrm{p}}(x) \sim e^{ik_x x}$.

We now count the number of electrons $\mathrm{d}N_{\mathrm{e}}$ in a small energy interval $\mathrm{d}E$ (Fig. 1.10). We start by recalling that the total number of k-states in $\mathrm{d}k$ is given by $\mathrm{d}k$ divided by the spacing between neighboring k-values, in our case given by $2\pi/L$. As in our example, each state can host one spin-up and one spin-down electron, the total number of right moving electrons $\mathrm{d}N_{\mathrm{R,e}}$ in the wave vector interval $\mathrm{d}k$ is

$$\mathrm{d}N_{\mathrm{R,e}} = 2 \cdot \frac{\mathrm{d}k}{2\pi/L} = 2 \cdot \frac{\mathrm{d}E}{2\pi/L} \left(\frac{\mathrm{d}E}{\mathrm{d}k} \right)^{-1} . \tag{1.22}$$

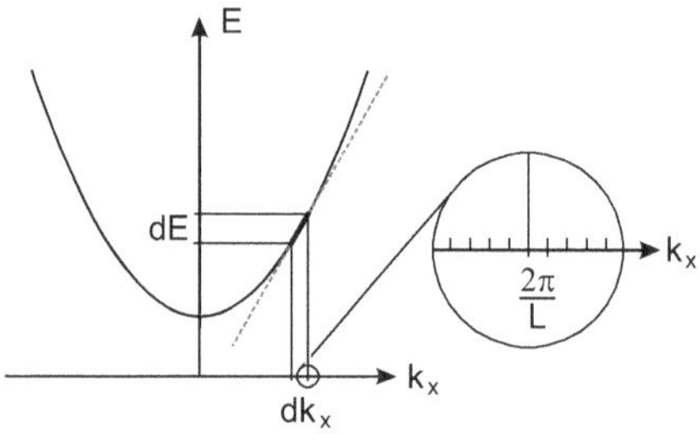

Fig. 1.10: Dispersion of a low energy mode in a quantum wire highlighting the allowed k-states in the k_x direction, which have a spacing of $2\pi/L$.

The prefactor of two originates from the spin-degeneracy. In some particular cases, the spin degeneracy might be lifted and in these cases the factor of two has to be replaced by a factor of one.

Next, we will express the velocity v_x for electrons in an one-dimensional band. As crystal electrons are described by Bloch waves, the group velocity of an electron wave packet is given by $v = \mathrm{d}\omega/\mathrm{d}k$, where $\omega = E/\hbar$ is the angular frequency of the Bloch wave. Thus, the group velocity in the x-direction is given by

$$v_x = \frac{1}{\hbar}\frac{\mathrm{d}E}{\mathrm{d}k} .$$

Therefore the current $\mathrm{d}I$ for a small energy interval $\mathrm{d}E$ can be expressed using eq. (1.21) by

$$\mathrm{d}I = \frac{e}{L}\mathrm{d}N_{\mathrm{R,e}}v_x = \frac{e}{L}\frac{2\mathrm{d}E}{2\pi/L}\left(\frac{\mathrm{d}E}{\mathrm{d}k}\right)^{-1}\frac{1}{\hbar}\frac{\mathrm{d}E}{\mathrm{d}k} = \frac{2e}{h}\mathrm{d}E .$$

The total current is finally given by integrating over the energy window ranging from μ_{R} to μ_{L}, since only these electrons contribute to the current, resulting in

$$I = \frac{2e}{h}\int_{\mu_{\mathrm{R}}}^{\mu_{\mathrm{L}}=\mu_{\mathrm{R}}+eV}\mathrm{d}E = \frac{2e^2}{h}V .$$

Hence the conductance is given by

$$G_0 = \frac{2e^2}{h} = 7.8\cdot 10^{-5}\,\mathrm{S} , \qquad (1.23)$$

which is called the conductance quantum.

In summary, we have shown that the conductance of a one-dimensional system with one occupied subband only depends on fundamental constants, namely the elementary charge e and the Planck constant h. Importantly, it does not depend on the dispersion relation $E(k)$ or the Fermi energy E_F.

1.4.1.1 Finite number of modes

Let us next examine the case of several partially occupied subbands. Such occupied subbands are also called open modes of the quantum wire. We consider a quantum wire with a circular cross-section of radius $\widetilde{R}$ (Fig. 1.11(a)). Assuming a separable solution of the time-independent, free-particle Schrödinger equation for an in-plane (i.e., z-y-plane) circular, infinitely high potential well, we express the total wave function ψ by a confined part ψ_{c} and a propagating plane-wave part $\psi_{\mathrm{p}}(x) \propto e^{-ik_x x}$ according to $\psi = \psi_{\mathrm{c}}(z, y)\psi_{\mathrm{p}}(x)$. Using a cylindrical coordinate system, one can find the following form of the confined wave functions, with radius r and azimuthal angle φ,

$$\psi_{\mathrm{c}}(z, y) = \psi_{\mathrm{c}}(r, \varphi) \sim J_m(k_{n,m}\, r)\, e^{im\varphi} ,$$

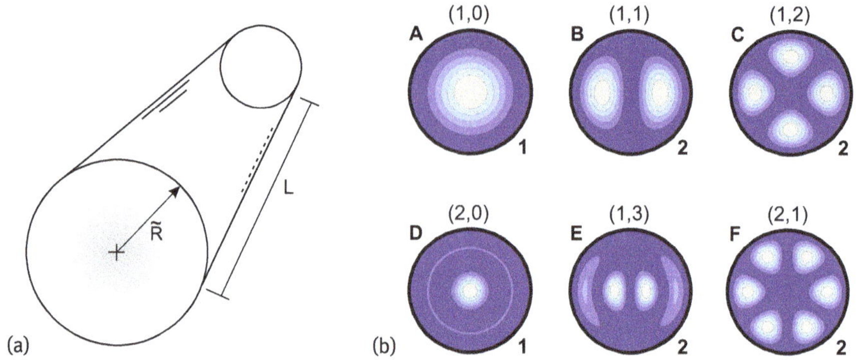

Fig. 1.11: Low energy modes in a quantum wire with a finite radius $\widetilde{R}$. (a) Schematic illustration of the wire. (b) The probability density of the confined wave functions $|\psi_c(r,\varphi)|^2$ for the lowest six energy eigenstates (A–F) labeled by the quantum number (n, m). The orbital degeneracy of each mode energy is given by the numbers in the lower right corner, which neglects the additional spin degeneracy.

where $m = 0, \pm 1, \pm 2, \ldots$ is the angular quantum number and $J_m(k_{n,m}\, r)$ are the regular Bessel functions, where $k_{n,m}$ is chosen such that it leads to the nth zero of $J_m(k\widetilde{R})$. The resulting energy subbands are given by

$$E_{n,m}(k_x) = \frac{\hbar^2 k_{n,m}^2}{2m^*} + \frac{\hbar^2 k_x^2}{2m^*} ,$$

where m^* is the effective mass. Figure 1.11(b) shows the probability densities $|\psi_c(r,\varphi)|^2$ of the six lowest energy eigenstates. Note, that the $|m| > 0$ states are doubly degenerate (labels below panels in Fig. 1.11(b)), since both positive and negative values of m exhibit the same energy. This corresponds physically to the equivalence of clockwise and counter-clockwise motion.

Figure 1.12 shows the corresponding one-dimensional subbands $E_{n,m}(k_x)$ for $\widetilde{R} =$ 50 nm and $m^* = 0.05 m_e$. Depending on the Fermi energy, one may have one subband (horizontal line at 4 meV) or several subbands (horizontal line at 11.5 meV) occupied by electrons. Only these bands contribute to the transport, since they are crossing E_F. Each subband provides a conductance value of G_0 times its orbital degeneracy. For example, we find for $E_F = 11.5$ meV a conductance value of $G = G_0(1 + 2 + 2 + 1) = 6G_0 = 12e^2/h$. In short, we say that this configuration of the quantum wire has six open modes. In general, the total conductance can therefore be written as

$$G = \frac{2e^2}{h} M ,$$

where M is the number of subbands crossing E_F, i.e., the number of open modes. Note that we assumed again that all subbands are spin degenerate and all modes are perfectly conducting.

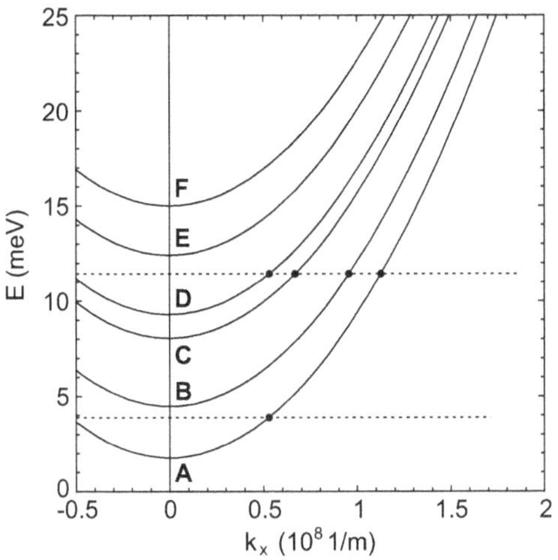

Fig. 1.12: One-dimensional subbands of a quantum wire with radius $\widetilde{R} = 50$ nm and hard wall confinement. Here, an effective mass of $m^* = 0.05 m_e$ is assumed. The subbands are labeled with A-F corresponding to Fig. 1.11(b). Dashed horizontal lines mark possible Fermi energies with resulting Fermi points (black dots).

1.4.2 Quantized conductance in a quantum point contact (QPC)

Let us next consider a two-dimensional electron gas (2DEG), e.g., formed on the interface between GaAs and AlGaAs (Section 1.3.1), with a small constriction of width W (Fig. 1.13(a), (top)) on the order of the size of the Fermi wave length ($W \sim \lambda_F$). Such systems form so-called quantum point contacts (QPCs), since – as we will show in this section – the conductance through the constriction is given by multiples of G_0. For simplicity, we model the constriction by a two-dimensional ribbon with width W and infinite length (Fig. 1.13(a), (bottom)). Then the eigenstates of the model system are obtained by solving the two-dimensional Schrödinger equation

$$-\frac{\hbar^2}{2m^*}\,\Delta\psi(x,y) = E\,\psi(x,y)$$

with the boundary condition

$$\psi(x,y) = \psi(x, \pm W/2) = 0\,.$$

When separating the wave function ψ into a lateral, propagating and a transverse, confined part, we obtain for the transverse direction quantized eigenstates of the standing wave form

$$\psi_n(y) = \sqrt{\frac{2}{W}}\,\sin\left[\frac{n\pi}{W}\left(y + \frac{W}{2}\right)\right]\,, \tag{1.24}$$

where $n = 1, 2, \ldots$ is the mode number. This leads to a reduced set of allowed k_y-values given by

$$k_{y,n} = \frac{n\pi}{W}\,.$$

This results in a so-called zone-folding of the original two-dimensional dispersion relation $E(k) = \hbar^2 k^2/(2\,\mathrm{m}^*)$ giving rise to one-dimensional subbands with band index

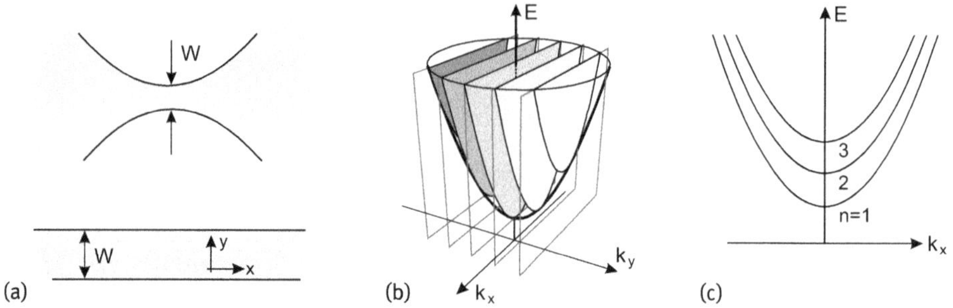

Fig. 1.13: (a) Schematic illustration of a constriction (upper panel) and a perfect ribbon (lower panel), both of width W in a 2DEG. The grey areas are conducting, while the white areas are insulating. (b) Two-dimensional band structure highlighting the allowed k-states in y-direction (planes) due to the finite size of the system. (c) Resulting one-dimensional subbands in k_x direction, which give rise to the quantized conductance steps.

n (Fig. 1.13(b), (c)). The confined, transverse wave functions are again multiplied by a plane wave $\psi_{\mathrm{p}}(x) \sim e^{-ik_x x}$ in the longitudinal direction with wave vector

$$k_{x,n} = \sqrt{k^2 - \left(\frac{n\pi}{W}\right)^2} \, . \tag{1.25}$$

If $n\pi/W < k_{\mathrm{F}}$, one obtains a real value of $k_{\mathrm{F},x,n}$ and the channel n is said to be open, because the states at E_{F} can carry charge in the $\vec{e}_x$-direction. For $n\pi/W > k_{\mathrm{F}}$, one gets an imaginary $k_{\mathrm{F},x,n}$ corresponding to an evanescent wave, which can not carry charge through the wire. The channel n is said to be closed. The number of open modes is then given by $M = n_{\max}$, where $n_{\max}$ is the largest n for which the condition $n_{\max}\pi/(Wk_{\mathrm{F}}) < 1$ holds. With $k_{\mathrm{F}} = 2\pi/\lambda_{\mathrm{F}}$, we therefore find

$$M = \left\lfloor \frac{W}{\lambda_F/2} \right\rfloor , \tag{1.26}$$

where $\lfloor x \rfloor$ is the floor integer value of x. The conductance is consequently given by

$$G = \frac{2e^2}{h} \left\lfloor \frac{2W}{\lambda_F} \right\rfloor = \frac{2e^2}{h} \left\lfloor \frac{W}{\pi} k_F \right\rfloor .$$

This directly implies that, in this case one can tune either the width W of the constriction or the Fermi wave length λ_{F}, e.g., by changing the electron density via gating, one can observe steps in the conductance, i.e., the current.

This indeed has been observed in 1988 by tuning the width W of a split-gate induced quantum point contact in a GaAs/AlGaAs heterostructure. In Fig. 1.14(a), we show a schematic of the device used by B. van Wees and colleagues [5]. By applying a negative voltage on the two top gates, the electron gas at the interface of the GaAs and AlGaAs (Fig. 1.5) is depleted leading to the constriction as depicted in Fig. 1.13(a). As the gates are several nanometers above the 2DEG the electric fringe-fields of the gate,

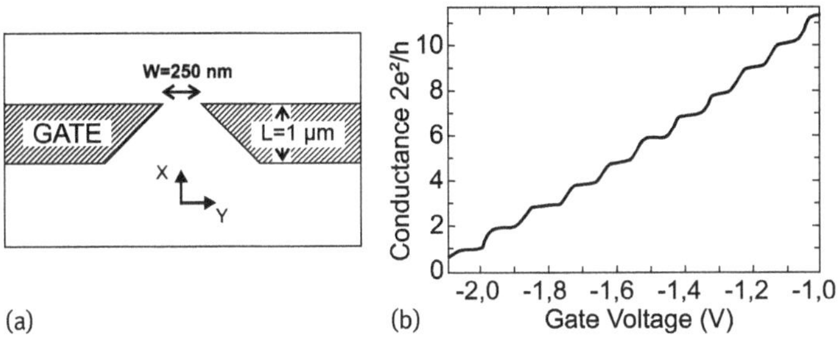

Fig. 1.14: (a) Schematic illustration of an experimentally realized split-gate geometry employing two gates (hatched areas) on top of a GaAs heterostructure (Fig. 1.5). (b) Low-temperature conductance of such a quantum point contact device as a function of gate voltage featuring conductance steps of $2e^2/h$. Adapted from [5].

which are controlled by the magnitude of the applied voltage V_{Gate}, we change the constriction width W. Thus, for large negative V_{Gate} the width W reduces to a value below $\lambda_F/2$ implying $G = 0\,\mathrm{S}$. With increasing V_{Gate}, the width increases and therefore the number of open modes $M = \lfloor 2\,\mathrm{W}/\lambda_F \rfloor$. This results in measurable conductance steps with step height $2e^2/h$ (Fig. 1.14(b)).

For larger systems, where the importance of the conductance steps becomes negligible, we can, in turn, approximate the ballistic conductance by

$$G \approx \frac{2e^2}{h}\frac{W}{\pi}k_F\,.$$

1.4.3 Ballistic transport through quantum billiards

Next, we extend our discussion to systems where the quantum wire is interrupted by a two-dimensional cavity region (Fig. 1.15(a)). This is often called a quantum billiard. Hence, we consider a two-dimensional region, which is connected to two 2D quantum wires with finite number of modes m and n. We introduce a local coordinate system (x_i, y_i) with $i = 1, 2$ for the exit and the entrance lead, respectively. Here, x_i denotes the longitudinal and y_i the transversal direction of lead i (Fig. 1.16). In the transverse direction, we obtain again quantized eigenstates as standing waves (eq. (1.24)),

$$\psi_n(y_i) = \sqrt{\frac{2}{W}}\sin\left[\frac{n\pi}{W}\left(y_i + \frac{W}{2}\right)\right]\,, \tag{1.27}$$

where W is the width of the left and right quantum wires[23]. The transverse wave

23 For simplicity, we assume that both wires have the same width W.

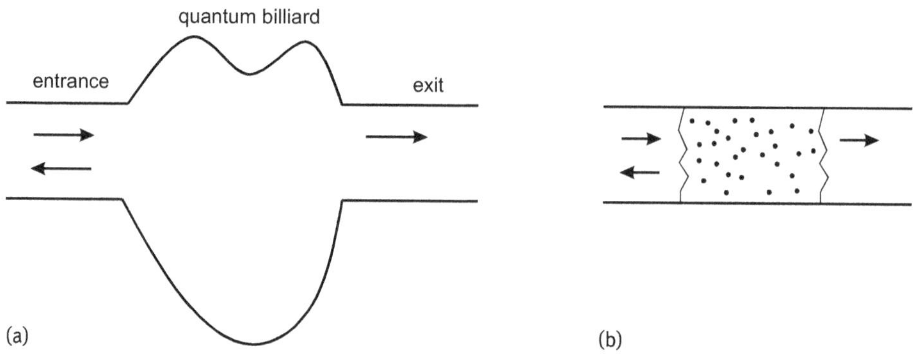

Fig. 1.15: (a) Schematic illustration of a quantum billiard connected to an entrance and exit lead. The arrows indicate that transport can be understood as a scattering problem via transmission into the exit lead and reflection back into the entrance lead. (b) Illustration of a two-dimensional wire with a central region of disorder. The black points are scatterers establishing diffusive electron transport.

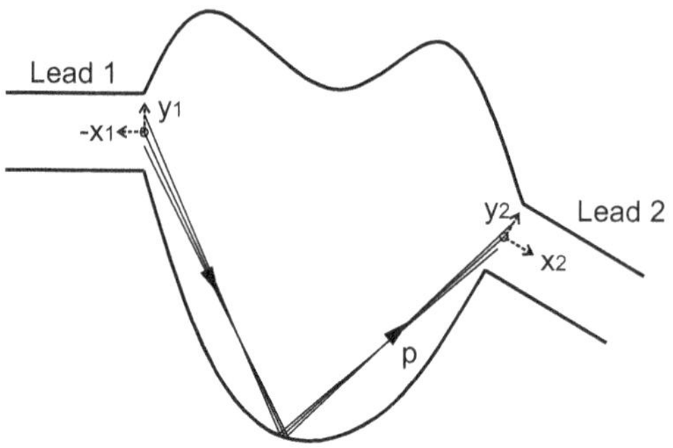

Fig. 1.16: Illustration of a quantum billiard with entrance lead 1 and exit lead 2 highlighting the local coordinate systems x_i, y_i. A path bundle p connecting lead 1 with lead 2 is shown as well.

function is multiplied by a plane wave in the longitudinal direction with wave vector (eq. (1.25))

$$k_{x_i,n}(y_i) = \sqrt{k^2 - \left(\frac{n\pi}{W}\right)^2} . \tag{1.28}$$

Again, if $n\pi/W < k_F$, $k_{F,x_i,n}$ is real and the channel n is said to be open.

For an electron approaching the cavity in the mth open mode in lead 1, the full entrance lead wave function is given by

$$\psi_1^{(m)}(x_1, y_1) = e^{ik_{x_1,m}x_1}\psi_m(y_1) + \sum_{n=1}^{M} r'_{nm} e^{-ik_{x_1,n}x_1}\psi_n(y_1) , \tag{1.29}$$

where r'_{nm} are the reflection amplitudes for reflection from mode m into mode n of lead 1 and M is the number of open channels. Here, we assume a lead length $L_1 = 1$ for the sake of simplicity. Notice that the same description can also be applied to the situation depicted in Fig. 1.15(b), where a region with impurities and defects but constant width is connected to two ideal quantum wire regions.

Similarly to the reflection, the wave function coming from mode m of lead 1 and representing the transmission into lead 2, again of length $L_2 = 1$, is

$$\psi_2^{(m)}(x_2, y_2) = \sum_{n=1}^{M} t'_{nm} e^{ik_{x_2,n}x_2} \psi_n(y_2) . \tag{1.30}$$

Here t'_{nm} denote the transmission amplitudes for transmission from mode m of the entrance lead into mode n of the exit lead. For simplicity, we assume that lead 2 has also M open modes. By making use of the definition of the probability current density $|\vec{j}|$ of the wave function ψ in one dimension, which is given by $|\vec{j}| = -i\hbar[\psi^*(\partial\psi) - \psi(\partial\psi^*)]/(2m^*)$, we can express the total probability current or electron flux through the entrance lead for an incoming wave of mode m by

$$J_{x_1}^{(m)} = \int_{-W/2}^{W/2} dy_1 j_{x_1}^{(m)}(x_1, y_1) .$$

Here $j_{x_1}^{(m)}$ is given by

$$j_{x_1}^{(m)} = -\frac{i\hbar}{2m^*}\left[\psi_1^{(m)*}(\partial_{x_1}\psi_1^{(m)}) - \psi_1^{(m)}(\partial_{x_1}\psi_1^{(m)*})\right] .$$

With this expression, we obtain, using eq. (1.29), the total electron flux $J_{x_1}^{(m)}$

$$J_{x_1}^{(m)} = \frac{\hbar k_{x_1,m}}{m^*} - \sum_{n=1}^{M} |r'_{nm}|^2 \left(\frac{\hbar k_{x_1,n}}{m^*}\right) . \tag{1.31}$$

This expression can be simplified by using the longitudinal velocity of the electron in the mth mode, given by $v_{x_1,m} = \hbar k_{x_1,m}/m^*$, leading to

$$J_{x_1}^{(m)} = v_{x_1,m} - \sum_{n=1}^{M} |r'_{nm}|^2 v_{x_1,n} . \tag{1.32}$$

Correspondingly, the flux in the exit lead is (eq. (1.30))

$$J_{x_2}^{(m)} = \sum_{n=1}^{M} |t'_{nm}|^2 v_{x_2,n} . \tag{1.33}$$

Since the electron flux is a conserved quantity, i.e., the incoming flux must be equal to the outgoing flux, it is convenient to normalize the channels to carry unit flux each, i.e., writing equations (1.29) and (1.30) in the following way

$$\frac{1}{\sqrt{v_{x_1,m}}}\psi_1^{(m)}(x_1, y_1) = \frac{1}{\sqrt{v_{x_1,m}}} e^{ik_{x_i,m}x_1}\psi_m(y_1) + \sum_{n=1}^{M} r_{nm}\frac{1}{\sqrt{v_{x_1,n}}} e^{-ik_{x_1,n}x_1}\psi_n(y_1) ,$$

and

$$\frac{1}{\sqrt{v_{x_2,m}}}\psi_2^{(m)}(x_2, y_2) = \sum_{n=1}^{M} t_{nm}\frac{1}{\sqrt{v_{x_2,n}}} e^{ik_{x_2,n}x_2}\psi_n(y_2) .$$

The new transmission matrix elements t_{nm} are related to the old ones by

$$t_{nm} = \sqrt{\frac{v_{x_2,n}}{v_{x_2,m}}} t'_{nm} \tag{1.34}$$

and the new reflection matrix elements are accordingly

$$r_{nm} = \sqrt{\frac{v_{x_1,n}}{v_{x_1,m}}} r'_{nm} . \tag{1.35}$$

The conservation of electron flux now reads

$$J_{x_1}^{(m)} = v_{x_1,m}\left(1 - \sum_{n=1}^{M} |r_{nm}|^2\right) = v_{x_1,m} \sum_{n=1}^{M} |t_{nm}|^2 = J_{x_2}^{(m)} . \tag{1.36}$$

This leads to the unitarity of the scattering matrix, also called S-matrix, which consists of the transmission amplitudes t_{nm} and the reflection amplitudes r_{nm} according to

$$1 = \sum_{n=1}^{M} (|r_{nm}|^2 + |t_{nm}|^2) . \tag{1.37}$$

The absolute squares of the transmission (reflection) amplitudes are the probabilities T_{nm} (R_{nm}) for the transmission (reflection) from mode m of the entrance lead to mode n of the exit lead (entrance lead).

1.4.3.1 Connection between conductance and transmission

In this subsection, we will derive the connection between the transmission matrix elements t_{nm} and the conductance G. Our starting point is the one-dimensional density of states per unit length moving into the right direction $\mathfrak{D}_{\mathrm{R}}(E) = 1/L \cdot \mathrm{d}N_{\mathrm{R,e}}/\mathrm{d}E = (\pi \mathrm{d}E/\mathrm{d}k)^{-1}$ (eq. (1.22)). For the mth mode of a parabolic band with dispersion $E - E_s^{(m)} = \hbar^2 k_m^2/2m^*$, onset energy $E_s^{(m)}$, and wave vector k_m, the density of states reads:

$$\mathfrak{D}_{\mathrm{R}}^{(m)}(E(k)) = \frac{m^*}{\pi \hbar^2 k_m} = \frac{2}{h v_{x_1,m}} .$$ [24]

We consider the billiard to be attached between two perfect electron reservoirs with electrochemical potentials μ_1 and $\mu_2 = \mu_1 - eV$, where V is the applied bias voltage. Electrons in lead 1 within the energy interval eV between μ_1 and μ_2, which are moving to the right, can be injected into the billiard and can be transmitted into lead 2. In order to calculate the resulting current, one has to multiply each injected $\mathfrak{D}_{\mathrm{R}}^{(m)}(E(k))$ with its velocity $v_{x_1,m}$, its charge e, and its transmission probability into all channels n of lead 2, before integrating across the energy interval between μ_1 and μ_2. Since

24 Inserting $k_m = \sqrt{2m^*(E - E_s^{(m)})}/\hbar$, one obtains $\mathfrak{D}_{\mathrm{R}}^{(m)}(E) = \sqrt{\frac{2m^*}{h^2(E - E_s^{(m)})}}$.

$e\mathcal{D}_R^{(m)}(E(k))v_{x_1,m}\sum_n |t_{nm}|^2 = 2e/h\sum_n |t_{nm}|^2$ is independent of E, the integration boils down to a multiplication by $\mu_1 - \mu_2 = eV$. Thus, the current $I^{(m)}$ emerging from channel m is given by

$$I^{(m)} = \frac{2e^2}{h}\left(\sum_{n=1}^{M} |t_{nm}|^2\right)V\,,$$

where we assumed that M channels of lead 2 are located below the Fermi level of lead 1. The total current from all open channels is accordingly

$$I = \sum_{m=1}^{M} I^{(m)} = \frac{2e^2}{h}\left(\sum_{n,m=1}^{M} |t_{nm}|^2\right)V = \frac{2e^2}{h}\left(\sum_{n,m=1}^{M} T_{nm}\right)V\,,$$

where $T_{nm} = |t_{nm}|^2$ are the transmission probabilities between the channels of the two leads. We assumed, reasonably, that the number of open channels in lead 1 equals the number of available channels in lead 2, which holds at low bias voltage V for identical materials of the two leads. By following Ohm's law $I = GV$, this leads to the so-called Landauer-Büttiker formula for the conductance G:

$$G = \frac{2e^2}{h}\sum_{n,m=1}^{M} T_{nm}\,. \tag{1.38}$$

In other words, the Landauer-Büttiker formula states that the conductance through a ballistic cavity is fully determined by the transmission between the incoming and the outgoing channels. This is quite different from the classical Drude–Sommerfeld theory, which describes conductance in a macroscopic sample in terms of electrons scattered by impurities, phonons and other electrons.

1.4.3.2 Finite temperature

For finite temperatures, we have to rewrite the current, which so far has been expressed as an integral over an energy window eV. Now, we have to use infinitely large integrals due to the smeared occupation of bands via the Fermi–Dirac distribution function, that reads

$$I = \frac{2e}{h}\sum_{n,m=1}^{\infty}\int_{-\infty}^{\infty} \mathrm{d}E\,T_{nm}(E)\,[f_1(E,T) - f_2(E,T)]\,.$$

Here, $f_{1,2}(E,T)$ are the Fermi–Dirac distribution functions in both contacts as given by

$$f_{1,2}(E,T) = \frac{1}{1 + e^{(E-\mu_{1,2})/(k_\mathrm{B}T)}}\,, \tag{1.39}$$

where $\mu_1 = \mu_2 + eV$. We added an energy dependence to the transmission probability T_{nm}, which is necessary due to the large energy window. In the frame of linear response, i.e., at negligibly small voltage, $f_1(E,T) - f_2(E,T)$ can be approximated by:

$$f_1(E,T) - f_2(E,T) \approx \frac{\mathrm{d}f_2(E,T)}{\mathrm{d}(eV)}eV = -\frac{\mathrm{d}f_2(E,T)}{\mathrm{d}E}eV\,.$$

Using this expression, we obtain

$$I = \frac{2e}{h} \sum_{n,m=1}^{\infty} \int_{-\infty}^{\infty} \mathrm{d}E T_{nm}(E) \left(-\frac{\mathrm{d}f_2(E,T)}{\mathrm{d}E} \right) eV$$

and the so-called linear conductance

$$G = \frac{2e^2}{h} \sum_{n,m=1}^{\infty} \int_{-\infty}^{\infty} \mathrm{d}E T_{nm}(E) \left(-\frac{\mathrm{d}f_2(E,T)}{\mathrm{d}E} \right) .$$

1.4.3.3 Semiclassical approximation for transmission and reflection

For the calculation of the transmission and reflection amplitudes, one could try to evaluate the full quantum wave function inside the cavity and match it with the mode wave functions of the leads. A more convenient way, however, serves as a starting point for our more qualitative discussion. In particular, we express the transmission amplitudes t_{nm} as the projection of the retarded Green's function, evaluated at the Fermi energy $E_\mathrm{F} = (\hbar k_\mathrm{F})^2/2m^*$, onto the transverse wave functions ψ_m and ψ_n of the incoming and outgoing modes, respectively. This reads

$$t_{nm}(k_\mathrm{F}) = -\mathrm{i}\hbar\sqrt{v_{\mathrm{F},x_2,n} v_{\mathrm{F},x_1,m}} \int_{-W/2}^{W/2} \mathrm{d}y_2 \int_{-W/2}^{W/2} \mathrm{d}y_1 \psi_n^*(y_2) G(y_2, y_1, k_\mathrm{F}) \psi_m(y_1) . \quad (1.40)$$

The retarded Green's function $G(y_2, y_1, k_\mathrm{F})$ describes the constant energy propagation at k_F between two points labeled y_1 and y_2. Hence, projected onto the transverse wave functions of the leads, it describes the propagation between the exit of lead 1 and the entrance of lead 2 (Fig. 1.16). The velocity prefactors are due to the quasi-one-dimensional densities of states in the entrance and exit leads.

A popular numerical method to calculate the retarded Green's function is the recursive Green's function algorithm, which allows solving of the scattering problem for arbitrary single-particle tight-binding Hamiltonians. The publicly available "Kwant" package is a good example of an implementation of such an algorithm [6]. Unfortunately, the numerical algorithms are not so helpful for gaining qualitative insights.

For a qualitative discussion, we will make use of Gutzwiller's semiclassical approximation to the Green's function. This basically sums the different paths, which can lead from an initial position y_1 to a final position y_2, including its phase factors. One then gets

$$G^\mathrm{SC}(y_2, y_1, k_\mathrm{F}) = \sum_{p:\, y_1 \to y_2} A_p \mathrm{e}^{\mathrm{i}S_p/\hbar - \mathrm{i}\pi\nu_p/2} , \quad (1.41)$$

where

$$A_p = \frac{-\mathrm{i}}{\hbar} \left(\frac{1}{2\pi\mathrm{i}\hbar} \right)^{1/2} \sqrt{D_p} = \frac{-\mathrm{i}}{\hbar} \left(\frac{1}{2\pi\mathrm{i}\hbar} \right)^{1/2} \sqrt{\frac{1}{|v_{x_1}||v_{x_2}|} \left| \frac{\partial^2 S_p(y_2, y_1, k_\mathrm{F})}{\partial y_2 \partial y_1} \right|} \quad (1.42)$$

is a prefactor that weights the probability of a certain path, S_p is the classical action of the path p, and D_p is the classical deflection factor, which is a measure of the divergence of nearby trajectories, also known as the stability of the path. For simplicity, we neglect, in the following, the Maslov index ν_p, which adds an additional phase factor depending on the topology (Section 5.4.1) of the phase space motion. The classical action S_p of path p is then given by

$$S_p(y_2, y_1, k_F) = \int\limits_{p:\, y_1 \to y_2} \mathrm{d}\vec{r}\vec{p} \,,$$

with $p\colon y_1 \to y_2$ being a path from y_1 to y_2. Without an external magnetic field, the action is

$$S_p(y_2, y_1, k_F) = \hbar k_F L_p(y_1, y_2) \,,$$

where L_p denotes the length of the classical path p. With an external magnetic field $\vec{B}$ and using the the so-called minimal substitution

$$\vec{p} \to \vec{p} + e\vec{A} \,,$$

where $\vec{A}$ is the electromagnetic vector potential, we obtain

$$S_p(y_2, y_1, k_F) = \hbar k_F L_p(y_1, y_2) + e \int\limits_{p:\, y_1 \to y_2} \mathrm{d}\vec{r}\vec{A} \,.$$

By making use of the Landau gauge $\vec{A}_j = B(-y_j, 0)$ and bringing $\psi_m(y_1)$ and $\psi_n(y_2)$ on the same reference gauge (Fig. 1.17) we find

$$S_p(y_2, y_1, k_F) = \hbar k_F L_p(y_1, y_2) + e \int\limits_{p - \Psi_2 + \Psi_1} \mathrm{d}\vec{r}\vec{A} = \hbar k_F L_p(y_1, y_2) + eBa_p \,, \tag{1.43}$$

where a_p is the oriented area associated with each path and $p - \Psi_2 + \Psi_1$ describes the integration path as sketched in Fig. 1.17.

Consequently, we find the following semiclassical Green's function,

$$G^{SC} = \sum_p A_p \mathrm{e}^{\mathrm{i}(k_F L_p + eBa_p/\hbar)} \,, \tag{1.44}$$

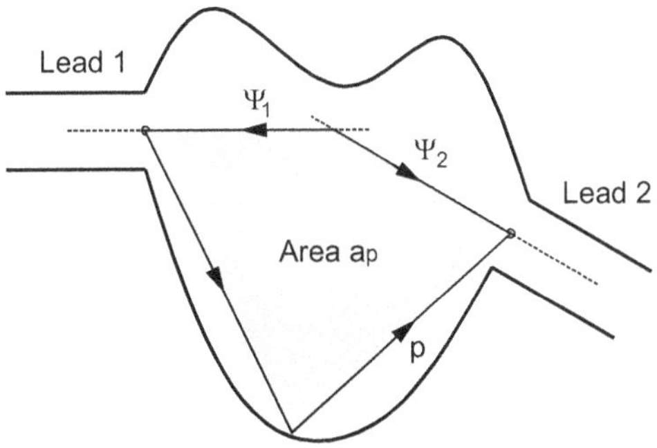

Fig. 1.17: Illustration of the area of a path p connecting the entrance (Lead 1) and the exit (Lead 2) of a quantum billiard. The area is closed by the straight extensions of the paths ψ_1 from the entrance lead and ψ_2 from the exit lead.

For simplicity, we can now rewrite the transmission amplitudes (eq. (1.40)) by

$$\begin{aligned} t_{nm} &\propto \int_{-W/2}^{W/2} dy_2 \int_{-W/2}^{W/2} dy_1 \psi_n^*(y_2) G^{SC}(y_2, y_1) \psi_m(y_1) \\ &= \int_{-W/2}^{W/2} dy_2 \int_{-W/2}^{W/2} dy_1 \sum_{p:\, y_1 \to y_2} \psi_n^*(y_2) A_p(y_2, y_1)\, e^{iS_p/\hbar} \psi m(y1)\,. \end{aligned} \tag{1.45}$$

with the geometrically determined phase factor $e^{iS_p/\hbar}$ given by eq. (1.43). For narrow leads, we next approximate the transmission amplitudes by

$$t_{nm} \approx \sum_{p:\, y_1^0 \to y_2^0} F_p^{nm} e^{iS_p/\hbar}\,, \tag{1.46}$$

where F_p^{nm} is an integrated weight of the path bundle p. Due to the integration over y_j used in order to determine F_p^{nm}, we must only consider the paths ending at single points $y_1 = y_1^0 \simeq 0$ and $y_2 = y_2^0 \simeq 0$ within the entrance and the exit lead, respectively. The sum over the path p, hence, runs over all path bundles, which connect lead 1 with lead 2.

The conductance can therefore be written as (eq. (1.38))

$$G = \frac{2e^2}{h} \sum_{n,m=1}^{M} \left| \sum_p F_p^{nm} e^{iS_p/\hbar} \right|^2 . \tag{1.47}$$

This leads to

$$G = \frac{2e^2}{h} \sum_{n,m=1}^{M} \sum_p \sum_{p'} F_p^{nm} \left(F_{p'}^{nm}\right)^* e^{i(S_p - S_{p'})/\hbar}\,. \tag{1.48}$$

Finally, we separate the two main contributions by multiplying either only equal or only unequal paths, respectively:

$$G = \frac{2e^2}{h} \sum_{n,m=1}^{M} \sum_p |F_p^{nm}|^2 + \frac{2e^2}{h} \sum_{n,m=1}^{M} \sum_{p \neq p'} F_p^{nm} \left(F_{p'}^{nm}\right)^* e^{i(S_p - S_{p'})/\hbar} = G_{cl} + \delta G_{qm}\,. \tag{1.49}$$

They are called the classical contribution G_{cl} without a phase factor with the quantum mechanical correction δG_{qm} providing a phase factor due to the addition of different phase factors from different paths between y_1^0 and y_2^0. In the following subsections, we will show that the quantum correction δG_{qm} can explain the origin of universal conductance fluctuations (Section 1.4.3.4), the Aharonov–Bohm effect (Section 1.4.3.5) as well as the Altshuler–Aronov–Spivak oscillations (Section 1.4.3.6).

1.4.3.4 The origin of universal conductance fluctuations (UCF)

For simplicity, we will assume in the following that we only have one open mode in lead 1 and lead 2 ($n = m = 1$), $B = 0\,\mathrm{T}$, and that the transport is dominated by two

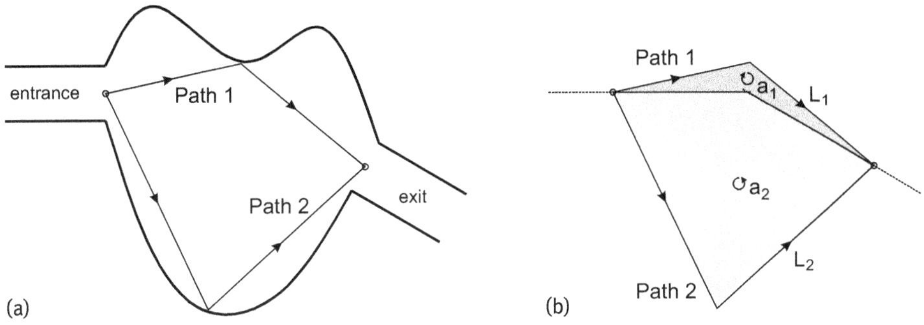

Fig. 1.18: Illustrations of the origin of universal conductance fluctuations. (a) Two possible different paths, paths 1 and path 2, connect the entrance and the exit channel. (b) These paths will interfere via their different lengths L_1 and L_2 and, in case of an applied B field, additionally via their different encircled areas a_1 and a_2 (eq. (1.52)).

paths (bundles) only ($p = 1, 2$). Path 1 is once reflected at the upper side of the cavity before hitting lead 2, while path 2 is reflected once at the bottom side of the cavity before reaching the exit lead (Fig. 1.18(a)). Crucially, the length of path 1, L_1, and path 2, L_2, are not identical (Fig. 1.18(b)). Following eq. (1.49), the conductance reads

$$G = \frac{2e^2}{h} \sum_{p=1}^{2} |F_p|^2 + \frac{2e^2}{h} \sum_{p \neq p'} F_p (F_{p'})^* \mathrm{e}^{\mathrm{i}k_\mathrm{F}(L_p - L_{p'})} . \tag{1.50}$$

This can be further simplified to

$$G = \frac{2e^2}{h} \left(|F_1|^2 + |F_2|^2 \right) + \frac{2e^2}{h} \Re \left\{ 2F_1 F_2^* \mathrm{e}^{\mathrm{i}k_\mathrm{F} \Delta L_{1,2}} \right\} , \tag{1.51}$$

where $\Delta L_{1,2} = L_1 - L_2$ and $\Re$ determines the real part of a complex number. The second term in eq. (1.51) is responsible for oscillations as a function of k_F with a period proportional to $1/\Delta L$. These oscillatory contributions to the conductance is the origin of the so-called *universal conductance fluctuation (UCF)*, which is on the order of e^2/h. In a more realistic scenario, we have many different paths p with different lengths L_p, which contribute to transport, resulting in many different differences $\Delta L_{p,p'} = L_p - L_{p'}$. This gives rise to a summation over many different phase factors $k_\mathrm{F}\Delta L_{p,p'}$. However, the amplitude of these oscillations remains on the order of e^2/h due to the fact that most phase factors cancel each other out via the negative and positive values of $\mathrm{e}^{\mathrm{i}k_\mathrm{F}\Delta L_{p,p'}}$. Hence, almost independently on the billiard shape, the sample size and the spatial dimension, an order of about one channel will change its contribution such that the UCFs are of the order e^2/h. Notice, that eq. (1.49) also holds to describe the conductance in a phase coherent sample with impurities (Fig. 1.15(b)). Thus, in such a system, the UCF appears similarly independent of the impurity configuration and the size L of the sample, as long as $L < l_\varphi$.

UCF can be observed as a function of k_F, which in an experiment is varied by a gate voltage V_G leading to a change in carrier density (Fig. 1.19). It can also appear

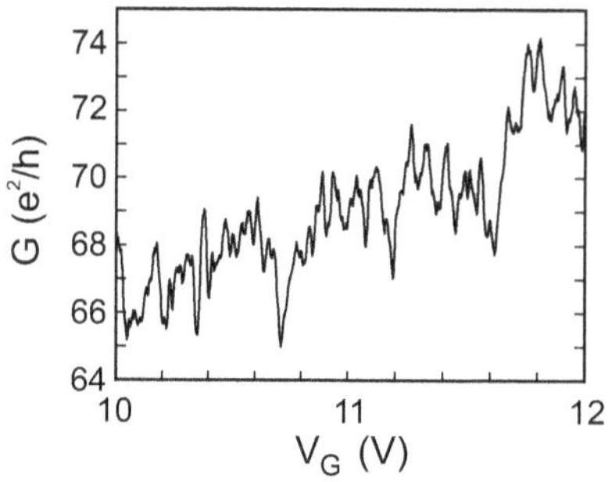

Fig. 1.19: Universal conductance fluctuations in a mesoscopic graphene device. The measurement has been taken at temperature $T < 10$ mK. Adapted from [7].

as a function of an externally applied B-field as is obvious from eq. (1.44). Universal conductance fluctuations are then often called magneto-conductance or magneto-resistance fluctuations. The presence of universal magneto-conductance fluctuations is caused by the different areas encircled by different paths, which leads to finite area differences $\Delta a_{p,p'}$ entering the quantum correction

$$\delta G_{\mathrm{qm}}(k_{\mathrm{F}}, B) = \frac{2e^2}{h} \sum_{p \neq p'} F_p (F_{p'})^* \mathrm{e}^{\mathrm{i}(k_{\mathrm{F}} \Delta L_{p,p'} + eB\Delta a_{p,p'}/\hbar)} \,. \tag{1.52}$$

Here, $\Delta L_{p,p'}$ are the length differences and $\Delta a_{p,p'}$ the corresponding area differences of the paths p and p' (Fig. 1.18). In short, we have shown that quantum coherence leads to sample-specific and reproducible conductance fluctuations in mesoscopic systems due to the many paths which are possible between two positions and which are able to interfere. This leads to a largely sample-independent amplitude of the universal conductance fluctuations of the order e^2/h.

1.4.3.5 The Aharonov–Bohm (AB) effect

A more controlled way to exploit the phase differences between different paths is to use a ring geometry as shown in Fig. 1.20(a). This leads to one of the most remarkable phenomena resulting from the wave nature of electrons, namely the Aharonov–Bohm effect. The effect is directly connected to the magneto-conductance fluctuations and demonstrates that in quantum theory the electromagnetic potentials, in particular the vector potential $\vec{A}$ plays a key role.

For simplicity, we again assume to have only one open mode in lead 1 and lead 2 ($n = m = 1$) governing the transport in the two semi-circle paths $p = 1, 2$, which both have the same lengths $L_1 = L_2$. However, the two paths have the opposite sign of the oriented path-area $a_2 = -a_1$. Furthermore, we assume that we have a finite magnetic field B penetrating the ring. The conductance then reads

$$G = \frac{2e^2}{h} \sum_{p=1}^{2} |F_p|^2 + \frac{2e^2}{h} \sum_{p \neq p'} F_p (F_{p'})^* \mathrm{e}^{\mathrm{i}k_F(L_p - L_{p'}) + \mathrm{i}eB(a_p - a_{p'})/\hbar} \,. \tag{1.53}$$

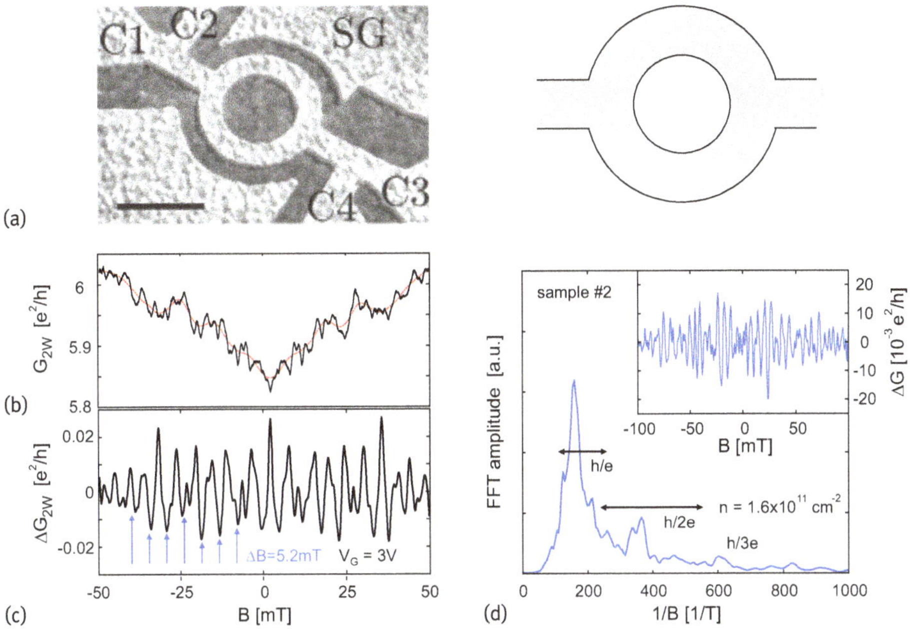

Fig. 1.20: (a) Left: Scanning force microscope image of a graphene ring on SiO_2 with leads C1-C4 and side gate (SG), which can tune the carrier density in both arms of the ring. Right: Schematic illustration of the ring structure, which offers two fundamentally different paths along the lower or along the upper half of the ring. (b) Two-terminal conductance of a graphene ring as shown in (a) as a function of B field. The red line marks the smoothly varying background mostly caused by weak localization (Section 1.4.3.7). (c) After subtracting the background signal (red line in (b)), regular oscillations with a period of 5.2 mT appear, which are the Aharonov–Bohm oscillations. (d) Fourier transformation (FFT) of the background subtracted conductance (inset) of another ring structure. The peaks of the Fourier transformation are labeled by the corresponding oscillation period in magnetic flux. All the measurements have been taken at a temperature $T < 100$ mK. (a) [8], (b)–(d) after [7].

This can be simplified by using $F_1 = F_2$ and $\Delta L = 0$, both since $L_1 = L_2$, (eq. (1.42) and (1.46)) to

$$G = \frac{2e^2}{h} 2|F_1|^2 + \frac{2e^2}{h} 2|F_1|^2 \cos\left(\frac{eBa_c}{\hbar}\right) , \tag{1.54}$$

where $a_c = 2a_1$ is the area encircled by the ring. This expression can be further simplified to

$$G(B) = \frac{4e^2}{h} |F_1|^2 \left[1 + \cos\left(\pi \frac{\Phi(B)}{\Phi_0}\right)\right] , \tag{1.55}$$

where $\Phi(B) = Ba_c$ is the magnetic flux through the ring structure and $\Phi_0 = h/2e = 2.07 \cdot 10^{15}\,\mathrm{J\,A^{-1}}$ is the magnetic flux quantum. The phase factor $\varphi_{AB} = \pi\Phi(B)/\Phi_0$ is known as the Aharonov–Bohm phase. It tells us that the conductance oscillates for every two magnetic flux quanta through the ring. These so-called Aharonov–Bohm (AB) oscillations have been first observed in 1985 in a normal metal ring [9].

Figure 1.20(a) shows a ring geometry with two contacts at the entrance and two contacts at the exit allowing a four-point measurement of the conductance, while slowly changing the B field. The resulting conductance is shown in Fig. 1.20(b). It consists of a general increase of conductance with $|B|$ due to weak localization as discussed below (Section 1.4.3.7). Removing this contribution (red line in Fig. 1.20(b)) via a high pass filter, Fig. 1.20(c) shows relatively regular oscillations with a period of $\Delta B = 5.2\,\text{mT}$. The deviations from the cosine-function are due to the fact that many different paths contribute to the interference, which pairwise encircle more or less the same area, but not exactly the same one. Moreover, the interfering paths from the upper and the lower half of the ring (Fig. 1.20(a)) can have slightly different lengths L_1 and L_2, which additionally contributes to the interference (eq. (1.44)). Another example of the conductance oscillations in such a ring geometry is shown in the inset of Fig. 1.20(d). The main figure exhibits the Fourier transformation of these data showing a clear peak at about $1/B = 170/\,\text{T}$, which corresponds to a period of the magnetoconductance oscillations by a magnetic flux of h/e as expected from eq. (1.54). However, further contributions with smaller periods as, e.g., $h/2e$ are visible, too.

1.4.3.6 The Altshuler–Aronov–Spivak (AAS) oscillations

The origin of these additional oscillations with a period of $\varphi_{\text{AB}}/2 = h/2e$ will now be discussed. It can be shown that, in the framework of our semiclassical approximation, these oscillations cannot be found in the transmission. It is therefore convenient to express the conductance in terms of the reflection,

$$G = \frac{2e^2}{h} \sum_{n,m=1}^{M} |t_{nm}|^2 = \frac{2e^2}{h} \left(M - \sum_{n,m=1}^{M} |r_{nm}|^2 \right), \tag{1.56}$$

where M is the total number of open modes and

$$r_{nm} \approx \sum_{p:\, y_1 \to y_1} F_p^{nm} e^{iS_p/\hbar} \propto \left\langle m \middle| G^{\text{SC}}(y_1, y_1, k_{\text{F}}) \middle| n \right\rangle .$$

Here, the sum over the paths p runs over all path bundles, which connect lead 1 back to the same lead 1.

By including the paths, which contribute to the reflection, we obtain a correction to the ballistic conductance $G = M \cdot G_0$ of

$$\delta G_{\text{R}} = -\frac{2e^2}{h} \sum_{n,m=1}^{M} |r_{nm}|^2 . \tag{1.57}$$

Let us assume again only one open mode in lead 1 ($n = m = 1$), such that the correction is dominated by two closed circle paths $p = 1, 2$, which are time-reversed, counter-propagating partners.[25] Consequently, both paths have the same length $L_1 = L_2$. However, they have the opposite sign of the oriented path-area $a_c = a_1 = -a_2$. Furthermore, we assume to have a small magnetic field penetrating the ring. The conductance correction due to reflection δG_R can then be written as,

$$\delta G_R = \frac{2e^2}{h}\sum_{p=1}^{2}|F_p|^2 + \frac{2e^2}{h}\sum_{p\neq p'} F_p(F_{p'})^* e^{ik_F(L_p-L_{p'})+ieB(a_p-a_{p'})/\hbar} . \tag{1.58}$$

This can be simplified by using $F_1 = F_2$ (since $L_1 = L_2$) and $\Delta L_{1,2} = 0$ to

$$\delta G_R = \frac{2e^2}{h}2|F_1|^2 + \frac{2e^2}{h}2|F_1|^2 \cos\left(\frac{2eBa_c}{\hbar}\right) , \tag{1.59}$$

where a_c is the area of the ring. This expression can be further simplified to

$$\delta G_R(B) = \frac{4e^2}{h}|F_1|^2 \left[1 + \cos\left(2\pi\frac{\Phi(B)}{\Phi_0}\right)\right] , \tag{1.60}$$

where $\Phi(B) = Ba_c$ is the magnetic flux through the ring and $\Phi_0 = h/2e$ is the magnetic flux quantum. These oscillations have exactly half the period in B with respect to the Aharonov–Bohm oscillations explaining the peak at $h/2e$ in Fig. 1.20(d). These so-called Altshuler[26]–Aronov[27]–Spivak[28] (AAS) oscillations were first observed in 1981, four years before the observation of the AB oscillations. Indeed, the AAS oscillations are easier to measure as they are robust against the contribution of interfering paths with different lengths, because the interfering paths are time-reversed partners encircling the ring clockwise and counterclockwise along the same path, hence, they automatically have the same length. Interestingly, the first AAS experiment by Sharvin was based on a human hair coated with Mg, such that the Mg forms a cylinder, which can be pierced by the magnetic flux [10].

1.4.3.7 Corrections to the average conductance

The constructive interference of closed paths with their time reversed, counter-propagating partners leads to a robust and very general reduction of the average conductance of a general, phase-coherent conductor. This can be seen by rewriting the conductance of a quantum billiard as (eq. (1.56))

$$G = \frac{2e^2}{h}\left(M - \sum_{n,m=1}^{M}|r_{nm}|^2\right) .$$

25 These time-reversed partners necessarily exist, in the absence of B, due to the time-reversal symmetry of the corresponding Schrödinger equation.

26 Boris L. Altshuler, 1955–, Russian-American physicist.

27 Arkady G. Aronov, 1939–1994, Russian physicist. He was the PhD supervisor of B. Altshuler and B. Spivak.

28 Boris Spivak, 1948–, Russian-American physicist.

This leads to

$$G = \frac{2e^2}{h}\left(M - \sum_{n,m=1}^{M}\sum_{p}|F_p^{nm}|^2 - \sum_{n,m=1}^{M}\sum_{p\neq p'} F_p^{nm}\left(F_{p'}^{nm}\right)^* \mathrm{e}^{\mathrm{i}(S_p - S_{p'})/\hbar}\right),$$

where all paths p and p' are closed orbits, such that they contribute to reflection.

Let us assume again only one open mode in lead 1 ($n = m = 1$). The conductance can then be written as

$$G = \frac{2e^2}{h}\left(1 - \sum_{p}|F_p|^2\right) - \frac{2e^2}{h}\sum_{p\neq p'} F_p(F_{p'})^* \mathrm{e}^{\mathrm{i}(S_p - S_{p'})/\hbar}\,. \tag{1.61}$$

The first term on the right hand side is the classical conductance, while the second term is the quantum correction δG_{qm}. The second term can be further divided into two contributions. The first contribution $\delta G_{\mathrm{qm}}^{(1)}$ is given by pairing the path p with its time-reversed partner $p' = -p$. Then, $L_p = L_{-p} = L_{p'}$ and, for $B = 0\,\mathrm{T}$, also $S_p = S_{-p}$. Therefore, we find that for $B = 0\,\mathrm{T}$,

$$\delta G_{\mathrm{qm}}^{(1)} = -\frac{2e^2}{h}\sum_{p}|F_p|^2\,. \tag{1.62}$$

For all the other $p \neq \pm p'$, we get, if we assume no additional symmetry in the device, $L_p - L_{p'} \neq 0$. Since we will have a continuous distribution of positive and negative values of $L_p - L_{p'}$, these terms will cancel each other for a large enough number of different paths p. Hence, the average quantum correction is only

$$\langle \delta G_{\mathrm{qm}} \rangle = \delta G_{\mathrm{qm}}^{(1)} = -\frac{2e^2}{h}\sum_{p}|F_p|^2\,. \tag{1.63}$$

This δG_{qm} is called the correction caused by weak localization, since the constructive interference of two time-reversed paths leads to an increased probability density $|\psi|^2$ at the starting point due to reflection. This starting point could be any point of the wave function, such that any point gets an increased probability density. Due to the normalization requirement of the wave function, the spatial extent of the wave function must decrease, i.e., the electronic state becomes more localized, hence, the term weak localization.

1.4.3.8 Suppression of weak localization

Applying a finite magnetic field leads to a suppression of the above mentioned quantum correction, which is due to the breaking of the time reversal symmetry of path p and $-p$. For a finite B-field, we find for $L_p = L_{-p}$ (eq. (1.44))

$$S_p - S_{-p} = \frac{2eBa_p}{\hbar}\,. \tag{1.64}$$

Therefore, the average quantum correction becomes

$$\langle \delta G_{\mathrm{qm}}(B) \rangle = -\frac{2e^2}{h} \sum_p |F_p|^2 \cos\left(\frac{2eBa_p}{\hbar}\right) . \tag{1.65}$$

To evaluate this expression, we substitute the sum over all closed paths p by an integration over all path areas a_p. This leads to

$$\langle \delta G_{\mathrm{qm}}(B) \rangle \propto -\int_0^\infty \mathrm{d}a_p P(a_p) \cos\left(\frac{2eBa_p}{\hbar}\right) , \tag{1.66}$$

where $P(a_p)$ is a weighted distribution of path areas a_p. It can be shown that for example the area distribution for chaotic quantum billiards is given by

$$P(a_p) \propto \mathrm{e}^{-\tilde{\alpha} a_p} \tag{1.67}$$

with sample specific factor $\tilde{\alpha}$. Therefore we find

$$\langle \delta G_{\mathrm{qm}}(B) \rangle \propto -\int_0^\infty \mathrm{d}a_p \mathrm{e}^{-\tilde{\alpha} a_p} \cos\left(\frac{2eBa_p}{\hbar}\right) . \tag{1.68}$$

This integral can be solved, resulting in a Lorentzian line shape,

$$\langle \delta G_{\mathrm{qm}}(B) \rangle \propto -\frac{1}{1 + \left(\frac{2eB}{\tilde{\alpha}\hbar}\right)^2} . \tag{1.69}$$

This is the so-called magnetic-field dependent weak-localization dip in the conductance. Note that only such paths can contribute to the correction, for which the phase coherence lengths l_φ is larger than L_p, since the constructive interference of different paths also gets destroyed by the loss of phase coherence, e.g., via inelastic scattering processes (Section 1.2.7). Since the quantum billiard can generally be replaced by a diffusive sample region (Fig. 1.15(b)), weak localization and its suppression by a magnetic field is also observed in macroscopic, diffusive samples. This will be discussed in Section 1.5.1 including experimental data (Fig. 1.23).

1.5 Phase coherent transport and weak localization

In this section, we will extend our discussion on phase coherent transport to the regime of diffusive transport. In particular, we revisit the quantum correction due to weak localization from the starting point of the classical Drude conductivity. Moreover, we will discuss the connection between the formalism of ballistic conductance (Section 1.4) and diffusive transport (Section 1.2).

1.5.1 Quantum correction to the Drude conductivity

In Section 1.4.3.7, we have seen that quantum coherence leads to the constructive interference of closed paths with their time reversed partners. This leads to an universal reduction of the average conductance due to quantum interference.

As the average quantum correction, given by eq. (1.63), is directly connected to the closed paths and their time reversed partners, we will now address the question how this effects diffusive transport.

Within the Drude–Sommerfeld model, quantum mechanical scattering at individual impurities is considered, but the coherent motion between scattering events, which leads to interference phenomena, is neglected. In the 1980s, scattering theories were developed, in which phase coherent multiple scattering was taken into account systematically. It was found that these processes lead to an enhanced backscattering of electrons and thereby to a logarithmic increase of the resistance. This effect, called weak localization (discovered in 1979), can be thought of as a precursor of strong localization of electrons (Section 1.6).

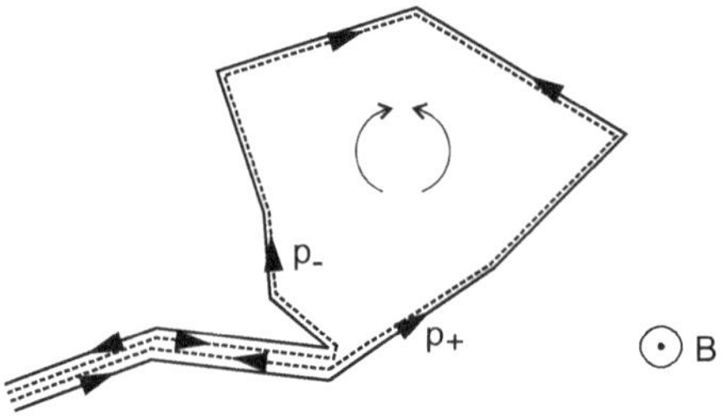

Fig. 1.21: Schematic illustration of two diffusion paths in a two-dimensional sample. The electron propagates on path, p_+ (solid line) and its time-reversed partner p_- (dashed line). At zero magnetic field such paths are always interfering constructively.

Before briefly revisiting diffusive transport, we start by considering two time-reversed paths (p_+, p_-) forming a loop (Fig. 1.21). The electron propagates on both paths and the two partial waves return after multiple scattering events to the starting point, but the loop is traveled by them in opposite directions. If we denote the complex quantum mechanical amplitudes of the two paths with F_+ and F_-, the probability of returning to the starting point is given by

$$P^{(\text{ret})} = |F_+ + F_-|^2 = |F_+|^2 + |F_-|^2 + F_+F_-^* + F_+^*F_- . \tag{1.70}$$

Here, the first two terms on the right hand side are the so-called classical contributions to backscattering which are contained in the Drude–Boltzmann theory. The last two terms are interferences which are neglected in the incoherent approximation of the Drude–Boltzmann theory. They manifest the weak localization correction (Section 1.4.3.7). At zero magnetic field, time-reversal symmetry ($L_+ = L_-$) requires that $F_+ = F_- = F$. The classical contribution to the return probability is therefore

$$P_{\text{cl}}^{(\text{ret})} = |F_+|^2 + |F_-|^2 = 2|F|^2 , \tag{1.71}$$

whereas the quantum mechanical (phase coherent) return probability is enhanced by a factor of two as a result of the interference terms

$$P_{\mathrm{qm}}^{(\mathrm{ret})} = 2|F|^2 + F_+F_-^* + F_+^*F_- = 2P_{\mathrm{cl}}^{(\mathrm{ret})} . \tag{1.72}$$

This result is called enhanced backscattering by quantum interference and leads to weak localization.

Importantly, the interference correction to the classical return probability is equal to the classical return probability itself. Note that the same result has been found for the quantum billiard (eq. (1.63)), where the scattering appears at the boundaries between the entrance and exit lead, instead at impurities in the crystal. Consequently, one can semi-quantitatively describe the weak localization correction to the conductivity at zero magnetic field by considering the classical return paths of electrons.

1.5.1.1 Revisiting diffusion

For simplicity, we restrict the following discussion to the two-dimensional case. Let $C(t)\mathrm{d}^2\vec{r}$ be the classical probability $\mathrm{d}P_{\mathrm{cl}}^{(\mathrm{ret})}(t)$ that a diffusing particle returns to the volume element $\mathrm{d}^2\vec{r}$ around its starting point ($\vec{r} = \vec{0}$) after time t. From the solution of the classical diffusion equation we find

$$C(t) = \frac{1}{4\pi Dt} , \tag{1.73}$$

for time scales on which the electron has traveled distances larger than the elastic mean free path l_{m}. The probability density of finding the electron at the time t at distance $|\vec{r}|$ is shown in Fig. 1.22(b) (solid line).[29] If we form semiclassical wave packets from the occupied electronic states in the system, we cannot localize an electron better than on the scale of λ_{F}. We are therefore interested in the return probability into the area $\lambda_{\mathrm{F}}^2 = v_{\mathrm{F}}\delta t\lambda_{\mathrm{F}} = h\delta t/m^*$ (eq. (1.11)), with δt being the time to travel a distance of λ_{F}. The return probability is, hence, given by

$$C(t)\lambda_{\mathrm{F}}^2 = \frac{h}{m^*}\frac{\delta t}{4\pi Dt} . \tag{1.74}$$

Next, we consider several paths returning within an interval of different times, since we have to take into account the phase coherence time τ_φ as an upper boundary for the constructive interference later on. This implies a time integral over the different return probabilities at different times leading to the classical return probability within a time period between t_1 and t_2,

$$P_{\mathrm{cl}}^{(\mathrm{ret})}(t_1, t_2) = \frac{h}{m^*}\int_{t_1}^{t_2} \frac{\mathrm{d}t}{4\pi Dt} = \frac{\hbar}{2m^*D}\ln\frac{t_2}{t_1} = \frac{1}{k_{\mathrm{F}}l_{\mathrm{m}}}\ln\frac{t_2}{t_1} . \tag{1.75}$$

For the last step, we have used the Einstein relation in two dimensions, i.e., $D = v_{\mathrm{F}}^2\tau_{\mathrm{sc}}/2$ (eq. (1.17)), $\hbar k_{\mathrm{F}} = m^* v_{\mathrm{F}}$, and $l_{\mathrm{m}} = v_{\mathrm{F}}\tau_{\mathrm{sc}}$.

29 The classical diffusion equation in 2D yields for the probability density $p(\vec{r}, t) = C(t)e^{-C(t)\vec{r}^2/\pi}$.

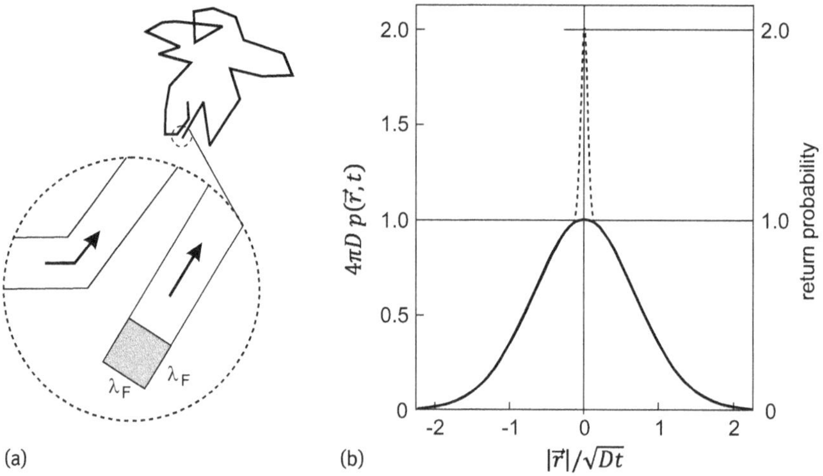

(a) (b)

Fig. 1.22: (a) Diffusion path of a conduction electron highlighting the finite width of λ_F of the ray-tube. (b) The probability distribution of a diffusing electron which starts at $\vec{r} = \vec{0}$ at the time $t = 0$. The classical solution is given by the solid line. In quantum diffusion, the return probability to $|\vec{r}| = 0$ is enhanced by a factor of two (dashed line). After [11].

From the discussion leading to eq. (1.72), we know that the quantum mechanical correction to the classical return probability is given by the classical return probability itself, $\delta P^{(\mathrm{ret})}_{\mathrm{qm}} = P^{(\mathrm{ret})}_{\mathrm{qm}} - P^{(\mathrm{ret})}_{\mathrm{cl}} = P^{(\mathrm{ret})}_{\mathrm{cl}}$. Thus, the classical return probability will be directly proportional to the normalized quantum correction of the resistivity. Moreover, the correction to the resistivity $\delta\rho_{\mathrm{qm}}$ is directly proportional to the resistivity ρ, since the ratio between return path probability and density of scattering centers, which determine the resistivity, does only depend on the dimensionality of the electron system. We get $\delta\rho_{\mathrm{qm}} \propto \rho P^{(\mathrm{ret})}_{\mathrm{cl}}$. Using the fact that $\delta\rho_{\mathrm{qm}}$ is small compared to ρ, we obtain, via adequate Taylor expansion of $\sigma = (\rho_{\mathrm{cl}} + \delta\rho_{\mathrm{qm}})^{-1}$, the quantum correction $\delta\sigma_{\mathrm{qm}}$ of the Drude conductivity σ, reading

$$\frac{\delta\sigma_{\mathrm{qm}}}{\sigma} \propto -P^{(\mathrm{ret})}_{\mathrm{cl}}(2\tau_{\mathrm{sc}}, \tau_\varphi)\,. \tag{1.76}$$

Here, we introduced $t_1 = 2\tau_{\mathrm{sc}}$ as the lower time, which is the minimum time needed to return (exactly one backscattering event) and $t_2 = \tau_\varphi$ as the upper time, since for longer times interference is suppressed due to the loss of phase coherence. To determine the proportionality factor, one has to determine the contribution of the return paths to the total resistivity, which goes beyond the scope of this book. One eventually obtains for the quantum mechanical conductivity correction[30]

$$\frac{\delta\sigma_{\mathrm{qm}}}{\sigma} = -\frac{1}{\pi k_{\mathrm{F}} l_{\mathrm{m}}} \ln\frac{\tau_\varphi}{2\tau_{\mathrm{sc}}}\,. \tag{1.77}$$

This is nearly identical to eq. (1.75). The additional factor $1/\pi$ is connected to the fact that in 2D systems the phase coherent units are discs and not squares (Fig. 1.25). It is also plausible that the contribution of return paths scales with λ_F/l_m, i.e., the larger l_m compared to $\lambda_F \propto 1/k_F$, the less likely is the return to the original area of size λ_F^2 (Fig. 1.22).

By expressing the time scales via the corresponding length scales, using $\tau_\varphi = l_\varphi^2/D$ and $\tau_{sc} = l_m/v_F = l_m^2/(2D)$ (eq. (1.17)), we find for the total conductivity

$$\sigma = \sigma_{cl} + \delta\sigma_{qm} = \frac{e^2}{h}k_F l_m - \frac{2e^2}{\pi h}\ln\frac{l_\varphi}{l_m}\,. \tag{1.78}$$

For the first term describing σ_{cl}, we have used eq. (1.12) reading $\sigma = \frac{e^2}{h}n_e l_m \lambda_F$ and $n_e = k_F^2/2\pi$. The logarithmic quantum correction $\delta\sigma_{qm}$ to the classical Drude conductivity σ_{cl} reduces the conductivity and is called the weak localization term. Its origin is the constructive interference of time-reversed paths.

1.5.1.2 Suppression of weak localization in a magnetic field

The microscopic picture of phase coherent backscattering leading to weak localization can be used straightforwardly to describe the magnetic field dependent suppression of weak localization (see also Section 1.4.3.8). The magnetic field influences the phases of the amplitudes F_+ and F_- in eq. (1.72) by adding the Aharonov–Bohm phase (Section 1.4.3.5). We recall that the magnetic field acts on the phase of the wave functions such that (eq. (1.44)ff.)

$$F_\pm(B) = Fe^{\pm i\varphi_{AB}}\,, \tag{1.79}$$

where the Aharonov–Bohm phase φ_{AB} is given by

$$\varphi_{AB} = 2\pi\frac{eBa_c}{h} = \pi\frac{\Phi}{\Phi_0}\,, \tag{1.80}$$

with a_c being the area enclosed by the two counter propagating paths, Φ the encircled magnetic flux, and Φ_0 the magnetic flux quantum (Fig. 1.21). This leads to a quantum mechanical, phase coherent return probability given by (see also eq. (1.55))

$$P_{qm}^{(ret)} = |F_+(B) + F_-(B)|^2 = 2|F|^2 + 2|F|^2\cos\left(2\pi\frac{2e\Phi}{h}\right)\,. \tag{1.81}$$

Crucially, the interference correction of the backscattering probability is $h/2e$-periodic in the magnetic flux encircled by the electron path. This interference term of a single path is identical to the $h/2e$-periodic Altshuler–Aronov–Spivak (AAS) oscillations (Section 1.4.3.6).

However, as in a macroscopic diffusive sample many paths encircle different areas a_c, we must introduce a probability density distribution $P(a_c)da_c$ denoting the relative

30 The following equation is only valid for $\tau_\varphi \gg \tau_{sc}$, i.e., when many different paths are contributing to the return probability. The term $\ln(\tau_\varphi/(2\tau_{sc}))$ can be approximated by $\ln(\tau_\varphi/\tau_{sc})$, leading to the expression of $\delta\sigma_{qm} = -e^2/(\pi h)\ln(\tau_\varphi/\tau_{sc})$, which is often found in literature.

contribution of areas of size a_c to the total conductivity. As a result, we can expect that the oscillatory contribution in B of individual time reversed paths averages out in σ_{qm} at magnetic fields, where a path surrounding an area with $\Phi > \Phi_0$ is traveled much faster than τ_φ. Closer to zero field, however, all time-reversed paths, which need less than τ_φ for traveling, contribute with similar phases, being different by less than π, such that they mostly still interfere constructively. This leads to a minimum in the conductivity $\sigma(B)$ or a maximum in the resistivity $\rho(B)$ around $B = 0$ T. The result is a quantum correction $\delta\sigma_{qm}(B)$ to the conductance which has the form

$$\delta\sigma_{qm}(B) = -\frac{\sigma}{k_F l_m} \int_0^\infty da_c P(a_c, l_\varphi) \cos\left(4\pi \frac{eBa_c}{h}\right), \tag{1.82}$$

where $P(a_c, l_\varphi)$ is the probability density to travel a path encircling a_c phase coherently given the phase coherence length l_φ. The integral is the Fourier cosine transform of the function $P(a_c, l_\varphi)$ and leads to a minimum of the conductance at $B = 0$ T in mesoscopic and macroscopic samples. The expression of $\delta\sigma_{qm}$ at zero magnetic fields is given in the previous subsection (eq. (1.77)).

The smallest areas a_c contributing to backscattering are of the order of l_m^2, whereas the largest encircled areas are of the order l_φ^2. The sharpness, i.e., the curvature, of the conductance minimum at $B = 0$ T will be determined by the phase coherence length l_φ, because the largest encircled areas lead to the contributions with the smallest periods in B (eq. (1.81)). To identify the quantity Dt with an effective area a_c, one writes heuristically for the probability distribution

$$P(a_c, l_\varphi) \propto \frac{1}{a_c}\left(1 - e^{-a_c/l_m^2}\right) e^{-a_c/l_\varphi^2}. \tag{1.83}$$

The first term $1/a_c$ describes geometrically that the return probability in 2D decreases with the encircled area a_c ($P \sim \lambda_F^2/a_c$). The second term eliminates the path lengths, which on average can not return since smaller than the scattering length l_m. Finally, the last term eliminates paths which are longer than the phase coherence length l_φ. Inserting this expression in eq. (1.82), we obtain

$$\delta\sigma_{qm} \propto -\int_0^\infty da_c \frac{1}{a_c}\left(1 - e^{-a_c/l_m^2}\right) e^{-a_c/l_\varphi^2} \cos\left(4\pi \frac{eBa_c}{h}\right). \tag{1.84}$$

Performing the integration numerically, one gets the typical magnetoresistance correction $\delta\sigma_{qm}(B)$, which has the shape of a dip in the conductivity around $B = 0$ T. This corresponds to a peak in resistivity as shown in the experimental data of Fig. 1.23. The peak gets narrower with decreasing temperature, since the phase coherence length increases such that larger areas a_c contribute to weak localization, which are prone to a faster oscillation in B.

The accurate quantitative theory of the weak localization effect has been worked out using diagrammatic methods [12] that are beyond the scope of this book. The result

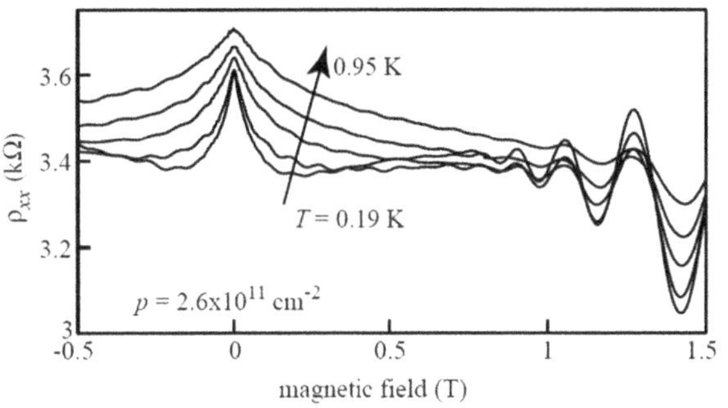

Fig. 1.23: Resistivity ρ_{xx} of a 2DEG as a function of magnetic field for different temperatures T highlighting the observed resistivity peak at $B = 0$ T, known as weak localization peak. The hole density $p = 2.6 \cdot 10^{11}$ cm^{-2} is marked. After [1].

for the magnetic field dependent correction of the Drude conductivity is

$$\delta\sigma_{\text{qm}}(B) - \delta\sigma_{\text{qm}}(0) = \frac{e^2}{\pi h}\left[\Psi_\Gamma\left(\frac{1}{2}+\frac{\tau_B}{2\tau_\varphi}\right) - \Psi_\Gamma\left(\frac{1}{2}+\frac{\tau_B}{2\tau_{\text{sc}}}\right) + \ln\left(\frac{\tau_\varphi}{\tau_{\text{sc}}}\right)\right]. \tag{1.85}$$

Here, $\tau_B = \hbar/(2eDB)$ and $\Psi_\Gamma(x)$ is the digamma function[31]. This relation is correct for $W, L \gg \tau_\varphi \gg \tau_{\text{sc}}$. The important parameter τ_B describes the traveling time after which a pair of returning, counter propagating paths interfere destructively at the origin for the first time due to an Ahoronov–Bohm phase difference of π.

1.5.2 From Landauer conductance to Drude conductivity

In Section 1.4, we derived the Landauer-Büttiker formula (eq. (1.38)), which incorporates the connection between conductance and transmission for ballistic samples. In particular, the Landauer formalism incorporates the correct properties of the resistance of small conductors, namely the length-independent interface resistance associated with the contacts (see below) and the discrete steps in coductance G as a function of open mode number M, which is related to the transverse modes in narrow conductors (Fig. 1.14).

We will now show that for large conductors, we recover the familiar Drude conductivity, i.e., the conductance given by $G = \sigma W/L$. For a wide conductor with many open modes the number of modes is proportional to the width: $M \approx Wk_F/\pi$ (eq. (1.26)). Therefore, the Landauer conductance can be written as (eq. (1.38)),

$$G = \frac{2e^2}{h}M\overline{T} \approx \frac{e^2}{h}k_F\frac{2\overline{T}W}{\pi} = \frac{e^2}{h}\left(k_F L\frac{2\overline{T}}{\pi}\right)\frac{W}{L}, \tag{1.86}$$

31 The digamma function is given by $\Psi_\Gamma(x) = \frac{1}{2}\frac{\mathrm{d}}{\mathrm{d}x}\ln\Gamma(x)$, where $\Gamma(x) = \int_0^\infty t^{x-1}e^{-t}\mathrm{d}t$ is the gamma function. For large arguments x, it can be approximated by $\Psi_\Gamma(x) \approx \ln(x) - \frac{1}{x} \approx \ln(x)$.

where $\overline{T} = \langle T_{nm} \rangle$ is the average transmission coefficient per mode of all open modes. The last expression on the right hand side has been brought in a form which can directly be compared with the conductance from the Drude model. Using eq. (1.15), we can write the conductance of the Drude model as

$$G = \frac{e^2}{h}(k_F l_m)\frac{W}{L} . \tag{1.87}$$

By comparing these two expressions, it becomes obvious that the transmission $\overline{T}$ must depend on the relation between mean free path l_m and sample length L. Thus, next we will show that the average transmission coefficient can be written as

$$\overline{T}(L) = \frac{L_0}{L + L_0} ,$$

where L_0 is a characteristic length scale of the order of the mean free path l_m.

We start by considering two conductors with transmission probabilities T_1 and T_2 connected in series (Fig. 1.24(a)). The problem is to find the probability of transmission T_{12} through the series combination. It is obtained by summing the probabilities for transmission with zero reflections, with two reflections, with four reflections and so on (Fig. 1.24(b)):

$$T_{12} = T_1T_2 + T_1T_2R_1R_2 + T_1T_2R_1^2R_2^2 + \cdots = \frac{T_1T_2}{1 - R_1R_2} . \tag{1.88}$$

We can rewrite this result, using $R_1 = 1 - T_1$ and $R_2 = 1 - T_2$,

$$\begin{aligned}\frac{1 - T_{12}}{T_{12}} = \frac{1 - R_1R_2 - T_1T_2}{T_1T_2} &= \frac{1 - (1 + T_1T_2 - T_1 - T_2) - T_1T_2}{T_1T_2} \\ &= \frac{T_1 + T_2 - 2T_1T_2}{T_1T_2} = \frac{1 - T_1}{T_1} + \frac{1 - T_2}{T_2} .\end{aligned}$$

This shows that, when we place two scatterers in cascade, the quantity $(1 - \overline{T})/\overline{T}$ is additive, meaning that the transmission probability $\overline{T}(N)$ of N scatterers in series, each

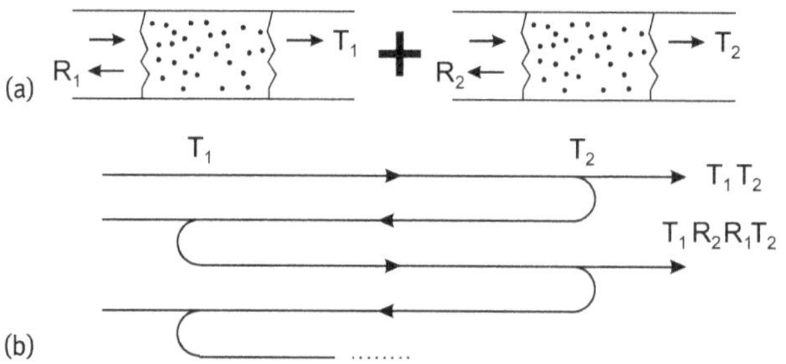

Fig. 1.24: (a) Connecting two regions of scatterers with transmission probabilities T_1 and T_2 and corresponding reflection probabilities R_1 and R_2. (b) Illustration of how the total transmission can be summed up by multiple paths featuring multiple reflections at the interfaces.

having a transmission probability $\overline{T}$, is given by,

$$\frac{1-\overline{T}(N)}{\overline{T}(N)} = N\frac{1-\overline{T}}{\overline{T}} \,.$$

This straightforwardly leads to

$$\overline{T}(N) = \frac{\overline{T}}{N(1-\overline{T})+\overline{T}} \,.$$

For N scatterers in series in a conductor of length L, we divide the length into parts of lengths L/N with transmission probability $\overline{T}$ including one scatterer each on average. Using additionally the definition

$$\frac{L}{L_0} = \frac{\overline{T}}{N(1-\overline{T})} \,,$$

we get a simple expression for the length dependent transmission probability

$$\overline{T}(L) = \frac{L_0}{L+L_0} \,, \tag{1.89}$$

Next, we insert this result into eq. (1.86) leading to

$$G = \frac{e^2}{h}k_F\frac{2W}{\pi}\frac{L_0}{L+L_0} = \frac{e^2}{h}k_F\left(\frac{2L_0}{\pi}\right)\frac{W}{L+L_0} \simeq \sigma\frac{W}{L+L_0} \,. \tag{1.90}$$

In the last step, we assumed $L_0 \ll L$ in order to connect the conductance G with the conductivity σ. Comparing with eq. (1.87), we get $L_0 \simeq l_m\pi/2$, i.e., L_0 is indeed about the mean free path l_m. Equation (1.90) can be also expressed as total resistance

$$R = G^{-1} = \sigma^{-1}\frac{L+L_0}{W} \,.$$

This is a series combination of an actual resistance $\sigma^{-1}L/W$ obeying Ohm's law and a contact or interface resistance $\sigma^{-1}L_0/W$. Alternatively, one can define a contact conductance $G_c = \sigma W/L_0$ and an intrinsic conductance

$$G_s = \sigma\frac{W}{L} = \frac{e^2}{h}k_F l_m\frac{W}{L} = \frac{e^2}{h}\pi\frac{l_m}{L}M \,,$$

giving

$$G_c = \sigma\frac{W}{L_0} = \frac{2e^2}{h}k_F\frac{l_m}{L_0}\frac{W}{\pi} \simeq \frac{2e^2}{h}M = G_0M$$

employing $M \simeq Wk_F/\pi$ and $L_0 \simeq l_m\pi/2$. Hence, the number of open channels M, defines the contact serial conductance of a device, providing a conductance quantum G_0 each, while the ratio of mean free path l_m and sample length L determines its intrinsic conductance.

1.5.2.1 Weak localization correction

In our discussion, leading to eq. (1.90), we again neglected quantum interference. This will be added now by considering the enhanced backscattering of electrons. Our starting point is eq. (1.86), respectively eq. (1.38) leading to

$$G = \frac{2e^2}{h} M\overline{T} = \frac{2e^2}{h} M(1 - \overline{R}) \, ,$$

where $\overline{R}$ is the average reflection probability of the open modes. Making use of the unitarity condition $\overline{T}+\overline{R} = 1$, we can express $\overline{R}(L)$ by repeating the calculation starting at eq. (1.88) and ending at eq. (1.89) as

$$\overline{R}(L) = \frac{L}{L + L_0} \, .$$

Assuming that the scatterers are isotropic, we would expect an incident electron in mode m to be reflected into all modes $n = 1, 2, \ldots, M$ with equal probability reading

$$R_{nm}(L) = \frac{1}{M} \frac{L}{L + L_0} \, .$$

When we take quantum interference into account this result still holds on the average for $m \neq n$. But the probability of reflection back into the incident mode is doubled from its classical value (eq. (1.72)). This leads to

$$R_{\mathrm{qm},nm}(L) = \frac{1}{M} \frac{L}{L + L_0} \quad \text{for} \quad m \neq n \tag{1.91}$$

$$= \frac{2}{M} \frac{L}{L + L_0} \quad \text{for} \quad m = n \, . \tag{1.92}$$

Hence, the average reflection probability of an electron is a little larger than the classical value

$$\overline{R}_{\mathrm{qm}}(L) = \sum_{n=1}^{M} R_{\mathrm{qm},nm}(L) = \frac{L}{L + L_0} + \frac{1}{M} \frac{L}{L + L_0} \, .$$

Thus, the transmission probability must be a little smaller than its classical value by the same amount (eq. (1.37)) reading

$$\overline{T}_{\mathrm{qm}}(L) = \overline{T}_{\mathrm{cl}}(L) - \frac{1}{M} \frac{L}{L + L_0} \, . \tag{1.93}$$

For large samples, where $L \gg L_0$ we can approximate this expression by

$$\overline{T}_{\mathrm{qm}}(L) \approx \overline{T}_{\mathrm{cl}}(L) - \frac{1}{M} \, . \tag{1.94}$$

This finally leads, via the Landauer-Büttiker formula (eq. (1.38)), to

$$G_{\mathrm{qm}} \approx G_{\mathrm{cl}} - \frac{2e^2}{h} \, .$$

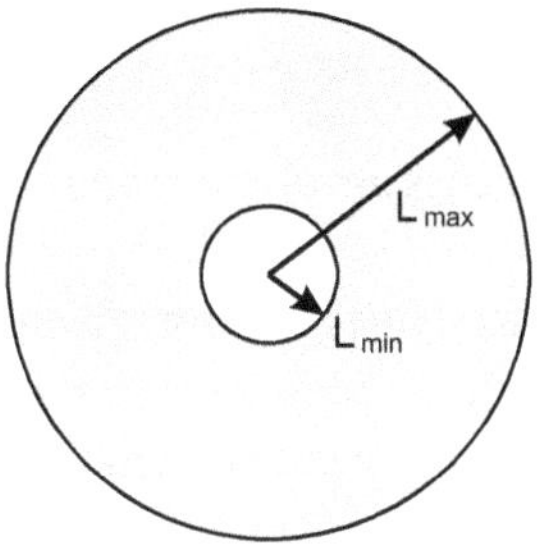

Fig. 1.25: Illustration of a two dimensional circular conductor with an inner radius of L_{min} and an outer radius of L_{max} featuring the area which is explored by a diffusive path propagating between the two lengths from an origin.

Due to the coherent backscattering, the average conductance of a phase coherent sample is reduced by $2e^2/h$ from its classical value. This effect is qualitatively observable even for samples, which are larger than the phase coherence length l_φ, since individual return paths can be smaller than l_φ.

Next, we calculate the corresponding reduction in the conductivity of a two-dimensional sample with width and length, both much larger than the phase coherence length l_φ. We consider a circular conducting region, which an electron can explore within τ_φ (Fig. 1.25). It is straightforward to show geometrically that the conductance G of such a circular conductor is related to its conductivity by the relation

$$G = \frac{\pi\sigma}{\ln(L_{max}/L_{min})},$$

where L_{max} and L_{min} are the outer and inner radii of the circular conductor (Fig. 1.25). It is natural to identify the outer radius for interfering paths with the phase coherence length, i.e., $L_{max} = l_\varphi$ (Section 1.5.1.1). The inner radius is not so clear-cut. It should be of the order of the mean free path l_m, but the precise value does not lead to any easily observable consequences as long as L_{min} is unaffected by temperature or magnetic field. It is common to set $L_{min} = l_m$ (Section 1.5.1.1). For this circular phase coherent conductor, we could argue as before that the conductance should be reduced by $(2e^2/h)$ due to coherent backscattering, so that

$$\frac{\pi\sigma_{qm}}{\ln(l_\varphi/l_m)} = \frac{\pi\sigma_{cl}}{\ln(l_\varphi/l_m)} - \frac{2e^2}{h}. \tag{1.95}$$

This finally leads to the same result as obtained in Section 1.5.1 (eq. (1.78)),

$$\sigma_{qm} = \sigma_{cl} - \frac{2e^2}{\pi h}\ln\left(\frac{l_\varphi}{l_m}\right). \tag{1.96}$$

Hence, considering the reflection probabilities of electrons at the boundaries of a mesoscopic sample area (eq. (1.96)) or the return probabilities via diffusion (eq. (1.78)) leads to the same value of the quantum correction to the conductivity.

1.6 Strong localization and scaling theory

In the previous section, we have discussed phase coherent transport phenomena in the regime of $k_F l_m \gg 1$. In particular, we have seen that, at low temperatures and correspondingly long phase coherence lengths, quantum interference phenomena are crucial. They give rise, e.g., to universal conductance fluctuations, Aharonov–Bohm oscillations, and Altshuler–Aronov–Spivak oscillations. Moreover, we have seen that weak localization leads to small quantum corrections to the classical conductivity. In this section, we will consider the case where $k_F l_m < 1$. For the sake of simplicity, we will discuss only phenomena where electron-electron interaction can be neglected. Such effects will be discussed in Chapters 4 and 5 of this book. In the actual section, we will find an insulating state without electron-electron interaction for the case that E_F is located within a band. This obviously deviates from the expectation of Bloch states, e.g., within the Boltzmann relaxation model, which implies metallic conduction, if E_F is located within an electron band. The insulating state in the presence of disorder is caused by the thermally activated hopping between localized states within this band, which is suppressed at low temperature.

1.6.1 Anderson localization

When the mean free path l_m is on the order of the Fermi wavelength ($\lambda_F \sim l_m$), the wave function tends to localize. In the case of

$$k_F l_m < 1 , \tag{1.97}$$

we talk about strong localization or Anderson[32] localization.

In 1958, Anderson pointed out that the electronic wave function in a random potential may be profoundly altered if the randomness is sufficiently strong. Most importantly, Anderson showed that if the disorder is very strong, the wave function becomes localized due to quantum interference (Fig. 1.26(a)). This leads to a wave function exhibiting an envelope, which decays exponentially from some point in space (Fig. 1.26(a)), i.e.,

$$|\psi(\vec{r})|^2 \sim e^{-|\vec{r}-\vec{r}_0|/\xi} , \tag{1.98}$$

where ξ is the localization length or localization radius (see also Section 5.3.1). This contrasts with the traditional view, where the wave functions remain extended throughout the sample and scattering by a random potential causes the Bloch waves only to change their quantum mechanical phase factor (Fig. 1.26(b)), also via changing the direction of $\vec{k}$. The presence of localized states introduces the new length scale ξ, which plays an important role for understanding of the corresponding metal-insulator transition.

32 Philip W. Anderson (1923–) is an American physicist.

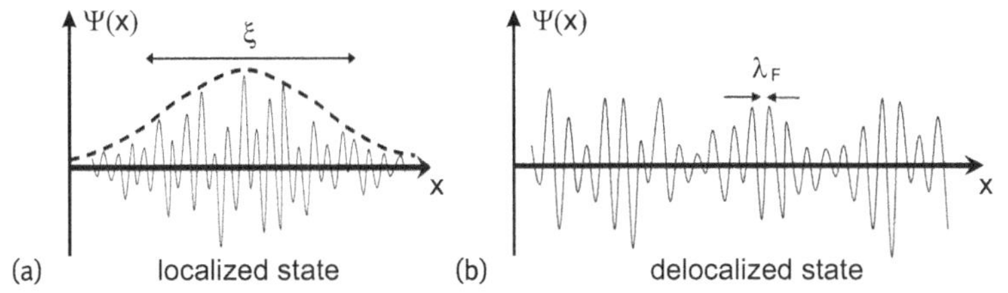

Fig. 1.26: Schematic illustration of (a) a localized state with a localization length ξ and (b) an extended state. Note that in both cases the wave function is oscillating on the length scale λ_F.

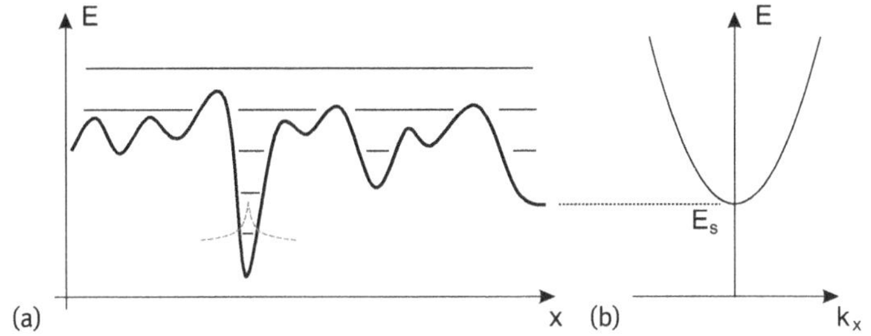

Fig. 1.27: (a) Illustration of a disorder potential exhibiting an extended state (highest horizontal line) and localized states (other horizontal lines). At the lowest state in energy, the electron wave function is sketched as a blue line. (b) Parabolic dispersion highlighting that the disorder preferentially localizes the states originating from the band edge E_S.

The existence of localized states can be well understood at the limit of very strong disorder. Then, a zeroth-order description of the eigenstate would be a bound state within the deepest fluctuation in the random disorder potential (blue line in Fig. 1.27). The admixture of this strongly localized state with other orbitals from different locations could be considered as a perturbation. Crucially, such admixtures will not produce an extended state composed of linear combinations of infinitely many localized orbitals. The main reason is that localized states, i.e., orbitals that are nearby in space, so that the wave functions overlap significantly, are in general very different in energy, so that the admixture is small, because of the large energy denominator appearing in a perturbation-type description. However, localized states that are nearly degenerate in energy are typically very far apart in space, so that the spatial overlap of its wave functions is exponentially small. Thus, in the strongly disordered limit, some wave functions will be exponentially localized. Actually, it is easier to establish localized states in the presence of disorder than to establish extended states. We will discuss below that, in one dimension, it can be shown rigorously that all states are localized, no matter how weak the disorder.

More generally, the solution for the electronic transport with strong disorder is called "Anderson localization theory." It predicts that infinitely large one- and two-dimensional systems are always insulators at $T = 0\,\mathrm{K}$ due to localization and gives useful indications on the nature of the metal-insulator transition in three dimensions.

1.6.2 Mobility edge and metal-insulator transition

In Sections 1.4 and 1.5, we have seen that the constructive interference of time-reversed paths gives rise to a quantum correction to the conductivity. This invites the question, how the accumulation of quantum interference corrections affects the transport in a strongly disordered environment. As pointed out above, the electronic states tend to localize in such an environment with the characteristic localization length ξ. This length limits the radius, which the electron can explore by diffusion, i.e., $\langle \vec{r}^2 \rangle = Dt \to \xi^2$ for $t \to \infty$. Hence, the diffusion constant D as well as the conductivity $\sigma = e^2 \mathcal{D} D$ (eq. (1.16)) are going to zero for large samples with length and width much larger than ξ. Thus, a strongly disordered system becomes insulating ($\sigma = 0\,\mathrm{S}$) at $T = 0\,\mathrm{K}$.

Importantly, the metal-insulator transition is directly connected to the localization length and takes place when $k_F l_m \approx 1$. The phase transition is called the Anderson transition, which exhibits the strength of the disorder as the order parameter. This is in contrast to the Mott[33] transition, where the metal-insulator transition is driven by changing the electron correlations relative to the band width, but at a fixed disorder potential (Section 4.5.3). Sometimes, a clear distinction between the two mechanisms is experimentally difficult and therefore *electronically-driven*[34] metal-insulator transitions are frequently called Mott–Anderson transitions.

According to its energy, a given state can either be extended or localized depending on the degree of disorder. Crucially, extended and localized states at the same energy do not coexist. This is due to the fact that the admixture of an extended state at

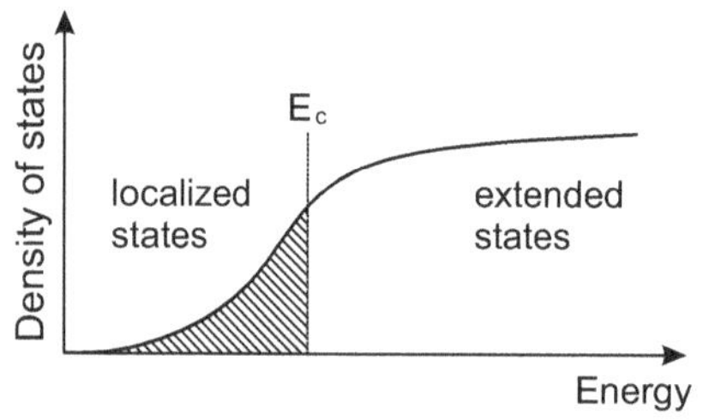

Fig. 1.28: Density of states of a three-dimensional electron system as a function of energy highlighting the mobility edge E_C, which separates energy regions of localized states (hatched area) and of extended states (white area).

33 Sir Nevill F. Mott (1905–1996) was a British physicist who shared the Nobel Prize for Physics in 1977 with P. Anderson and J. H. Van Vleck.

34 The *electronically-driven* transitions are distinct from *crystallographically-driven* metal-insulator transitions, where due to structural changes of the atomic lattice the band structure gets modified. The latter play an important role in so-called phase change materials, which for example are used for optical data storage in DVDs and Blue-ray Discs.

the same energy by arbitrarily small perturbations would lead to the delocalization of a formerly localized state. Moreover, it is natural that states at the band edge are more susceptible to the formation of localized states (Fig. 1.27). These considerations suggest a profile for the density of states, in which the low-lying localized states are separated from higher-energy extended states by the so-called mobility edge. This is illustrated in Fig. 1.28, where the localized states (hatched area) are separated from the extend states (white area) and the energy of the mobility edge is given by the critical energy, E_C (see also Fig. 5.7). The transition signaled by the mobility edge is the Anderson transition.

1.6.3 Thermally activated conduction and hopping transport

In this section, we will extend the discussion on charge transport to insulators and systems with localized states. So far, we discussed transport in metals and doped semiconductors with extended states.

If the states at the Fermi energy E_F are localized, $E_F < E_C$, the quantum conduction at $T = 0$ K is zero. At low but finite temperatures, the electrons can gain thermal energy typically from other excitations, in particular phonons, enabling charge transport via extended states above E_C or via hopping between localized states. This allows for different transport mechanisms, including pure thermal activation and various hopping processes (Fig. 1.29).

For localized states energetically close to the mobility edge or at elevated temperatures, the activation to and above the mobility edge (Fig. 1.29(a)) yields a conductivity

$$\sigma = \sigma_0 e^{-(E_C - E_F)/k_B T} . \tag{1.99}$$

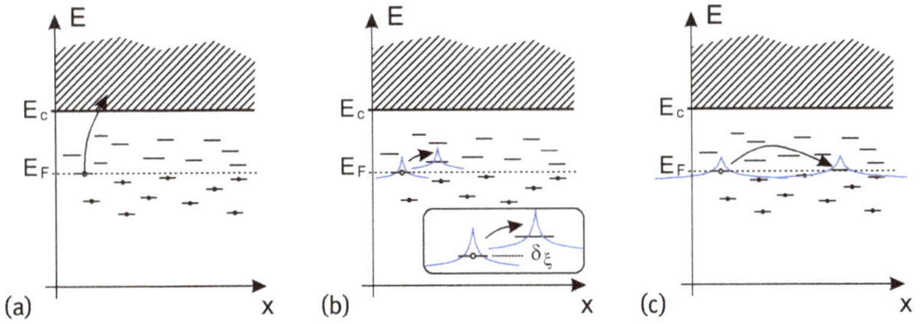

Fig. 1.29: Illustrations of the different hopping processes which contribute to transport: (a) Thermal activation of localized electrons into extended states. (b) Nearest neighbor hopping between localized states. The inset highlights the required energy penalty δ_ξ. (c) Variable range hopping between localized states at similar energy but larger distance. The black horizontal lines are energy levels, the blue lines depict electron wave functions and the hatched area marks the energy region of extended states.

The coefficient σ_0 is proportional to the square of the electron-phonon coupling strength (eq. (5.5)). The temperature dependence of σ, hence, enters via the thermal smearing of the Fermi–Dirac distribution, which is expressed by the Boltzmann factor $\exp(-(E_C - E_F)/k_B T)$.

For localized states at lower energies and in particular at lower temperatures, the transport is dominated by so-called hopping transport (Fig. 1.29(b), (c)). The conductivity in this case is then proportional to the possibility of hopping of electrons between localized states. By assuming that the transition from a localized state on site i to one on site j is accompanied by the adsorption of a phonon, we can express the hopping probability by

$$P_{ij} = F\left(1 - f_j(E, T), f_{ij}^{\text{Bose}}, f_i(E, T)\right) \left|\langle\psi_j| e^{i\vec{q}\cdot\vec{r}} |\psi_i\rangle\right|^2 \delta(\hbar\omega_{\text{ph}} - \Delta E_{ij}), \qquad (1.100)$$

where $\omega_{\text{ph}} = v_a|\vec{q}|$ is the frequency of the acoustic phonon with momentum $\vec{q}$ and speed v_a. The term $\Delta E_{ij} = E_j - E_i$ is the energy difference between the localized states on site j and i. The statistical factor F takes into account that for a hopping process the initial state needs to be filled (Fermi distribution $f_i(E, T)$), the final state needs to be empty $(1 - f_j(E, T))$ and phonon emission or adsorption processes have to be accounted for (f_{ij}^{Bose}). The leading term of the statistical factor can be approximated by,

$$F(1 - f_j, f_{ij}^{\text{Bose}}, f_i) \propto e^{-\epsilon_{ij}/(k_B T)}. \qquad (1.101)$$

Here, $\epsilon_{ij} \simeq \Delta E_{ij} \ll k_B T$ depends on the specific transport scenario as discussed below. Moreover, in a first approximation the squared matrix element can be simplified to

$$\left|\langle\psi_j| e^{i\vec{q}\cdot\vec{r}} |\psi_i\rangle\right|^2 \propto e^{-2r_{ij}/\xi}, \qquad (1.102)$$

where r_{ij} is the distance between site i and site j. The factor two in the exponent results from the squaring on the left hand side of eq. (1.102). Consequently, we can approximate the hopping probability by

$$P_{ij} \approx P_0 e^{-\epsilon_{ij}/(k_B T)} e^{-2r_{ij}/\xi}. \qquad (1.103)$$

Notice that the hopping probability is directly connected to the conductivity, since hopping of electrons transports charge. In the following, we will discuss the temperature dependence of the conductivity and we will distinguish between nearest neighbor hopping (Fig. 1.29(b)) and variable range hopping (Fig. 1.29(c)).

1.6.3.1 Nearest neighbor hopping (NNH)

If the localization length is ξ and the density of states per volume at the Fermi energy is $\mathcal{D}(E_F)$, the number of states per energy in a volume of linear dimension ξ in d dimensions is given by $\mathcal{D}(E_F)\xi^d$. Hence, the typical energy separation between overlapping states on the scale of ξ is

$$\delta_\xi \approx \frac{1}{\mathcal{D}(E_F)\xi^d}. \qquad (1.104)$$

This yields an activated nearest neighbor hopping process between localized states proportional to $e^{-\delta_\xi/k_B T}$ and, thus, a conductivity of the form[35]

$$\sigma_2 = \sigma_0 e^{-\delta_\xi/k_B T} . \tag{1.105}$$

Hence, one expects an Arrhenius type of temperature dependence for the conductivity and, respectively, the resistivity $\rho(T)$, which provides an experimental access to ξ, if $\mathcal{D}(E_F)$ is known. Exemplary experimental $\rho(T)$ data of a three-dimensional indium phosphide (InP) sample are shown in Fig. 1.30 [13]. The Arrhenius representation shows ρ in log-scale as a function of $1/T$. At high temperatures, i.e., small $1/T$, the experimental data indeed follow the Arrhenius type of the temperature dependence, $\rho_2 = \sigma_2^{-1} \propto e^{\delta_\xi/k_B T}$ (dash-dotted curve labeled NNH).

1.6.3.2 Variable range hopping (VRH)

However, as suggested by Mott, it pays sometimes for the electron to hop a larger distance, thereby reducing the necessary inelastic energy transfer. This introduces the next type of activated conductivity called variable range hopping, particularly important at low temperatures. We assume that the contribution of the hopping conductivity to a state localized a distance $L \gg \xi$ away from the starting state is proportional to the overlap matrix element squared, which is proportional to $e^{-2L/\xi}$ (eq. (1.103)). On the other hand, the energy distance between these two spatially distant states δ_L, is obtained by generalizing the argument leading to eq. (1.105) and noting that δ_L is proportional to L^{-d} (d: dimensionality) while $\mathcal{D}(E_F) \simeq 1/(\xi^d \delta_\xi)$. Hence, we get

$$\delta_L \approx \frac{1}{\mathcal{D}(E_F)L^d} \approx \delta_\xi \left(\frac{\xi}{L}\right)^d , \tag{1.106}$$

for $L \gg \xi$. The hopping probability over a length L is then controlled by $P_{ij} \propto e^{-2L/\xi - \delta_L/k_B T}$ (eq. (1.103)). At low temperatures, it pays to make hops with $L \gg \xi$.

The optimal length for such jumps L^* is given by minimizing the exponent with respect to L leading to

$$0 = \frac{2}{\xi} + \frac{\mathrm{d}}{\mathrm{d}L}\left(\frac{\delta_L}{k_B T}\right) = \frac{2}{\xi} - \frac{d}{\mathcal{D}(E_F)L^{d+1}k_B T} \tag{1.107}$$

Solving for L, gives the optimal hopping length

$$L^* \approx \left(\frac{d\xi}{2\mathcal{D}(E_F)k_B T}\right)^{1/(d+1)} . \tag{1.108}$$

Such a long-range hopping mechanism, selecting an optimal hopping length L^*, is relevant as long as $L^* \gg \xi$, i.e., when the temperature $T \ll T_{VRH}$ with

$$T_{VRH} \approx d\delta_\xi / k_B . \tag{1.109}$$

35 We use eq. (1.101) with $\epsilon_{ij} = \delta_\xi$.

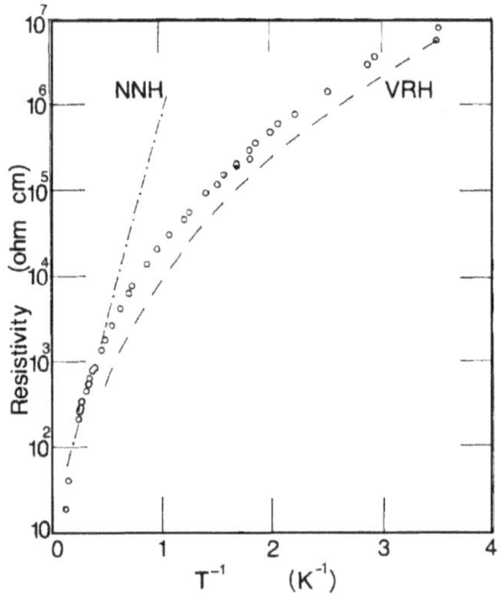

Fig. 1.30: Comparison of theoretical and experimental resistivity (data points) of a three-dimensional indium phosphide sample as function of temperature. The dash-dotted curve is the theoretical resistivity deduced for nearest neighbor hopping (NNH) (eq. (1.105)). The dashed curve is the theoretical resistivity for variable range hopping (VRH) (eq. (1.110)) [13].

At such low temperatures, the variable range hopping (VRH) conductivity σ_3 is given by

$$\sigma_3 = \tilde{\sigma}_0 e^{-C(T_{\text{VRH}}/T)^{1/(d+1)}} , \tag{1.110}$$

where C is a dimensionless constant and $\tilde{\sigma}_0$ is a prefactor depending on details of the wave functions and the band structure. Indeed, the inverse equation for $d = 3$, reading $\rho_3 = \sigma_3^{-1} \propto e^{C(T_{\text{VRH}}/T)^{1/4}}$, largely fits the low temperature data (large $1/T$) of the three-dimensional InP sample (Fig. 1.30, dashed curve labeled VRH).

1.6.4 Thouless energy

Remembering that the expression of the Drude conductivity can be written in the form of the zero-temperature Einstein relation (eq. (1.16))

$$\sigma = \frac{e^2 n_e \tau_{\text{sc}}}{m^*} = e^2 \mathcal{D}_d(E_F) D , \tag{1.111}$$

where the diffusion constant is $D = v_F^2 \tau_{\text{sc}}/d$ for a d-dimensional system and the density of states at the Fermi energy E_F, $\mathcal{D}_d(E_F)$ in d-dimensions ($d = 1, 2, 3$), we now determine the conductance $G(L)$ of a cubic ($d = 3$), square ($d = 2$), or stretched ($d = 1$) piece of material with side length L, which is much bigger than the mean free path l_m of the electrons. With the length of the sample L and its area perpendicular to the current direction being L^{d-1}, we get

$$G(L) = \sigma \frac{L^{d-1}}{L} = \frac{e^2}{\hbar} \left(\mathcal{D}_d(E_F) L^d \right) \frac{\hbar D}{L^2} . \tag{1.112}$$

In this expression, the quantity $\mathcal{D}_d(E_F) L^d = 1/\delta E$ is the number of quantum states at the Fermi energy per unit energy interval within the whole sample, which introduces the typical energy level spacing δE. The quantity $\hbar D/L^2$, hence, has the dimension of an energy. It is called the Thouless[36] energy

$$E_{\mathrm{Th}} = \frac{\hbar D}{L^2} \,. \tag{1.113}$$

The physical meaning of the Thouless energy becomes intuitive, when considering it as the inverse of a time scale $\tau_{\mathrm{Th}} = \hbar/E_{\mathrm{Th}}$. This time scale is a measure for the time that an electron needs in order to explore the area L^2 diffusively. In other words, τ_{Th} is the time, at which on average the diffusing electron reaches the boundary of the sample. In the picture of wave functions, E_{Th} tells us how sensitive the energy of a particular wave function is to a change of the boundary conditions at the sample edge. It is intuitive that a strongly localized state will hardly change when the conditions at the sample boundary are changed (E_{Th} small) (see also Section 5.4.4). In contrast, an extended state changes strongly (E_{Th} large). From eq. (1.112), the conductance of the square of material can be written as

$$G(L) = \frac{e^2}{\hbar}\frac{E_{\mathrm{Th}}(L)}{\delta E(L)} \,. \tag{1.114}$$

The conductance is equal to the conductance quantum times the number of energy levels within the energy interval E_{Th}. The dimensionless conductance $\tilde{g}(L) = G(L)/(2e^2/h)$ is frequently called the Thouless number. In the range of validity of the Drude–Boltzmann theory we have $\tilde{g}(L) \gg 1$, i.e., $E_{\mathrm{Th}} \gg \delta E$. In the regime of strong localization $\tilde{g}(L) \ll 1$ and correspondingly $E_{\mathrm{Th}} \ll \delta E$. We can estimate the localization length via the condition $\tilde{g}(\xi) \approx 1$, if an expression for $\tilde{g}(L)$ is known. For two limiting cases, $\tilde{g}(L)$ is indeed known:

- For $\tilde{g}(L) \gg 1$, we have the expression from the Drude model (eqs. (1.112))

$$\tilde{g}(L) = \frac{h}{2e^2}\sigma L^{d-2} \,. \tag{1.115}$$

- For $\tilde{g}(L) \ll 1$, we use the expression for the strong localization regime at $T = 0\,\mathrm{K}$ (Section 1.6.3.2), where the following expression is found

$$\tilde{g}(L) = \tilde{g}_0 \mathrm{e}^{-2L/\xi} \tag{1.116}$$

  with $\tilde{g}_0$ being a constant.

1.6.5 Scaling theory of localization

The conductance of a macroscopic disordered sample can usually neither be calculated analytically nor numerically, if one starts from the Schrödinger equation. This is

36 David J. Thouless (1934–) is a British-US condensed-matter physicist. He is a laureate of the 2016 Nobel Prize for physics.

due to the huge number of electrons implying a huge number of degrees of freedom. The goal of the scaling theory of localization is the calculation of the conductance by other means. The basic idea is the following: the sample is considered to be divided in smaller blocks of material with side length $L \gg l_{\mathrm{m}}$, for which a numerical calculation of $G(L)$ is possible. The conductance of the macroscopic sample is then deduced from that of the small cube by subsequent doubling of the cube's side length. This procedure can be applied repetitively in one, two, or three dimensions. The important ingredient is that the conductance $G(2L)$ of a piece of material with side length $2L$ depends only on the conductance $G(L)$. Mathematically, this can be expressed by the relation

$$\frac{\mathrm{d}\ln\tilde{g}}{\mathrm{d}\ln L} = \beta(\tilde{g}) , \tag{1.117}$$

employing the dimensionless conductivity $\tilde{g}$ (Thouless number), where $\beta(\tilde{g})$ is called the scaling function, and $\tilde{g} = E_{\mathrm{Th}}/\delta E$ is called the scaling parameter.

Next, we argue that $\tilde{g}$ is indeed the only scaling parameter. Assuming that we have solved Schrödinger's eigenvalue problem for two material blocks of size L, such that the energy levels and wave functions are known, we find the solution for the combined system by matching the wave functions at the boundaries where the two blocks touch. States of the two blocks will mix to form extended states, if there are many levels in an energy interval E_{Th} in both blocks, i.e., if $E_{\mathrm{Th}}/\delta E \gg 1$. In turn, states of the two blocks will hardly mix, if there are few states within E_{Th}, i.e., if $E_{\mathrm{Th}}/\delta E \ll 1$. Therefore, it is reasonable to assume that $\tilde{g}(2L)$ is essentially depending on $E_{\mathrm{Th}}/\delta E = \tilde{g}(L)$.

The question is now, how the scaling function $\beta(\tilde{g})$ can be determined. In order to get a feeling for this function, we consider the well-known limiting cases:

- If $\tilde{g}(L) \gg 1$, we are in the limit of the Drude–Boltzmann theory and have

$$\tilde{g}(L) = \frac{h}{2e^2}\sigma L^{d-2} . \tag{1.118}$$

  In this limit, we find

$$\lim_{\tilde{g}\to\infty} \beta(\tilde{g}) = \frac{\mathrm{d}\ln\tilde{g}}{\mathrm{d}\ln L} = d - 2 , \tag{1.119}$$

  i.e., $\beta(\tilde{g})$ tends to a constant in all dimensions d.
- If $\tilde{g}(L) \ll 1$, we are in the limit of strong localization. In this regime, the conductance depends exponentially on the sample size, i.e., we have

$$\tilde{g}(L) = \tilde{g}_0 \mathrm{e}^{-2L/\xi} . \tag{1.120}$$

  This leads to

$$\lim_{\tilde{g}\to 0} \beta(\tilde{g}) = \frac{\mathrm{d}\ln\tilde{g}}{\mathrm{d}\ln L} = -\frac{2L}{\xi} = \ln\left(\frac{\tilde{g}}{\tilde{g}_0}\right) , \tag{1.121}$$

  in all dimensions.

From these two examples, we can see that the scaling function $\beta(\tilde{g})$ depends in both limiting cases only on $\tilde{g}$ as a sample specific measure. Scaling theory assumes that between these two limiting cases, the scaling function $\beta(\tilde{g})$ does also depend only on the scaling parameter $\tilde{g}$.

We now discuss qualitatively, what the suggested functional form of $\beta(\tilde{g})$ means for systems of different dimensions. Suppose, we have calculated $\tilde{g}(L)$ for a small system of size L. We can obtain the dimensionless conductance $\tilde{g}(L_{\mathrm{M}})$ of a macroscopic system of size L_{M} by considering the integral equation

$$\int\limits_{\ln \tilde{g}(L)}^{\ln \tilde{g}(L_{\mathrm{M}})} \frac{d \ln \tilde{g}}{\beta(\tilde{g})} = \int\limits_{\ln L}^{\ln L_{\mathrm{M}}} d \ln L = \ln\left(\frac{L_{\mathrm{M}}}{L}\right) . \tag{1.122}$$

Since $L_{\mathrm{M}} > L$, we find that for (i) $\beta(\tilde{g}) > 0$ follows $\tilde{g}(L_{\mathrm{M}}) > \tilde{g}(L)$ and (ii) $\beta(\tilde{g}) < 0$ follows $\tilde{g}(L_{\mathrm{M}}) < \tilde{g}(L)$.

1D-systems: For a one-dimensional system, $\beta(\tilde{g}) < 0$ for all values of $\tilde{g}$ (eq. (1.119) and (1.121)). This means that $\tilde{g}(L_{\mathrm{M}}) < \tilde{g}(L)$, no matter from which value $\tilde{g}(L)$ we start. The scaling theory predicts that macroscopic one-dimensional systems are always insulators, because for arbitrary large L_{M}, $\tilde{g}(L_{\mathrm{M}})$ becomes arbitrarily small.

3D-systems: The situation is different in the case of a three-dimensional system. For certain values of $\tilde{g}(L) > \tilde{g}_{\mathrm{c}}$, $\beta(\tilde{g})$ is a positive number (eq. 1.119), whereas for $\tilde{g}(L) < \tilde{g}_{\mathrm{c}}$ it is negative (eq. 1.121). This means that the conductance of a macroscopic three-dimensional system can either become very large with increasing system size, if $\tilde{g}(L)$ is bigger than $\tilde{g}_{\mathrm{c}}$, which is called metallic behavior. On the other hand, the conductance can become arbitrarily small, if $\tilde{g}(L) < \tilde{g}_{\mathrm{c}}$, which is called insulating behavior. In three-dimensional systems, we, consequently, have a metal-insulator transition as a function of the strength of potential fluctuations in the sample.

2D-systems: The case of the two-dimensional system is the most interesting one, because the limit of $\beta(\tilde{g})$ is zero for $\tilde{g} \to \infty$ (eq. 1.119). It is therefore very important to find out, whether $\beta(\tilde{g})$ is positive or negative for large $\tilde{g}$. In the first case, we would expect a metal-insulator transition in two-dimensional systems, whereas in the other case, all macroscopic two-dimensional systems would be insulating. In 1979, Abrahams et al. [14] suggested a result that was later proven to be correct. According to their theory, the scaling function in two-dimensions has for large $\tilde{g}$ the asymptotic form

$$\beta(\tilde{g})(L) = -\frac{\tilde{g}_1}{\tilde{g}} , \tag{1.123}$$

where $\tilde{g}_1 \approx 1$. The prediction is therefore that in two-dimensions, $\beta(\tilde{g}) < 0$, i.e., all macroscopic systems are insulators and there is no metal-insulator transition. By using eq. (1.122) and plugging in $\beta(\tilde{g})(L) = -\tilde{g}_1/\tilde{g}$, we find

$$\tilde{g}(L_{\mathrm{M}}) = \tilde{g}(L) - \tilde{g}_1 \ln\left(\frac{L_{\mathrm{M}}}{L}\right) . \tag{1.124}$$

The logarithmic term makes the result different from the Drude–Boltzmann limit for $d = 2$. In fact, this term reminds us of the logarithmic quantum mechanical correction $\delta\sigma_{qm}$ of the Drude conductivity found as a result of enhanced coherent backscattering (eq. (1.78)). Indeed, we can rewrite this expression using $\tilde{g}_1 \simeq 1$

$$\sigma(L_M) \simeq \sigma - \frac{2e^2}{h} \ln\left(\frac{L_M}{L}\right) . \tag{1.125}$$

In two-dimensional systems, there is thus a logarithmic correction to the Drude–Boltzmann conductivity with its origin in the quantum diffusion of electrons. The above equation can, however, only be used as long as the logarithmic term is smaller than the Drude–Boltzmann σ, since negative σ do not exist. From eq. (1.124), we can again estimate the localization length ξ for a two-dimensional system using the relation $\tilde{g}(\xi) \approx 1$. Employing eq. (1.15), one of which reads $\sigma = e^2/h \cdot k_F l_m$, we obtain

$$1 = \frac{k_F l_m}{2} - \tilde{g}_1 \ln\left(\frac{\xi}{l_m}\right) \Longrightarrow \xi \approx l_m e^{k_F l_m/2} \tag{1.126}$$

with $\tilde{g}_1 \simeq 1$. Moreover, we have assumed that $k_F l_m \gg 1$, and we have set $L \approx l_m$ which gives only a small error, because of the logarithm. We see that the localization length grows exponentially with $k_F l_m$. If we consider a high mobility two-dimensional electron gas with a density $n_e = k_F^2/2\pi = 3 \cdot 10^{11}\ \text{cm}^2$ (text following eq. (1.14)) and a low-temperature mobility $\mu = e\tau_{sc}/m^* = el_m/v_F m^* = el_m/\hbar k_F = 10^6\ \text{cm}^2/(\text{Vs})$ (eq. (1.10)), we find $k_F l_m = h\mu n_e/e = 1241$ and $l_m = \hbar\mu\sqrt{2\pi n_e}/e = 9\ \mu\text{m}$. This leads to $\xi \approx 2 \cdot 10^{264}$ m which is an astronomically large length scale. In turn, for an electron gas with the same density, but a mobility $\mu = 10^4\ \text{cm}^2/(\text{Vs})$, we find $k_F l_m = 12.4$, $l_m = 90$ nm and $\xi \approx 44\ \mu\text{m}$, such that localization becomes relevant for mesoscopic samples. For even smaller mobilities, the regime of strong localization is reached even more quickly.

1.7 Summary

This first chapter of the book was dedicated to the description of electronic transport at mesoscopic length scales, which are scales well above the atomic limit but below the limit, where phase coherence of the electron wave functions is lost. We found a number of surprising transport phenomena, which evolved from a description of ballistic transport in terms of open modes via the Landauer-Büttiker formalism. This description was extended up to a macroscopic sample size, where it leads to important corrections to the conductivity such as the contact resistance, universal conductance fluctuations and weak localization. The formalism also predicts straightforwardly oscillatory quantum effects at smaller length scales such as the Aharonov–Bohm effect or the Altshuler–Aronov–Spivak oscillations, which are both verified experimentally. Hence, the mesoscopic length scale is the scale, where quantum effects become apparent due to a subtle interplay between phase preserving and phase disturbing scattering processes, both caused by static or dynamic disorder within a crystalline sample.

The distinction between such a more diffusive quantum description and a quantum description via tunneling processes between localized states has been additionally analyzed leading to the so-called scaling theory of conductance. For the sake of simplicity, we neglected electron-electron interactions so far, which, however, can be implemented into the formalisms via modifications of the scattering events discussed so far and via introducing of additional scattering channels [12]. This hampers the transparency of corresponding calculations, but often contains open scientific questions. More fundamental and well understood consequences of the electron-electron interaction will be discussed in chapters 4 and 5 of this book.

Recent ongoing experiments in mesoscopic physics tackle, e.g., transport phenomena of samples including interfaces between different materials. Partly, these materials are strongly influenced by electron-electron interaction effects such as superconductors, Mott-insulators or ferromagnets. Other experiments explore the role of additional degrees of freedom of the electrons, such as, e.g., different valleys of the band structure at E_F or a sublattice degree of freedom within the unit cell, which are present, e.g., in graphene or in similar 2D materials. We believe that this introduction into basic principles of mesoscopic physics is also an excellent starting point for approaching these scientific questions.

Gero von Plessen

2 Interaction of light with electrons

2.1 Introduction

In the previous chapter of this book, we have described transport properties of electrons in solids, in particular at mesoscopic length scales. The quantum mechanical phase of the electron wave function, which is decisive for the interference of electrons, was crucial for the measurable conductivity in different types of devices. Hence, quantum mechanics was central. However, we mostly did not solve the Schrödinger or Dirac equation explicitly. Instead, we have built on a more intuitive description of general wave properties in order to explain the different experimentally relevant effects.

This points to a very general approach in experimental solid state physics, justified by the impossibility of solving the fundamental equations such as the Schrödinger equation rigorously for the many degrees of freedom present in a solid. Instead, one needs to operate with adequate approximations, often chosen by intuition or guided by experimental results and justified in detail later. Thus, one often switches between different model languages, appropriate for simple intuition, but consistent with respect to the subsequent detailed justification.

In the present chapter of the book, which deals with the optical properties of solids, we will switch even more extensively between different model languages employing mechanical, electrodynamical, and quantum mechanical explanations. Concretely, we will describe the interaction of electrons with light, which, e.g., determines how we see solids and provides us with means to distinguish between different solids. The detailed understanding of the resulting optical properties has enabled many technical applications of tremendous importance in daily life such as lasers, photodiodes, and solar cells. The interaction of light with electrons is also of fundamental interest as it gives us access to complex many-body interactions of the electrons in solids, whose understanding is still a central challenge in solid state physics (Chapters 4 and 5). While some of these issues can be understood by referring to classical, electrodynamic models describing light as an electromagnetic field and the electrons as classical particles, some of them require treating either the electrons or light or both quantum mechanically.

In order to deal with these different levels of complexity, we will first review some of the electrodynamics of light in solids as well as quantum mechanical descriptions of the transition between different energy levels of an electron due to an electromagnetic field. We will then focus on a special type of optical excitations in solids, namely interband transitions, in which an optical transition of an electron from one band to another occurs. There, we treat the electrons quantum mechanically, namely as Bloch states located in electron bands, but introduce the action of the light as purely provid-

https://doi.org/10.1515/9783110438321-002

ing the energy for the transition, and keep their quantum mechanical photon character in mind. The interband transitions are of particular importance for the optical properties of semiconductors and are the base for several optoelectronic devices, namely photodiodes, photovoltaic solar cells, light-emitting diodes and semiconductor lasers. After discussing the interband transitions, we will treat the way in which the resulting excited electron and hole can be bound to each other in a state called an exciton. We will see how interband electron-hole excitations can, in turn, emit light, a light emission called luminescence. Finally, we will deal with the free-electron currents in metals and doped semiconductors that give rise to their characteristic reflection spectra.

Along the way, we will familiarize ourselves with some of the most common spectroscopic methods used to measure the optical properties of solids, namely absorption spectroscopy, photoluminescence spectroscopy and reflection spectroscopy.

2.2 Fundamentals of light-matter interaction

In this section, we will summarize some basic aspects of light-matter interaction that are important for treating the optical properties of solids. The light-matter interaction will be considered at different formal levels:

- exclusively using methods from classical physics,
- semi-classically, i.e., matter will be treated quantum mechanically and the light field classically,
- quantum optically, i.e., also the light field is treated quantum physically.

Each level will be chosen according to what is required in order to describe the physical situation correctly and to be as transparent as possible at the same time.

We start by briefly reviewing what should be known a priori from introductory lecture courses or textbooks, namely the central macroscopic terms which describe the light-solid interaction:

Propagation, i.e., the process of light traveling in a medium, usually at a phase velocity lower than in vacuum,

light scattering, i.e., a process in which incident light changes its propagation direction due to an interaction with an optical inhomogeneity in its propagation path,

reflection, i.e., a special case of light scattering in which a smooth solid surface scatters an incident light beam in a single direction into the half-space from which it was incident,

refraction, i.e., a special case of light scattering in which a smooth solid surface transmits an incident light beam in a single direction different from the direction of incidence,

absorption, i.e., an uptake of light energy by the medium without coherent re-emission of this energy at the same frequency.

Microscopically, all of these phenomena result from the interaction of light with electrons in the medium. The high-frequency oscillations of the electromagnetic field (in the visible, near-infrared and ultraviolet frequency ranges) drive periodic motions of these electrons. The resulting charge oscillations lead to the emission of light at the same frequencies and, in non-transparent media, also result in the generation of heat.

2.2.1 Light waves in vacuum

In this section, the properties of light without interaction with matter, i.e., in vacuum, will be briefly summarized. This is needed for an understanding of the following Section 2.2.2.

Light can be described as an electromagnetic wave. Its wave nature is concluded from the characteristic wave behavior that it shows in experiments (propagation, refraction, dispersion, interference, diffraction). Its electromagnetic nature can be concluded from the fact that it is generated by rapidly oscillating charges and exerts forces on charges and currents. Visible light differs from other electromagnetic waves by its wavelength, which ranges from approximately 390 nm to approximately 780 nm, i.e., from the color blue via green, yellow and orange to red. At shorter wavelengths, we have the ultraviolet (UV) regime and at longer wavelengths, the infrared (IR) regime. The term light is also often used for radiation in these wavelength ranges. Light from natural light sources such as incandescent lamps can have many different wavelengths. If it has only one wavelength, e.g., generated by a laser, it is called monochromatic. The spectral distribution of light can be determined experimentally by analyzing it in a grating spectrometer and measuring the light power at the individual wavelengths with a detector. Light waves can theoretically be described using classical electrodynamics. In this approach, light waves represent oscillations of the $\vec{B}$ and $\vec{E}$ field which travel in space, in a process called wave propagation. They obey the electromagnetic wave equation

$$\nabla^2 \vec{E} = \frac{1}{c^2}\frac{\partial^2 \vec{E}}{\partial t^2} \quad \text{and} \quad \nabla^2 \vec{B} = \frac{1}{c^2}\frac{\partial^2 \vec{B}}{\partial t^2} \tag{2.1}$$

for the electric field $\vec{E}$ and the magnetic induction $\vec{B}$. ∇ is the Nabla operator and c the phase velocity of the waves. In vacuum, it has the value $c_0 = 3.00 \times 10^8\ \mathrm{m\,s^{-1}}$. One class of solutions to the wave equation are plane harmonic waves of the form

$$\vec{E}(\vec{r}, t) = \vec{E}_0 \cos(\vec{k} \cdot \vec{r} - \omega t)\ , \tag{2.2}$$

with an analogous form for $\vec{B}$. Here, the wave vector $\vec{k}$ describes the propagation direction of the wave. Its magnitude is connected to the wavelength λ via $|\vec{k}| = 2\pi/\lambda$. Here, ω is the angular frequency. The direction of the amplitude vector $\vec{E}_0$ is the direction of the electric field of the wave, the so-called polarization direction. The magnetic field

$\vec{B}$ obeys the relation

$$\vec{B} = \frac{1}{\omega}\left(\vec{k} \times \vec{E}\right) , \tag{2.3}$$

i.e., $\vec{B}$, $\vec{E}$ and the propagation direction of the wave along $\vec{k}$ are pairwise perpendicular to each other. The plane that is spanned by $\vec{E}_0$ and the propagation direction is called the polarization plane. If the polarization plane of the wave remains the same independently of time, the wave is called linearly polarized. For instance, the light from many laser sources is linearly polarized. Light from natural light sources such as the sun or glowing bodies is unpolarized, i.e., this light consists of a superposition of many wave trains with statistically distributed polarization directions. The argument $\vec{k} \cdot \vec{r} - \omega t$ of the cosine function (eq. (2.2)) is called the phase of the wave. All space points of equal phase at a fixed time t are part of a phase front or phase plane. The velocity at which a phase front travels through space is given by the phase velocity of the light wave

$$c = \frac{\lambda}{T_\omega} = \frac{\omega}{k} , \tag{2.4}$$

where T_ω is the time period of the oscillation. The relation derived from this expression,

$$\omega(k) = c \cdot k , \tag{2.5}$$

is often called the dispersion relation of the light wave. If two waves exhibit a spatially or temporally constant difference between their phases, they are called coherent with each other. We define the intensity or energy flow density I_p as the power P transported by the light wave through a unit area A perpendicular to its propagation direction:

$$I_\mathrm{p} = \frac{P}{A} . \tag{2.6}$$

The intensity of a plane light wave at a given time t and position $\vec{r}$ can be calculated using eq. (2.2), yielding

$$I_\mathrm{p}(\vec{r}, t) = c\varepsilon_0 \vec{E}^2(\vec{r}, t) = c\varepsilon_0 \vec{E}_0^2 \cos^2(\vec{k} \cdot \vec{r} - \omega t) , \tag{2.7}$$

where ε_0 is the vacuum permittivity. Integrating over a time period T_ω of the oscillation and dividing by T_ω gives the time-averaged intensity

$$\bar{I}_\mathrm{p} = \frac{c}{2}\varepsilon_0 \vec{E}_0^2 . \tag{2.8}$$

This relates the measurable physical quantity $\bar{I}_\mathrm{p}$ to the amplitude $\vec{E}_0$ of the electric field from eq. (2.2). An alternative representation of the electric field is the complex notation

$$\tilde{\vec{E}}(\vec{r}, t) = \vec{E}_0 \mathrm{e}^{\mathrm{i}\left(\vec{k} \cdot \vec{r} - \omega t\right)} , \tag{2.9}$$

for which

$$\vec{E}(\vec{r}, t) = \mathrm{Re}\big[\tilde{\vec{E}}(\vec{r}, t)\big] = \mathrm{Re}\big[\vec{E}_0 \mathrm{e}^{\mathrm{i}(\vec{k} \cdot \vec{r} - \omega t)}\big] , \tag{2.10}$$

holds. From the complex representation, we thus get back the form of the electric field in eq. (2.2). The complex notation is helpful because in many cases, it allows one to shorten calculations considerably as compared to approaches based on eq. (2.2).

We emphasize that this classical description neglects the corpuscular character of light as described by photons. Hence, it is only applicable at sufficiently high intensities.

2.2.2 Light waves in media

In the previous section, we have dealt with the propagation of light in vacuum, which is an environment that exerts no back effect on the light waves. This applies no longer to the light propagation in media, because the light field causes charge oscillations in the atoms, which can take up energy from the wave, resulting in its gradual attenuation. Moreover, the atoms can, in turn, emit electromagnetic waves, thus giving rise to new contributions to the light field. The medium thus changes the light wave during its propagation. This applies most strongly to condensed matter, which differs from gaseous media by its high number density of atoms. In order to build the basis for the following sections, we will have a closer look at light-matter interaction in media and discuss how one can formally describe the changes experienced by light waves in the course of this interaction.

2.2.2.1 Light absorption and propagation in media

The alternating electric field $\vec{E}(\vec{r}, t)$ can excite forced oscillations of an electron in an atom at the position $\vec{r}$. From a quantum mechanical point of view, such an oscillation consists of a periodic displacement of the probability density of the electron relative to the positively charged atom core. The cause of this displacement is the force $-e\vec{E}(\vec{r}, t)$ exerted on the electron by the alternating electric field of the light wave. For many purposes, it has proven useful to consider the oscillation not from a quantum mechanical perspective but rather classically as a displacement $\vec{x}(t)$ of a point-like electron (Fig. 2.1). The displacement can be treated using the concepts of mechanics and electrodynamics. Moreover, we assume the restoring force exerted on the electron by the atom core is, in linear approximation, proportional to the displacement $\vec{x}(t)$. The system is thus assumed to behave as if the electron and the atom core were connected by a spring. Together, these assumptions are called the Lorentz oscillator model and the system consisting of electron, atom core, and spring is called a Lorentz oscillator. Despite its simplicity, this model yields essential insights into the light matter interaction.

In the Lorentz model, we describe the dynamics of an electron of mass m_e in an atom under the influence of light through an equation of motion for a forced oscillation:

$$m_e\ddot{\vec{x}} + m_e\gamma\dot{\vec{x}} + m_e\omega_0^2\vec{x} = -e\vec{E}_0 e^{i(\vec{k}\cdot\vec{r}-\omega t)} . \tag{2.11}$$

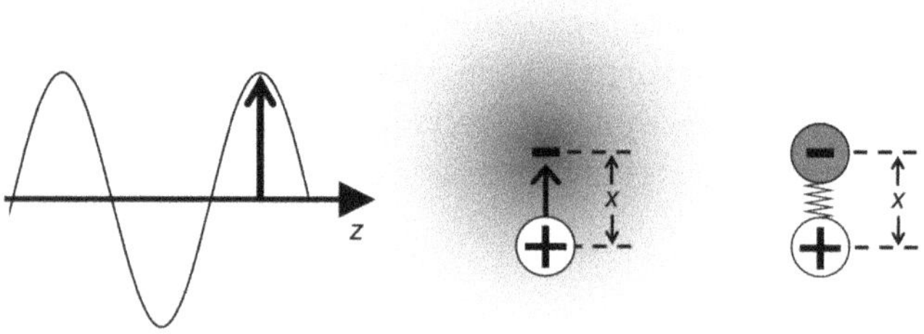

Fig. 2.1: Charge oscillation in an atom under the effect of the alternating electric field of a light wave (left). Center: schematic drawing of the instantaneous displacement of the electronic probability density with respect to the atom core at a given moment in time. Right: Simplified model for the situation depicted in the middle, the so-called Lorentz oscillator. The negative electron charge has its equilibrium position at the position of the positive ion core and is attracted towards the core as if pulled by a spring attached to the electron, which is extended by x.

Here, γ describes a possible damping of the oscillation due to energy losses (for example, due to light emission or collisions of atoms with each other), and ω_0 is the eigenfrequency of the undamped oscillation of the electron. The term on the right side of the equation represents a time-dependent force which drives the oscillation with $\vec{E}_0 e^{i(\vec{k}\cdot\vec{r}-\omega t)}$ being the electric field in the complex notation (eq. (2.9)). Recall that $\vec{r}$ is the position of the atom while $\vec{x}$ describes the displacement of the electron charge with respect to the atom core. The complex-valued solution to the differential equation (2.11) for the displacement of the electron has the form

$$\tilde{\vec{x}}(t) = \vec{x}_0 e^{i\left(\vec{k}\cdot\vec{r}-\omega t+\varphi\right)}, \tag{2.12}$$

where $\vec{r}$ is again the position of the atom,

$$\vec{x}_0 = \frac{-e\vec{E}_0}{m_e} \frac{1}{\sqrt{\left(\omega_0^2-\omega^2\right)^2+\left(\gamma\omega^2\right)}} \tag{2.13}$$

is the (real-valued) amplitude of the electron oscillation and φ is the phase delay of the displacement relative to the external force $-e\vec{E}_0 e^{i(\vec{k}\cdot\vec{r}-\omega t)}$. For φ we obtain

$$\tan\varphi = \frac{\gamma\omega}{\omega_0^2-\omega^2}. \tag{2.14}$$

For not too strong damping, the resonance frequency, i.e., the frequency of the exciting light field for which the amplitude becomes maximal (Fig. 2.2(a)), agrees well with the eigenfrequency ω_0. The displacement $\vec{x}(t)$ of the electron is obtained as the real part of the complex-valued function $\tilde{\vec{x}}(t)$ given by eqs. (2.12), (2.13), and (2.14). We see from eq. (2.12) that the time dependence of the displacement for a given position $\vec{r}$ of the atom represents a harmonic oscillation. Hence the Lorentz oscillator

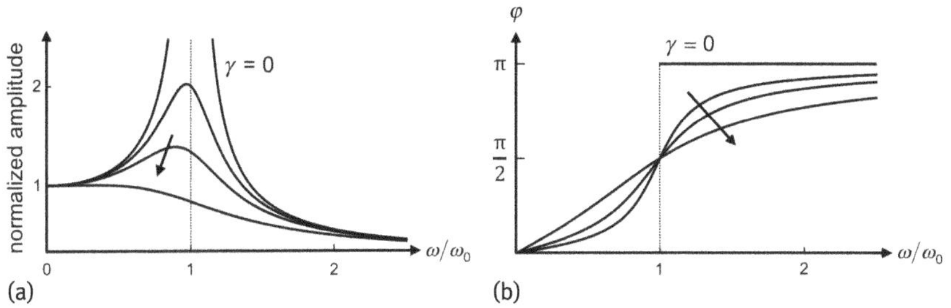

Fig. 2.2: (a) Amplitude of a Lorentz oscillator plotted versus the driving frequency ω for different dampings γ, which increase in the direction of the arrow. The amplitude is normalized to its value at zero frequency (static displacement), and the frequency to the eigenfrequency ω_0. In the undamped case, the resonance frequency is equal to ω_0. It decreases with increased damping. (b) Phase shift φ of the displacement relative to the phase of the external force for the same dampings γ as in (a). As in (a), the arrow shows the direction of increased γ.

model describes in a simple way, via eqs. (2.13) and (2.14), the frequency dependence of the electron oscillation. How this frequency dependence is reflected in measurable macroscopic quantities will be considered in the next subsection.

The oscillation energy of the excited electron originates from the light wave. It may lose that energy through the following processes:

a) Via collisions (for example with other atoms) or – provided that the atom is part of a molecule – via coupling of the electron oscillation to mechanical oscillations of the molecule. In these processes, the oscillation energy is ultimately transformed into heat. Such an energy uptake with subsequent complete or partial transformation into heat is called absorption. For $\omega \approx \omega_0$, we have resonant absorption.
b) Via emission of the energy taken up from the light wave by the atom. The oscillating electron and the corresponding atom core form an oscillating dipole with electric dipole moment

$$\vec{p}_E(t) = -e\vec{x}(t) \qquad (2.15)$$

which, in turn, radiates a light wave, called a secondary wave. If the electron executes a purely forced oscillation and no other degrees of freedom are involved, this wave has the same frequency as the exciting light wave. We then call the reradiation elastic light scattering. Due to the phase delay between the exciting field and the charge oscillation (eq. (2.14)), the emitted light wave also has a phase delay with respect to the exciting wave.

Because a medium through which the light wave propagates is composed of many atoms, the light absorption in the individual atoms leads to macroscopically measurable consequences. Because a part of the light energy is transformed into heat during the propagation of the light wave through the medium, the intensity decays with increased propagation distance. If the medium is homogeneous, the relative decrease

$\frac{dI_p}{dz}\frac{1}{I_p(z)}$ along a small interval dz of the light path is independent of z:

$$\frac{dI_p}{dz}\frac{1}{I_p(z)} = -\alpha\,. \tag{2.16}$$

The constant α is called absorption coefficient. It is measured in m^{-1} or cm^{-1}. Integrating eq. (2.16) yields an exponential decay of intensity with increased propagation distance z in the medium:

$$I_p(z) = I_{p,0}e^{-\alpha z} \quad \text{(Beer's law)}\,. \tag{2.17}$$

In general, the absorption coefficient is frequency-dependent. It exhibits maxima at frequencies for which the absorption is resonant. Because the resonance frequencies are controlled by the electronic structure of the medium, the frequency dependence of α is specific for the material. For instance, while some materials, e.g., glass, are transparent ($\alpha \approx 0$) in the visible spectral range, most materials absorb ($\alpha > 0$) there. Similarly, the re-radiation of light by the individual atoms has consequences that are macroscopically measurable. At each point in space, the portion of the exciting light wave that has not been absorbed and the secondary waves are superimposed. Because of the phase delay φ between the exciting field and the charge displacement, the phase fronts created by this superposition also experience a delay with respect to the exciting wave. Because this delay takes place in each individual layer of the medium in the direction of propagation, the accumulated effect is that the phase velocity c in the medium is reduced with respect to the vacuum light velocity c_0. The factor by which c is reduced is called the refractive index:

$$c = \frac{c_0}{n} \quad \text{(speed of light in a medium)}\,. \tag{2.18}$$

The refractive index changes the phase velocity, but not the angular frequency ω of the light wave. Therefore and due to eq. (2.4),

$$k = \frac{\omega}{c} = \frac{n\omega}{c_0} = nk_0 \tag{2.19}$$

and

$$\lambda = \frac{2\pi}{k} = \frac{2\pi}{nk_0} = \frac{\lambda_0}{n}\,. \tag{2.20}$$

Here, k_0 and λ_0 denote the values of the wave number and wavelength in vacuum, respectively. When light transits from vacuum into a medium, its velocity is, thus, reduced by a factor of n, the wave vector is stretched by n and the wavelength is shortened by the same factor. Similarly to the absorption coefficient, the refractive index is specific for the material (e.g., $n \approx 1.5$ for glass in the visible spectrum range) and, in general, frequency-dependent. The different refractive indices of media and vacuum show themselves macroscopically, for instance, in the phenomenon of light refraction during the transition of a light wave from the vacuum into the medium or in the opposite direction.

So far, we have considered light scattering and absorption separately from each other. Both effects can be formally summarized in complex notation by extending eq. (2.9) by a factor that describes the absorption-induced amplitude decay with propagation distance z (eq. (2.17)), and by replacing k from eq. (2.19) by nk_0:

$$\tilde{\vec{E}}(z,t) = \vec{E}_0 e^{-\frac{\alpha}{2}z} e^{i(nk_0 \cdot z - \omega t)} . \tag{2.21}$$

This representation becomes more concise when we define a complex refractive index $\tilde{n}$ through

$$\tilde{n} := n + i\kappa \tag{2.22}$$

with the extinction coefficient

$$\kappa := \frac{\alpha}{2k_0} . \tag{2.23}$$

From eq. (2.21) and using eqs. (2.22) and (2.23), we obtain the more compact expression

$$\tilde{\vec{E}}(z,t) = \vec{E}_0 e^{i(\tilde{n}k_0 \cdot z - \omega t)} . \tag{2.24}$$

When, in analogy to eq. (2.19), a complex wavenumber $\tilde{k}$ is defined by

$$\tilde{k} = \tilde{n}\frac{\omega}{c_0} = \tilde{n}k_0 , \tag{2.25}$$

the representation becomes still shorter:

$$\tilde{\vec{E}}(z,t) = \vec{E}_0 e^{i(\tilde{k} \cdot z - \omega t)} . \tag{2.26}$$

Here, the absorption-induced decay of the amplitude is described by the imaginary part of the complex wavenumber $\tilde{k}$.

2.2.2.2 Microscopic and macroscopic optical responses

In the previous subsection, we have argued that the electron oscillation in the atoms, which is a microscopic response to the optical field, leads to a retardation and absorption of the light wave in the medium, i.e., macroscopically measurable effects. In the present subsection, we will study the connection between the microscopic and macroscopic optical responses of the medium in more detail. To this end, we will start from the equation of motion (eq. (2.11)) of the Lorentz oscillator. We slightly rewrite eq. (2.12) to obtain compact expressions later on:

$$\tilde{\vec{x}}(t) = \tilde{\vec{x}}_0 e^{i(\vec{k} \cdot \vec{r} - \omega t)} \tag{2.27}$$

with the complex-valued amplitude $\tilde{\vec{x}}_0 = \vec{x}_0 e^{i\varphi}$. The phase factor $e^{i\varphi}$ is now included in the amplitude and $\vec{r}$ is again the position of the atom. When we enter the ansatz from eq. (2.27) into the equation of motion (2.11), we obtain the complex-valued amplitude

$$\tilde{\vec{x}}_0 = \frac{-e\vec{E}_0}{m_e} \frac{1}{\omega_0^2 - \omega^2 - i\gamma\omega} . \tag{2.28}$$

The displacement $\vec{x}(t)$ of the electron is then obtained as the real part of the complex-valued function $\tilde{\vec{x}}(t)$ calculated using eqs. (2.27) and (2.28). Like eq. (2.13), eq. (2.28) shows that the oscillation amplitude is especially large for $\omega \approx \omega_0$, i.e., near the resonance frequency, and small away from it. The displacements $\vec{x}(t)$ of the negatively charged electron relative to the position of the atom core give rise to an oscillating dipole moment $\vec{p}_{E,\mathrm{res}}(t) = -e\vec{x}(t)$. This can also be written in complex notation as

$$\tilde{\vec{p}}_{E,\mathrm{res}}(t) = -e\tilde{\vec{x}}(t)\,. \tag{2.29}$$

The index "res" is supposed to remind us that the amplitude of this dipole moment becomes, according to eq. (2.28), maximal at the resonance frequency. In what follows, n_A denotes the number of atoms per unit volume. The contribution made to the dielectric polarization $\tilde{\vec{P}}$ by the dipole moments $\tilde{\vec{p}}_{E,\mathrm{res}}$ generated by the oscillation is:

$$\tilde{\vec{P}}_\mathrm{res} = n_\mathrm{A}\tilde{\vec{p}}_{E,\mathrm{res}} = -n_\mathrm{A}e\tilde{\vec{x}}\,. \tag{2.30}$$

We assume that all atoms oscillate in phase, i.e., we consider a volume smaller than λ of the light wave. Additionally, we subsume into a term $\tilde{\vec{P}}_\mathrm{nonres}$ the contributions from charge oscillations that are nonresonant at $\omega \approx \omega_0$, e.g., of ions of the crystal lattice. The total dielectric polarization can thus be written as:

$$\tilde{\vec{P}} = \tilde{\vec{P}}_\mathrm{nonres} + \tilde{\vec{P}}_\mathrm{res}\,. \tag{2.31}$$

If we express the polarizations by the associated susceptibilities χ and the electric field according to

$$\tilde{\vec{P}} = \varepsilon_0\chi\tilde{\vec{E}}\,, \tag{2.32}$$

we obtain

$$\chi = \chi_\mathrm{nonres} + \chi_\mathrm{res}\,, \tag{2.33}$$

where χ_res, because of eqs. (2.28)–(2.33), has the frequency dependence

$$\chi_\mathrm{res}(\omega) = \frac{n_\mathrm{A}e^2}{\varepsilon_0 m_\mathrm{e}}\frac{1}{\omega_0^2 - \omega^2 - \mathrm{i}\gamma\omega}\,. \tag{2.34}$$

As known from electrodynamics, the dielectric polarization is connected to the electric displacement $\vec{D}$ through the equation

$$\tilde{\vec{D}} = \varepsilon_0\tilde{\vec{E}} + \tilde{\vec{P}}\,. \tag{2.35}$$

Inserting eq. (2.32) gives

$$\tilde{\vec{D}} = \varepsilon_0\varepsilon\tilde{\vec{E}}\,, \tag{2.36}$$

where

$$\varepsilon = 1 + \chi \tag{2.37}$$

is the relative permittivity, also called the dielectric constant of the medium. From eq. (2.36) and the fact that in the absence of free charges a $\vec{D}$ field perpendicular to the surface of the medium has the same value inside and outside the medium, we obtain

$$\tilde{\vec{E}} = \frac{\tilde{\vec{E}}_{\text{ext}}}{\varepsilon}, \tag{2.38}$$

where $\tilde{\vec{E}}_{\text{ext}}$ is the electric field outside the medium. The term $1/\varepsilon$ thus gives the factor by which the electric field in the medium differs from that outside, if the fields are directed perpendicular to the surface of the medium. This difference is a consequence of the electric counter field that is generated by the charge displacements in the medium, which is superimposed on the external electric field. Like χ, ε is also frequency-dependent. For instance, for the Lorentz oscillators described by eq. (2.34), it has the form:

$$\varepsilon(\omega) = 1 + \chi_{\text{nonres}} + \frac{n_{\text{A}} e^2}{\varepsilon_0 m_{\text{e}}} \frac{1}{\left(\omega_0^2 - \omega^2 - \text{i}\gamma\omega\right)}. \tag{2.39}$$

To stress the frequency dependence of ε, $\varepsilon(\omega)$ is often called the dielectric function. The frequency dependence described in eq. (2.39) is plotted in Fig. 2.3 for the example of the D_2 line of sodium. As expected, it shows a resonance at the eigenfrequency ω_0, with a width determined by the damping rate γ.

Real solids have more than one optical resonance as they support vibrational and electronic excitations at various frequencies. For many solids, their dielectric functions can be looked up in tables. Alternatively, they can be modelled as superpositions

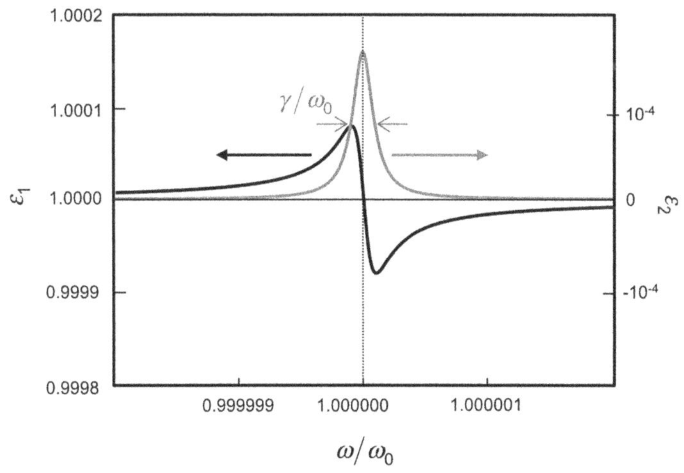

Fig. 2.3: Frequency dependence of the dielectric function $\varepsilon(\omega)$ of a medium with Lorentz oscillators with density $n_A = 10^{17}\,\text{m}^{-3}$, eigenfrequency $\omega_0 = 3.20 \times 10^{15}\,\text{s}^{-1}$, damping rate $\gamma = 6.28 \times 10^8\,\text{s}^{-1}$, and $\chi_{\text{nonres}} = 0$ (eq. (2.39)). These numbers are valid for the strongest hyperfine component of the D_2 line of a gas of sodium atoms [15]. ε_1 and ε_2 are the real and imaginary parts of the dielectric function $\varepsilon(\omega)$, respectively.

of Lorentz oscillator terms, one for each optical excitation in the solid, and each with their own resonance frequency ω_{0_j}, damping rate γ_j and spectral weight f_j:

$$\varepsilon(\omega) = 1 + \frac{n_A e^2}{\varepsilon_0 m_e} \sum_j \frac{f_j}{\left(\omega_{0_j}^2 - \omega^2 - i\gamma_j \omega\right)} . \tag{2.40}$$

A comparison with eq. (2.39) reveals the origin of the term χ_{nonres}. It stands for those Lorentz oscillator terms in eq. (2.40) that are off-resonance at a given wavelength ω.

Once the dielectric function has been determined from tables or through a model calculation, the refractive index n and the absorption coefficient α of the respective materials can be calculated. This useful connection can be derived from Maxwell's equations

$$\nabla \cdot \vec{D} = \rho_c \tag{2.41}$$

$$\nabla \cdot \vec{B} = 0 \tag{2.42}$$

$$\nabla \times \vec{E} = -\frac{\partial \vec{B}}{\partial t} \tag{2.43}$$

$$\nabla \times \vec{H} = \vec{j} + \frac{\partial \vec{D}}{\partial t} , \tag{2.44}$$

where $\vec{H}$ is the magnetic field, ρ_c the density of free charges, and $\vec{j}$ the electric current density. Let us assume that our medium is a dielectric, i.e., no currents can flow in it (current density $\vec{j} = \vec{0}$). Furthermore, the medium is assumed to be nonmagnetic with permeability $\mu_M = 1$ and, hence, we have $\vec{B} = \mu_0 \vec{H}$, where μ_0 is the vacuum permeability. By combining eqs. (2.43) and (2.44), we obtain:

$$\nabla \times (\nabla \times \vec{E}) = -\mu_0 \varepsilon_0 \varepsilon \frac{\partial^2 \vec{E}}{\partial t^2} . \tag{2.45}$$

If we take into account the vector identity $\nabla \times (\nabla \times E) = \nabla(\nabla \cdot \vec{E}) - \nabla^2 \vec{E}$ and assume no free charges ($\rho_c = 0$), we obtain the equation

$$\nabla^2 \vec{E} = \mu_0 \varepsilon_0 \varepsilon \frac{\partial^2 \vec{E}}{\partial t^2} . \tag{2.46}$$

Entering a plane harmonic wave of the form in eq. (2.26) into eq. (2.46) and exploiting $c_0 = 1/\sqrt{\mu_0 \varepsilon_0}$ gives:

$$\tilde{k}^2 = \varepsilon \frac{\omega^2}{c_0^2} . \tag{2.47}$$

A comparison with eq. (2.25) gives for the refractive index

$$\varepsilon = \tilde{n}^2 . \tag{2.48}$$

Equation (2.48) makes an interesting physical statement: The change of the electric field caused by the medium as described by ε is directly linked to the phase velocity and the absorption of the light wave in this medium, as described by $\tilde{n} = n + \mathrm{i}\kappa$. From eq. (2.48), the following relations are derived between the real and imaginary parts of the two quantities $\varepsilon = \varepsilon_1 + \mathrm{i}\varepsilon_2$ and $\tilde{n} = n + \mathrm{i}\kappa$ (eq. (2.22)):

$$\varepsilon_1 = n^2 - \kappa^2 \tag{2.49}$$

and

$$\varepsilon_2 = 2n\kappa\,. \tag{2.50}$$

The conversion formulas in the opposite direction are

$$n = \frac{1}{\sqrt{2}}\sqrt{\varepsilon_1 + \sqrt{\varepsilon_1^2 + \varepsilon_2^2}} \tag{2.51}$$

and

$$\kappa = \frac{1}{\sqrt{2}}\sqrt{-\varepsilon_1 + \sqrt{\varepsilon_1^2 + \varepsilon_2^2}}\,. \tag{2.52}$$

Using these two equations and eq. (2.23), the refractive index n and the absorption coefficient α can be calculated from the dielectric function $\varepsilon(\omega)$. An example is shown in Fig. 2.4, in which n and κ have been calculated from the dielectric function shown in Fig. 2.3.

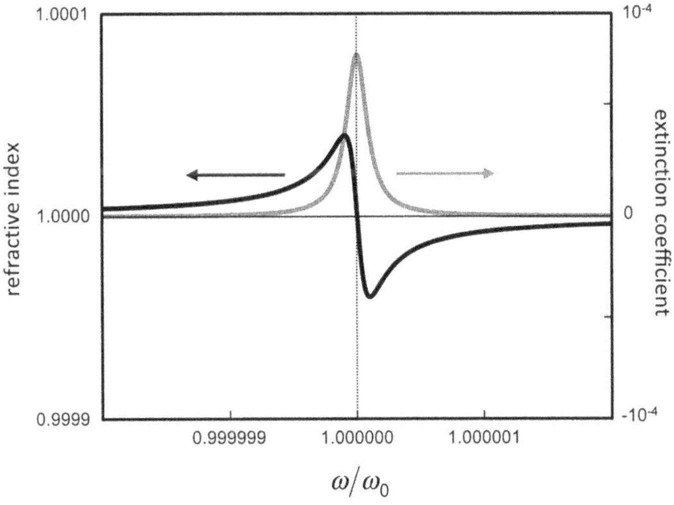

Fig. 2.4: Real refractive index n and extinction coefficient κ calculated from the dielectric function according to eqs. (2.51) and (2.52) for the same medium of Lorentz oscillators as in Fig. 2.3

2.2.2.3 Light reflection

So far, we have dealt with light propagation and absorption in homogeneous media. We know from electrodynamics that light incident on the interface between two media

1 and 2 is partially reflected by the interface. For instance, a fraction of approximately 4% of the light power is reflected for perpendicular incidence of a light wave from air onto a glass surface. In what follows, we will deal with the question how one can calculate the reflected fraction. The ratio of the time-averaged reflected light power $\overline{P}_\mathrm{r}$ to the time-averaged total incident power $\overline{P}_\mathrm{i}$

$$R_y := \frac{\overline{P}_\mathrm{r}}{\overline{P}_\mathrm{i}} \tag{2.53}$$

is called reflectivity. The corresponding fraction transmitted through the interface

$$T_y := \frac{\overline{P}_\mathrm{t}}{\overline{P}_\mathrm{i}} \tag{2.54}$$

is called transmissivity. If the participating media are transparent and thus no absorption takes place,

$$R_y + T_y = 1 \tag{2.55}$$

must hold due to energy conservation. Because of eqs. (2.6) and (2.8), the reflectivity is connected to the electric field amplitudes of the incident and reflected wave by

$$R_y = \left(\frac{E_{0,\mathrm{r}}}{E_{0,\mathrm{i}}}\right)^2 . \tag{2.56}$$

The amplitude ratio appearing in eq. (2.56) can also be derived from Maxwell's equations for arbitrary angles of incidence β of the light wave. For perpendicular incidence ($\beta = 0$) from air onto the surface of a medium with refractive index $\tilde{n} = n + \mathrm{i}\kappa$, one obtains:

$$R_y(\beta = 0) = \left|\frac{\tilde{n} - 1}{\tilde{n} + 1}\right|^2 = \left|\frac{n + \mathrm{i}\kappa - 1}{n + \mathrm{i}\kappa + 1}\right|^2 = \frac{(n-1)^2 + \kappa^2}{(n+1)^2 + \kappa^2} . \tag{2.57}$$

Equation (2.57) shows that $R_y \approx 1$ for $\kappa \gg n$. Therefore, strongly absorbing media also have a high reflectivity. Furthermore, if the absorption coefficient α and thus the extinction coefficient κ are strongly dependent on the wavelength, this also applies to the reflectivity.

2.2.2.4 Classical light emission

In Subsection 2.2.2.1, it was mentioned that an excited atom can re-radiate the energy taken up from the light wave. In this way, it behaves like every oscillating dipole which radiates electromagnetic waves due to its charge oscillation. The oscillation energy is thus gradually transformed into electromagnetic field energy, which is transported away from the atom. The loss of oscillation energy limits the time during which an atom can oscillate after the passage of the exciting electromagnetic wave train. In what follows, we will calculate the decay time of the oscillation. The mechanical oscillation energy of the Lorentz oscillator (Section 2.2.2.1) is given by:

$$E_\mathrm{osc} = \frac{1}{2} m_\mathrm{e} \omega_0^2 x_0^2 . \tag{2.58}$$

As known from electrodynamics, the power that is radiated by an oscillating dipole with charges $\pm e$, averaged over one oscillation period, is given by

$$\overline{P}_{\mathrm{em}} = \frac{e^2 \omega_0^4 x_0^2}{12\pi\varepsilon_0 c^3} . \tag{2.59}$$

If the lost energy is not replaced from the outside, the oscillation energy of the dipole decreases at the rate described by eq. (2.59). According to eq. (2.58), this means that the oscillation amplitude x_0 decays. The relative reduction of the oscillation energy is

$$\frac{\mathrm{d}E_{\mathrm{osc}}}{\mathrm{d}t}\frac{1}{E_{\mathrm{osc}}} = \frac{\overline{P}_{\mathrm{em}}}{E_{\mathrm{osc}}} = -\frac{e^2 \omega_0^2}{6\pi\varepsilon_0 m_{\mathrm{e}} c^3} . \tag{2.60}$$

Solving the differential eq. (2.60) gives an exponential decay of the oscillation energy:

$$E_{\mathrm{osc}}(t) = E_{\mathrm{osc}}(t=0) \cdot e^{-\frac{t}{\tau_{\mathrm{R}}^{(\mathrm{cl})}}} . \tag{2.61}$$

Here, $\tau_{\mathrm{R}}^{(\mathrm{cl})} = \frac{e^2 \omega_0^2}{6\pi\varepsilon_0 m_{\mathrm{e}} c^3}$ is called the radiative lifetime of the classical dipolar oscillation.

As mentioned at the end of Section 2.2.1, the classical description neglects the quantization of the light field into photons, which becomes relevant at lower light intensity and if one considers the energy exchange between quasiparticles, such as crystal electrons, and the light field. The latter is the subject of the following section.

2.2.3 Quantum physics of optical transitions

Our classical treatment of the microscopic optical properties of solids in the previous section has enabled us to achieve a simple understanding of important phenomena such as light absorption. However, there are optical phenomena in solids which cannot be described in this way and which can only be explained using quantum physics. One example is stimulated emission as employed by lasers. In the present section, we will discuss the quantum physics of optical transitions using a simple model system, the two-level system. For simplicity, we will initially do this in a semi-classical approximation, in which matter is treated quantum-mechanically and the light field classically. Subsequently, we will describe spontaneous emission in a simplified quantum-optical picture.

2.2.3.1 The two-level system

As mentioned in Section 2.2.2.1, an optically driven electron oscillation in an atom represents, from a quantum physical perspective, an oscillating spatial displacement of the electron wave function relative to the positively charged atom core. In what follows, we will show that this displacement can be described by an appropriate superposition of the energy eigenfunctions of the electron. We keep this treatment as simple as possible by assuming that only two states of the electron in the atom contribute to this superposition and that no damping processes are involved. The two-level system is composed of a state ψ_1 with energy E_1 and a second state ψ_2 with energy $E_2 > E_1$ (Fig. 2.5). In the absence of light, ψ_1 and ψ_2 are energy eigenstates of the system and thus obey the time-independent Schrödinger equation:[1]

$$\widehat{H}_0\psi_j = E_j\psi_j\,, \tag{2.62}$$

where $\widehat{H}_0$ is the Hamilton operator of the unperturbed two-level system without incident light. ψ_1 and ψ_2 have the time-dependent form

$$\widetilde{\psi}_j(\vec{r}, t) = \psi_j(\vec{r})\exp\left(-\frac{\mathrm{i}E_j t}{\hbar}\right), \quad j = 1, 2\,. \tag{2.63}$$

When ψ_1 und ψ_2 are coupled by the alternating electric field of a light wave at frequency ω, they cease to be true eigenstates. The time evolution of the system is then described by the time-dependent Schrödinger equation:

$$\mathrm{i}\hbar\frac{\partial\widetilde{\psi}}{\partial t} = \left(\widehat{H}_0 + \widehat{H}'\right)\widetilde{\psi}\,, \tag{2.64}$$

where $\widetilde{\psi}$ is a time- and position-dependent wave function. $\widehat{H}'$ from eq. (2.64) is the operator of the perturbation by light. In classical electrodynamics, the electric field $\vec{E}$ causes an energy shift of a dipole $\vec{p}_E$ by $-\vec{p}_E\cdot\vec{E}$. The perturbation operator $\widehat{H}'$ can be expressed in terms of this energy shift by replacing the two involved classical quantities, $\vec{p}_E$ and $\vec{E}$, with their associated operators. In what follows, we will use the semiclassical approximation, in which particles are treated quantum mechanically while the field is that of a classical electromagnetic wave. The perturbation operator is then:

$$\widehat{H}' = -\vec{E}\cdot\widehat{\vec{p}}_E\,, \tag{2.65}$$

where

$$\widehat{\vec{p}}_E = -e\widehat{\vec{x}} \tag{2.66}$$

is the electric dipole operator. In agreement with eq. (2.2), we write the field at the position $\vec{r} = \vec{0}$ in an atom as

$$\vec{E}(t) = \frac{1}{2}\vec{E}_0\left(\mathrm{e}^{\mathrm{i}\omega t} + \mathrm{e}^{-\mathrm{i}\omega t}\right). \tag{2.67}$$

$\widetilde{\psi}$ can be written as a superposition of the two states ψ_1 and ψ_2 of the unperturbed two-level system:

$$\widetilde{\psi} = c_1\psi_1 + c_2\psi_2 \tag{2.68}$$

with the time-dependent coefficients $c_1(t)$ and $c_2(t)$. Entering the ansatz of eq. (2.68) into the Schrödinger eq. (2.64), multiplying with $\widetilde{\psi}^*$ from the left side, integrating over

---

1 We use a hat above a symbol (such as in $\widehat{A}$) to indicate quantum mechanical operators.

the volume of the system and taking into account eq. (2.62) yields a system of two coupled differential equations:

$$\mathrm{i}\hbar\dot{c}_1 = \left(E_1 + \widehat{H}'_{11}\right) c_1 + \widehat{H}'_{12} c_2 \quad \text{and} \quad \mathrm{i}\hbar\dot{c}_2 = \widehat{H}'_{21} c_1 + \left(E_2 + \widehat{H}'_{22}\right) c_2 \tag{2.69}$$

with the matrix elements

$$\widehat{H}'_{jk} = \vec{E} \cdot \vec{\mu}_{jk} , \tag{2.70}$$

where

$$\vec{\mu}_{jk} = \int \psi_j^* e\vec{x}\psi_k \mathrm{d}^3\vec{x} \tag{2.71}$$

is the dipole matrix element between the states ψ_j and ψ_k for $j, k = 1, 2$ (also called the transition dipole moment). In many cases, the diagonal elements $\vec{\mu}_{11}$ and $\vec{\mu}_{22}$ vanish for symmetry reasons, as we assume in the following. Of course, we have to assume additionally that the selection rules allow the transitions $\psi_1 \rightarrow \psi_2$ and $\psi_2 \rightarrow \psi_1$ to occur, which implies, in particular, $\vec{\mu}_{12} \neq \vec{0}$ and $\vec{\mu}_{21} \neq \vec{0}$. Using, moreover, the transformation $c_j =: d_j \exp\left(\frac{-\mathrm{i}E_j t}{\hbar}\right)$, eqs. (2.69) read

$$\mathrm{i}\hbar\dot{d}_1 = \widehat{H}'_{12} d_2 \mathrm{e}^{-\mathrm{i}\frac{(E_2-E_1)t}{\hbar}} \quad \text{and} \quad \mathrm{i}\hbar\dot{d}_2 = \widehat{H}'_{21} d_1 \mathrm{e}^{-\mathrm{i}\frac{(E_1-E_2)t}{\hbar}} . \tag{2.72}$$

By entering the eqs. (2.70) and (2.67) into eq. (2.72), restricting ourselves to resonant excitation $\hbar\omega = E_2 - E_1$, neglecting rapidly rotating terms at frequency 2ω (rotating wave approximation) and transforming back to the coefficients c_1 and c_2, one obtains the following solutions:

$$c_1 = \mathrm{e}^{\frac{-\mathrm{i}E_1 t}{\hbar}} \cos\left(\omega_{\mathrm{Rabi}} t\right) \tag{2.73}$$

and

$$c_2 = \mathrm{e}^{\frac{-\mathrm{i}E_2 t}{\hbar}} \sin\left(\omega_{\mathrm{Rabi}} t\right) , \tag{2.74}$$

where $\omega_{\mathrm{Rabi}} = \frac{\vec{E}_0 \cdot \vec{\mu}_{12}}{\hbar}$ is called the Rabi frequency. Thus we have determined the time dependence of the electron state $\widetilde{\psi}$ from eq. (2.68). The probability density oscillates between state ψ_1 and state ψ_2. Particularly instructive is the expectation value $\langle\vec{x}\rangle$ of the electron position, which can be calculated using eqs. (2.68), (2.71), (2.73) and (2.74):

$$\langle\vec{x}\rangle = \int \widetilde{\psi}^* \vec{x} \widetilde{\psi} \mathrm{d}^3\vec{x} = -\frac{\vec{\mu}_{12}}{e} \sin\left(\omega_{\mathrm{Rabi}} t\right) \sin\left(\frac{(E_2 - E_1)\, t}{\hbar}\right) . \tag{2.75}$$

This equation tells us that the superposition of the two wave functions ψ_1 and ψ_2 leads to an oscillation of the average electron position back and forth in space at the frequency $\frac{(E_2-E_1)}{\hbar}$. This corresponds to the frequency ω_0 at which the Lorentz oscillator of eq. (2.12) oscillates back and forth. In addition, the amplitude of the quantum mechanical oscillation is modulated at the Rabi frequency ω_{Rabi}, which is typically much smaller than $\frac{E_2-E_1}{\hbar}$ and tunable by the electric field strength $\vec{E}_0$ of the light wave. The physical meaning of this frequency becomes clear when we calculate the probabilities $|c_j|^2$ of finding the electron in the lower or upper state via eqs. (2.73) and (2.74):

$$|c_1|^2 = \cos^2\left(\omega_{\mathrm{Rabi}} t\right) \tag{2.76}$$

and

$$|c_2|^2 = \sin^2(\omega_{\text{Rabi}} t) \ . \tag{2.77}$$

In the absence of damping processes, the electron goes back and forth between the lower and the upper state at the frequency $2 \cdot \omega_{\text{Rabi}}$ – a process called Rabi oscillation. This sequence of light-induced excitation and de-excitation between the two states is a purely quantum-physical effect and is caused by the light-induced coupling of the two states. It enables one to deliberately transfer the ground state ψ_1 into a superposition of both states ψ_1 and ψ_2 via choosing the time t that the light wave acts on the two-level system (see also Section 3.4.3). It is also responsible for the oscillation of the average charge position in eq. (2.75).

The oscillation vanishes at those moments at which the electron is found with certainty in one of the two states. In contrast, it is maximal when $|c_1(t)|^2 = |c_2(t)|^2 = \frac{1}{2}$. This behavior is distinct from the behavior of the classical Lorentz oscillator, whose oscillation amplitude would, in the absence of damping, increase monotonically with time, taking up energy from the light field in every oscillation cycle of the field. The light-induced transitions between the states in the Rabi oscillation act back on the light field due to energy conservation. During the transition $\psi_1 \rightarrow \psi_2$, energy is absorbed from the light field, thus reducing the light power (Fig. 2.5). In the reverse transition $\psi_2 \rightarrow \psi_1$, energy is restored to the light field, a phenomenon that is called stimulated emission as it does not occur spontaneously but is induced by the light field.

Normally, damping processes prevent an observation of complete cycles of a Rabi oscillation. In this case, the damping rate is much larger than the Rabi frequency and the damping-induced de-excitation of the two-level system dominates. It can be shown that for stationary excitation conditions, the occupation of the upper state then becomes

$$|c_2|^2 \approx \left(\frac{\omega_{\text{Rabi}}}{2\gamma}\right)^2 , \tag{2.78}$$

i.e., $|c_2|^2 \ll 1 \approx |c_1|^2$. The system always tends to this equilibrium value, if it has either been in the lower or in the upper level at the beginning of the light illumination. In addition, it follows from eq. (2.78) that in stationary excitation conditions $|c_2|^2$ is approximately proportional to ω^2_{Rabi} and thus to $|\vec{E}_0|^2$, which is also true for the oscillation energy of the classical Lorentz oscillator as described in eq (2.58). Hence, strong damping restores the classical behavior of electrons in an atom.

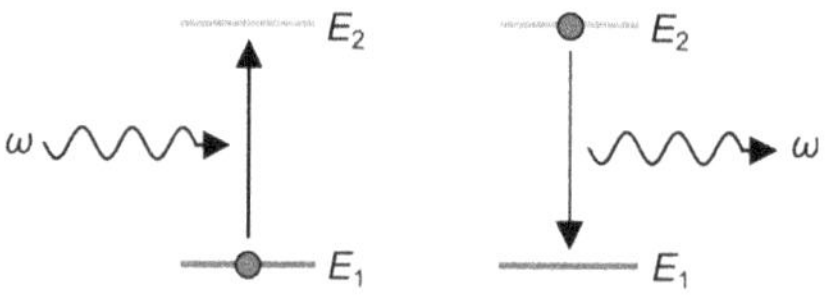

Fig. 2.5: Optical transitions of an electron between two energy levels in an atom due to absorption (left) and emission (right) of light.

2.2.3.2 Spontaneous and stimulated emission

As known from quantum physics, light can be decomposed into tiny portions called light quanta or photons. They have the photon energy

$$E_\gamma = \hbar\omega \tag{2.79}$$

and photon momentum

$$\vec{p}_\gamma = \hbar\vec{k}_\gamma \,. \tag{2.80}$$

The research field that deals with the properties of photons and their interaction with matter is called quantum optics. Here, the photons are the excitations of the light field. In the quantum optical treatment of light absorption and emission, it is assumed that these two processes involve transitions of atoms between two quantum states. An atom absorbs a photon by jumping into a higher energy level with energy E_2 (Fig. 2.5). The emission of a photon involves a jump into a lower level with energy E_1. If we neglect – in good approximation – the effects of photon absorption or emission on the center of mass motion of the atom, we can identify the energies E_1 and E_2 with those of the jumping electron in the atom. Energy conservation requires that the energy of the photon equals the energy difference between the initial and the final level for both directions of a jump:

$$\hbar\omega = E_2 - E_1 \,. \tag{2.81}$$

In order to determine the temporal rates at which electron jumps occur, we choose a simple description using Einstein coefficients. This description can be derived from a more demanding quantum theoretical treatment. We express the number of absorption events by the rate at which the occupation N_1 of the lower level changes due to upward transitions:

$$\frac{\mathrm{d}N_1}{\mathrm{d}t} = -B_{12}N_1u(\omega) \,. \tag{2.82}$$

Here B_{12} is the Einstein B coefficient which expresses how sensitively an electron in the lower level reacts to a perturbation by a photon. The function $u(\omega)$ is the spectral energy density of light at frequency ω. It is proportional to the intensity of light or – in quantum optical language – to the number density of photons at frequency ω per volume. Here, we consider only the case of resonant absorption, meaning that ω is determined by the resonance condition (2.81). In an analogous fashion, we can express the rate at which the occupation N_2 of the upper level changes due to stimulated emission:

$$\frac{\mathrm{d}N_2}{\mathrm{d}t} = -B_{21}N_2u(\omega) \,. \tag{2.83}$$

Here B_{21} is the Einstein B coefficient for stimulated emission, which expresses how sensitively an electron in the upper level reacts to a perturbation by a photon. It is possible to show that the two Einstein B coefficients are closely linked, i.e., the ratio B_{21}/B_{12} is independent of the material. In stimulated emission, light that induces the downward transition adds another photon and thus increases the number of photons. It can be shown that the light added by stimulated emission always has the same optical phase as the incident light, which means that it oscillates coherently with the latter. In Section 2.2.2.4, we have seen that a classical oscillating dipole spontaneously loses energy through electromagnetic radiation. This phenomenon also exists in quantum optics and is known here as spontaneous emission. It can be understood as stimulated emission caused by vacuum fluctuations of the electromagnetic field. In the Einstein description, the rate at which the occupation N_2 of the upper level changes due to spontaneous emission has the form

$$\frac{\mathrm{d}N_2}{\mathrm{d}t} = -A_{21}N_2 \,. \tag{2.84}$$

Here, A_{21} is the Einstein A coefficient for spontaneous emission, which expresses how sensitively an electron in the upper level reacts to the perturbation by the vacuum fluctuations. As the vacuum fluctuations are not included in the spectral energy density, $u(\omega)$ is absent in eq. (2.84). The solution of the equation shows that spontaneous emission leads to an exponential decay of the occupation of the upper level:

$$N_2(t) = N_2(0)\mathrm{e}^{-A_{21}t} = N_2(0)\mathrm{e}^{\frac{-t}{\tau_R}} \,, \tag{2.85}$$

where $\tau_R = A_{21}^{-1}$ is called the radiative lifetime of the transition. One thus obtains an exponential decay of the same form as in the classical decay of the oscillation energy (eq. (2.61)). It can be shown that $A_{21}(\omega)/B_{21}(\omega)$ is also independent of the material, i.e., the Einstein coefficient for spontaneous emission is related to that for stimulated emission.

An important application of stimulated emission are lasers. The word LASER is an acronym for "light amplification by stimulated emission of radiation." Lasers are based on an excitation of electrons (called pumping) within a medium, so that the occupation N_2 becomes larger than the occupation N_1. This is called inversion. Provided that the values of B_{21} and B_{12} are not too different, the consequence of the inversion is, according to eqs. (2.82) and (2.83), that the stimulated emission rate exceeds the absorption rate, resulting in an amplification of the light field created by the first spontaneous emission processes rather than a light attenuation as in ordinary absorbing media. In this case, the medium consisting of excited atoms is called a gain medium. Moreover, for lasers, a resonator is set up by placing the gain medium between two mirrors (Fig. 2.6). One of the mirrors, the high reflector, is in general highly reflective, and the other one, the output coupler, is not as highly reflective, but slightly transparent for the photons. Because the photons circulate between the mirrors several times before leaving the resonator through the output coupler, the spectral energy density in the gain medium is high and, hence, the stimulated emission is much larger than the spontaneous one. If the stimulated emission dominates, the light waves emitted at various positions in the gain medium are coherent resulting in laser emission.

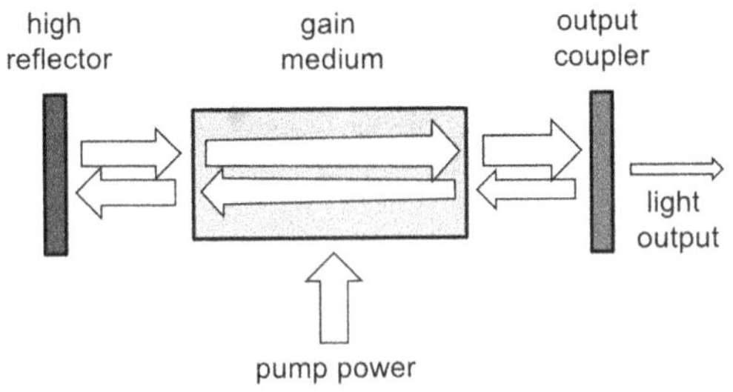

Fig. 2.6: Laser resonator. Light circulates in the resonator by repeatedly passing through a gain medium and being reflected by mirrors. The gain medium is optically or electrically excited by an external energy source (a process called pumping) and amplifies the light during each of its passages. One of the mirrors has a high reflectivity (high reflector). A fraction of the light leaves the resonator through the other mirror with the lower reflectivity (output coupler).

In a fashion analogous to the light attenuation in an absorbing medium, which is expressed by the absorption coefficient α (eq. 2.16), we can express the optical amplification at the frequency ω using the gain coefficient g_ω:

$$dI_\mathrm{p} = +g_\omega I_\mathrm{p}(z)\mathrm{d}z\,. \tag{2.86}$$

The positive sign in the differential reflects that the intensity grows with distance. Integration of the equation gives:

$$I_\mathrm{p}(z) = I_{\mathrm{p},0}\mathrm{e}^{g_\omega z}\,. \tag{2.87}$$

A condition for stable laser emission is that the light intensity in the resonator does not change over time. This is the case if the intensity increase caused by the amplification exactly compensates the light losses due to the non-perfect reflectivities of the mirrors and other losses. For a complete round-trip in the resonator through the gain medium of length L with mirrors having reflectivities $R_{\gamma,1}$ and $R_{\gamma,2}$, this condition reads

$$\mathrm{e}^{(g_\omega-\alpha_\mathrm{loss})\cdot 2L}\cdot R_{\gamma,1}\cdot R_{\gamma,2} = 1\,. \tag{2.88}$$

The attenuation coefficient α_loss takes into account scattering losses as well as absorption by impurities in the medium. The condition in eq. (2.88) can be written in the form

$$g_\mathrm{th} := g_\omega = \alpha_\mathrm{loss} - \frac{1}{2L}\ln(R_{\gamma,1}R_{\gamma,2})\,. \tag{2.89}$$

This equation defines the threshold gain g_th, which is the value where laser emission sets in.

2.3 Interband transitions in solids

In Subsections 2.2.2.1 and 2.2.3.1, we have dealt with optical transitions between the electron states in atoms. These states are spatially confined to the individual atoms and energetically discrete. In contrast, the electron states in crystalline solids are spatially extended and arranged in quasi-continuous energy continua called energy bands. If the initial and final state of an optical transition lie in different bands, the transition is called an interband transition. The present section deals with interband transitions as they appear in absorption and emission spectra of solids.[2]

2.3.1 Interband absorption

Absorption of light by interband transitions of electrons in solids is called interband absorption. In the present section, we will learn that the properties of interband absorption are controlled by properties of the initial and final states of the transition. Measurements of interband absorption thus reveal important information about electronic properties such as excited-state energies and the densities of states of those bands.

2.3.1.1 Interband transitions of single electrons

Similarly to the case of single atoms, the alternating electric field of an incident light wave can excite an electron oscillation in solids. Let us consider a single electron occupying a state in a solid. In many cases, the light field couples this initial state predominantly to one state of higher energy. As in Section 2.2.3.1, the resulting oscillation is described quantum physically by a superposition of the two states. Again, the field-induced superposition is accompanied by a transition of the electron from the occupied initial state to the higher empty state. In crystalline solids, these states are extended Bloch waves:

$$\psi_{\vec{k}}(\vec{r}) = \frac{1}{\sqrt{V}} u_{\vec{k}}(\vec{r}) e^{i\vec{k}\cdot\vec{r}} . \tag{2.90}$$

Here, V is the volume of the solid, $u_{\vec{k}}(\vec{r})$ is a function with the same periodicity as the lattice and $\vec{r}$ is the position of the electron. We denote the initial state as $\psi_{\vec{k}_i}$ and the final state as $\psi_{\vec{k}_f}$. The energy required for the excitation is absorbed from the light field as a photon with energy $\hbar\omega$. For instance, Fig. 2.7 shows a transition that promotes an electron from an initial state with an energy E_i in a fully occupied lower band into a final state with the energy E_f in an initially empty upper band. Since both states are Bloch states, they are both spatially extended.

The possible combinations of initial and final states are not arbitrary. As in the atomic case, this has several reasons. First, the initial state must be occupied and the final state must be empty for an optical transition to occur. The law of energy conservation and the selection rules must be fulfilled. Due to the lattice periodicity of the Bloch states, also the law of momentum conservation must be fulfilled. This law reads

2 In this section, we will follow the excellent presentation of Ref. [15].

for the absorption of a photon:[3]

$$\hbar\vec{k}_\mathrm{f} = \hbar\vec{k}_\mathrm{i} + \hbar\vec{k}_\gamma \,. \tag{2.91}$$

Here, $\hbar\vec{k}_\gamma$ is the momentum of the exciting photon (eq. (2.80)). Equation (2.91) holds if the electron is not scattered by a phonon during the photon absorption. An optical transition for which eq. (2.91) holds is called a direct transition, and the corresponding absorption process is known as direct absorption. Because the wavelength of light is much larger than the lattice constant of the crystal, the photon momentum can, in general, be neglected compared to typical electron momenta. We therefore get, to a good approximation,

$$\hbar\vec{k}_\mathrm{f} = \hbar\vec{k}_\mathrm{i} \,, \tag{2.92}$$

i.e., direct transitions are almost vertical in the $E(\vec{k})$ diagram. For direct absorption, the law of energy conservation reads

$$E_\mathrm{f} = E_\mathrm{i} + \hbar\omega \,, \tag{2.93}$$

i.e., the energy required for the excitation originates completely from the absorption of the photon. These conditions for direct absorption (appropriate state occupation, selection rules, momentum and energy conservation) usually forbid direct transitions of single electrons within the same band at frequencies in the visible and near-infrared spectral ranges. In contrast, an excitation of the electron from an initial state in a lower band into a final state in a higher band is often allowed. An example of such an interband absorption has been shown schematically in real space in Fig. 2.7. The simultaneous momentum and energy conservation can best be illustrated in an $E(\vec{k})$ diagram of a solid such as Fig. 2.8. As in Fig. 2.7, the shading of the bands indicates that the lower band is occupied by electrons and the upper band is unoccupied. The arrow depicts

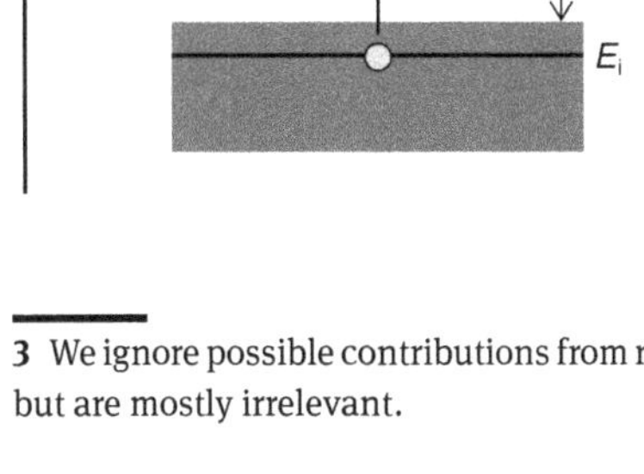

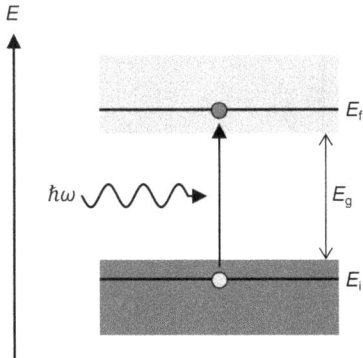

Fig. 2.7: Schematic diagram of an interband absorption process in a crystalline solid in real space. Absorption of a photon with energy $\hbar\omega$ promotes an electron from an initial state with energy E_i in a fully occupied lower band (shaded dark grey) to a final state with energy E_f in an initially empty upper band (shaded light grey). E_g is the band gap between the two bands.

3 We ignore possible contributions from reciprocal lattice vectors $\vec{G}_i$, which could add momenta $\pm\hbar\vec{G}_i$, but are mostly irrelevant.

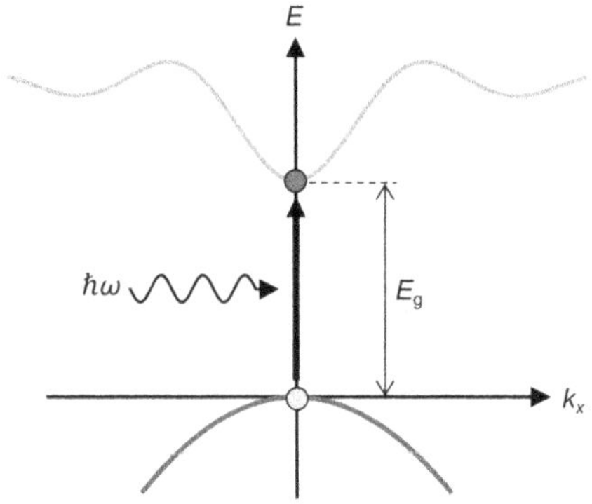

Fig. 2.8: An interband absorption process (arrow) displayed in an $E(\vec{k})$ diagram of a crystalline solid. An electron is optically excited from an initial state in the occupied lower (valence) band (dark grey line) to a final state in the unoccupied upper (conduction) band (light grey line).

the momentum and energy conservation for the special case of monochromatic excitation at the photon energy $\hbar\omega = E_g$. Moreover, we assume that the symmetry properties of the involved band states satisfy the selection rules for optical transitions.

Figure 2.8 depicts a situation in which the maximum of the lower occupied band is at the same point in $\vec{k}$ space as the minimum of the higher empty band. This means that the lowest photon energy at which direct transitions can occur is the band gap energy, $\hbar\omega = E_g$. Such a situation is found in direct semiconductors, in which the global maximum of the highest valence band and the global minimum of the lowest conduction band both lie at the same point in $\vec{k}$ space, usually at $\vec{k} = \vec{0}$. An example is the semiconductor gallium arsenide shown in Fig. 2.9. Direct transitions from states of the valence band maximum into empty states of the conduction band minimum can be excited at the Γ point of the Brillouin zone, i.e., at $\vec{k} = \vec{0}$.

In contrast to direct semiconductors, in indirect semiconductors, such as silicon or germanium, the conduction band minimum lies far away from the valence band

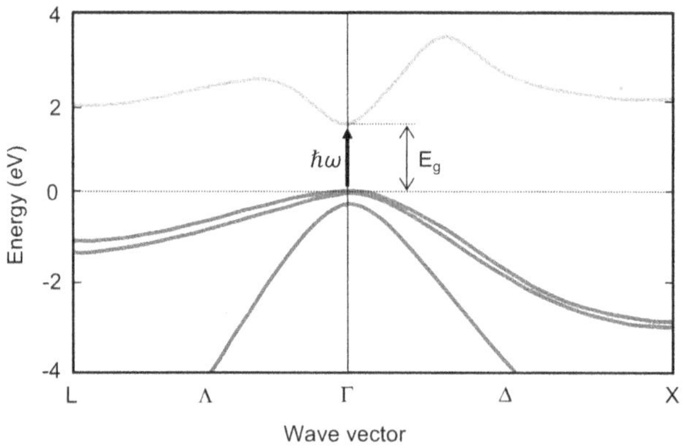

Fig. 2.9: Direct interband absorption at the Γ point of the Brillouin zone in gallium arsenide. For the absorption process to take place at this point, the photon energy $\hbar\omega$ of the excitation light must match the band gap E_g. After Fig. 3.4 of [15], there after [16].

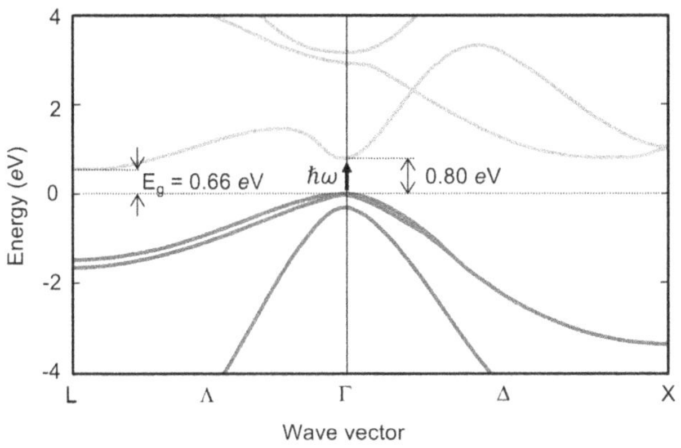

Fig. 2.10: Direct interband absorption at the Γ point of the Brillouin zone in germanium. For the absorption process to take place at this point, the photon energy $\hbar\omega$ of the excitation light must be equal to 0.80 eV, larger than the band gap $E_g = 0.66$ eV. After Fig. 3.10 of [15], there after [17].

maximum, which is at $\vec{k} = \vec{0}$. For instance, the conduction band minimum in germanium lies near the L-point of the Brillouin zone, far away from the Γ-point (Fig. 2.10).

Therefore, in indirect semiconductors, a direct transition is not possible at a photon energy of E_g. In contrast, indirect interband transitions are possible, in which a phonon is additionally absorbed or emitted during the transition. As depicted in Fig. 2.11, the phonon provides the necessary difference in $\vec{k}$ between the initial and the final state of the transition. In an extension of eq. (2.92), the law of momentum conservation for such an indirect interband absorption reads:

$$\hbar\vec{k}_f = \hbar\vec{k}_i \pm \hbar\vec{q}\ , \tag{2.94}$$

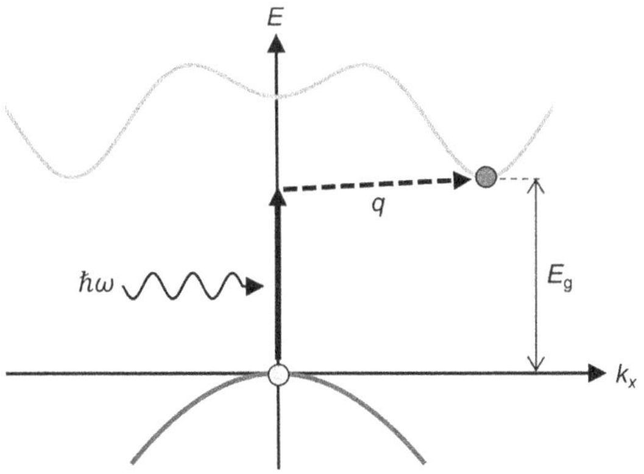

Fig. 2.11: An indirect interband absorption process in an $E(\vec{k})$ diagram. An electron is optically excited from an initial state in an occupied lower band into a final state in an unoccupied upper band. The transition involves a phonon (dashed arrow), which provides the needed momentum $\hbar q$. The energy of the phonon is much smaller than the photon energy.

where $\vec{q}$ is the wave vector of the phonon. As in eq. (2.92), we have neglected the momentum of the photon. The law of energy conservation for the transition reads:

$$E_\mathrm{f} = E_\mathrm{i} + \hbar\omega \pm \hbar\omega_\mathrm{ph} \,, \tag{2.95}$$

where the + sign stands for the absorption and the – sign for the emission of a phonon. Typical phonon energies $\hbar\omega_\mathrm{ph}$ in solids are a few meV to a few tens of meV and, therefore, small in comparison to typical photon energies for interband transitions (several hundred meV to a few eV).

When an electron is excited into a higher band, it leaves behind a positively charged hole in the initial state in the original band. Conceptually, we can think of these two charge carriers as an electron-hole pair. Such an electron-hole pair is electrically neutral. The creation of an electron-hole pair due to the absorption of a photon is called excitation of the electron-hole pair. The total energy acquired by the electron and hole during the excitation is said to be the energy of the electron-hole pair. After the excitation, the electron-hole pair can lose this energy, partially or entirely, due to energy relaxation of the electron and the hole, e.g., by phonon emission. Electron-hole pairs will be discussed in more detail in Subsection 2.3.3.

2.3.1.2 Absorption coefficient for direct interband transitions

According to Subsection 2.2.2.2, the frequency dependence of optical absorption in a medium is controlled by its electronic structure. A special property of crystals is that the electronic energy levels are arranged in continuous bands with forbidden energy gaps between them. The effects of this energetic arrangement on the frequency dependence of the absorption coefficient can be understood from Fig. 2.7. For a transition from the lower band, there exist no suitable final states at photon energies $\hbar\omega$ below E_g. Hence, no absorption takes place, and α is zero at these photon energies. In contrast, transitions between states in the lower and upper band are possible for $\hbar\omega \geq E_\mathrm{g}$. Thus, interband absorption sets in at $\hbar\omega = E_\mathrm{g}$. The absorption due to interband transitions between states that lie close to the band gap is called band edge absorption. For very large photon energies, no initial and final states lying far enough apart in energy can be found in the two bands, and the absorption vanishes again. Between these limits, the absorption coefficient has a continuous frequency dependence, because the energies of the initial and final states in the bands are continuously distributed.

The goal of this and the following subsections is to derive expressions for the absorption coefficient associated with direct interband transitions. According to eq. (2.16), the absorption coefficient is defined by

$$\alpha = \frac{\mathrm{d}I_\mathrm{p}}{\mathrm{d}z}\left[-\frac{1}{I_\mathrm{p}(z)}\right] . \tag{2.96}$$

A light beam can be imagined as a flow of photons that move in the beam direction at the velocity of light. From eqs. (2.6) and (2.79), we obtain the relation between the intensity of this light and the number N_γ of photons per time t and cross-sectional area A of the beam:

$$I_p = \frac{N_\gamma E_\gamma}{At} . \tag{2.97}$$

The term α in eq. (2.96) thus describes the relative reduction of the photon flow due to absorption in a layer of thickness dz. This relative reduction is proportional to the transition rate $W_{i\to f}$ of electrons from the electronic ground state to the excited state in the medium:

$$\alpha \propto W_{i\to f} . \tag{2.98}$$

For a direct transition, the transition rate can be calculated quantum physically using Fermi's Golden Rule:

$$W_{i\to f} = \frac{2\pi}{\hbar}|\widetilde{M}|^2 g(\hbar\omega) . \tag{2.99}$$

Here $\widetilde{M}$ is the so-called optical matrix element. It describes the effect of the perturbation exerted by the alternating electric field of the light wave on the electrons in the initial states. Moreover, $g(\hbar\omega)$ is the joint density of states, which tells us how many initial and final states that are $\hbar\omega$ apart exist per volume. For the sake of simplicity, we assume that all pairs of states contributing to $g(\hbar\omega)$ have the same matrix elements $\widetilde{M}$, which is reasonable for transitions involving only two bands. In what follows, we will derive expressions for this optical matrix element and the joint density of states. We will insert these expressions into eq. (2.99) in order to calculate the dependence of the absorption coefficient on $\hbar\omega$ using eq. (2.98). The optical matrix element has the form:

$$\begin{aligned}\widetilde{M} = \langle f|\widehat{H}'|i\rangle &= \int \widetilde{\psi}_f^*(\vec{r},t)\widehat{H}'(\vec{r},t)\widetilde{\psi}_i(\vec{r},t)\mathrm{d}^3\vec{r} \\ &= e^{\frac{i(E_f-E_i)t}{\hbar}} \int \psi_f^*(\vec{r})\widehat{H}'(\vec{r},t)\psi_i(\vec{r})\mathrm{d}^3\vec{r} ,\end{aligned} \tag{2.100}$$

where the operator $\widehat{H}'$ describes the perturbation by the alternating electric field of the light wave (eqs. (2.65) and (2.66)), $\widetilde{\psi}_{i,f}(\vec{r},t)$ are the time-dependent wave functions of the initial and final states, respectively, and $\psi_{i,f}(\vec{r})$ are their time-independent counterparts. The integration is over the entire crystal volume. For the electric field of the light wave, we use the complex notation from eq. (2.9). Inserting this ansatz into eq. (2.100) gives

$$\widetilde{M} = \int \psi_f^*(\vec{r})e\vec{r}\cdot\vec{E}_0 e^{i\vec{k}\vec{r}}\psi_i(\vec{r})\mathrm{d}^3\vec{r} , \tag{2.101}$$

where we restrict ourselves to the case of a resonant excitation,

$$\omega = \frac{E_f - E_i}{\hbar} . \tag{2.102}$$

The initial and final state of the optical transition are described by Bloch functions (eq. (2.90)). Inserting them into eq. (2.101) and reordering gives:

$$\begin{aligned}\widetilde{M} &= \frac{e}{V}\int u_f^*(\vec{r})e^{-\vec{k}_f\cdot\vec{r}}\vec{r}\cdot\vec{E}_0 e^{i\vec{k}\vec{r}}u_i(\vec{r})e^{i\vec{k}_i\cdot\vec{r}}\mathrm{d}^3\vec{r} \\ &= \frac{e}{V}\int u_f^*(\vec{r})\vec{r}\cdot\vec{E}_0 u_i(\vec{r})\left(e^{-\vec{k}_f\cdot\vec{r}}e^{i(\vec{k}_i+\vec{k})\cdot\vec{r}}\right)\mathrm{d}^3\vec{r} .\end{aligned} \tag{2.103}$$

Taking into account eq. (2.91) and assuming, without loss of generality, that the light is polarized along the x-direction, we obtain

$$\widetilde{M} = \frac{e|\vec{E}_0|}{V}\int u_f^*(\vec{r})xu_i(\vec{r})d^3\vec{r} = \frac{N_{uc}|\vec{E}_0|}{V}\int\limits_{\text{unit cell}} u_f^*(\vec{r})exu_i(\vec{r})d^3\vec{r}\,. \tag{2.104}$$

Here, N_{uc} is the number of unit cells in the crystal and the integration on the right part is over the volume of a single unit cell. The functions $u_i(\vec{r})$ and $u_f(\vec{r})$ are the lattice periodic parts of the Bloch wave functions, which are closely related to the electron orbitals of the individual atoms and are equal in all unit cells. The integral needs to be calculated from the material-specific functions $u_i(\vec{r})$ and $u_f(\vec{r})$. Typically, these functions do not change too strongly within a band. A single matrix element $\widetilde{M}$ is thus a good approximation for all transitions between two bands, which, a posteriori, justifies eq. (2.99). In order to evaluate eq. (2.99), we still need an expression for the joint density of states $g(\hbar\omega)$. Because the initial and final states lie in bands, the joint density of states is a continuous function of $\hbar\omega$. The precise $\hbar\omega$ dependence of this quantity describes how many initial and final states are $\hbar\omega$ apart and additionally obey momentum conservation. It thus depends on the dispersions of the two bands involved in the optical transitions. Here, we will derive it for direct transitions between two isotropic parabolic bands (sketched in Fig. 2.8 and realized, e.g., in the vicinity of the band gap of direct semiconductors). We obtain from eqs. (2.93) and (2.92)

$$\hbar\omega = E_g + \frac{\hbar^2\vec{k}^2}{2m_e^*} + \frac{\hbar^2\vec{k}^2}{2m_h^*}\,, \tag{2.105}$$

where m_e^* and m_h^* are the effective masses of the conduction and valence bands of the direct semiconductor, respectively. Using the reduced mass $\tilde{\mu}$

$$\frac{1}{\tilde{\mu}} = \frac{1}{m_e^*} + \frac{1}{m_h^*}\,, \tag{2.106}$$

this relation reads

$$\hbar\omega - E_g = \frac{\hbar^2 k^2}{2\tilde{\mu}}\,. \tag{2.107}$$

This equation tells us that the excess energy $\hbar\omega - E_g$ that the electron-hole pair possesses with respect to the band gap has a quadratic $\vec{k}$-dependence similar to the dispersion $E(\vec{k})$ of a single parabolic isotropic band. On the basis of this similarity, one can derive an expression for the joint density of states by performing the same steps as in the derivation of the electronic density of states of a parabolic isotropic band (see eq. (1.13) and footnote 16, p. 10). We obtain:

$$g(\hbar\omega) = \frac{1}{2\pi^2}\left(\frac{2\tilde{\mu}}{\hbar^2}\right)^{\frac{3}{2}}(\hbar\omega - E_g)^{\frac{1}{2}} \quad \text{for} \quad \hbar\omega \geq E_g\,. \tag{2.108}$$

Moreover, it is clear that

$$g(\hbar\omega) = 0 \quad \text{for} \quad \hbar\omega < E_g\,. \tag{2.109}$$

Inserting these results into eq. (2.99) and making use of eq. (2.98), we obtain the frequency dependence of the absorption coefficient:

$$\alpha(\hbar\omega) \propto (\hbar\omega - E_g)^{\frac{1}{2}} \quad \text{for} \quad \hbar\omega \geq E_g \tag{2.110}$$

$$\alpha(\hbar\omega) = 0 \quad \text{for} \quad \hbar\omega < E_g\,. \tag{2.111}$$

For a direct semiconductor with isotropic parabolic bands, our treatment thus gives a vanishing absorption coefficient within the band gap (as argued at the start of this subsection) and a square-root dependence on photon energy above the band gap. This prediction is confirmed by experimental measurements. Figure 2.12 shows the absorption coefficient of the direct semiconductor indium arsenide (InAs) at room temperature. Its bands have an isotropic parabolic dispersion in the vicinity of the band gap. The photon energy dependence of α^2 follows a straight line, in agreement with eq. (2.110). The photon energy at which α goes to zero (0.35 eV) represents the band gap energy E_g at room temperature.

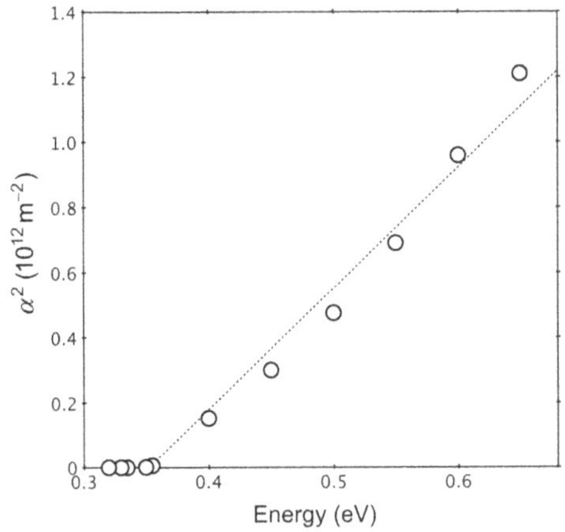

Fig. 2.12: Square of the measured absorption coefficient α versus photon energy $\hbar\omega$ for the direct semiconductor indium arsenide at room temperature. The dashed line is a guide to the eye. After Fig. 3.6 of [15], there after [18].

In many direct semiconductors, the photon energy dependence of the absorption coefficient is found to significantly deviate from eq. (2.110). For instance, at low temperatures and for a sufficiently large band gap, excitonic effects appear, which stem from the Coulomb interaction between the excited electron and hole. The Coulomb interaction enhances the absorption in the spectral vicinity of the band gap. We will consider excitonic effects in Section 2.3.3.1. However, these effects are negligible for the case of room-temperature InAs considered in Fig. 2.12. Moreover, deviations from the parabolic band dispersion that may appear away from the Brillouin zone center may also lead to deviations from the square-root dependence of the absorption coefficient (eq. (2.110)). In such cases, the joint density of states cannot be calculated using an effective mass model, but must be determined numerically on the basis of the real bandstructure.

2.3.1.3 Absorption coefficient for indirect interband transitions

In the previous subsection, we have derived the absorption coefficient for the band-edge absorption in a direct semiconductor. A similar treatment for an indirect semiconductor would be too lengthy to be presented here. It yields the following photon-energy dependence for a valence band maximum at the Γ-point and a conduction band minimum at the L-point with parabolic band dispersions at both band extrema:

$$\alpha(\hbar\omega) \propto \left(\hbar\omega - E_g \mp \hbar\omega_{ph}\right)^2 , \tag{2.112}$$

where, as in eq. (2.95), $\hbar\omega_{ph}$ is the phonon energy.

It is clear from eq. (2.112) that the indirect interband absorption must also show a threshold behavior in the spectral vicinity of E_g, albeit shifted by one phonon energy. The sign of the shift depends on whether the phonon involved in the transition is absorbed (−) or emitted (+). Such a shift is indeed observed in Fig. 2.13, which shows data from absorption measurements on the indirect semiconductor germanium. The solid circles represent the square root of the absorption coefficient versus the photon energy in the spectral vicinity of the band-gap energy at 0.66 eV. Above 0.65 eV, the points follow a straight line, as expected from eq. (2.112). This line intersects the photon energy axis at 0.65 eV. Therefore, the absorption sets in already 0.01 eV below the band-gap energy of 0.66 eV. This is attributed to a phonon of energy 0.01 eV, which is absorbed together with the photon. The energy agrees with the energy of the so-called transversal acoustic phonons in germanium. Moreover, the solid circles show an absorption tail below 0.65 eV. This tail can be explained by the absorption of higher-frequency phonons and multiple phonons.

Besides the shift of the absorption threshold, the difference between indirect and direct interband absorption is that the exponent in eq. (2.112) is larger than in

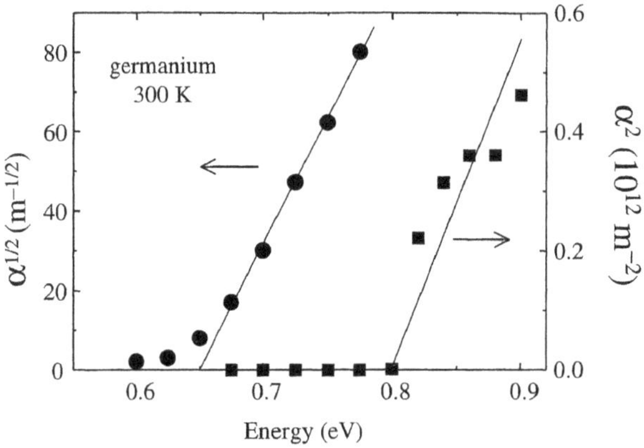

Fig. 2.13: Measured data for the absorption coefficient α of germanium, which is once plotted as $\alpha^{1/2}$ according to eq. (2.112) for indirect transitions and once plotted as α^2 according to eq. (2.110) for direct transitions. Fig. 3.10 of [19], there after [18].

eq. (2.110). The larger exponent is a consequence of the fact that the momentum transfer provided by the phonon enables other combinations of initial and final states than those possible for direct transitions. This difference in absorption can be used to determine whether the band gap of the given semiconductor is of a direct or indirect nature. For instance, a pronounced additional increase of the absorption coefficient is observed for germanium above 0.80 eV. This is shown by the black squares in Fig. 2.13, which exhibit the square of the absorption coefficient versus the photon energy. The increase starting at 0.80 eV follows approximately a straight line and thus obeys the power law of eq. (2.110), which demonstrates that the additional increase is caused by direct absorption. This absorption corresponds to a direct transition at the Γ-point with an energy of 0.80 eV (Fig. 2.10).

Notice that the absolute value of the absorption coefficient above 0.80 eV is orders of magnitude larger than immediately above 0.65 eV. The reason is that an indirect absorption needs a phonon to be absorbed (or emitted) at the same time as the photon is absorbed. The probability for such a transition depends on the probability that these processes occur simultaneously. The process is, hence, a second order process, which has a much lower probability than the first order process of a direct transition, which requires only the absorption of a photon. This difference can also be seen in Fig. 2.14, which shows the measured absorption coefficients of the semiconductors silicon (Si) and gallium arsenide (GaAs). Si possesses an indirect band gap at an energy of 1.12 eV, and GaAs a direct band gap at 1.42 eV. Fig. 2.14 shows that the absorption coefficient of the indirect semiconductor Si at photon energies slightly above 1.1 eV is orders of magnitude smaller than that of the direct semiconductor GaAs at 1.42 eV.

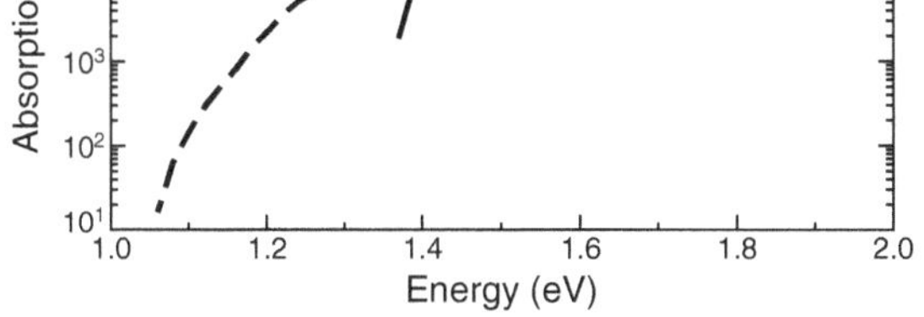

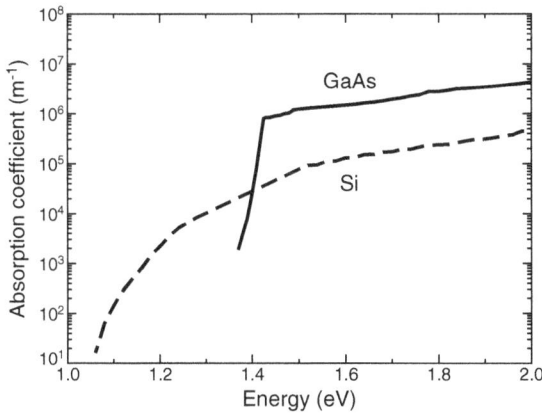

Fig. 2.14: Comparison between the absorption coefficients of the direct semiconductor gallium arsenide and the indirect semiconductor silicon at room temperature. After Fig. 3.9 of [15], there after [18].

2.3.1.4 Absorption spectroscopy

A plot of the absorption coefficient α versus the photon energy $\hbar\omega$ or wavelength λ of the light is called an absorption spectrum. Measuring such spectra and the methods applied in the measurement are called absorption spectroscopy. As we have seen in the context of Fig. 2.13, absorption spectra may reveal valuable information about the optical and electronic properties of a solid. Absorption spectroscopy is therefore a widespread method for the characterization of solids. Experimentally, absorption spectra are taken by measuring, in a spectrally resolved way, the intensity of light transmitted through a thin layer of the investigated material. A typical experimental set-up is shown in Fig. 2.15.

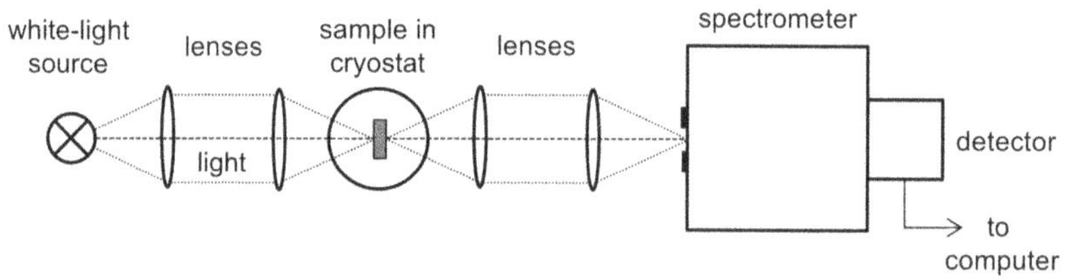

Fig. 2.15: Experimental set up for the measurement of absorption spectra. Light from a white-light source is transmitted through a sample, which is mounted in an optical cryostat for temperature control. The light is spectrally dispersed in a spectrometer and measured with a detector, e.g., a silicon photodiode array detector. The spectrum is recorded with a computer.

Light from a white-light source is transmitted through a sample of known thickness L and spectrally dispersed in a grating spectrometer. The wavelength-dependent light power is measured with a detector, e.g., a silicon photodiode array detector, and recorded with a computer. The ratio of the transmitted light power with and without the sample in the beam path is calculated at each wavelength λ to determine the spectrally resolved transmissivity $T_\gamma(\lambda)$ according to eq. (2.54). The absorption coefficient α is calculated from this data using

$$T_\gamma = (1 - R_{\gamma,1})\, e^{-\alpha L} \,(1 - R_{\gamma,2}) \,. \tag{2.113}$$

Here, $R_{\gamma,1}$ and $R_{\gamma,2}$ are the reflectivities of the front and rear surface of the sample, respectively, which can be determined in a reflection experiment (Section 2.4.3). The sample can be mounted in an optical cryostat in order to control its temperature. The silicon photodiode array detector is suited to measurements in the sensitivity range of silicon, i.e., roughly $\lambda = 200$–1000 nm. In the wavelength range from 1000 nm to 1600 nm, it needs to be replaced with an InGaAs array detector. At still greater wavelengths in the infrared, scanning monochromators or Fourier transform spectrometers with suitable infrared detectors can be employed.

2.3.1.5 Semiconductor photodiodes

One of the most important technical applications of interband absorption is the photodiode. This is a semiconductor diode in which photo-generated carriers generate a current that flows in an electric circuit connected to the diode. Photodiodes can be employed as photodetectors and as photovoltaic cells. Figure 2.16 shows the principle of a photodiode that is used as a photodetector, i.e., for the detection of light. It consists of a p-i-n semiconductor structure with an intrinsic, i.e., undoped, layer i between a p-doped and an n-doped zone. This results in a band profile where the built-in voltage difference between the p-doped region and the n-doped region drops across the intrinsic zone.[4] The photodiode is operated using an applied reverse bias voltage V_0 that, hence, adds to the built-in voltage. Consequently, there is a strong electric field in the intrinsic zone, but, in the absence of light, only a very small electric current (dark current). When light with a photon energy above the band gap E_g is incident on the diode, the interband absorption of photons generates free electrons and holes in the i-zone. These carriers are pulled in opposite directions by the strong electric field, i.e., electrons are pulled towards the n-zone and holes towards the p-zone. Carriers that reach the contacts are measurable as a photocurrent I_{pc} in the external circuit. This photocurrent can be used to detect light.

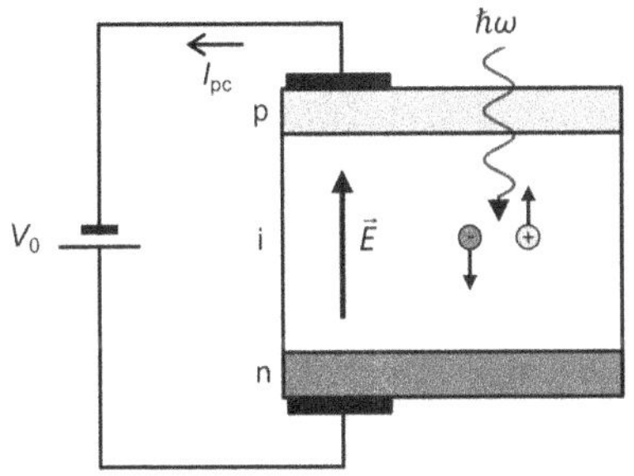

Fig. 2.16: A p-i-n photodiode in an external electric circuit. A reverse bias voltage V_0 is applied to the diode, resulting in an electric field $\vec{E}$ in the intrinsic (i) zone. The interband absorption of photons of energy $\hbar\omega$ in the intrinsic zone generates free electrons and holes, which are pulled into opposite directions by the field. Carriers that reach the contact regions p and n are measurable as a photocurrent I_{pc} in the external circuit.

In what follows, we will estimate the photocurrent. We consider a photodiode whose intrinsic layer has a length L. Light with a photon energy $\hbar\omega > E_g$ and power P is incident on the diode, corresponding to a photon current of $\frac{P}{\hbar\omega}$ photons per second. We assume that the surface of the photodiode is oriented towards the light source and is anti-reflection coated to prevent a reflection of the incident light. Moreover, we assume that the absorption in layers above the intrinsic layer is negligible. According to Beer's law (eq. (2.17)), the portion of the light power absorbed along the distance L is equal to $(1 - e^{-\alpha L})$. Here α is again the absorption coefficient at the photon energy $\hbar\omega$. As each absorbed photon generates one electron-hole pair, of which only one charge carrier contributes to the current in the external circuit, the photocurrent is then given

4 Corresponding band profiles can be found, e.g., in [2, Chapter 12.6] or in [15, appendix E].

by

$$I_{\mathrm{pc}} = e\eta_{\mathrm{q}}\frac{P}{\hbar\omega}\left(1 - \mathrm{e}^{-\alpha L}\right) . \tag{2.114}$$

Here, the quantum efficiency η_{q} denotes the fraction of photogenerated electrons and holes that are extracted into the external circuit. A quantity derived from eq. (2.114) that is important for the suitability of the diode as a photodetector is its responsivity R_{ph}. It is defined as the ratio

$$R_{\mathrm{ph}} = \frac{I_{\mathrm{pc}}}{P} \tag{2.115}$$

and tells us how efficient the photodiode transforms an incident optical power P into a photocurrent I_{pc}. According to eq. (2.114), the responsivity is

$$R_{\mathrm{ph}} = \frac{e\eta_{\mathrm{q}}}{\hbar\omega}\left(1 - \mathrm{e}^{-\alpha L}\right) . \tag{2.116}$$

For a use of the photodiode as a photodetector, essentially three properties are desirable:
- high responsivity
- low noise
- fast response

According to eq. (2.116), a high absorption coefficient α and a high quantum efficiency η_{q} are required to achieve a high responsivity of the photodiode. An upper limit to R_{ph} will be reached if both $(1 - \mathrm{e}^{-\alpha L})$ and η_{q} are equal to unity. In this case, the responsivity will be $\frac{e}{\hbar\omega}$. To achieve a high absorption, it is, of course, necessary that the semiconductor material has a band gap $E_{\mathrm{g}} < \hbar\omega$ in the entire spectral region of interest. This criterion makes the choice of the semiconductor dependent on the spectral range to be covered. Because of their higher absorption, direct semiconductors such as gallium arsenide are better suited than indirect ones such as silicon. For many applications, the disadvantage of indirect semiconductors can be compensated by a larger layer thickness, as, e.g., in the case of silicon. For a high responsivity, it is also useful to make the contact that faces the light source as transparent as possible. Therefore, one chooses a small thickness and a suitable material such as, e.g., indium tin oxide (ITO). The second desirable property, i.e., low noise, is achieved by making the band gap E_{g} as large as possible. This reduces the noisy current contribution that results from carriers thermally excited across the band gap (dark current). For the third property, i.e., a fast response, small layer thicknesses and large carrier mobilities μ of the semiconductor material are advantageous, as they allow a fast extraction of the photogenerated carriers from the photodiode.

Figure 2.17 shows the principle of a photodiode that is used as a photovoltaic cell, i.e., for the transformation of light energy into electric energy. Here, the voltage source from Fig. 2.16 has been replaced by an electric load with resistance R. No photocurrent flows in the dark, meaning that the voltage drop across the load, and thus also the voltage at the contacts of the diode, is zero. When the diode is illuminated, the light

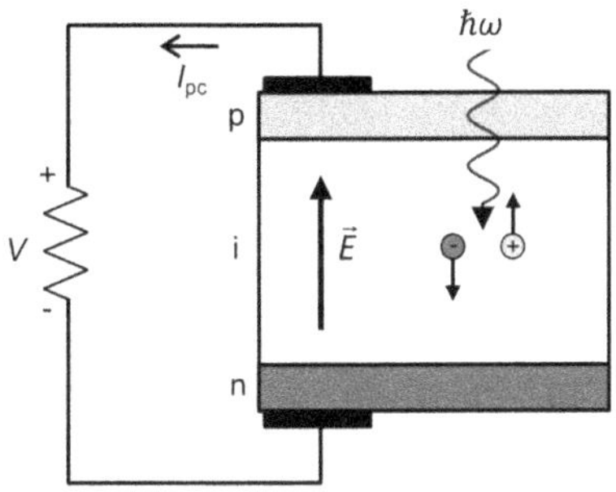

Fig. 2.17: p-i-n photodiode operated as a photovoltaic cell. The photocurrent I_{pc} generated by the illumination results in a power $I_{pc}^2 R$ consumed in the load resistance R marked by its voltage drop V. The voltage $V = I_{pc}R$ dropped across R due to the current flow opposes the built-in field across the intrinsic zone of the diode.

absorption in the intrinsic zone generates electron-hole pairs. The electrons and holes are separated by the built-in field of the diode and swept to the n- and p-zones, respectively. The photocurrent I_{pc} resulting from this motion flows through the load, where it can do work or generate heat. Hence, the photovoltaic cell transforms the absorbed light power into an electrical power RI_{pc}^2. The voltage RI_{pc} dropped across the load and thus applied to the diode biases the diode in the forward direction. This voltage opposes the built-in field $\vec{E}$ and thus reduces the field transporting the carriers across the intrinsic zone. Therefore, when the illumination power is increased from zero, the photocurrent saturates as soon as $I_{pc}R$ approaches the value of the built-in voltage V_{bi} across the i-zone of the diode. The photocurrent saturation limits the electrical power that can be generated.

2.3.2 Interband luminescence

In Section 2.3.1, we have dealt with the excitation of electrons in solids due to light absorption. The reverse process is known for atoms in which electrons perform transitions from excited levels into lower ones while emitting light (Section 2.2.3). When such a radiative emission process occurs in a solid, it is called luminescence. One distinguishes between photoluminescence and electroluminescence, according to whether the carriers in the initial levels have been excited optically or electrically. Moreover, it is possible to generate stimulated emission (Section 2.2.3.2). This phenomenon is used in semiconductor lasers. In the present section, we will study the features of different kinds of light emission in solids.

2.3.2.1 Optical emission from electron-hole pair recombination

Generally, in luminescence, electrons jump from excited levels into lower ones by spontaneously emitting a photon. As shown schematically in Fig. 2.18, the two levels involved in visible or near-infrared luminescence in solids lie, in general, in two different energy bands. This process is known as interband luminescence. Such a downward transition shares with the upward transitions involved in absorption (Sec-

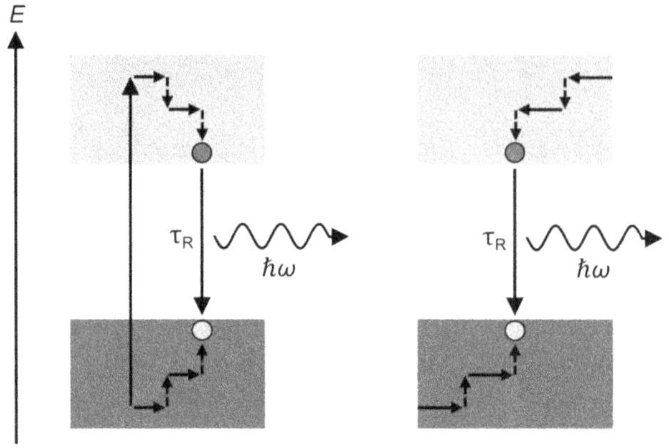

Fig. 2.18: Schematic representations of luminescence processes in a solid in real space. In photoluminescence, carriers are injected into the bands optically (left), and in electroluminescence, electrically (right). In both cases, electrons are injected into the previously empty upper band, while holes are injected into the lower band, which was initially fully occupied with electrons. The carriers rapidly relax into states at or near the band edge via phonon emission (dashed arrows). The electrons subsequently jump into the empty states formed by the holes while emitting photons with energy $\hbar\omega$. τ_R: radiative recombination lifetime.

tion 2.3.1.1) that the initial state must be occupied and the final state empty. The initial occupation is achieved by injecting electrons into states of the previously empty upper band optically, in the case of photoluminescence, or electrically, in the case of electroluminescence. The upper band may be the conduction band of a semiconductor. In general, the injection is into states high up in the band, from where they rapidly relax into states at or near the band edge via phonon emission. In situations with interband luminescence, usually, the lower band is initially fully occupied with electrons, as, e.g., the valence band of semiconductors. Empty states are created in this band by injecting holes simultaneously with the injection of electrons into the upper band. The holes also rapidly relax towards the edge of their band. In the end of the relaxation process, the electron and hole distributions become quasi-stationary. On longer time scales (in general, nanoseconds and longer), the electrons jump into the empty states formed by the holes in the lower band, while emitting photons. This process is called radiative electron-hole recombination.

Let us consider the two levels involved in such a jump. As summarized in Section 2.2.3.2, the spontaneous emission rate in such a two-level system is described by the Einstein A coefficient. If, therefore, the upper level has an occupation N_2 at time t, this occupation decreases at a rate (eq. (2.84))

$$\left(\frac{\mathrm{d}N_2}{\mathrm{d}t}\right)_{\mathrm{R}} = -A_{21}N_2\,, \tag{2.117}$$

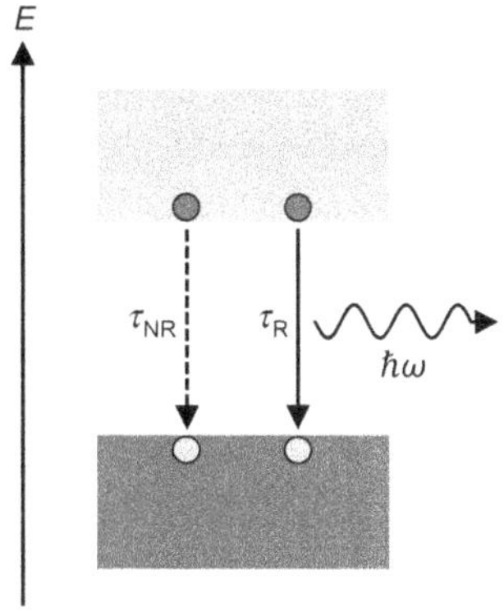

Fig. 2.19: Nonradiative (left) and radiative (right) recombination processes of carriers in band-edge states. τ_R: radiative recombination lifetime, τ_{NR}: nonradiative recombination lifetime.

where „R“ denotes the radiative nature of the occupation change. With the radiative lifetime $\tau_R = A_{21}^{-1}$ (eq. (2.85)), this becomes

$$\left(\frac{dN_2}{dt}\right)_R = -\frac{N_2}{\tau_R}\,. \tag{2.118}$$

Because the rate at which photons are emitted is $|(\frac{dN_2}{dt})_R|$, eq. (2.118) tells us that the emitted power is proportional to the occupation N_2 of the upper level.

Radiative emission is not the only process by which electrons can make a transition from an excited state into the ground state. The carriers may also lose their excitation energy via emission of phonons, in particular if states within the band gap are available as intermediate states for a sequential phonon emission (Fig. 2.19). Such a nonradiative recombination competes with radiative recombination, i.e., carriers that recombine nonradiatively do not contribute to the emitted light. The total rate of occupation change in the excited level is, hence, obtained by extending eq. (2.118):

$$\left(\frac{dN_2}{dt}\right)_{\text{total}} = -\frac{N_2}{\tau_R} - \frac{N_2}{\tau_{NR}} = -\left(\frac{1}{\tau_R} + \frac{1}{\tau_{NR}}\right) N_2\,. \tag{2.119}$$

Here, τ_{NR} is called the nonradiative lifetime. The fraction of radiatively recombining electron-hole pairs is called the radiative quantum efficiency η_R:

$$\eta_R = \frac{\left(\frac{dN_2}{dt}\right)_R}{\left(\frac{dN_2}{dt}\right)_{\text{total}}}\,. \tag{2.120}$$

According to eqs. (2.118) and (2.119), this implies

$$\eta_R = \frac{1}{1 + \frac{\tau_R}{\tau_{NR}}}\,, \tag{2.121}$$

which is independent of N_2. If τ_R and τ_{NR} both have identical values for all electron-hole pairs in the vicinity of the band edges, eq. (2.121) holds for all of them and η_R is called the luminescence efficiency or luminescence yield.

If $\tau_{NR} \ll \tau_R$, η_R becomes small and the luminescence is very inefficient. For instance, the luminescence efficiency of the interband luminescence of gold is typically on the order of 10^{-6}, meaning that luminescence is hardly observable. Conversely, an efficient luminescence (i.e., $\eta_R \to 100\%$) requires that $\tau_{NR} \gg \tau_R$. Direct and indirect transitions behave differently, not only in absorption, but also in luminescence. The law of energy conservation reads for a direct photoluminescent transition, i.e., not involving a phonon:

$$E_f = E_i - \hbar\omega \,. \tag{2.122}$$

This reflects the fact that the final level is lower than the initial level by the energy of the emitted photon. The corresponding law of momentum conservation reads:

$$\hbar\vec{k}_f = \hbar\vec{k}_i - \hbar\vec{k}_y \approx \hbar\vec{k}_i \,, \tag{2.123}$$

again exploiting that the momentum of the photon $\hbar\vec{k}_y$ is negligibly small compared to the Brillouin zone (Section 2.3.1.1).

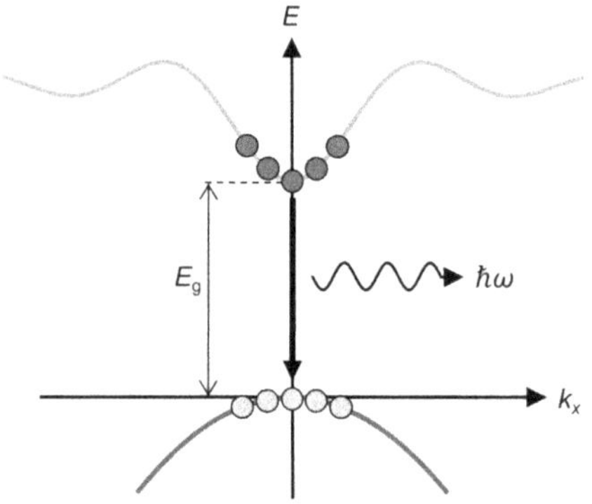

Fig. 2.20: An interband emission process in the $E(\vec{k})$ diagram of a direct semiconductor. An electron jumps from one of the occupied initial states in the conduction band (dark grey circles) across the band gap into one of the empty final states in the valence band (light grey circles) while emitting a photon with energy $\hbar\omega \approx E_g$.

The simultaneous conservation of energy and momentum can again best be illustrated in an $E(\vec{k})$ diagram. For instance, Fig. 2.20 shows a schematic of an interband luminescence in a direct semiconductor such as GaAs. Here, the circles indicate that the lowest part of the conduction band is occupied with electrons, while the topmost part of the valence band is unoccupied (or occupied with holes). These carriers originate from injection into states high up in the respective bands. After their injection, they rapidly relax to the band minima via phonon emission. They form quasistationary distributions, before they recombine radiatively near $\vec{k} = \vec{0}$. Each recombining electron-hole pair hence emits a photon with energy $\hbar\omega \approx E_g$.

This means that independently of how we originally injected the electrons and holes, we obtain luminescence at energies near the band gap. An example is shown in Fig. 2.21. It depicts a photoluminescence spectrum and an absorption spectrum of the direct semiconductor gallium nitride (GaN) at $T = 4$ K. At this temperature, the band gap energy is 3.5 eV. The absorption shows a steep increase close to the band

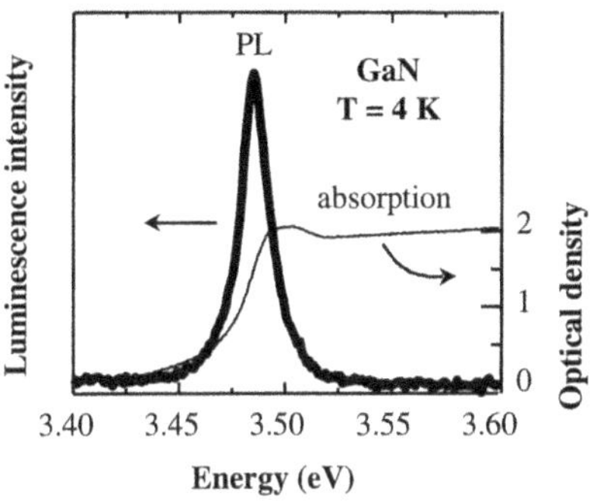

Fig. 2.21: Photoluminescence (PL) spectrum (thick line) and optical density spectrum (thin line) of gallium nitride at $T = 4\,\mathrm{K}$. The optical density is proportional to the absorption coefficient. The luminescence is excited with light at a photon energy of 4.9 eV. Fig. 5.3 of [19], from unpublished data from K. S. Kyhm and R. A. Taylor.

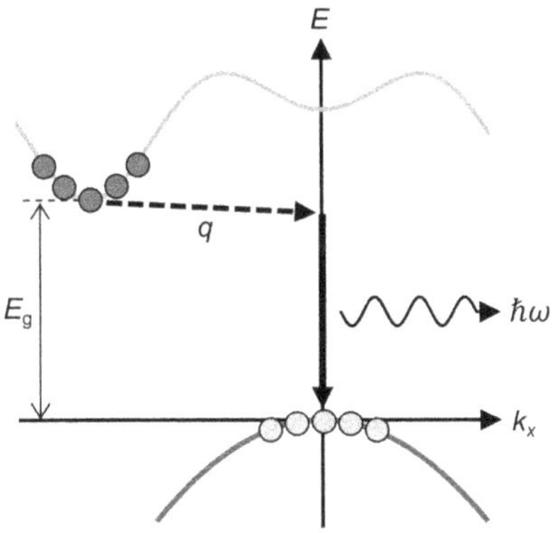

Fig. 2.22: Schematic representation of an interband luminescence process in the $E(\vec{k})$ diagram of an indirect semiconductor. An electron jumps from an occupied initial state in the conduction band across the band gap into an empty final state in the valence band. The transition involves a phonon (dashed arrow), which provides the needed momentum $\hbar q$. The energy of the phonon is much smaller than the photon energy.

gap, which is caused by the so-called Sommerfeld enhancement to be described in Section 2.3.3. Above, the absorption is nearly constant. In contrast, the luminescence consists of a narrow emission line near the band gap energy only.

Figure 2.22 shows a sketch of an interband luminescence in an indirect semiconductor such as silicon. After their injection, the electrons relax into the conduction band minimum away from $\vec{k} = \vec{0}$, and the holes into the valence band maximum at $\vec{k} = \vec{0}$. Then, momentum conservation requires that each emission of a photon is accompanied by the simultaneous emission or absorption of a phonon. The probability for a radiative recombination to occur is determined by the probability of the corresponding second order process, and thus is relatively small. Therefore, the radiative lifetime τ_R is much longer than for direct transitions. According to eq. (2.121), this means that the photoluminescence efficiency η_R is small, as a consequence of the competition with nonradiative recombination processes. For this reason, indirect semiconductors such as silicon and germanium are, in general, poor light emitters.

2.3.2.2 Photoluminescence in solids

In this subsection, we will deal specifically with photoluminescence, which appears after the optical excitation of a solid. Fig. 2.23 shows a schematic drawing of a photoluminescence process in a direct semiconductor such as gallium arsenide. In addition

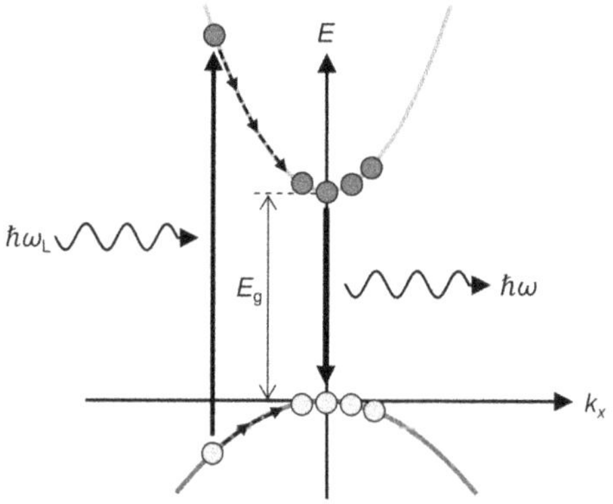

Fig. 2.23: (a) Sketch of photoluminescence process in a direct semiconductor after optical excitation by a laser photon of energy $\hbar\omega_L$. The excitation occurs into high band states, followed by rapid relaxation via sequential inelastic scattering processes (dashed arrows). The carriers accumulate in the vicinity of the band edges and form quasi-stationary thermal distributions, until they recombine radiatively by emission of photons with energy $\hbar\omega$ or non radiatively (not shown).

to the emission process depicted in Fig. 2.20, Fig. 2.23 includes the preceding optical excitation of the electrons and the relaxation of the carriers within both bands. Here, the excitation occurs between states far away from the band extrema, followed by a relaxation via sequential inelastic scattering processes (dashed arrows) by phonons towards the conduction band minimum and the valence band maximum. Each of these scattering events occurs on a timescale on the order of 10^{-13} s. The relaxed carriers accumulate in the vicinity of the band edges and reach thermal equilibrium with the other carriers in the same band via inelastic electron-electron or hole-hole scattering. As a consequence, electrons and holes each form a quasi-stationary thermal distribution, before they recombine radiatively or nonradiatively.

These quasi-stationary thermal distributions can be calculated as follows. The number density of electrons in the conduction band must satisfy the following equation:

$$n_e = \int_0^\infty \mathcal{D}_c(E_e) f_e(E_e, T) \mathrm{d}E_e \,, \tag{2.124}$$

where $\mathcal{D}_c(E_e)$ is the density of states and $f_e(E_e, T)$ the Fermi–Dirac distribution function of the excited, quasi-stationary electrons in the conduction band with temperature T. This quasi-stationary distribution is indicated by the subscript e. The electron energy E_e is given with respect to the conduction band minimum. For a parabolic conduction band, $\mathcal{D}_c(E_e)$ reads (eq. (1.13)):

$$\mathcal{D}_c(E_e) = \frac{1}{2\pi^2} \left(\frac{2m_e^*}{\hbar^2} \right)^{\frac{3}{2}} E_e^{\frac{1}{2}} \,. \tag{2.125}$$

The Fermi distribution function is given by

$$f_e(E_e, T) = \left[\exp\left(\frac{E_e - E_{F,e}}{k_B T} \right) + 1 \right]^{-1} \,. \tag{2.126}$$

Here, $E_{F,e}$ denotes the Fermi energy of the thermal distribution of the excited electrons in the conduction band. This energy is different from the Fermi level of all carriers in equilibrium. The distinction is important, because the electrons are in thermal equilibrium with each other within the conduction band, but not with the holes in the valence band. Hence, $E_{F,e}$ is called quasi Fermi energy of the electrons. Similarly, the holes are in thermal equilibrium with each other, but not with the electrons. Correspondingly, there is a quasi-Fermi energy $E_{F,h}$ of the holes. The absence of a thermal equilibrium of all carriers with each other, which could be characterized by a single Fermi energy E_F, is a consequence of the fact that there are more electrons and holes after an optical excitation than would be present for a purely thermal excitation across the band gap. Summarizing eqs. (2.124), (2.125), and (2.126), we obtain:

$$n_e = \int_0^\infty \frac{1}{2\pi^2}\left(\frac{2m_e^*}{\hbar^2}\right)^{\frac{3}{2}} E_e^{\frac{1}{2}} \left[\exp\left(\frac{E_e - E_{F,e}}{k_B T}\right) + 1\right]^{-1} dE_e \,. \tag{2.127}$$

Analogously, we write for the number density of the holes in the valence band:

$$n_h = \int_0^\infty \frac{1}{2\pi^2}\left(\frac{2m_h^*}{\hbar^2}\right)^{\frac{3}{2}} E_h^{\frac{1}{2}} \left[\exp\left(\frac{E_h - E_{F,h}}{k_B T}\right) + 1\right]^{-1} dE_h \,. \tag{2.128}$$

In this integral, we have chosen the maximum of the valence band as the zero of energy, and both E_h and $E_{F,h}$ are counted from the valence band maximum downwards. Because the excitation process generates equal numbers of electrons and holes, $n_e = n_h$ must hold. Both number densities are determined by the intensity of the exciting light. The quasi-Fermi energies can then be determined from a given carrier density n_e or n_h using eqs. (2.127) and (2.128), if T is known. Here, we assume that T is the sample temperature for both electrons and holes.

In the limit of low carrier densities, $E_{F,e}$ lies far below the conduction band minimum and we can neglect the +1 term in eq. (2.126). The electron distribution can then be approximated by a Boltzmann distribution function:

$$f_e(E_e, T) \propto \exp\left(\frac{-E_e}{k_B T}\right) . \tag{2.129}$$

This also holds for the hole distribution:

$$f_h(E_h, T) \propto \exp\left(\frac{-E_h}{k_B T}\right) . \tag{2.130}$$

The probability that a given two-level system with the energy $E_e + E_h = \hbar\omega - E_g$ is occupied both with an electron and a hole is, hence, given by

$$f_e(E_e, T) f_h(E_h, T) \propto \exp\left(\frac{-E_e}{k_B T}\right)\exp\left(\frac{-E_h}{k_B T}\right) = \exp\left(-\frac{\hbar\omega - E_g}{k_B T}\right) . \tag{2.131}$$

Multiplying with the joint density of states from eq. (2.108) yields the number density of two-level systems which are occupied both with an electron and a hole. The intensity of the photoluminescence emission is proportional to this number density.

Therefore, the frequency dependence of the photoluminescence intensity for a direct semiconductor reads for $\hbar\omega > E_g$

$$I_P(\omega) \propto (\hbar\omega - E_g)^{\frac{1}{2}} \exp\left(-\frac{\hbar\omega - E_g}{k_B T}\right) . \tag{2.132}$$

The behavior can be seen in Fig. 2.24. It shows a photoluminescence spectrum of gallium arsenide at 100 K. The spectrum exhibits a steep rise near E_g due to the density of states factor in eq. (2.132) and then drops exponentially due to the Boltzmann factor. The exponential character of the decay is most easily visible in a semilogarithmic plot of the same data (inset of Fig. 2.24). The slope of the photoluminescence line at high photon energies is in reasonable agreement with the temperature of 100 K at which the measurement was performed. The width (full width at half maximum) of the photoluminescence line is close to $k_B T$, as expected from the combination of steep rise and exponential decay.

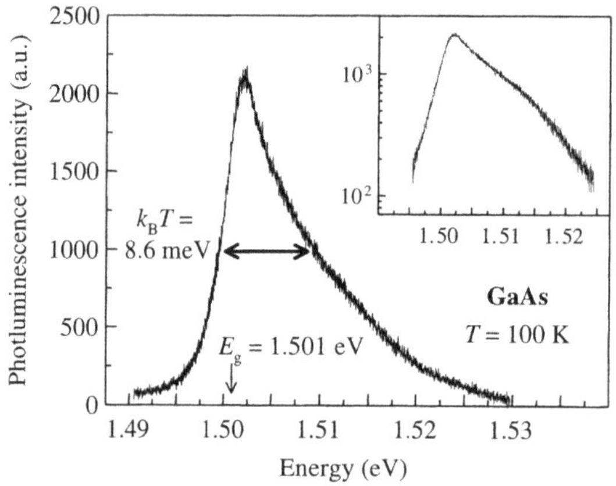

Fig. 2.24: Photoluminescence spectrum of GaAs at a sample temperature of T = 100 K. The excitation was performed at a photon energy of 1.96 eV. The inset shows the same spectrum in a semilogarithmic plot. Fig. 5.6 of [15], there from unpublished data from A. D. Ashmore and M. Hopkinson.

2.3.2.3 Photoluminescence spectroscopy

The measurement of photoluminescence spectra is called photoluminescence spectroscopy. It can give valuable information about the optoelectronic properties of a solid, for instance the energies of states, the existence of defects and the rates of relaxation processes. Photoluminescence spectroscopy is mainly used to characterize semiconductor materials and structures. For instance, it is useful in the development of light-emitting diodes and lasers. Experimentally, the photoluminescence spectra are taken by measuring, in a spectrally resolved way, the intensity of light that has been emitted by a sample after an optical excitation at a high photon energy. A typical experimental setup is shown in Fig. 2.25. The sample is mounted in a cryostat, which makes it possible to control its temperature. The photoluminescence is excited with laser or lamp light at a photon energy above the band gap. A portion of the photoluminescence light emitted by the sample is collected with a lens and guided to a grating spectrometer, in which it is dispersed spectrally. The wavelength dependent light power is measured with an array detector, e.g., a CCD array. Alternatively, the spectrometer can be replaced with a monochromator and the intensity is measured separately at each wavelength with a more sensitive detector, e.g., a photomultiplier tube. In both cases, the measured data is read out with a computer and saved for evaluation.

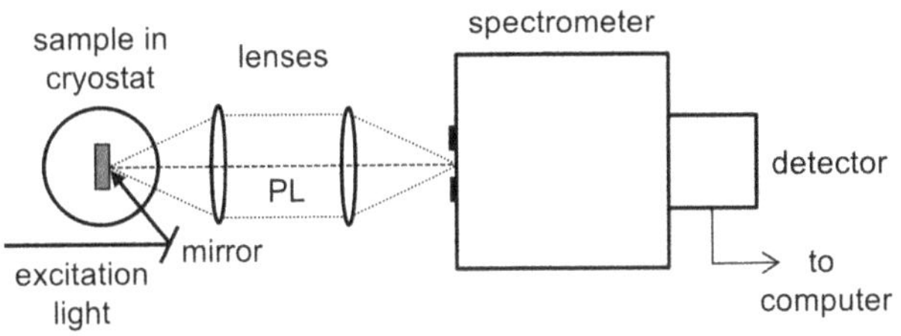

Fig. 2.25: Experimental setup for the measurement of photoluminescence spectra. The sample, which is mounted in a variable temperature cryostat, is excited with light at a photon energy above the band gap of the sample material. The photoluminescence spectrum is obtained by measuring the emission (PL) of the sample as a function of wavelength using a spectrometer and a detector.

In a version of this spectroscopy method, called photoluminescence excitation spectroscopy (PLE), the luminescence intensity is measured at the wavelength of the emission maximum, while the wavelength of the excitation light is varied. The spectra consist of plots of the luminescence intensity versus the photon energy of the exciting light. A feature of this measurement is that the signal is proportional to the density of excited carriers, which, in turn, is proportional to the absorption coefficient at the excitation wavelength. Therefore, the technique can be used to determine the absorption coefficient of samples which are too thick for transmission measurements.

The two varieties of photoluminescence spectroscopy presented here provide a spectral, but not a temporal resolution. The latter can be achieved if the luminescence intensity at the selected wavelength is measured as a function of time after the absorption of a short laser pulse which excites the sample. The measurements are taken using the setup depicted in Fig. 2.25, but with a pulsed laser as the light source and a monochromator as the spectrometer. The time resolution is limited by the pulse duration and the response time of the detection technique. Time resolutions down to 0.1 picoseconds (10^{-13} s) are possible. The time dependence of the measured photoluminescence transients enables the determination of radiative lifetimes and yields information about the dynamics and efficiencies of carrier relaxation and recombination processes.

2.3.2.4 Electroluminescence

Electroluminescence is luminescence generated by electrical injection of electrons and holes into the crystal. Two optoelectronic devices based on electroluminescence, the light-emitting diode (LED) and the laser diode, have found important technical applications. We will study these two devices in this subsection.

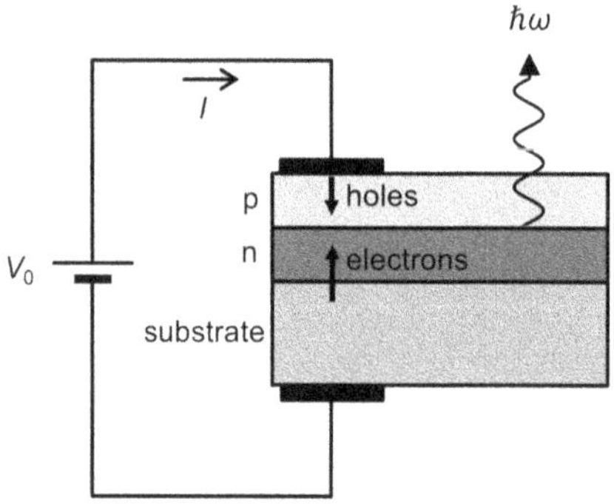

Fig. 2.26: Sketch of a light-emitting diode (LED). The diode is operated in the forward direction, which gives rise to a strong current flow through the layers. The current is composed of carriers that are injected into the transition region between the n- and p-layers. In this zone, the electrons recombine with the holes, emitting photons ($\hbar\omega$) via interband luminescence. The recombining carriers are continuously replaced by the current I that flows through the diode from the external circuit.

Figure 2.26 shows the principle of an LED. It is composed of various thin semiconductor layers, which have been deposited on a thick crystalline semiconductor substrate (Section 1.3.1). The task of the substrate is to support the thin layers above it. These layers form a p-n diode or p-i-n diode. By connecting it to a voltage source, the diode is operated in forward direction, which gives rise to a strong current flow through the layers. The current is composed of carriers that are injected into the transition region between the two layers, the active zone. The electrons are injected through the n-layer, and the holes through the p-layer. In the active zone, the electrons recombine with the holes, emitting photons with energy E_g via interband luminescence. The recombining carriers are continuously replaced by the current that flows through the diode from the external circuit as depicted in Fig. 2.26.

Figure 2.27 shows the electroluminescence spectrum of a p-i-n diode made from gallium arsenide and biased in the forward direction. Light is generated in the thin intrinsic layer between the p-and n-layers. The emission line is at a photon energy of 1.42 eV, which is the band gap energy of GaAs at room temperature. The width of the

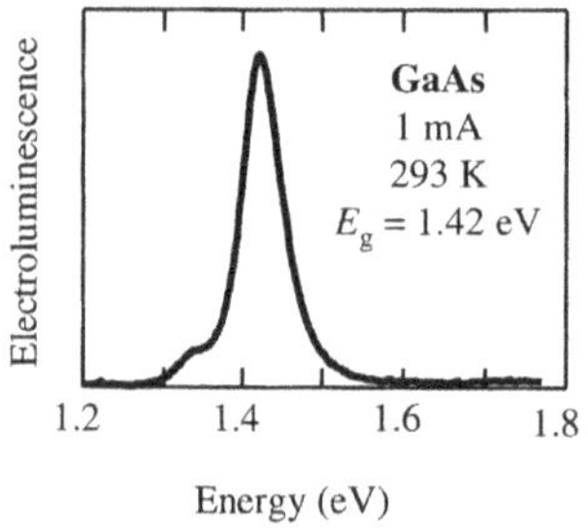

Fig. 2.27: Room-temperature electroluminescence spectrum of a light-emitting diode made from gallium arsenide. Fig. 5.13 of [15], there from unpublished data from A. D. Ashmore.

line (full width at half maximum) is close to 60 meV. This approximately corresponds to twice $k_B T$ and thus is in the order of magnitude expected from the discussion of photoluminescence linewidths in Section 2.3.2.2 (eq. (2.132)). LEDs have found widespread use as light sources for lighting, displays, signals and data communication.

Not only LEDs, but also lasers can be built from semiconductor diodes. They are called laser diodes or semiconductor lasers. For this purpose, the diode structure must be modified in two respects (Section 2.2.3.2). First, a high photon density is required for stimulated emission. Therefore, a laser resonator with appropriate mirrors must be incorporated. This can be achieved by increasing the reflectivities R_γ of the semiconductor-air interfaces at the crystal edges (typically $R_\gamma \approx 30\,\%$) by suitable coatings. Second, an inversion must be generated in the semiconductor medium. This is achieved by injecting strong currents I_{in}, which make the carrier density in the active zone high. The gain coefficient g_ω (eq. (2.86)) then reaches high values. If we assume that g_ω grows linearly with I_{in}, as shown in Fig. 2.28(a), there will be a value of I_{in} at which the gain reaches a large enough value of g_ω, as determined by eq. (2.89), and laser emission sets in. This value is called the threshold current I_{th}.

For all $I_{\text{in}} \geq I_{\text{th}}$, the value of g_ω is fixed by the equilibrium condition in eq. (2.88), and thus independent of I_{in} (an effect called gain clamping). At currents beyond the threshold current, the additional electrons and holes injected into the active zone cause no additional gain, but give rise to a stronger light field, which is amplified by the same factor per length due to the correspondingly larger density of electron-hole pairs. Above the laser threshold, the output power P_{out} then grows according to $P_{\text{out}} \propto (I_{\text{in}} - I_{\text{th}})$, as depicted in Fig. 2.28(b). If η_{L} denotes the fraction of injected electrons that emit laser photons, one gets:

$$P_{\text{out}} = \eta_{\text{L}} \frac{\hbar\omega}{e} (I_{\text{in}} - I_{\text{th}}) \,. \tag{2.133}$$

In an ideal laser diode η_{L} would be unity, i.e., each electron passing through the device emits a photon. The best laser diodes get very close to this value.

As semiconductor lasers have, due to the built-in resonator, a more complex structure than LEDs, their fabrication is more expensive. However, they have important advantages for applications, such as narrower linewidths, higher output efficiencies, better beam quality and higher response speeds than LEDs. They have found widespread use as light sources for various applications, e.g., fiber optics communication, laser pointers, CD and DVD players, barcode scanners, laser printers and laser surgery.

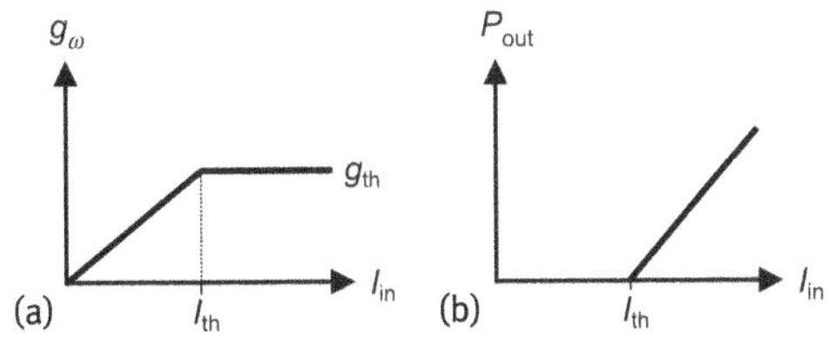

Fig. 2.28: Behavior of a semiconductor laser diode in dependence of injection current I_{in}. Gain g_ω(a) and output power P_{out}(b) versus injection current I_{in}. I_{th} is the threshold current, and g_{th} the threshold gain. After Fig. 5.15 of [15].

Commercial LEDs and laser diodes are fabricated from direct band-gap semiconductors such as GaAs, because the radiative efficiency of materials with an indirect band gap is too low. The choice of direct semiconductors is mainly guided by the following criteria:
- size of the band gap and, hence, emission frequency ω.
- restrictions due to the lattice matching between the semiconductor layers.
- ability to dope the semiconductor material.

The size of the band gap is important as it determines the emission wavelength. The term lattice matching refers to the necessity of matching the lattice constants of subsequent semiconductor layers of the device as well as possible. This is necessary to avoid dislocations due to distorting forces between the lattices (Section 1.3.1). Such dislocations typically reduce the optical quality by enabling non-radiative recombination and decrease the electrical quality through a reduction of the conductivity due to additional scattering. The doping of the semiconductor materials is required to enable the production of the n- and p-layers of the diode.

2.3.3 Excitons

Absorption and photoluminescence spectra of semiconductors at low temperatures show lines in the spectral vicinity of the band gap that are not predicted by the simple picture of interband transitions introduced above. These lines can be understood from the following considerations. When an electron-hole pair is created, e.g., as a consequence of an interband absorption, the electron and hole attract each other via the Coulomb interaction. In certain conditions, this attraction is able to bind the electron and hole to each other. The resulting carrier pair looks electrically neutral to its environment and is called an exciton. Quantum physically, it represents a quasiparticle like a phonon. Its lifetime is finite, because, after its excitation, it can either decay by emitting a photon or it may get destroyed by scattering processes that separate the electron and the hole. Excitons are observed in low temperature absorption and photoluminescence spectra of semiconductors and insulators. In principle, they can also be created in metals; however, the carrier scattering and the screening are too strong to preserve the excitons for a relevant period of time. Essentially, there are two kinds of excitons: Wannier–Mott excitons and Frenkel excitons. In this section, we will study the properties of both kinds in various materials and for different excitation conditions.

2.3.3.1 Wannier–Mott excitons

Wannier–Mott excitons typically appear in direct semiconductors at low temperatures. An important property of these excitons is that the mean distance between electron and hole is much larger than the lattice constant of the crystal, as shown schematically in Fig. 2.29.

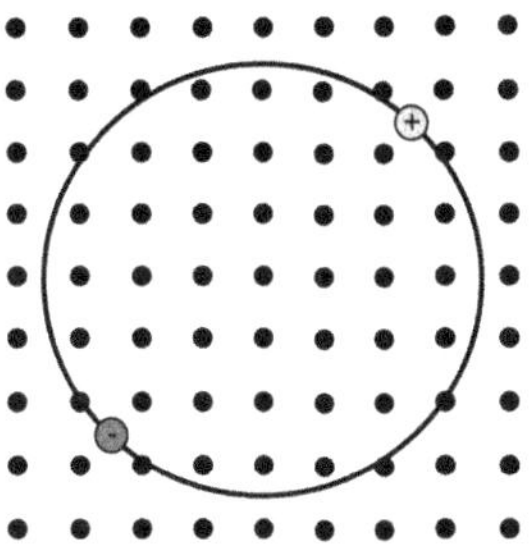

Fig. 2.29: Schematic representation of a Wannier–Mott exciton in a crystal. Black dots represent the atoms. The relative motion of the electron (-) and hole (+) is classically pictured as a common orbit around their shared center of mass. After Fig. 4.1 of [15].

The mean electron-hole distance is thus orders of magnitude larger than the mean distance between the electron and the proton in a hydrogen atom. The difference is a consequence of field screening. The Coulomb attraction force, which binds the negatively charged electron and the positively charged hole, also generates polarization charges in the surrounding medium, which, in turn, partially screen the Coulomb fields of the electron and hole, i.e., partially balance their attraction force. Therefore, the electron and hole are bound less tightly than the electron and proton in the hydrogen atom.

Because the distance between electron and hole is so large, we can average over the atomic structure of the lattice and describe the field screening by the position-independent dielectric constant ε (Section 2.2.2.2). In particular, the Coulomb attraction force in the crystal is of a similar $\frac{1}{r^2}$ form as in vacuum. Therefore, the Schrödinger equation for the two-particle wavefunction $\Psi(\vec{r}_\mathrm{e}, \vec{r}_\mathrm{h})$ of the electron-hole pair has the same form as for the electron-proton pair of the hydrogen atom – with the exception that the Coulomb interaction term in the Hamilton operator is a factor of $\frac{1}{\varepsilon}$ smaller and that we use as particle masses the effective masses m_e^* and m_h^* of electron and hole (Section 1.2.2, eq. (1.6)):

$$\left[\frac{\hat{\vec{p}}_\mathrm{e}^2}{m_\mathrm{e}^*} + \frac{\hat{\vec{p}}_\mathrm{h}^2}{m_\mathrm{h}^*} - \frac{e^2}{4\pi\varepsilon_0\varepsilon|\vec{r}_\mathrm{e} - \vec{r}_\mathrm{h}|}\right]\Psi = E\Psi\,, \tag{2.134}$$

where $\hat{\vec{p}}_\mathrm{e}$ and $\hat{\vec{p}}_\mathrm{h}$ are the momentum operators for the electron and hole, respectively. In this form of the Schrödinger equation, we assume that the electron-hole pair already exists and only the kinetic energies of both carriers as well as the interaction between them must be included – precisely as in the case of the hydrogen atom, in which the electron and proton already exist before their mutual binding. This means that we will have to consider, a posteriori, the energy invested for the creation of the electron-hole pair. When we use the separation ansatz

$$\Psi(\vec{r}_\mathrm{e}, \vec{r}_\mathrm{h}) = F(\vec{R})G(\vec{r})\,, \tag{2.135}$$

rewriting the two-particle wave function in terms of the center-of-mass coordinate

$$\vec{R} = \frac{m_\mathrm{e}^*\vec{r}_\mathrm{e} + m_\mathrm{h}^*\vec{r}_\mathrm{h}}{m_\mathrm{e}^* + m_\mathrm{h}^*} \tag{2.136}$$

and the relative coordinate

$$\vec{r} = \vec{r}_\mathrm{e} - \vec{r}_\mathrm{h} , \tag{2.137}$$

the Schrödinger equation breaks up – as in the case of the hydrogen atom – into an equation for the center-of-mass motion and an equation for the relative motion, which can be solved separately. As part of the solution for the center-of-mass motion, we obtain the energy eigenvalues

$$E^\mathrm{center} = \frac{\hbar^2 \vec{K}^2}{2\,(m_\mathrm{e}^* + m_\mathrm{h}^*)} , \tag{2.138}$$

where $\hbar\vec{K}$ is the center-of-mass momentum of the exciton. The quadratic energy dispersion appearing in eq. (2.138) is characteristic of free particles (Section 1.2.2) and shows that the Wannier–Mott exciton can freely move as a unit through the crystal. Therefore, it is also called a free exciton. As part of the solution for the relative motion we obtain energy eigenvalues of an analogous form as in the case of the hydrogen atom:

$$E^\mathrm{rel} = -\frac{E_\mathrm{Ryd,X}}{n^2} , \tag{2.139}$$

where

$$E_\mathrm{Ryd,X} = \frac{\tilde{\mu}}{m_\mathrm{e}\varepsilon^2} E_\mathrm{Ryd} \tag{2.140}$$

is known as the excitonic Rydberg constant or excitonic binding energy. The principal quantum number is called n, $\tilde{\mu}$ is the reduced mass of the electron-hole pair (eq. (2.106)), m_e the electron mass in vacuum, and $E_\mathrm{Ryd} = 13.6\,\mathrm{eV}$ the Rydberg constant of the hydrogen atom. The negative sign in eq. (2.139) indicates that the eigenfunctions of the relative motion are bound states, just like the bound electron states in the hydrogen atom. In a classical picture, the relative motion of an electron and hole can be imagined as an orbit around their shared center of mass, as illustrated in Fig. 2.29. The radius of this orbit is given by

$$r_n = n^2 a_\mathrm{X} , \tag{2.141}$$

where

$$a_\mathrm{X} = \frac{m_\mathrm{e}\varepsilon}{\tilde{\mu}} a_\mathrm{B} \tag{2.142}$$

is called the exciton Bohr radius and $a_\mathrm{B} = 0.5\,\text{Å}$ is the Bohr radius of the H atom. According to eqs. (2.139) and (2.141), the $n = 1$ state has the largest binding energy and the smallest orbital radius. The higher states with $n > 1$ have smaller binding energies and larger orbital radii. The possible energies that the exciton can have are consequently given as a sum of the above energies:

$$E_{n,\vec{K}} = E_\mathrm{g} + \frac{\hbar^2 \vec{K}^2}{2\,(m_\mathrm{e}^* + m_\mathrm{h}^*)} - \frac{E_\mathrm{Ryd,X}}{n^2} . \tag{2.143}$$

Here, the first term takes into account that an excitation across the band gap of the direct semiconductor is required for an electron-hole pair to be created. This energy term was not included in eq. (2.134). The second term represents the energy eigenvalues of the center-of-mass motion, and the third one those of the bound states of the relative motion. The energy-level scheme described by eq. (2.143) is shown in Fig. 2.30. Here, the energies of the bound electron-hole states are depicted as parabolas. The continuously distributed energy levels of the unbound electron-hole states (shaded gray) are denoted as an electron-hole continuum. In a classical picture, these states can be imagined as non-closed trajectories, in which the electron and hole pass each other with so much kinetic energy that they cannot trap each other into a bound state. Similarly to the bound states, these states are solutions to the Schrödinger equation (eq. (2.134)), but have energies exceeding those described by eq. (2.139) due to the kinetic energies of unbound electrons and holes.

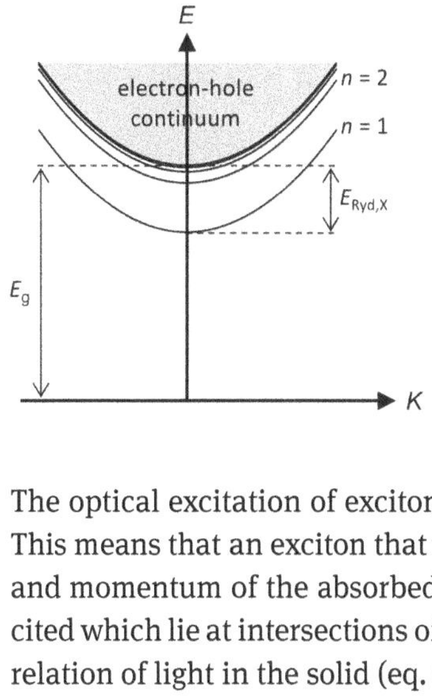

Fig. 2.30: Energy level scheme of Wannier–Mott excitons according to eq. (2.143). After Fig. 4.1 of [15].

The optical excitation of excitons is subject to energy and momentum conservation. This means that an exciton that is in the process of being created takes up the energy and momentum of the absorbed photon. Hence, only those exciton states can be excited which lie at intersections of the level scheme in Fig. 2.30 with the $E(\vec{k})$ dispersion relation of light in the solid (eq. (2.5)):

$$\hbar\omega = \hbar c|\vec{k}| \,. \tag{2.144}$$

However, as discussed in Section 2.3.1.1, the photon wave number $|\vec{k}|$ in the visible range is much smaller than the size of the Brillouin zone, which implies transitions at $\vec{K} \simeq \vec{0}$ in Fig. 2.30. We thus may simplify eq. (2.143) to

$$E_n = E_g - \frac{E_{\mathrm{Ryd,X}}}{n^2} \,. \tag{2.145}$$

We therefore expect absorption lines at the photon energies given by eq. (2.145). An example of such an absorption is given in Fig. 2.31, which shows a measured absorption spectrum of gallium arsenide at low temperature (1.2 K). The three lowest excitonic lines ($n = 1$, $n = 2$, $n = 3$) are visible. As expected from eq. (2.145), the two lowest lines are the furthest apart. The higher ones lie increasingly closer, such that lines with $n > 3$ can not be resolved due to their lifetime broadenings. Above $E_g = 1.5191$ eV, the absorption is caused by the excitation of unbound electron-hole pair states (electron-hole continuum). Immediately above E_g, much higher values of α are observed than those predicted by eq. (2.110). This enhancement is called the Sommerfeld factor. It is a consequence of the enhanced overlap of the electron and hole wave functions. This overlap is caused by the Coulomb attraction force between the two carriers. The overlap enhances the transition probability compared to independent carriers via the matrix element $\widetilde{M}$ (eqs. (2.103)). This enhancement is also visible in Fig. 2.21.

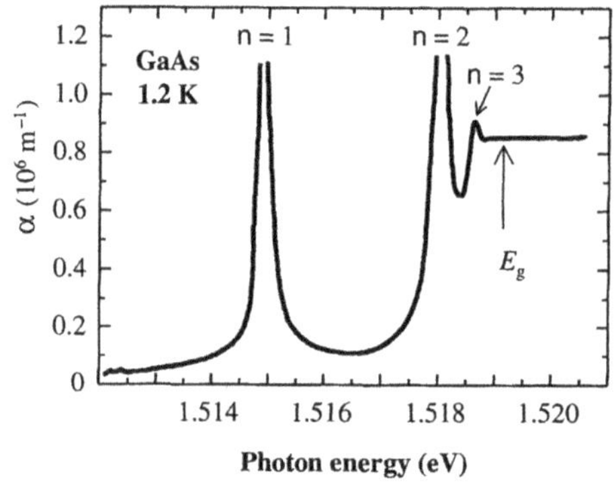

Fig. 2.31: Measured absorption spectrum of GaAs at a temperature of 1.2 K. The labels mark the three lowest excitonic absorption lines ($n = 1$, $n = 2$, $n = 3$) and the energy of the band gap E_g. Fig. 4.4 of [19], there after [20].

According to eq. (2.145), the exciton binding energy in gallium arsenide can be determined from Fig. 2.31 as the difference between the energy of the ($n = 1$)-line (1.5149 eV) and the band gap energy (1.5191 eV) revealing $E_{Ryd,X} = 4.2$ meV. Whether any exciton lines can be observed in a given semiconductor at a given temperature depends on this exciton binding energy. Stable excitons only exist if the binding energy is large enough to preserve the bound state from destructive collisions with phonons. If the temperature rises so that the thermal energy $k_B T$ increases beyond $E_{Ryd,X}$, it becomes likely that the electron-hole pair is excited to an unbound state by phonon absorption, and is, hence, dissociated. Consequently, the exciton line in the spectrum disappears as shown in Fig. 2.32. This process is called thermal ionization of the exciton. Because the electron and hole in a Wannier–Mott exciton attract each other relatively weakly due to the high ε, Wannier–Mott excitons have small binding energies on the order of 10 meV. As $k_B T$ is 25 meV at room temperature, exciton lines in direct semiconductors are observed only at low temperatures. In contrast to the lines themselves, the Sommerfeld enhancement is more persistent, appearing even at room temperature, as shown in Fig. 2.32.

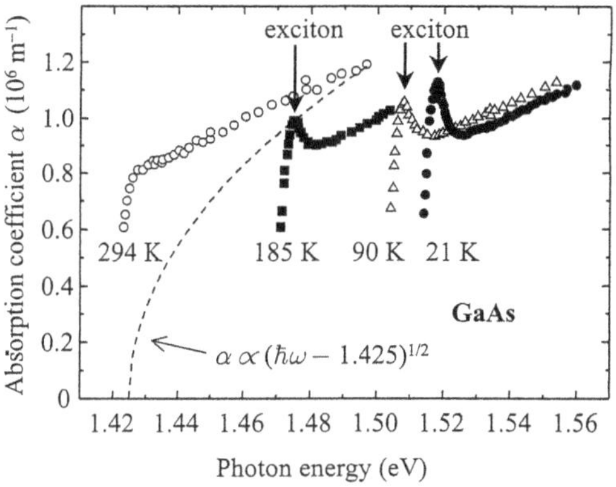

Fig. 2.32: Measured absorption spectra of gallium arsenide at different temperatures. In addition to a band edge shift typical of semiconductors and caused by the thermal expansion of the crystal, the spectra show an increasing thermal ionization of the excitons with increased temperature via the disappearance of the exciton line. The dashed line shows the result of a fit to the measured spectrum at 294 K using eq. (2.110). At this temperature, the band-gap energy is $E_g = 1.425$ eV. Fig. 4.3 of [15], there after [21].

Excitons may also get ionized as a consequence of an external electric field. When a strong, time-independent, electric field acts on an exciton, the electron and hole are pulled apart. It is possible to show that the field strength required for this phenomenon is

$$|\vec{E}_{\text{critical}}| \approx \frac{2E_{\text{Ryd,X}}}{ea_{\text{X}}} \,. \tag{2.146}$$

This exciton dissociation is called field ionization of the exciton. It can be observed in absorption spectra as a lifetime broadening of the exciton ($n = 1$)-line or as a complete disappearance of the line. Thermal and field ionizations both contribute to the fact that excitonic effects only play a minor role in most semiconductor optoelectronic devices. The value of the binding energy $E_{\text{Ryd,X}}$ is critical for the stability of the exciton against thermal and field ionizations. A systematic comparison of exciton binding energies for direct semiconductors shows that $E_{\text{Ryd,X}}$ generally grows with increased band gap. The reason is that the effective masses of the free carriers increase with increasing E_g and that polarization charges are less easily induced at large E_g. Consequently, the reduced mass $\tilde{\mu}$ of the electron-hole pairs becomes larger and the dielectric constant ε smaller. Both trends increase the exciton binding energy (eq. 2.140). Additionally, the exciton Bohr radius a_{X} decreases with increasing E_g (eq. (2.142)). In insulators with band gaps above 5 eV, a_{X} is similar to the lattice constant. Consequently, the model of the Wannier–Mott exciton loses its meaning. Conversely, in semiconductors with small band gaps such as indium arsenide ($E_g = 0.35$ eV), it is difficult to observe any effects of Wannier–Mott excitons due to ionization, as shown in Fig. 2.12. Therefore, Wannier–Mott excitons can best be observed in semiconductors with medium-sized band gaps (1 eV to 3 eV).

2.3.3.2 Frenkel excitons

Frenkel excitons appear in insulators with large E_g, in molecular crystals and conjugated polymers. A characteristic property of these excitons is that the mean distance between electron and hole is similar to the lattice constant (Fig. 2.33). As explained

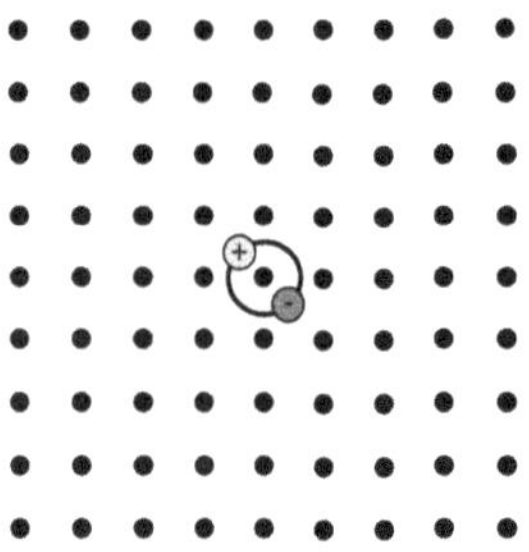

Fig. 2.33: Schematic representation of a Frenkel exciton in a crystal. The relative motion of the electron (e) and hole (h) is pictured as a common orbit of electron and hole around a single atom or molecule of the crystal.

in the previous subsection, a large E_g implies a weak screening of the Coulomb interaction between electron and hole resulting in a strong attraction. Frenkel excitons, hence, have small radii and large binding energies, which typically range from 0.1 eV to several eV. Their binding energies are thus much larger than the thermal energy $k_B T$, and the Frenkel excitons are normally stable at room temperature. Large band gaps are also associated with large effective masses of the carriers. This explains why Frenkel excitons are much less mobile than Wannier–Mott excitons. They are bound to single atoms or molecules of the crystal lattice and can only move by phonon-assisted transport between lattice sites, a process called exciton hopping (see also Section 1.6.3).

For the theoretical treatment of Frenkel excitons, there is no simple model. The calculation usually follows a tight-binding approach, which emphasizes the analogy to atomic and molecular states.

A class of materials in which Frenkel excitons have been demonstrated are the alkali halogenide crystals, e.g., sodium chloride. They have large band gaps, i.e., their excitonic lines lie in the ultraviolet. Their excitonic binding energies are between 0.3 eV for sodium iodide and 0.9 eV for lithium fluoride. These values are far larger

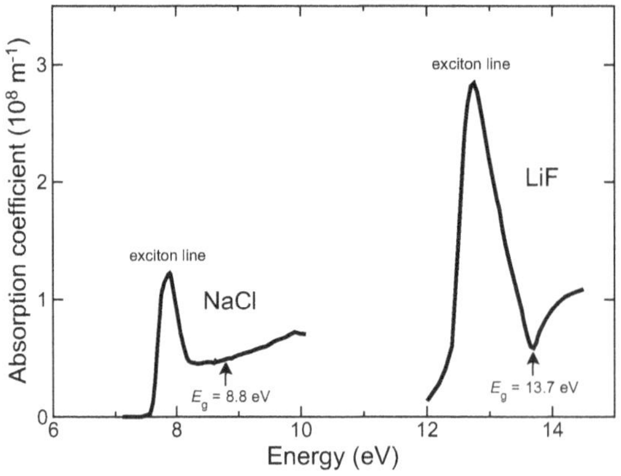

Fig. 2.34: Absorption spectra of sodium chloride and lithium fluoride at room temperature. The exciton lines and the band gap absorption at E_g are marked. After Fig. 4.8 of [15], there after [18].

than $k_B T = 25$ meV at room temperature. Therefore, the exciton lines do not disappear at this temperature, in contrast to those in gallium arsenide. Figure 2.34 shows absorption spectra of sodium chloride and lithium fluoride. From the energy differences between the $(n = 1)$ exciton lines and the band gaps E_g, the excitonic binding energies are determined to be $E_{Ryd,X} = 0.8$ eV for sodium chloride and $E_{Ryd,X} = 0.9$ eV for lithium fluoride.

2.4 Free-electron excitations in solids

The electric field of a light wave is able to induce interband transitions as described in the previous sections, but also transitions between states within the same band, so-called intraband transitions. Such intraband transitions are known from electric transport, where they describe the transition of electrons between states of different crystal momentum $\hbar\vec{k}$ within the same band, corresponding to a change of the velocity of the electrons in real space (Section 1.2.2, Fig. 1.3). Without scattering, the instantaneous crystal momentum of the electron at time t is given by Bloch's acceleration theorem

$$\hbar\vec{k}(t) = \hbar\vec{k}(t = 0) - e\int_0^t \vec{E}(\tilde{t})d\tilde{t}\,. \tag{2.147}$$

The corresponding momentum changes are, however, minute at the high frequencies ω of the light, which imply a fast change of the electric field $\vec{E}(t)$. Nevertheless, they have non-negligible effects if a large number of electrons occupy the same band, e.g., the conduction band of a doped semiconductor or of a metal. The collective motion resulting from the momentum changes may lead to a considerable electrical current. If the frequency of the driving electric field is in the optical range, the rapidly oscillating current will influence the optical properties of the crystal. In the present section, we will study this influence in electrical conductors such as metals.

2.4.1 Optical properties of metals

As an example of a good electrical conductor, we consider a typical non-ferromagnetic metal, e.g., silver. Here, we base most of our treatment on methods from classical physics. The time-dependent electric and magnetic fields at the surface and in the interior of the metal are described by the Maxwell equations (eqs. (2.41)–(2.44)). In contrast to a dielectric, a conductor such as a metal possesses free electrons, which are not bound to individual atoms but are able to move freely through the medium and contribute to electrical currents. Therefore, the electrical current densities $\vec{j}(t)$ induced in the medium by the alternating electric field of the light wave are non-negligible. Unlike in Section 2.2.2.2, we are thus not allowed to set $\vec{j}$ to zero. In analogy to eq. (2.46), we obtain from the Maxwell equations:

$$\nabla^2 \vec{E} = \mu_0 \varepsilon_0 \varepsilon_\mathrm{d} \frac{\partial^2 \vec{E}}{\partial t^2} + \mu_0 \frac{\partial \vec{j}}{\partial t} , \tag{2.148}$$

where ε_d is the dielectric constant that the material would have without electric currents. It describes how far the bound charges in the medium can be displaced by the electric field. Inserting into eq. (2.148) the expression from eq. (1.1)

$$\vec{j} = \sigma \vec{E} \tag{2.149}$$

yields

$$\nabla^2 \vec{E} = \mu_0 \varepsilon_0 \varepsilon_\mathrm{d} \frac{\partial^2 \vec{E}}{\partial t^2} + \mu_0 \sigma \frac{\partial \vec{E}}{\partial t} . \tag{2.150}$$

This extends the wave equation, eq. (2.46), by the electrical conductivity σ. Inserting the ansatz of a plane harmonic wave according to eq. (2.26), we obtain

$$\tilde{k}^2 = \varepsilon_\mathrm{d} \frac{\omega^2}{c_0^2} + \mathrm{i} \mu_0 \sigma \omega . \tag{2.151}$$

A comparison with eq. (2.25) and use of $c_0 = 1/\sqrt{\varepsilon_0 \mu_0}$ yields for the complex refractive index

$$\tilde{n}^2 = \varepsilon_\mathrm{d} + \mathrm{i} \frac{\sigma}{\varepsilon_0 \omega} , \tag{2.152}$$

which extends eq. (2.48) by a conductivity term. Equation (2.152) reveals that the macroscopic optical properties of the medium described by $\tilde{n}$ (phase velocity and absorption) are determined by two microscopic properties, namely the polarizability of the bound charges (described by ε_d) and the ability to conduct electrical currents (described by σ). In the frequency range of light, σ differs from its value in the direct current case ($\omega = 0$). To emphasize this difference, $\sigma(\omega)$ is often called optical conductivity for frequencies in the visible range. To derive an expression for its frequency dependence, we consider each of the conduction electrons as a classical free point-like particle with mass m_e^* and charge $-e$ that moves under the effect of the periodic force $-e\tilde{\vec{E}}(\vec{r}, t) = -e\vec{E}_0 \mathrm{e}^{\mathrm{i}(\tilde{\vec{k}} \cdot \vec{r} - \omega t)}$ and is subject to a frictional force due to scattering at a rate γ. This approach is inspired by the Drude model, which describes the electric transport of free electrons in metals (Section 1.2.2). The equation of motion for the field-induced electron displacement $\vec{x}$ from its initial position reads

$$m_\mathrm{e}^* \ddot{\vec{x}} + m_\mathrm{e}^* \gamma \dot{\vec{x}} = -e\tilde{\vec{E}}(\vec{r}, t) . \tag{2.153}$$

This equation differs from the equation of motion for the Lorentz oscillator (eq. (2.11)) in that the restoring force of the oscillation is absent. The reason is that the electrons considered here are conduction electrons of the metal, which are not bound to individual atoms, but are able to move freely through the medium. Equation (2.153) can also be expressed by the velocity $\vec{v} = \dot{\vec{x}}$ of the electron (see also eq. 1.5):[5]

$$m_e^* \dot{\vec{v}} + m_e^* \gamma \vec{v} = -e\tilde{\vec{E}}(\vec{r}, t)\,. \tag{2.154}$$

The term $\vec{v}$, in turn, can be expressed by the electric field $\vec{E}$ in the following way. The relation between the electric current density $\vec{j}$ and the number N_e of electrons flowing through the cross-sectional area A of the conductor per time t is (eq. (1.2)):

$$|\vec{j}| = \frac{N_e e}{At} = \frac{N_e}{V}(-e)|\vec{v}| = -n_e e|\vec{v}|\,, \tag{2.155}$$

where $V = A|\vec{v}|t$ is the volume filled by the N_e electrons contributing to the current during time t and, hence, n_e is the electron density. Combining eq. (2.155) with eq. (2.149), taking into account that $\vec{j}$ is antiparallel to $\vec{v}$, solving for $\vec{v}$, and inserting the resulting expression into eq. (2.153) gives

$$m_e^* \frac{d}{dt}\tilde{\vec{E}}(\vec{r}, t) + m_e^* \gamma \tilde{\vec{E}}(\vec{r}, t) = n_e \frac{e^2}{\sigma}\tilde{\vec{E}}(\vec{r}, t)\,. \tag{2.156}$$

Entering a plane wave ansatz for $\vec{E}$ according to eq. (2.9), differentiating and solving for σ yields

$$\sigma(\omega) = \frac{n_e e^2}{m_e^*} \frac{1}{\gamma - i\omega}\,. \tag{2.157}$$

When we enter this expression into eq. (2.152), we obtain

$$\tilde{n}^2 = \varepsilon_d - \frac{n_e e^2}{m_e^* \varepsilon_0} \frac{1}{\omega(\omega + i\gamma)}\,. \tag{2.158}$$

Here, the first term on the right side represents the contribution from the bound electrons and the second one that of the conduction (or free) electrons. The first contribution can be described by the Lorentz oscillator model and the second one by the Drude model. Their combination, which has led us to eq. (2.158), is known as the Drude–Lorentz model. We can simplify eq. (2.158) by introducing the plasma frequency

$$\omega_p := \sqrt{\frac{n_e e^2}{m_e^* \varepsilon_0 \varepsilon_d}}\,. \tag{2.159}$$

Then eq. (2.158) becomes

$$\tilde{n}^2 = \varepsilon_d \left(1 - \frac{\omega_p^2}{\omega(\omega + i\gamma)}\right)\,. \tag{2.160}$$

The meaning of this formula becomes evident when we evaluate it for different frequency ranges:

5 Here, we do not use $\vec{v}_d$ to describe the velocity of the electrons, since the dominating scattering processes at the higher frequencies of light are distinct from the ones in the electronic transport, which has been discussed in Chapter 1 of the book.

Case a) Very low frequencies: $\omega \ll \omega_p, \gamma$:
Equation (2.160) becomes

$$\tilde{n}^2 \approx \varepsilon_d \left(1 + \frac{i\omega_p^2}{\omega\gamma}\right) . \tag{2.161}$$

In general, the damping rate γ is smaller than the plasma frequency ω_p, which implies

$$\tilde{n}^2 \approx \varepsilon_d \frac{i\omega_p^2}{\omega\gamma} . \tag{2.162}$$

The complex refractive index, hence, is

$$\tilde{n} \approx \sqrt{i}\sqrt{\varepsilon_d \frac{\omega_p^2}{\omega\gamma}} = \frac{1+i}{\sqrt{2}} \cdot \sqrt{\varepsilon_d \frac{\omega_p^2}{\omega\gamma}} . \tag{2.163}$$

At low frequencies, the bound electrons, in general, show no absorption, which implies that ε_d in eqs. (2.161), (2.162), and (2.163) is real. We obtain for the real refractive index and the extinction coefficient (eq. 2.22)

$$n, \kappa \approx \sqrt{\varepsilon_d \frac{\omega_p^2}{2\omega\gamma}} . \tag{2.164}$$

This means that at low frequencies, the values of n and κ are almost equal. Because κ is large due to $\omega \ll \omega_p, \gamma$, eq. (2.23) implies that the absorption coefficient α is large, too. Thus, metals absorb strongly at low frequencies. The penetration depth or skin depth of the wave,

$$\delta_{skin} := \frac{1}{\alpha} , \tag{2.165}$$

is then small (typically 20 nm to 50 nm at low frequencies). In other words, the light can penetrate only a very small distance into the metal. Because of the large values of κ and n, also the reflectivity R_γ is large (eq. (2.57)). For both reasons, a metal layer barely transmits light at low frequencies.

Case b) High frequencies below the plasma frequency: $\gamma \ll \omega < \omega_p$:
At high frequencies, eq. (2.160) becomes

$$\tilde{n} \approx \sqrt{\varepsilon_d \left(1 - \frac{\omega_p^2}{\omega^2}\right)} . \tag{2.166}$$

Provided that ε_d is real, eq. (2.166) implies that $\tilde{n}$ becomes purely imaginary. The real refractive index n then is zero, which means that the wave cannot propagate in the metal. Equation (2.57) implies that $R_\gamma \approx 1$, i.e., the metal surface is highly reflective. This gives metals their typical shine. However, the above condition that ε_d is real is not completely fulfilled for metals in the visible frequency range, due to the light absorption by bound electrons. Hence, the reflectivity drops below unity. For instance,

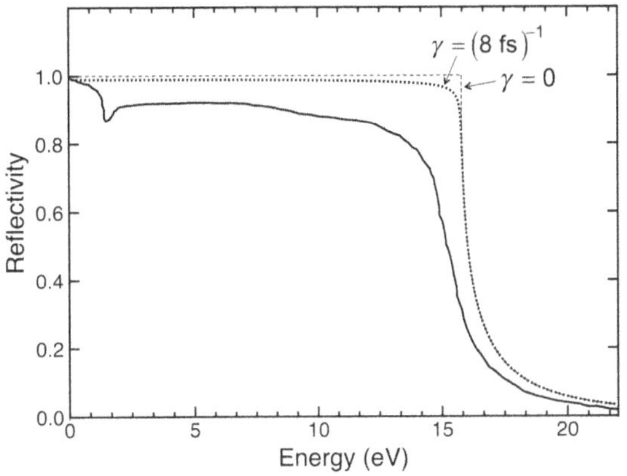

Fig. 2.35: Measured reflectivity spectrum of aluminium (solid line). For comparison, the dotted and dashed lines show spectra calculated using the free-electron model (eq. (2.158)) for $\varepsilon_d = 1$, $\hbar\omega_p = 15.8$ eV and different damping rates γ. After Fig. 7.2 of [15], there after [22].

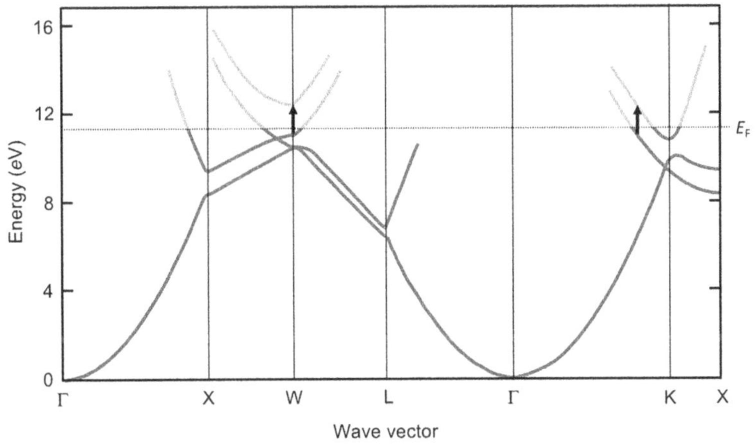

Fig. 2.36: Band structure of aluminium. States occupied with electrons are shaded dark grey, and empty states light grey. The dotted horizontal line marks the Fermi energy E_F. The interband transitions causing the reflection minimum at 1.5 eV in Fig. 2.35 are indicated by the arrows. After Fig. 7.3 of [15], there after [23].

the reflection spectrum of aluminium in Fig. 2.35 exhibits a minimum at 1.5 eV. At this photon energy, interband transitions between occupied states below the Fermi energy into empty states above it occur as indicated in Fig. 2.36 by arrows. In these positions, the marked pairs of bands run almost parallel to each other about 1.5 eV apart. This gives rise to a high joint density of states $g(\omega)$ at this photon energy and, hence, to a large absorption (eq. (2.110)). While the effect is smaller above 1.5 eV due to non-parallel band dispersions, R_γ is still considerably reduced there. The characteristic color differences of metals, such as aluminium, copper, silver, and gold, are hence caused by differences between their interband transitions in the visible range, leading to a frequency dependent reflection.

Case c) High frequencies above the plasma frequency: $\omega > \omega_p$:
Equation (2.166) holds also in this case. In metals in which no transitions of bound electrons occur at these high frequencies, ε_d is real, so that $\tilde{n}$ is also real, resulting in a vanishing extinction coefficient κ (eq. 2.22). Indeed, metals such as the alkali metals and aluminium are non-absorbing for $\omega > \omega_p$. Their conduction electrons do not absorb because they have too much inertia to keep up with the frequent changes of direction of the electric field. Thus, they cannot take up energy from the light wave. According to eq. (2.57), the vanishing extinction coefficient for $\omega > \omega_p$ implies that at the plasma frequency, the reflectivity R_γ will continuously drop from near unity to a smaller value determined by ε_d. An example of the reflectivity drop for aluminium is shown in Fig. 2.35 (solid line). For comparison, the results of an ideal free-electron model with and without damping (dotted and dashed line) are added. In the ideal metal case, ε_d is assumed to be unity so that R_γ drops to zero. The fact that a similar drop is observed experimentally demonstrates that, near ω_p, the value of ε_d is indeed close to unity. The small or vanishing values of κ and R_γ for $\omega > \omega_p$ imply that metals transmit light at ultraviolet wavelengths, an effect known as the ultraviolet transmission of metals.[6]

2.4.2 Particle plasmons

In the previous section, we have dealt with light reflection by extended metal surfaces. The high reflectivity of these surfaces gives metals their typical shine. The behavior of nanoparticles, i.e., particles with sizes between a few nm and ca. 100 nm, is very different, even if made of the same metals. As the extension of their surfaces is smaller than the wavelength of light, they do not reflect but scatter the light. This light can have brilliantly blue, green, or yellow colors, in contrast to the rather subtle colors of light reflected by extended metal surfaces. The colors of scattered light from metal nanoparticles are generated by collective oscillations of their conduction electrons relative to the positively charged ion background. These electron oscillations are excited by the alternating electric field of an incident light wave. An instantaneous picture of such an oscillation is shown in Fig. 2.37.

As the diameter of the nanoparticle is smaller than or similar to the penetration depth of the light, the alternating electric field of the light wave penetrates into the

6 Plasmons have also been observed in doped semiconductors, where the much lower charge carrier density n_e in the conduction or valence band leads to much lower plasma frequencies corresponding to $\hbar\omega_p \simeq 0.1$ eV (eq. (2.159)) [24].

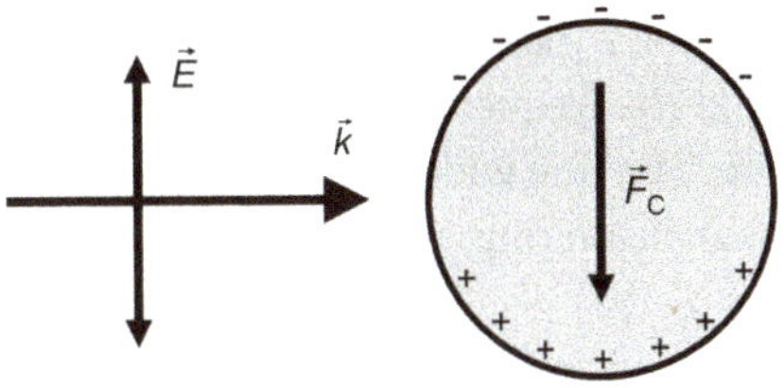

Fig. 2.37: Instantaneous distribution of polarization charges in a small spherical metal nanoparticle caused by the alternating electric field $\vec{E}$ of a light wave incident along $\vec{k}$. $\vec{F}_C$ is the restoring force exerted on the electrons by the positively charged ion cores.

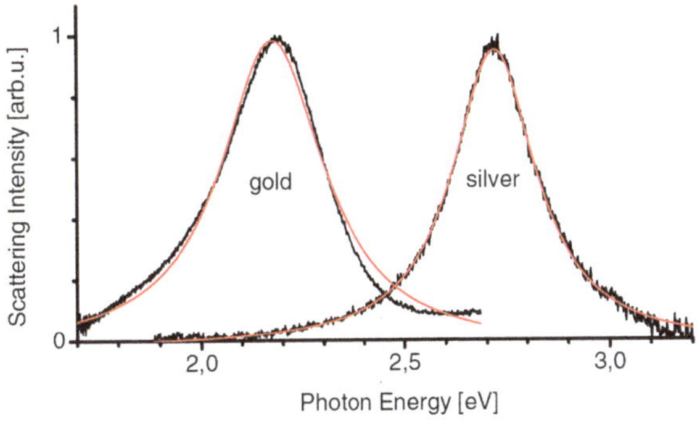

Fig. 2.38: The black curves show the measured light scattering spectra of two single nanoparticles made of gold and silver, respectively, as measured by microspectroscopy. The particles are spherical in shape and have a diameter of 60 nm. The red curves represent Lorentz fits to the data. Fig. 1 of [25].

metal and periodically displaces the conduction electrons. At the moment depicted in Fig. 2.37, the field has displaced all conduction electrons towards the top of the nanoparticle. This is followed by a displacement in the opposite direction in the next half cycle of the light field. The electron displacement results in polarization charges of opposite signs at the opposite surfaces of the nanoparticle and in a restoring force $\vec{F}_C$ exerted by the positive ion cores on the displaced electrons. The restoring force leads to an optical resonance, the particle plasmon resonance. Correspondingly, the energy quanta of the collective electron oscillation are called particle plasmons. Two examples of such resonances are shown in Fig. 2.38.

The spectral position of the resonance determines the color of the scattered light from the particles. This position is controlled by the size and shape of the particle as well as by the dielectric properties of the metal and its environment. The reason for these control parameters is that all these properties enter into the boundary conditions of the electromagnetic fields at the interface between the metal and its environment. Thus, they influence the force $\vec{F}_C$ exerted on the electrons. For instance, by solving the boundary problem, it is possible to show the following proportionality for the light power P_{scat} scattered by a small spherical metal nanoparticle of radius $\widetilde{R}$:

$$P_{\text{scat}}(\omega) \propto \left| 4\pi\widetilde{R}^3 \frac{\varepsilon_{\text{m}}(\omega) - \varepsilon_{\text{u}}(\omega)}{\varepsilon_{\text{m}}(\omega) + 2\varepsilon_{\text{u}}(\omega)} \right|^2 . \tag{2.167}$$

Here, $\varepsilon_m(\omega)$ is the dielectric function of the metal, and $\varepsilon_u(\omega)$ is the dielectric function of the surrounding material. The resonance lies at the frequency ω at which the denominator in eq. (2.167) goes through a minimum. The line width of the resonance is determined by damping processes such as electron scattering, which are included in $\varepsilon_m(\omega)$.

The unusual optical properties of metal nanoparticles and other metallic nanostructures have been investigated thoroughly over the last decades. This interest has been motivated by the wide range in which the spectral resonance positions of particle plasmons and related collective electron excitations can be tuned, and by the strong optical near-fields which are generated by the polarization charges at the metal surfaces and which are of interest for novel technical applications.

2.4.3 Reflection spectroscopy

A plot of the reflectivity versus the photon energy $\hbar\omega$ or wavelength λ of the light is called a reflection spectrum. Measuring such spectra and the methods applied in the measurement are known as reflection spectroscopy. Similarly to absorption spectra, reflection spectra can give valuable information about optical and electronic properties of a solid. Reflection spectroscopy can be especially helpful when the sample has a transmittance too small for absorption measurements. Figure 2.39 shows a typical experimental setup for reflection spectroscopy. Light from a white light source is spectrally filtered using a monochromator, and is incident on the sample. The reflected light power is measured with a detector, while the wavelength is varied using the monochromator. The detector should be positioned so that the angle between the incident and the reflected light is as small as possible to achieve nearly perpendicular incidence. To determine the reflectivity R_γ from the reflected light power, often a comparison with the power reflected by a calibrated mirror is performed, instead of a normalization to the incident light power as in eq. (2.53). This approach avoids an interruption of the incident light beam to measure the incident power. An example of a measured reflection spectrum is shown in Fig. 2.35. An important variant of the re-

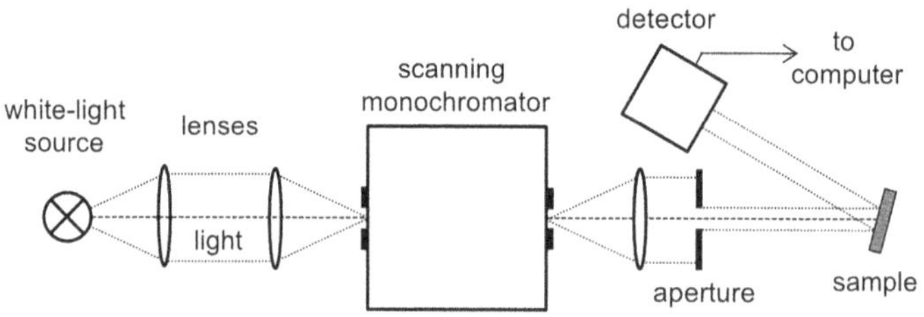

Fig. 2.39: Experimental set-up for the measurement of reflection spectra. Light from a white-light source is spectrally filtered with a monochromator and incident on the sample. The reflected light power is measured with a detector, while the wavelength is varied using the monochromator.

flection spectroscopy is ellipsometry, which is used to determine dielectric properties such as $\tilde{n}(\omega)$ and the thicknesses of thin layers. Its principle is based on determining the change of the polarization state of light upon reflection by a sample. For this purpose, light is obliquely incident on the sample with a polarization that is neither perpendicular nor parallel to the plane of incidence. The polarization properties of the reflected light are measured. As they depend on the optical constants and the thicknesses of the thin layers, both aspects can be determined.

2.5 Summary and outlook

This chapter of the book has treated the optical properties of solids. Microscopically, these properties result from the interaction of light with electrons in the medium. The goal has been to go into more depth on various microscopic aspects of the light-solid interaction, using quantum-mechanical and quantum-optical concepts when necessary and focusing on the important material classes of semiconductors and metals. We started off by reviewing some of the electrodynamics of light in solids. We then focused on a special kind of optical excitations in solids, namely interband transitions, in which an optical transition of an electron from one band to another occurs. They are of particular importance for the optical properties of semiconductors. Many optoelectronic devices are based on them, e.g., photodiodes, photovoltaic solar cells, light-emitting diodes, and semiconductor lasers. In particular, we have dealt with ways in which the electron and hole excited during an interband transition are bound to each other in a state called an exciton. We have seen how interband electron-hole excitations can, in turn, emit light, a light emission called luminescence. Subsequently, we have dealt with the free-electron currents in metals and doped semiconductors that give rise to their characteristic reflection spectra. Along the way, we have familiarized ourselves with some of the most common spectroscopic methods used to measure the optical properties of solids, namely absorption spectroscopy, photoluminescence spectroscopy and reflection spectroscopy.

The information provided here is regarded as a starting point to study more novel topics in optical research such as:

- quantization effects, which are based on electron confinement in semiconductor nanostructures (Section 1.3.1.1) and enable applications such as quantum-dot luminescence for color-television screens;
- nonlinear optics of solids, which is based on nonlinear responses at high light intensities and enables interesting applications such as optical frequency conversion;
- quantum-optical effects in solids, which are based on the quantum nature of the light field and give rise to phenomena such as photon entanglement;
- high-density excitation effects, which are based on many-body interactions in solids and can enable insights into different states of matter;

- optical metamaterials, which are based on sophisticated nanostructuring of optical materials and enable the manipulation of light in ways not available in nature, e.g., to fabricate optical cloaking devices.

Hendrik Bluhm

3 Quantum computing, qubits and decoherence

3.1 Motivation and introduction

3.1.1 Overview

While the previous chapters of this book largely used semiclassical descriptions for the properties of electrons in solids, we will now deal with a topic that requires a quantum mechanical description of the electron dynamics. Namely, we will introduce quantum computation, which exploits the dynamics of superposition states to more efficiently solve certain computational problems. The basic idea is that a superposition state represents several eigenstates at the same time and, thus, an ensemble of data sets, which are manipulated simultaneously. A computation with N quantum bits (qubits) then, in principle, allows a 2^N-fold parallelism compared to using N classical bits. However, due to the final projection of the result via a measurement, the gain can only be realized for certain computational problems. It is obvious that dealing with superposition states requires a quantum mechanical description of the system. Moreover, it builds on an enhanced control of the physical implementations of the quantum system. Both are described in detail in this chapter of the book. The presentation aims at providing a largely intuitive understanding of basic concepts, often using visualization via so-called Bloch spheres.

The significantly improved control of quantum mechanical degrees of freedom in solids also enables one to experimentally explore the fundamental question of how the quantum mechanical description of microscopic systems is related to the classical behavior of the macroscopic world consisting of many degrees of freedom. Is the reason that phase coherence and superposition states cannot be experienced in the macroscopic world merely a matter of complexity or is there some subtle change to the laws of physics? While discussing the few existing hypotheses and even more scarce experimental data on the latter possibility is beyond the scope of this course, we will study how interactions of a well prepared small quantum system with an environment consisting of many uncontrolled degrees of freedom can destroy phase coherence and thus leads to a transition to classical behavior.

The classic example for this puzzle is Schrödinger's cat [26]:

> A cat is penned up in a steel chamber, along with the following device (which must be secured against direct interference by the cat): in a Geiger counter, there is a tiny bit of radioactive substance, so small that perhaps in the course of the hour, one of the atoms decays, but also, with equal probability, perhaps none; if it happens, the counter tube discharges, and through a relay releases a hammer that shatters a small flask of hydrocyanic acid. If one has left this entire system to itself for an hour, one would say that the cat still lives if meanwhile no atom has decayed.

https://doi.org/10.1515/9783110438321-003

> The psi-function of the entire system would express this by having in it the living and dead cat (pardon the expression) mixed or smeared out in equal parts. It is typical of these cases that an indeterminacy originally restricted to the atomic domain becomes transformed into macroscopic indeterminacy, which can then be resolved by direct observation.

The notion of a cat being in a superposition of dead and alive is certainly strange. How can we reconcile that with our intuitive notions of cats? Studying the physics and concepts of quantum computing also helps us to answer this fundamental question.

The control of dynamical quantum effects, which has recently been achieved in solids, includes the deterministic creation of superposition states, control of the time evolution of such states under the Schrödinger equation, detailed investigations and manipulation of phase coherence, the deliberate creation of entanglement between quantum mechanical objects, and the projective measurement of quantum states. While the pursuit of such experiments was initially largely driven by curiosity, substantial efforts are nowadays fueled by the potential practical use. Major companies such as Google, Intel, Microsoft, and IBM as well as many smaller companies have started initiatives to bring these effects into practical use. The most prominent, in some cases already implemented applications are computing, communication and sensing. This chapter of the book will focus on quantum computing as one concrete example, but many of the outlined concepts are equally useful for other kinds of quantum technologies.

3.1.2 Digital vs. quantum computing

According to Wikipedia, the first Sumerian abacus appeared as early as 2700 B.C. Counting pebbles presumably emerged in the stone age. Hence, from the start of human civilization, computing has been based on discrete information. We are now at the stage of learning how to go beyond this principle by using quantum effects such as superposition states and entanglement[1] for information processing, thus fundamentally changing the way in which we encode information.

This new paradigm is not a mere continuation of Moore's law according to which the performance of computers doubles every two years, but opens up qualitatively new approaches that would render some computational problems solvable that were previously intractable in practice. This section will give an overview of how such quantum effects can be harnessed to achieve qualitatively superior computing.

Section 3.2 will sketch the main information-theoretical concepts of quantum computing. The main intention is to provide an overview of how quantum devices can be used for information processing, and what requirements they need to fulfill.

1 These terms, which might be known from previous courses in quantum mechanics, will be described in detail below.

The basic building blocks of such devices are quantum bits or qubits, which are fully controllable quantum mechanical two-state systems. Section 3.3 will introduce the two solid state approaches for implementing qubits that are generally considered most promising for large-scale computing applications. Demonstrator devices using these qubits have now reached a complexity of several tens of qubits. Section 3.4 will then revisit the quantum dynamics described by the Schrödinger equation in the context of qubits in order to build an intuitive understanding of their behavior and operation. In particular, we will introduce the concept of Bloch spheres as a vehicle for an intuitive geometric description of the quantum dynamics. Section 3.5 will finally focus on the modeling and understanding of decoherence processes, which are responsible for the loss of the predictable evolution of the quantum-mechanical phase. This subject is crucial for an understanding of the performance-limiting imperfections of qubits, and, at the same time, helps one to grasp how the classical macroscopic world arises from the laws of quantum mechanics.

3.1.3 Review of concepts of quantum mechanics

This section lists the quantum mechanical concepts that one should be familiar with. It is not written as a crash course, so one has to review the concepts in detail elsewhere (e.g., [27]), if necessary. Introductory courses of quantum mechanics often focus on wave mechanics, where the wave function $\psi(\vec{x})$ is a function of position $\vec{x}$. This approach is also partly employed in the previous sections. For the purpose of quantum computing, the equivalent matrix formulation is more suitable.

States are elements of a Hilbert space. For our purposes, the dimension d of the Hilbert space will usually be finite. States are denoted by a *ket* $|a\rangle$, where a is some useful label. The scalar product of two states is written as $\langle a|b\rangle$. By convention, states are usually normalized so that $\langle a|a\rangle = 1$.

Qubit states: For a single qubit (i.e., two level system), $d = 2$, and one typically uses $a = 0, 1$. Multi-qubit states are represented by the tensor product of the single qubit Hilbert spaces. A basis of a N-qubit Hilbert space is given by the product states $|a_1 \dots a_N\rangle = \otimes_{i=1}^{N} |a_i\rangle$. The corresponding Hilbert state thus has $d = 2^N$.

Superposition states are superpositions of basis states. The basis states are typically chosen as the eigenstates of a Hamiltonian, hence representing the corresponding conservation laws via quantum numbers. In this case, the superposition states are non-stationary and more fragile than the basis states, since they are not protected by a conservation law.

Entanglement: States that can be written as a product of individual qubit states (not necessarily basis states) are called **product states**. An example would be:

$$|0\rangle_{\mathrm{A}} \otimes \frac{1}{\sqrt{2}}(|0\rangle_{\mathrm{B}} + |1\rangle_{\mathrm{B}}) = \frac{1}{\sqrt{2}}(|00\rangle + |01\rangle)\,, \tag{3.1}$$

where A and B mark the two qubits. States where factorization is not possible are called **entangled states**. These entangled states can only be written as a sum of different product states. The most popular example is the Bell state $\frac{1}{\sqrt{2}}(|10\rangle+|01\rangle)$, which cannot be separated into a product of two independent states for qubit A and qubit B. The most peculiar consequence of an entangled state is that the measurement result of one qubit can set the state of the other qubit. If qubit A, prepared in the Bell state, reveals the result $|1\rangle_A$, qubit B is necessarily in state $|0\rangle_B$, even if the two qubits are separated by a distance that would require a superluminal velocity for an exchange of causal effects.[2]

Basis states: Any state can be written as a linear superposition of d orthonormal basis states: $|a\rangle = \sum_{n=1}^{d} |n\rangle\langle n|a\rangle$. A basis change corresponds to the multiplication of the expansion coefficients $\langle n|a\rangle$ with a unitary matrix.

Density matrices can be understood as classical probability distributions of pure states. A density matrix $\widehat{\rho} = \sum_i p_i |\psi_i\rangle\langle\psi_i|$ means that state $|\psi_i\rangle$ is encountered with probability p_i. Any pure state $|\psi\rangle$ can be written as a density matrix $|\psi\rangle\langle\psi|$. The advantage of density matrices is that they describe classical probabilities and quantum superpositions in a unified way.

Physical observables are represented by Hermitian operators, and can thus be written as Hermitian matrices. Any Hermitian operator has real eigenvalues and its eigenstates form an orthonormal basis of the Hilbert space.

Projective measurements of an observable $\widehat{A}$ yield as a result one of the eigenvalues λ_n of $\widehat{A}$ and leaves the system in the associated eigenstate $|n\rangle$ (or the eigenspace in the case of several eigenstates with the same eigenvalue). The probability of measuring λ_n for a state $|\psi\rangle$ is given by $|\langle n|\psi\rangle|^2$.

Time evolution is governed by the **Schrödinger equation** $i\hbar\frac{d}{dt}|\psi(t)\rangle = \widehat{H}|\psi(t)\rangle$, where the Hamiltonian $\widehat{H}$ is the energy operator. It can be straightforwardly solved in the energy basis if $\widehat{H}$ is constant in time.

Unitary evolution: It is useful to write the time evolution in terms of a **unitary time evolution operator** $\widehat{U}(t)$: $|\psi(t)\rangle = \widehat{U}(t)|\psi(t=0)\rangle$. The Schrödinger equation then reads $i\hbar\frac{d}{dt}\widehat{U}(t) = \widehat{H}\widehat{U}(t)$.

3.2 Quantum computing

This section will give an overview of the basic ideas and requirements for quantum computing. It is not intended as an in-depth study. Rather, it is supposed to provide an intuitive context for the subsequent section and a basic intuition for the theoretical foundations of quantum information processing.

2 Around 1930, Einstein called this result “spukhafte Fernwirkung” (spooky action at a distance) in order to challenge the theory of quantum mechanics, without success.

3.2.1 Basic ideas

There are two qualitative arguments that provide some intuition as to how quantum computers can outperform classical computers. First, they can represent a dramatically larger memory space. While a classical register consisting of N bits can store 2^N different integers, an N-qubit quantum register can store (up to normalization) all possible vectors of a 2^N dimensional Hilbert space. This significantly increased memory depth is the root of a quantum computer's superior capability to simulate quantum systems. Second, a quantum computer can store and process superposition states. Crudely speaking, it is thus able to process several states representing data at once. Contrary to today's parallel computers, employing many processors each working on a different part of a task, this quantum parallelism is achieved with a single piece of hardware.

Both of the above arguments show that the number of possible states and hence the computing power increases exponentially with the number of qubits. This scaling leads to the ability of quantum computers to execute certain algorithms exponentially faster than their classical counterparts. It would however be a major misconception to think that quantum computing could parallelize any known classical algorithm. The difficulty is that, while the execution is parallel in the sense discussed above, the outcomes encoded as superpositions in the final quantum state cannot be read out simultaneously. Reading a single qubit projects the state, thus destroying the information contained in other components of the multi-qubit superposition state. This problem is a direct consequence of the fact that it is not possible to fully determine a quantum state from only a single copy. It would of course be possible to circumvent this by repeating the algorithm many times, however this would eliminate the speedup as the number of measurements needed to fully characterize a multi-qubit state also grows exponentially with the number of qubits.

Together with the fact that we are lacking a natural intuition for quantum effects, this limitation makes it rather difficult to find algorithms where quantum mechanics offers a substantial speedup. Nevertheless, several such algorithms are known. The key feature of these algorithms is that they use superposition states only at intermediate stages, and then finish in a product state representing a *single* final answer. In the next section, we will summarize some of the most important quantum algorithms currently known.

3.2.2 Known algorithms

One well-known quantum algorithm that spurred much of the initial interest in quantum computing is Shor's algorithm for prime factorization. This algorithm terminates after a number of steps that scales polynomially with the length of the number to be factorized. In contrast, classical factorization procedures exhibit an exponential scaling. As a result, it is practically impossible for a classical algorithm to factorize numbers of more than a few hundred bits. Part of the interest in quantum computers comes from the fact that this difficulty of factorizing large numbers is at the root of the widespread RSA encryption algorithm.[3] With a quantum computer, RSA encryption could thus be broken. The Shor algorithm is based on a quantum mechanical version of the Fourier transform, which can also be applied to a few other problems.

Another prominent example is Grover's search algorithm, which finds a targeted object within a group of randomly ordered objects. More formally, given a function $f: 1, \ldots, N \to 0, 1$ with $f(n) = 1$ for $n = n_0$ for some unknown n_0 and $f(n) = 0$ for $n \neq n_0$, it can find n_0 with $\mathcal{O}(\sqrt{N})$ evaluations of f. This demonstrates the puzzling power of quantum computing quite clearly: classically, one would need on average $N/2$ evaluations, which for $N \gg 1$ is much larger than $\sqrt{N}$.

Another class of quantum algorithms relates to the ability of efficiently solving systems of linear equations.[4] Under certain conditions, it also offers an exponential speedup. However, both, equation solving and Grover's algorithm, illustrate a common problem with quantum computing: they require that f or the system of equations are available as quantum operators acting on qubit states, which in itself is not a trivial task. If one would like to search for an entry in a data base using Grover's search, the data base would probably have to be encoded in a quantum memory.

Another potential use for quantum computers, which might turn out to be most useful, is their ability to simulate other quantum systems.[5] Here, the difficulty for classical computers arises from the exponential growth of the Hilbert space of a quantum system with the number of particles it contains. For example, the 2^{50} dimensional state space for only 50 spin-½ particles exhausts the storage capacity of the largest supercomputers at the time of writing (2018), and any doubling of the available memory size would be offset by adding just one more spin. A quantum computer could circumvent this problem by using a quantum memory, whose depth also increases exponentially with the number of qubits. Hence, a doubling of the memory would enable a doubling of the size of the studied quantum mechanical system. One could then discretize the Schrödinger equation in a similar way to that in which differential equations are treated. Curiously, it was this at the time still unproven notion that led Feynman in 1982 to hypothesize that using quantum mechanics for data processing might be a powerful approach [28].

---

3 This asymmetric RSA encryption code is based on the ideas of Rivest, Shamir and Adleman, which provide the initials for the term RSA. It employs a combination of private and public keys. It is difficult to decode the encrypted information by a classical computer without knowing the private key, since factorization of a number into its prime factors scales exponentially with the number of digits, so that the private key enables an exponential speedup in decoding (see Wikipedia).

4 This can be employed, e.g., in all types of finite element simulations or for solving coupled equations of classical motion.

5 This is dubbed quantum simulations.

3.2.3 Other reasons to care about quantum information science

Quantum computing of major practical use is one of the most distant promises of quantum technology. There are a couple of other reasons researchers are interested in quantum information and related technologies and concepts. We will briefly discuss a few examples.

Some computer scientists and mathematical physicists study quantum algorithms in order to gain insight into their complexity class and the computing power of quantum computers. For example, they ask what kind of problems can be solved on quantum computers efficiently, i.e., with a run time or memory need that scales, at most, polynomially with the size of the problem rather than exponentially. Establishing equivalence of a certain task with such a so-called quantum algorithm indicates that the problem cannot be solved efficiently on a classical computer. Hence, quantum algorithms can serve as a tool for proving complexity theorems.

Recently, there has also been a growing interest in using quantum effects for sensing applications. For example, clouds of ultracold atoms are among the most sensitive magnetometers known to date [29] enabling sensitivities better than 1 nT. Such magnetometers can be based, e.g., on measuring the precession rate of atomic spins in the magnetic field to be determined. Preparing these atomic spins in an entangled state with a reduced uncertainty of the spin component along a particular axis can reduce the measurement noise below the value expected based on classical statistics. Other approaches employ the dependence of the dynamics of qubits on external fields to measure these fields.

3.2.4 First steps towards commercialization

Quantum mechanics was constitutional for the experimental realization of the transistor (~ 1945) and the laser (~ 1953). These kind of inventions based on single-particle concepts of quantum mechanics initiated the so-called "first quantum revolution." Corresponding devices are nowadays implemented in almost any technical product. The use of superposition states and entanglement for technical products goes conceptually beyond this stage and is thus sometimes dubbed the emerging "second quantum revolution." The first commercially available technology of this second revolution is quantum encryption. Here, superposition states of the transmitted qubits, usually encoded in the polarization of photons, are used to create a communication channel that makes undetected eavesdropping physically impossible. Any attempt to read the information transmitted would disturb the quantum states in which it is encoded and thus could be detected by the communicating parties. Such quantum cryptography systems that operate via standard optical fibers over distances up to about 100 km are offered by a few companies, e.g., ID Quantique, but there is not yet a large demand for such devices. Quantum-secured links via satellites have already been implemented in China.

Another early product are quantum number generators, which employ the measurement of photons in a polarization basis rotated by 45° with respect to the original polarization. They generate true random numbers rather than pseudorandom numbers that are based on a deterministic algorithm or classical physics, which is in principle also deterministic and hence could be predicted by simulation.

There are now also companies offering commercial quantum computing services and devices. However, at the time of writing (2018), these are not powerful enough to compete with conventional computers.

3.2.5 Requirements for qubits: DiVincenzo criteria

As we have discussed, the basic building blocks of a quantum computer are quantum mechanical two-level systems called qubits. A set of five criteria that qubits should fulfill in order to be a viable option for quantum information processing was laid down by David DiVincenzo. In brief, these are:

1. Qubits should rely on a scalable technology: it must be possible to fabricate large scale processors with many reliably functioning qubits.
2. It should be possible to initialize qubits into a known state.
3. The decoherence time, i.e., the time over which quantum information is retained, should be long compared to the time required for an operation.
4. It must be possible to execute a universal set of quantum gates. In other words, one must have full control over the qubits. Furthermore, these gates need to fulfill certain accuracy requirements related to criterion 3.
5. One must be able to read out the state of a qubit with high fidelity.

An accessible discussion of these criteria is given in Section 3 of [30]. Note that the above criteria are not absolute requirements since it is possible to trade deficiencies in some against strengths in others. For example, an initialization can be replaced by a fast projective measurement followed by an inversion conditional on the outcome. The measurement will project the qubit into a known basis state. If the latter is not the desired state, it can be transformed into the desired state via a suitable gate. Several other, more subtle trade offs are possible. The meaning of "long" in (3) and the accuracy requirements for (4) will be discussed in the following Section 3.2.6.

3.2.6 Sensitivity to errors and quantum error correction (QEC)

The above criteria impose partially conflicting requirements on a qubit. On the one hand, one needs to be able to control a qubit with external fields and to read out its state, while on the other hand, any fluctuation of a control field (or other field coupling to the qubit) will change its state in an unpredictable way. In practice, no control field is completely free of noise.

In addition, quantum mechanics implies that any leakage of information about the qubit's state to its environment will act like a measurement, thus erasing the phase information of superposition and causing a loss of information. This peculiarity is related to the no-cloning theorem, which says that quantum information cannot be copied [31]. Transferring it to another system implies loss of information in the original system. In principle, this problem could be overcome by completely decoupling a qubit from the readout mechanism when readout is not required, but in practice such switching on and off of an interaction is never perfect. We can conclude that any qubit will be subject to unpredictable kicks that will randomize its state. This process, which will be the topic of Section 3.5, is referred to as decoherence (see also Section 1.2.6).

While classical bits are also subject to environmental noise, they are inherently more robust against this noise because they have only two logical states. Hence, small changes in the state of a physical bit, e.g., leakage of a few electrons from the capacitor of a DRAM cell, do not change its logical state. This insensitivity is the reason why digitally stored music or movies maintain their quality over time and are not degraded by copying. The state of a qubit on the other hand is defined by the complex coefficients of its two basis states, which are continuous variables. Hence, even a slight perturbation changes the state, and at first sight there is no protection against errors.

Although most classical digital circuits are reliable enough to be used without error correction procedures, it is quite straightforward to detect and remove bit flip errors by creating and storing multiple copies. To check for and remove an error, all instances are read out and a majority vote is taken to determine the logical state. Evidently, at least three bits are needed, in which case a single bit flip error can be detected. More refined procedures employing more bits are robust against a larger number of bit flips.

For preserving arbitrary quantum states in qubits, this procedure is inadequate. The unitary evolution of quantum mechanics forbids the copying of quantum states (no cloning theorem), and reading out a full qubit state from a single copy is impossible because any projective measurement destroys the phase information. At first sight, it might thus seem that any attempt to build a quantum information processor is doomed to fail because of the importance and propagation of small errors.

Nevertheless, it was discovered that even quantum errors can be corrected, provided they occur infrequently enough. Depending on the assumptions and methods applied, one finds that 10^2 to 10^6 error free coherent operations per error are required. This insight was a central trigger for today's widespread experimental pursuit of quantum information processing. The basic idea of fault tolerant quantum computing is to encode a logical qubit state in several physical qubits. However, one does not create several copies, but uses a two dimensional subspace of a multi-qubit Hilbert space to encode a single logical qubit, the so-called logical subspace. The simplest example is the so called three-qubit code employing the states

$$|0_L\rangle = |000\rangle \quad \text{and}$$
$$|1_L\rangle = |111\rangle\,,$$

where $|0_L\rangle$ and $|1_L\rangle$ denote the two logical basis states. Note that mapping the state of an additional single physical qubit to such an encoded state is not forbidden by the no-cloning-theorem and can be realized quite straightforwardly using two controlled not (CNOT) operations.

The detection of errors then employs measurements that only determine whether the collective, entangled state of the physical qubits still lies within the logical subspace and what kind of error (e.g., a flip of which qubit) caused the leakage, but that do not distinguish the two logical basis states. The latter have to be chosen such that any error to be protected against leaks out of the subspace they span. Since no which-state information is obtained between the two logical states, the phase of superposition states can be maintained. Hence, it is possible to detect errors and to determine their nature without disturbing the stored quantum information. Subsequently, the error can be corrected, again without changing the logical state.

Note that in practice, the three-qubit code is not very useful because it protects only against a single bit flip error but not against errors affecting the phase of a qubit. This limitation can be remedied by using more advanced codes. An example is the Steane code with

$$\begin{aligned}|0_L\rangle = \frac{1}{\sqrt{8}}(&|0000000\rangle + |1010101\rangle + |0110011\rangle + |1100110\rangle\\ &+ |0001111\rangle + |1011010\rangle + |0111100\rangle + |1101001\rangle)\end{aligned}$$

and

$$\begin{aligned}|1_L\rangle = \frac{1}{\sqrt{8}}(&|1111111\rangle + |0101010\rangle + |1001100\rangle + |0011001\rangle\\ &+ |1110000\rangle + |0100101\rangle + |1000011\rangle + |0010110\rangle)\,.\end{aligned}$$

A detailed treatment shows that such an error correction procedure is only beneficial if the error probability per gate operation P_{error} is less than the so called error correction threshold. The reason is that the detection and correction of errors requires a number of additional operations, which will be subject to errors themselves. Hence, one has to make sure that on average no more errors are introduced in the correction procedure than were originally present. Roughly speaking, any single error will be corrected, while two or more errors in one complete error correction cycle will not be corrected but instead propagate to the next processing step. The probability for two errors to occur is approximately $(P_{\text{error}} \times (\text{number of gate operations per correction step}))^2$, which must be smaller than the bare error probability without quantum error correction. If this condition is fulfilled, an overall reduction of the error probability can be achieved with error correction.

We see that rather than completely eliminating all errors, error correction only reduces the error probability. Further error reduction can be achieved by hierarchically concatenating error correction codes: the logical qubits of any level are used as the physical qubits of the next higher level. As a result, the error reduction factors are exponentiated. It is also clear that error correction generates a significant overhead, because more physical qubits and gate operations are required. The longer an algorithm needs to run, the lower the error probability must be for reliable results, and hence the larger the overhead becomes. An important result is that the overhead does not grow exponentially with the size of the problem, so that the speedup offered by quantum information processing can be maintained.

Depending on the code and the assumptions made for the hardware implementation, the thresholds computed for different codes lie typically between 10^{-6} and 10^{-1} errors per quantum gate operation. These most advanced and promising codes, referred to as surface codes, employ a 2D array of qubits and topological ideas (Section 5.4) and achieve error correction thresholds as high as 10^{-2} to 10^{-1}. To avoid an excessive overhead leading to a very large number of qubits, the error rate should practically not be larger than 10^{-3}. Even then, the overhead is about 10^4, i.e., about 10.000 physical qubits are required for a single logical qubit.

This large overhead results from the fact that about 1000 physical qubits with an error rate of 0.1% are needed to represent a single logical qubit with an error rate of $\sim 10^{-15}$, which is required by the number of gate operations needed to solve practically relevant problems. Another factor of 10 arises from the need to iteratively implement certain gates that are not directly supported by the surface code. For an algorithm requiring 10^4 logical qubits, one arrives at a total number of 10^8 physical qubits. The exact number depends on the algorithm and the actual error rates, however in a sublinear way. Consequently, substantially lower error rates are needed to achieve a significant reduction in the number of physical qubits. For example, reducing the error rate from 10^{-3} to 10^{-4} only saves a factor of four in the number of physical qubits.

3.3 Experimental realization of qubits

Any implementation of a quantum information processor needs a physical system to represent its qubits. A large variety of different types of qubits have been explored to various degrees (Fig. 3.1). The challenge is to identify a truly scalable architecture that allows the controlled realization of thousands or even millions of interconnected qubits.

Approaches based on photons, atoms or ions (left bubble in Fig. 3.1(a)) make qubits with very impressive properties, e.g., with respect to error rates or entanglement fidelities. For these experiments, atoms or ions are trapped far away from disturbing influences in ultrahigh vacuum (UHV) chambers. Such macroscopic ion traps have been shown to accommodate as much as 53 ions [35], but seem funda-

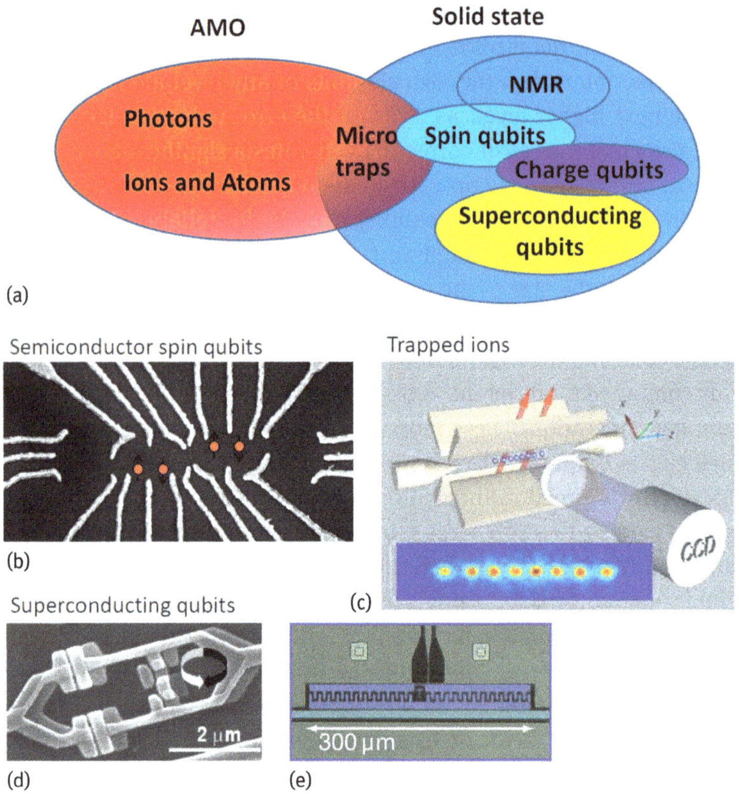

Fig. 3.1: Overview of different approaches for the realization of qubits. (a) Rough classification of qubit types including the overlap between them (AMO: atomic, molecular and optical physics, NMR: nuclear magnetic resonance). (b)–(e) The most promising types of qubits. (b) Scanning electron microscope (SEM) image of a gate structure as used for a controlled not (CNOT) gate (Section 3.4.4). It is made out of spin qubits (Section 3.3.2) consisting of four quantum dots with one electron spin each (red balls with arrows). (c) Main image shows a sketch of an elongated ion trap, which can host well-controlled ions (blue balls) in a row observable by the CCD chip via optical magnification. Each ion acts as a qubit and interacts with the other ions via Coulomb repulsion. A map of the ions obtained via the optics is shown in the inset. (d) SEM image of a circuit with three Josephson junctions as used for a flux qubit (Section 3.3.5.2). (e) Superconducting resonator as used to read out so-called transmon qubits (Section 3.3.5.4). (c) After [32]. (d) [33]. (e) [34].

mentally difficult to push to much larger numbers. Their size and costly fabrication including the use of multiple lasers are further major impediments to realizing very large qubit numbers. One possibility is to use microfabricated traps on which ions can even be moved around. However, according to a recent proposal, a large-scale quantum computer would still fill a football field [36].

Photons interact very weakly with each other, which implies poor prospects for multi-qubit operations. The interaction between qubits can be replaced by measurements and the preparation of highly entangled initial states, but the latter, so far, is only possible for a few photons with unpractically small rates.

Solid state systems tend to have complementary properties. Microfabrication techniques, in principle, allow the inexpensive fabrication of millions of qubits. However, these qubits are coupled to a rather complex environment consisting of crystal defects, phonons, free spins, dangling bonds, etc. This can cause rather strong decoherence, i.e., a loss of the quantum information of the superposition states, that has to be carefully avoided. These solid state qubits will be the main focus of this introduction to quantum computation. In particular, we will discuss spin, charge and superconducting qubits.

In principle, a pure spin 1/2 could be considered as the most natural qubit, since it has only two states. However, real spins are attached to particles, which also have orbital degrees of freedom that can interact with the spin via spin-orbit coupling and other mechanisms. Other solid state qubits, in particular superconducting ones, do not make use of the spin at all, but employ electromagnetic excitations instead. Hence, a physical qubit will usually have more than two states, and we need to consider all its degrees of freedom. From the complete state space of the solid state device, one can then select two eigenstates that span the computational subspace. Most importantly, leakage out of that subspace has to be avoided. It is often prevented via some form of energy selectivity. Using conservation of energy, one can avoid states other than the desired ones being populated.

In the following, we survey a selection of solid state qubits in order to convey the most important concepts.

3.3.1 Charge qubits

Charge qubits, whose state is represented by two possible locations of an electron, are conceptually among the most simple types of qubits. Although their coherence is rather poor due to strong coupling to electric noise, we will discuss them as an instructive example. Many of the principles encountered will help to understand spin and superconducting qubits, which provide better coherence properties.

Figure 3.2 shows a schematic representation of a charge qubit. Its state is encoded in whether the electron is located in the left or in the right box. To realize transitions between these two basis states, one needs to transfer the electron between left and right. This transfer cannot be done with some sort of wire, since interactions with other electrons in the wire would immediately destroy coherence. The remedy is to use tunnel coupling through a finite height barrier between two quantum dots (Section 1.3.1.1) as shown in Fig. 3.2 (right). Note that this approach is very similar to classical bits, where one also stores or removes charge, in this case on capacitor plates, in order to represent the 1 and the 0 of the bit, respectively. The main difference is that a qubit can also exist in a superposition of the two states.

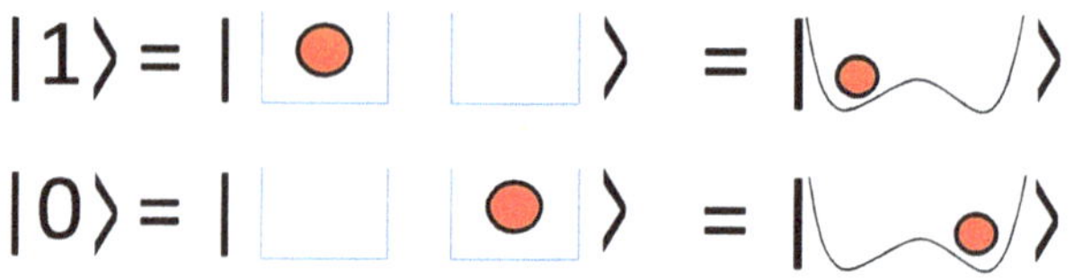

Fig. 3.2: Schematic representation of the states of a charge qubit, defined as an electron being located in the left or the right box (middle column), respectively in the left or right potential well (right column). Changes of the state can be achieved by tunneling across the finite-size barrier of the double potential well on the right.

3.3.1.1 Quantum dots as charge qubits

We will now shortly discuss one way to create a suitable confining potential for charge qubits, which can also be used to create spin qubits. We will build on the methods of heterostructure growth and lateral confinement by gates as described in Sections 1.3.1 and 1.3.1.1, respectively, i.e., we focus on electrostatically defined quantum dots based on a 2DEG, which is located at an epitaxially grown heterointerface between GaAs and AlGaAs (Fig. 1.6). Using such semiconductor structures raises the hope of leveraging the evidently very successful semiconductor technology. Hence, good prospects for scalability of these type of quantum devices are foreseen. Many of the key concepts also apply to other semiconductor qubit systems. Most prominently, Si-based structures have been used with great success to achieve a longer coherence time of spin superposition states (Section 3.5.5.3.1).

The remote doping with respect to the qubit plane (2DEG plane), as described in Section 1.3.1, is advantageous, since charge noise from the doping layer or from the surface would deteriorate the coherence of the qubits significantly. For qubit devices, the depth of the 2DEG is typically $\sim 100\,\text{nm}$ below the surface. Adequate gating of the heterostructure by gates from the surface can lead, e.g., to the occupation of two quantum dots with one electron each (red dots in Fig. 1.6) as required for certain types of spin qubits (Section 3.3.2.2).

The multiple gates on the surface are also used to tune the tunnel barriers. The barrier between two quantum dots is tuned by applying a negative potential to gate fingers in between the two quantum dots. The tunnel barriers between the quantum dots and the surrounding leads can be tuned via additional gates on the left and on the right of the respective quantum dot. The tuning of the tunnel barriers is always effected by small adjustments of all the different gate voltages relative to each other, which are not completely independent. Importantly, it is possible to define a suitable confinement potential including fine-tuned tunnel barriers for individual electrons as required for a charge qubit (Fig. 3.2) or a spin qubit (Section 3.3.2).

Recall that the small effective mass of GaAs implies that one can neglect excitations into higher confinement levels of the quantum dot at $T < 1\,\text{K}$, which is quite helpful for a controlled qubit operation. Moreover, the wave function envelope, neglecting the atomically periodic part of the wave function, is governed by a Schrödinger equation as for free particles with a modified mass m_e^* and an enhanced dielectric constant $\varepsilon \simeq 13$. This makes its properties rather intuitive and simple to understand.

3.3.1.2 Transport through quantum dots

Before describing the operation of quantum dots as qubits, we will discuss the transport of electrons through quantum dots. The focus will be on the phenomenon of Coulomb blockade, which arises from the nonzero charging energy. The corresponding Coulomb blockade measurements are widely used to extract information on quantum dot properties such as the charging energy, confinement energies and other types of electron-electron interaction energies, much beyond the field of quantum computing (Section 5.5.1). Furthermore, transport through quantum dots is widely used for qubit readout (Section 3.3.1.6).

A single quantum dot (QD) or metallic island connected to two leads as depicted in Fig. 3.3(a) on the right is also called a single electron transistor (SET), because its conductance depends on whether individual electrons can hop on and off the island. A SET can be formed by a true quantum dot with resolvable confinement levels as well as by a metallic island whose confinement energy splittings are so small that the states can be modelled as a continuum via the energy density of states. In the device of Fig. 3.3(a), the central gate electrode on the very right can be used to vary the electrostatic potential of the QD, while the outer gate electrodes on the very right are predominantly used to tune the tunnel coupling to the surrounding 2DEG acting as source (S) and drain (D) leads. Figure 3.3(b) shows a schematic representation of the dot.

Applying a bias voltage V_{SD} between source and drain can only induce a current through the quantum dot if the addition energy ΔE required to add an electron to the quantum dot lies between the Fermi levels of source and drain. In the Coulomb blockade regime, the most important contribution to the addition energy is the electrostatic contribution referred to as charging energy E_{C}. For typical devices larger than $\sim 100\,\mathrm{nm}$, E_{C} is larger than the single particle level spacing arising from the confinement energy.

The addition energy ΔE can be computed from a classical model of electrostatic energies. Details of the geometry and resulting electric fields are accounted for in terms of capacitances as shown in the equivalent circuit diagram in Fig. 3.3(c). The potential of the quantum dot, which for this purpose can be thought of as a conductor with floating potential, is given by

$$V_{\mathrm{QD}} = Q/C_\Sigma + V_{\mathrm{QD},0}\,. \tag{3.2}$$

Here, Q is the total charge on the QD, $C_\Sigma = \Sigma_i C_i$ is the total capacitance between the QD and its surroundings, and $V_{\mathrm{QD},0} = \frac{\Sigma_i C_i V_i}{C_\Sigma}$ is the average potential of the quantum dot as produced by the gates indexed by i, which are kept at potentials V_i. The term Q/C_Σ in eq. (3.2) is simply the voltage drop between the dot and the surroundings that arises

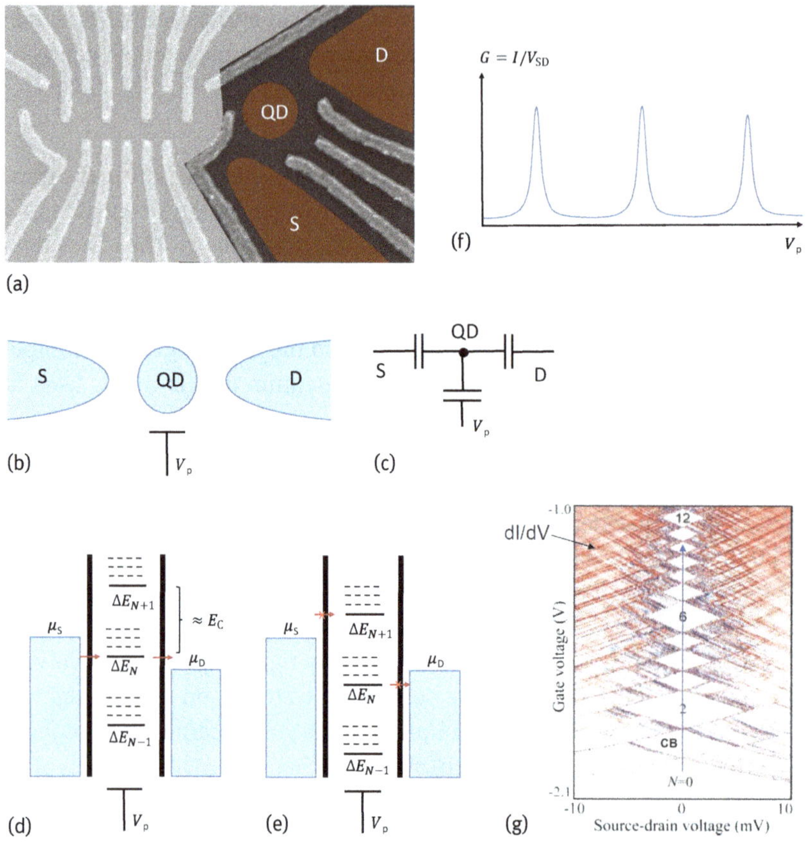

Fig. 3.3: Coulomb blockade in single electron transistor (SET). (a) SEM micrograph of a quantum dot operated as charge sensor to read out the qubit in the adjacent double quantum dot (light grey area). The brown regions represent the island (marked QD) and the 2DEG serving as source (S) and drain (D) leads. (b) Schematic of the SET. A quantum dot (QD) or metallic island is tunnel coupled to source (S) and drain (D) leads. The potential on the island can be controlled by capacitive coupling to the plunger gate with voltage V_p. (c) Equivalent circuit diagram used to compute the electrostatic energies. (d), (e) Energy diagram of a SET. The chemical potentials $\mu_{S,D}$ of source and drain are shifted by the applied bias voltage V_{SD} (not shown). The levels for the dots reflect the addition energies ΔE_{N_e} for N_e electrons already on the dots. Dashed lines represent excited states. For metallic islands, these excited states can be considered as a continuum described by a density of states. The difference of ΔE_{N_e} and ΔE_{N_e+1} is dominated by the charging energy E_c. In (d), ΔE_{N_e} lies in the bias window between μ_S and μ_D so that current flow is possible as indicated by the red arrows. Panel (e) corresponds to a different plunger gate voltage V_p where no addition energy is available in the bias window and current is blocked, since no electrons (holes) are available at the red arrows at source (drain). (f) Conductance $G = I/V_{SD}$ as a function of V_p at low $V_{SD} < \hbar\Gamma/e$, $k_B T_e/e$. The larger of the latter two quantities determines the width of the peaks. (g) Example of a Coulomb spectroscopy data set displaying dI/dV_{SD} in color scale as a function of V_p and V_{SD}. Numbers along the central blue arrow mark the number of electrons in the QD within the corresponding Coulomb diamond (CB). (g) [37].

from the charge stored on it. The energy ΔE_{N_e} required to add the $(N_e + 1)^{\text{th}}$ electron to the QD with N_e electrons already present is obtained by integrating eq. (3.2) with respect to Q. Employing

$$\int_{N_e e}^{(N_e+1)e} \frac{Q}{C_\Sigma} dQ = \frac{1}{2}\left[\frac{Q^2}{C_\Sigma}\right]_{N_e e}^{(N_e+1)e} = (N_e + 1/2)\frac{e^2}{C_\Sigma}, \tag{3.3}$$

this leads to

$$\Delta E_{N_e} = eV_{\text{QD},0} + \frac{e^2}{C_\Sigma}(N_e + 1/2) + \mu_{\text{QD},N_e}, \tag{3.4}$$

where the additional term μ_{QD,N_e} is the N_e-dependent chemical potential of the quantum dot. For a metallic island, it is simply the Fermi energy. For a dot with discrete levels, it is the energy of the lowest unoccupied level arising from the quantum confinement.

We see that ΔE_{N_e} depends on the gate voltages via $V_{\text{QD},0}$ and the number of electrons N_e already present: the more electrons are on the dot, the more energy is needed to add an additional one because of the Coulomb repulsion. The characteristic energy scale per electron is given by the charging energy $E_c = \frac{e^2}{C_\Sigma}$ as ΔE_{N_e} increases by E_c for each additional electron on the dot. By tunneling of electrons on or off the QD via the source and drain leads, N_e will adjust itself such that $\Delta E_{N_e+1} > \mu_{\text{S,D}}$ and $\Delta E_{N_e} < \mu_{\text{S,D}}$, where μ_S and μ_D are the chemical potentials of the leads, which we have for now assumed to be equal. If N_e electrons are on the dot in equilibrium, it must have been energetically favorable to add the N_eth electron, but must be unfavorable to add another one.

For a nonzero V_{SD}, $\mu_\text{S} = \mu_\text{D} + eV_{\text{SD}}$ and it becomes possible for electrons to flow through the dot. If $\mu_\text{S} > \Delta E_{N_e} > \mu_\text{D}$, the N_eth electron can enter from the source and subsequently can leave to the drain. This situation is illustrated in Fig. 3.3(d). If, on the other hand, there is no ΔE_{N_e} between μ_S and μ_D as depicted in Fig. 3.3(e), no current flow is possible. This effect is called Coulomb blockade. Due to the gate voltage dependence of ΔE_{N_e} via $V_{\text{QD},0}$, the plunger gate voltage V_p can be used to sweep the addition energies ΔE_{N_e} through the bias window given by eV_{SD}. Measuring the current as a function of V_p reveals so called Coulomb peaks of finite conductance $G = I/V_{\text{SD}}$, which are separated by regions without current flow (Fig. 3.3(f)). The largest V_{SD} for which current is suppressed, because no addition energy is located in the bias window, is evidently given by the charging energy, hence Coulomb blockade requires $eV_{\text{SD}} < E_c$.

In practice, the Coulomb peaks have a finite width even at $V_{\text{SD}} \approx 0$ due to two effects. First, a nonzero electron temperature T_e in the source and drain broadens the Fermi distributions and thus the condition for current flow.[6] Electrons become available, e.g., at the red arrows on the source side of Fig. 3.3(e), and holes become available

6 The electron temperature T_e is often larger than the lattice temperature T, typically at $T \leq 100$ mK, due to high-frequency electromagnetic noise, which couples to the electrons.

at the red arrows on the drain side due to the broadened Fermi distribution at finite T_e. Second, a finite tunneling rate Γ leads to a broadening of the dot levels according to the energy-time uncertainty relation. The width of a Coulomb peak is thus determined by the largest among V_{SD}, $\hbar\Gamma/e$ and $k_B T_e/e$. If any of these voltages exceeds E_c/e, the regions of blocked current become completely washed out. These conditions explain why the observation of Coulomb blockade requires very small structures in order to maximize E_c and/or low temperatures, and why the dot has to be separated from the leads by tunnel barriers with sufficiently small Γ. The decrease of the Coulomb blockade region with increasing V_{SD} leads to the characteristic so-called Coulomb diamonds visible as white, diamond-shaped regions in Fig. 3.3(g).

Other energy contributions, such as the confinement energy or more complex Coulomb interaction terms, e.g., exchange and correlation terms also contribute to ΔE_{N_e} via μ_{QD,N_e}. While they are typically small compared to E_c, the resulting variation of the spacing of the Coulomb peaks as a function of V_p can provide information about them (Section 5.5.1). Furthermore, the additional steps in the current as a function of V_{SD} can reveal the energies of excited states of the QD. Whenever an excited state with energy larger than the ground state of the QD with N_e electrons becomes energetically accessible in the bias window, it increases the rate at which electrons can tunnel into the dot. Hence, the current I through the dot increases, which is visible as lines in the differential conductance $\mathrm{d}I/\mathrm{d}V_{SD}$ parallel to the edges of the Coulomb diamonds (Fig. 3.3(g)). Hence, Coulomb blockade measurements are not only employed for a charge detector for qubits as described below, but also to study the physics of electrons in quantum dots in detail (Section 5.5.1).

3.3.1.3 Charge detectors

The SET as described in the previous Section 3.3.1.2 as well as a quantum point contact (QPC) (Section 1.4.2) can be used as a charge detector. They are employed, e.g., to distinguish the two eigenstates of a charge qubit, which exhibit a different spatial charge distribution. The most simple charge detector is the QPC, i.e., a small, gate controlled constriction between two areas of a 2DEG (Fig. 3.6). As discussed in Section 1.4.2, a QPC exhibits relatively sharp step-like changes of conductance by $2e^2/h$ as a function of gate voltage (Fig. 1.14). The repulsive potential of the electron to be detected adds to the electrostatic potential of the QPC and hence slightly shifts the gate voltage range of the conductance steps. Electrons would shift them upwards and holes would shift them downwards. Tuning the QPC into such a conductance step makes the conductance of the QPC dependent on the position or presence of the electron in the adjacent quantum dot (Fig. 3.7(d)). Hence, the sensor conductance depends on the position of the charge and thus on the state of the charge qubit (Section 3.3.1.6).

A better sensitivity was demonstrated by replacing the quantum point contact with a SET (Section 3.3.1.2). Since localized charges in the environment of the SET contribute to the electrostatic potential of its QD area, the position of the Coulomb peaks depends on the state of an adjacent charge qubit. Tuning V_p to the slope of a Coulomb peak leads to a strong change of I if an electron moves in the surrounding of the SET. Hence, the SET also acts as a sensitive charge detector. With modern amplification techniques, sensitivities well below $10^{-8}e/\sqrt{Hz}$ have been demonstrated, albeit for electrons directly above or below the SET. Of course, it is also possible to detect electrons for purposes other than qubit readout, for example to improve the performance of current standards based on a chain of SETs acting as charge pumps.

3.3.1.4 Energy scales

To further understand the behavior of electrons in the quantum dots acting as charge qubits, it is helpful to think of the QDs in terms of a potential whose shape can be controlled with gate voltages, which is drawn in Fig. 1.6 at the interface between GaAs and $Al_{0.3}Ga_{0.7}As$. For double quantum dots, the tunnel barrier linking the two dots is most important. From the shape of the double well potential, one can compute the energy scales, which are key to identify suitable qubit states and to understand the qubit dynamics.

The single particle level splitting between the orbital states in a single dot, as already encountered in Section 3.3.1.2, depends on the curvature of the confining potential. It can be estimated by approximating the system as a harmonic oscillator. For GaAs devices, typical values for level splittings are about 1 meV. The charging energy is typically slightly larger ($E_c \approx 3$ meV). Another important energy scale is the interdot charge coupling, which is the change of the addition energy of one dot when an electron is added to the other dot. It is typically about 0.5 meV and is decisive for electrostatically transferring an electron from one dot to the other, without allowing it to pass through the leads. Both of these Coulomb energies can be estimated from simple electrostatic models analogously to the description of the SET (Section 1.3.1.1). Note that all of these energy scales are considerably larger than the thermal energy $k_B T = 8.6\,\mu eV$ at $T = 100$ mK, which is a typical temperature used in qubit experiments employing dilution refrigerators. Hence, the system can quite easily be initialized into its ground state.

Finally, we also need to consider the tunnel coupling between the two dots, which can be identified with the energy difference between the odd and even eigenstates in a symmetric double well potential (Fig. 3.4(a)). The tunnel coupling between the dots depends exponentially on the barrier height and can thus be varied over a wide range. If the potential is tilted, the two lowest single particle eigenstates change to the electron being on the left and on the right quantum dot, respectively (Fig. 3.4(b)). The energy difference ϵ between these two states is called detuning and is (to a very good approximation) proportional to the tilt of the potential.

3.3.1.5 Hamiltonian and energy diagram

To understand how a single electron in a double dot can be used as a qubit, we need to understand its quantum dynamics. We will for now neglect its spin and focus on the

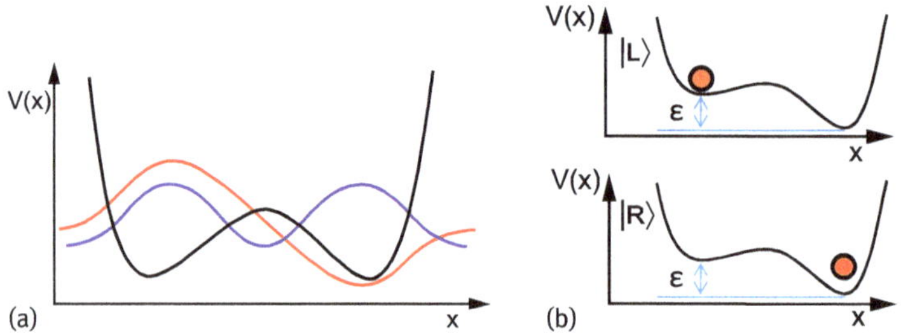

Fig. 3.4: (a) Double-well potential (black line) with wave functions of symmetric (blue line) and anti-symmetric (red line) eigenstate of the Hamiltonian. (b) Sketch of the two states of the charge qubit $|L\rangle$ and $|R\rangle$ for an asymmetric potential well with marked detuning energy ϵ between the states.

orbital dynamics. A complete picture could be obtained with a brute force solution of the Schrödinger equation. However, it is for most purposes sufficient to consider only the two lowest states and to describe the dynamics by an effective Hamiltonian, which we will construct in the following.

We choose as basis states $|L\rangle$ and $|R\rangle$, which correspond to the electron being on the left or on the right dot, respectively. By definition, the detuning energy, ϵ, is then described by a term $\frac{\epsilon}{2}\widehat{\sigma}_z$ in the Hamiltonian (Fig 3.4(b)). Here, $\widehat{\sigma}_z$ is the Pauli matrix, which describes the two state system consisting of $|L\rangle$ and $|R\rangle$.[7] The tunnel coupling, whose magnitude we denote by Δ, on the other hand causes transitions between the two basis states and can thus be described by a term $\frac{\Delta}{2}\widehat{\sigma}_x$. ($\widehat{\sigma}_y$ could also be used, but one usually prefers $\widehat{\sigma}_x$ because it is real). The total Hamiltonian is thus

$$\widehat{H} = \frac{\epsilon}{2}\widehat{\sigma}_z + \frac{\Delta}{2}\widehat{\sigma}_x \,. \tag{3.5}$$

As we will see later, this Hamiltonian also describes other types of qubits. Its eigenenergies are $E_\pm = \pm\frac{\sqrt{\epsilon^2+\Delta^2}}{2}$. A useful tool is the so called energy diagram, which plots the eigenenergies as a function of detuning ϵ (Fig. 3.5). It also provides information about the nature of the corresponding eigenstates because their dominating $|L\rangle$ or $|R\rangle$ component is proportional to $\frac{\partial(E_+-E_-)}{\partial\epsilon}$. At large detuning, the near-unity slope shows that the eigenstates are nearly $|L\rangle$ or $|R\rangle$, which would be exact for $\Delta = 0$. The avoided crossing of the two states with reduced slopes near $\epsilon = 0$ reflects the hybridization of $|L\rangle$ and $|R\rangle$ due to the tunnel coupling when they are nearly degenerate. Having constructed a Hamiltonian, we are in principle now in a position to describe how the qubit can be controlled. This is often done by varying ϵ, for example by applying voltage pulses to the outer gates as depicted in Fig. 3.6. We will develop a formalism to gain

7 The Pauli matrices are given explicitly in the list of symbols. Although they may be most familiar as operators acting on spin 1/2 particles, they are equally useful for describing any two-state system. In particular, they form a basis for the space of Hamiltonians of such systems. We use the basis $\{|1\rangle, |0\rangle\}$, so that $\widehat{\sigma}_z|1\rangle = |1\rangle$ and $\widehat{\sigma}_z|0\rangle = -|0\rangle$.

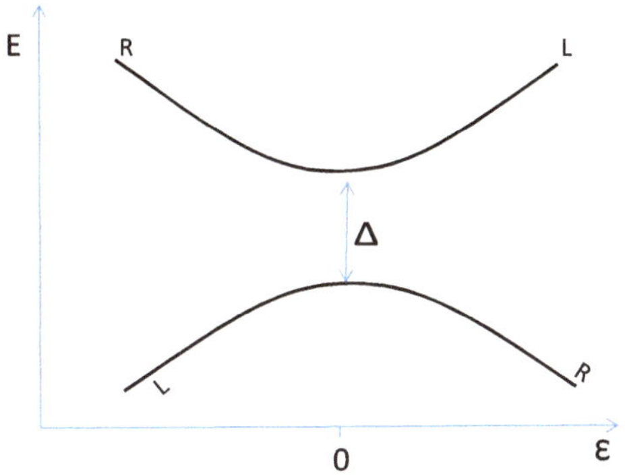

Fig. 3.5: Eigenenergies of a charge qubit as a function of detuning energy ϵ. The tunnel coupling Δ lifts the degeneracy at $\epsilon = 0$ and leads to a so-called anti-crossing whose size reflects the value of Δ. Away from the transition, the slopes of the states reflect their charge distribution with the dominant component of the bare states $|L\rangle$ and $|R\rangle$ marked.

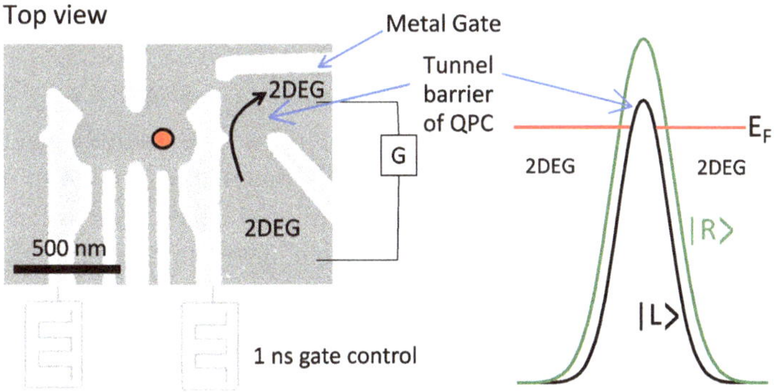

Fig. 3.6: Left: Electron microscopy image of a double quantum dot. The two broad outer gates are used to apply electric pulses for manipulation as sketched below the image (1 ns gate control). The state of the charge qubit can be measured by the quantum point contact (QPC, Section 1.4.2) on the right of the double-dot. An electron moving in the double dot between $|R\rangle$ and $|L\rangle$ raises or lowers the potential barrier of the QPC (right), respectively, thus changing its conductance. Hence, the position of the electron (state of the charge qubit) can be measured without removing the electron from the double quantum dot.

an intuitive understanding of the resulting qubit dynamics in Section 3.4. We note that initialization can be achieved by letting the qubit relax to its ground state, which will happen eventually as long as the level splitting is much larger than the temperature of the relevant bath that couples to the qubit subspace. Understanding the relaxation rates requires a detailed consideration of the relaxation mechanism.

3.3.1.6 Readout

A convenient way to read out a charge qubit is to use the charge detectors as described in Section 3.3.1.3. They can easily distinguish the two eigenstates at large detuning, where the spatial charge distribution of $|L\rangle$ or $|R\rangle$ differs by nearly a full electron. Using a quantum point contact (QPC) as charge detector, the conductance G of the QPC determines the qubit state (Fig. 3.6) via a projective measurement. With modern amplification techniques, the measurement can be performed in the μs range.

3.3.2 Electron spin qubits

Instead of using the charge degree of freedom of electrons in quantum dots, it is also possible to use their spin to encode a qubit. Such spin qubits can be realized in the same type of devices. This approach has been much more successful to date, because the weak coupling of spins to the environment leads to much longer coherence times.

3.3.2.1 Single-spin qubit

The conceptually most simple approach is to use single $s = 1/2$ spins, which form a natural two level system. To control these spins, one employs an AC magnetic field (effective or real, see Section 3.4.3). In this section we will discuss how a single spin can be initialized and read out [38]. A single quantum dot (Fig. 3.7(a)) is tuned to contain either zero or one electron. It is placed into a magnetic field of a few Tesla, so that the Zeeman energy is much larger than $k_B T$. Using voltage pulses V_p on one of the gates, the energy levels of the electron can be moved above and below the Fermi energy (E_F) of the reservoir, to which the dot is tunnel-coupled. A sequence of voltage pulses used to initialize and read out the spin qubit is shown in Fig. 3.7(b) and (c). First, V_p is made negative enough so that both the spin-up and the spin-down states are raised above E_F causing the dot to be emptied of the electron regardless of its spin state (i). Subsequently, pulling the two levels below E_F makes both of them energetically accessible, so that after some time either a spin-up or a spin-down electron will tunnel into the dot with approximately equal probability (ii). The Coulomb blockade will prevent the tunneling of further electrons to the QD.

To determine the spin state of the captured electron, V_p can be raised to bring only the spin-down state above E_F (iii). If the electron is in the higher energy spin-down state, it can tunnel out of the dot (iii, top). Another electron from the lead can now tunnel into the dot because it is no longer repelled by the presence of an electron already inside the dot via the Coulomb blockade. This incoming electron will have a spin-up state because only this state is below E_F. If on the other hand the dot was originally occupied by a spin-up electron, the electron cannot leave the dot so that nothing happens (iii, bottom). Hence, one can read out the electron spin by monitoring the tunneling of the electrons. The time scale of these tunneling events is determined by the tunnel rate Γ to the reservoir, which can be tuned by changing the gate voltages that determine the tunnel barrier.

During the whole process, the occupation of the dot can be monitored with the QPC acting as a charge detector (Section 3.3.1.6). The tunneling out of the quantum dot and subsequent tunneling back into the quantum dot, which is associated with a spin

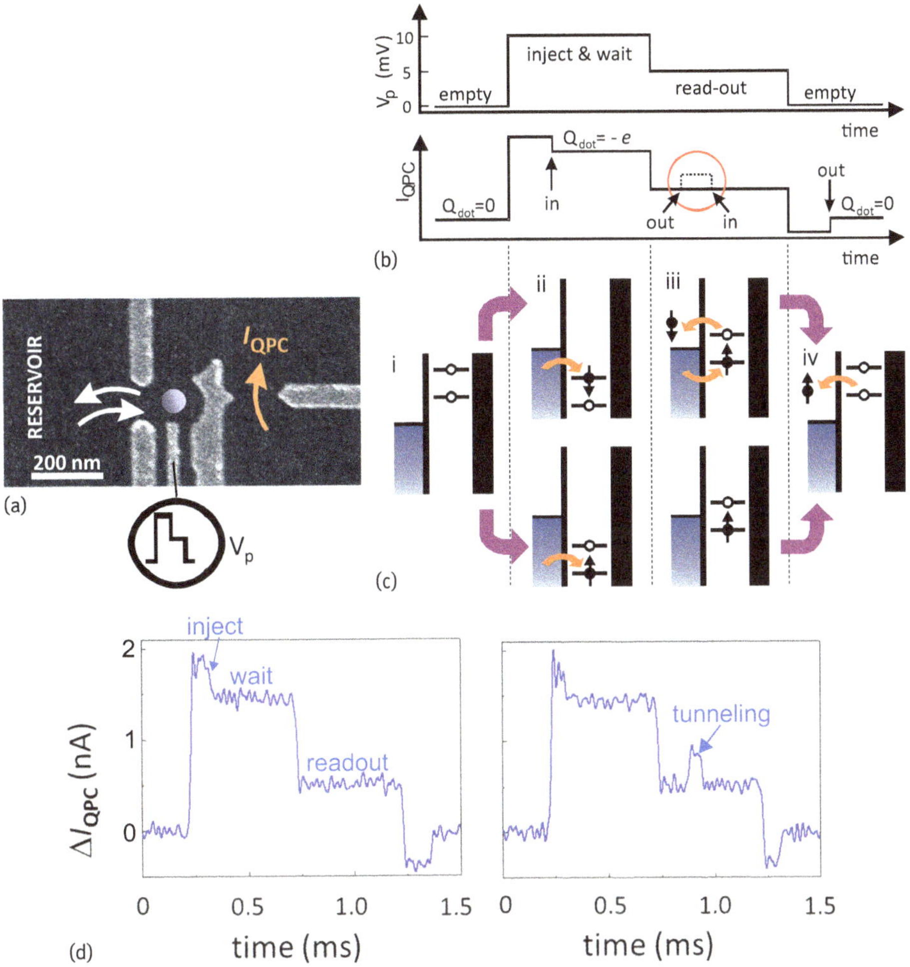

Fig. 3.7: Single-shot readout of a single spin. (a) Electron microscopy image of a single electron quantum dot, which is tunnel coupled to a reservoir (white arrows) and read out by a QPC (orange arrow). Voltage pulses V_p are applied to the central gate electrode of the quantum dot as marked. (b) Readout mechanism for a single-dot qubit. Top: gate voltage trace with injection and readout stage. Bottom: schematic of the corresponding readout signal, i.e., of the current through the QPC. The dashed blip in the red circle reflects an excited spin state tunneling out of the dot and a ground-state spin tunneling back in. (c) Electron transitions between reservoir and dot for spin-down and spin-up states: (i) emptying of the dot, (ii) filling of the dot with a random spin state, (iii) spin selective tunneling for readout leading to the presence or absence of the blip in (b, bottom), (iv) emptying of the dot. (d) Measured current through the QPC during the V_P sequence shown in (b, top). In the left trace no signal appears during readout, indicating a spin-up occupation, while the right trace shows a blip in the QPC signal during readout marked tunneling, which indicates a spin-down electron tunneling from the dot to the reservoir and another electron tunneling back into the dot. The other steps in ΔI_{QPC} reflect the response of the sensor to gate voltage pulses, respectively, the loading of the quantum dot by an electron with arbitrary spin (inject). Based on material provided by Lars Schreiber and Lieven Vandersypen, original publication in [38].

down electron, leads to an increase and subsequent decrease in the detector current because of the short absence and subsequently restored presence of an electron in the dot. This presence or absence of the additional electron makes it harder or easier for other electrons to flow through the detector, respectively. The signature of a spin down electron is hence a step up and then a step down of the charge detector current sketched as a dashed blip in Fig. 3.7(b), bottom (marked by red circle). Figure 3.7(d) shows measured data for both spin states, where the presence (absence) of the blip in the QPC current during readout implies a spin down (spin up) state. Note that at the end of the experimental cycle, i.e., before emptying the dot again, the electron is in a spin-up state independent of its initial state and so the process can also serve to initialize the spin state. Thus, two of the DiVincenzo criteria (Section 3.2.5), namely for initialization and readout of the spin information, are fulfilled.

By introducing a variable wait time between the preparation of the spin-up or spin-down state and the readout, it is also possible to measure the spin relaxation time. As the wait time is increased, the excited spin-down component will relax into the spin-up state, so that the probability of detecting a spin-down electron $P_{\downarrow}$ decreases, i.e., the blip in the QPC current during the readout time appears less often. The decay of $P_{\downarrow}$ directly reflects the spin relaxation. Depending on the external magnetic field, relaxation times between a few milliseconds and up to a second have been demonstrated by this method for GaAs quantum dots [39]. Even longer spin relaxation times are possible in Si devices [40].

3.3.2.2 Two-electron spin qubits

One potential disadvantage of single-electron spin qubits is the need to manipulate them via AC-magnetic fields. This requires microwave signals of fairly large amplitude that can cause heating and make it difficult to avoid crosstalk. A possible remedy is to use two spins in a double quantum dot (Fig. 3.8(a)) to encode a single qubit. Of the four possible spin states, only the $m = 0$ subspace spanned by the basis states $|\uparrow\downarrow\rangle$ and $|\downarrow\uparrow\rangle$ is used to store information. The main advantage of this encoding is that the exchange interaction[8] between the two spins can be used to manipulate the qubit very rapidly using only electric fields (see below and Section 3.4.2.1).

While the device in Fig. 3.8(a) is typically operated as a qubit with each of the electrons located in separate dots, it is possible to move both electrons into a single dot by changing the voltages on the large outer gates. How these gate voltages can be used to control the device can be understood by defining the detuning energy ϵ as the energy difference between the states with separated electrons, denoted as $(1, 1)$, and the $(0, 2)$ state with both electrons in one dot (Fig. 3.8(c)). This definition is analogous to that used for a charge qubit, except that an additional electron is present. The detuning is proportional to the difference between the voltages on the outer gates

8 A detailed introduction into the exchange interaction between two electrons is given in Section 4.4.

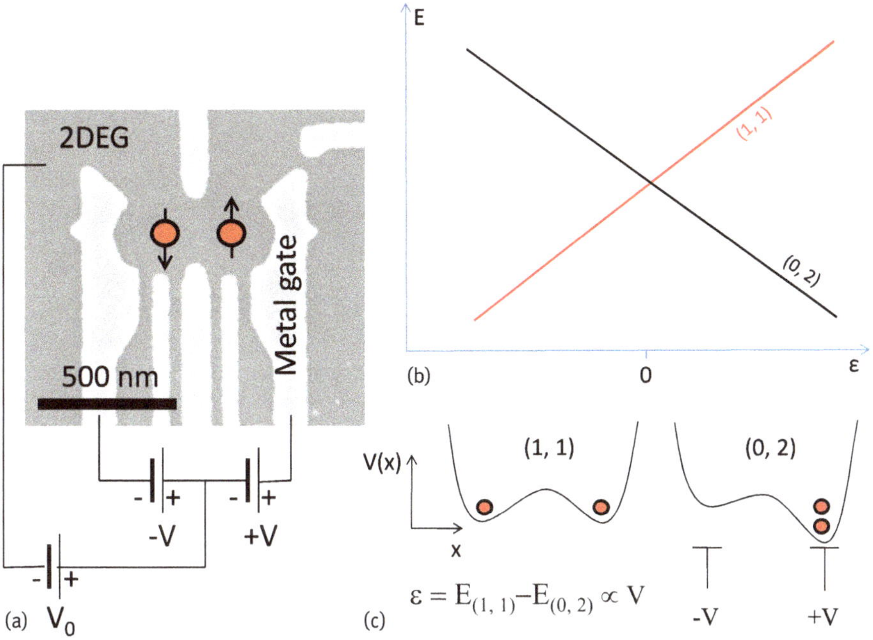

Fig. 3.8: Electrically controllable double quantum dot. (a) Electron microscopy image of gate structure for a double quantum dot realizing a two-electron singlet-triplet spin-qubit. (b) Energy diagram for a singlet-triplet qubit in the absence of tunnel-coupling and singlet-triplet splitting, thus distinguishing only the (1, 1) and the (0, 2) state. (c) Sketch of the double well potential with corresponding ground states in the (1, 1) and in the (0, 2) configuration, i.e., on the left and on the right of (b). A description of the detuning energy ϵ is included below the (1, 1) state.

(Fig. 3.8(a)). As for the charge qubit, the effect of ϵ can be described via the energy diagram, which we will now construct. Due to the presence of additional states, it is more complicated than that of the charge qubit.

In the absence of tunnel coupling and disregarding spin, the energy diagram would contain two straight lines for the (1, 1) and (0, 2) states, as shown in Fig. 3.8(b). The corresponding Hamiltonian is $\widehat{H} = \frac{\epsilon}{2}\widehat{\sigma}_z$.

An important effect to be considered is the singlet-triplet splitting of the (0, 2) state. Due to the Fermi-statistics (Pauli principle), the total two-electron wave-function must be antisymmetric under exchange of the two electrons. Hence, for an antisymmetric spin-singlet state $|S\rangle = 1/\sqrt{2}(|\uparrow\downarrow\rangle - |\downarrow\uparrow\rangle)$, the orbital wave function $\Psi(x_1, x_2)$ has to be symmetric with respect to exchange of the two position variables x_1 and x_2, i.e., $\Psi(x_1, x_2) = \Psi(x_2, x_1)$. This can be realized by the product of two single particle ground state wave functions ψ_0: $\Psi(x_1, x_2) = \psi_0(x_1)\psi_0(x_2)|S\rangle$.

In contrast, for a spin-triplet $|T\rangle$, the orbital wave function has to be anti-symmetric, which requires a second single-particle state ψ_1 according to $\Psi(x_1, x_2) = 1/\sqrt{2}(\psi_0(x_1)\psi_1(x_2) - \psi_0(x_2)\psi_1(x_1))|T\rangle$. Neglecting interaction effects, the energy of

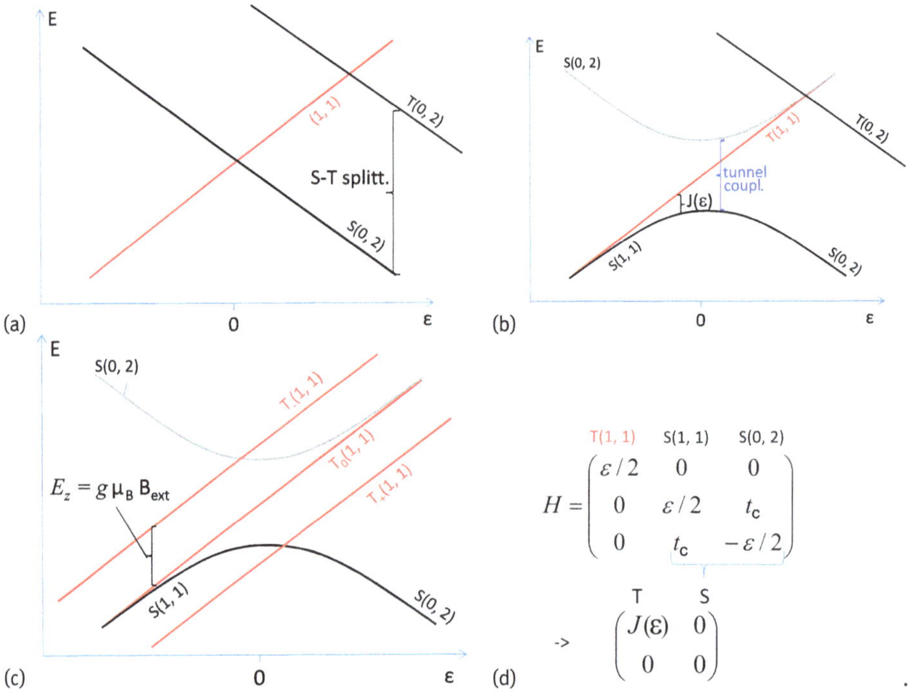

$$H = \begin{pmatrix} \varepsilon/2 & 0 & 0 \\ 0 & \varepsilon/2 & t_c \\ 0 & t_c & -\varepsilon/2 \end{pmatrix} \rightarrow \begin{pmatrix} J(\varepsilon) & 0 \\ 0 & 0 \end{pmatrix}$$

Fig. 3.9: Energy diagrams of the spin states of a singlet-triplet qubit in a double quantum dot. (a) Energies of the (0, 2)-singlet $S(0, 2)$, the (0, 2)-triplet $T(0, 2)$ and the (1, 1) state as a function of detuning ϵ for zero tunnel coupling and zero magnetic field. Here, (n, m) indicates n electrons in the left dot and m electrons in the right dot. (b) Effect of introducing a finite tunnel coupling between the dots. For simplicity, the anticrossing due to the tunnel coupling is only shown for the singlet states and not for the triplet states. (c) Effect of an additional Zeeman splitting, which energetically separates the three triplet states by the Zeeman energy E_Z. (d) Qubit Hamiltonian for a tunnel coupled double dot, which is reduced to an effective 2-state subspace by diagonalizing the lower singlet block of the upper matrix.

the triplet is thus larger by the single-particle level splitting of the two states ψ_1 and ψ_0. For a (1, 1) state, where the two electrons are in different quantum dots, the single particle splitting between the symmetric and antisymmetric state of Fig. 3.4 and hence the singlet-triplet splitting remains small as long as the two dots are sufficiently far apart. The resulting energy diagram, neglecting the singlet-triplet splitting in the (1, 1) state, is shown in Fig. 3.9(a). While interactions between the two electrons reduce the singlet-triplet splitting in the (0, 2) state via exchange contributions, one can show that it can never be completely eliminated at zero magnetic field.

Next, we consider the effect of tunnel coupling between the dots, which will induce transitions between the (1, 1) and (0, 2) states. This leads to avoided level crossings, as for the charge qubit (Fig. 3.5), which is depicted in Fig. 3.9(b). At the degeneracy point ($\epsilon \simeq 0$), one gets symmetric and antisymmetric superpositions of the (0, 2) and the $(1, 1)$ state, since the electron pair can tunnel between these two configurations. Since the two combinations imply a different charge distribution of the two electrons, they differ in energy, analogous to the symmetric and antisymmetric state of a single electron in the double dot of Fig. 3.4(a). Importantly, the tunnel coupling barely mixes S and T states, since this would require an additional spin flip during tunneling, so that only $S(0, 2)$ and $S(1, 1)$ as well as $T(0, 2)$ and $T(1, 1)$ are significantly hybridized by the tunnel coupling.

Due to the singlet-triplet splitting, the anticrossing of the triplet state occurs at a different detuning ϵ. For simplicity, this anticrossing is not shown in Fig. 3.9(b). As $T(0, 2)$ has a much higher energy, one often assumes that the triplet will always stay in the $T(1, 1)$ configuration.

We thus obtain an ϵ-dependent effective *exchange splitting* between the hybridized S-ground state, consisting of $S(1, 1)$ and $S(0, 2)$, and the $T(1, 1)$ state. This splitting is usually denoted as $J(\epsilon)$ (Fig. 3.9(b)). It can be formally derived by introducing the tunnel coupling t_c in the the three level Hamiltonian for $T(1, 1)$, $S(1, 1)$ and $S(0, 2)$. From this, one gets $J(\epsilon)$ by diagonalizing the S-block. Finally, one keeps only $T(1, 1)$ and the lower of the two resulting S-states as a basis of the computational subspace (Fig. 3.9(d)).

Next, we add a magnetic field to the energy diagram, which introduces a Zeeman splitting with Hamiltonian $\widehat{H}_Z = g\mu_B\widehat{S}_z$. This splits the three triplet states $|T_+\rangle = |\uparrow\uparrow\rangle$ $(m = 1)$, $|T_0\rangle = 1/\sqrt{2}(|\uparrow\downarrow\rangle + |\downarrow\uparrow\rangle)$ $(m = 0)$, and $|T_-\rangle = |\downarrow\downarrow\rangle$ $(m = -1)$ in energy, while the $|S\rangle$ state $(m = 0)$ remains unaffected. The complete energy diagram including the Zeeman splitting is shown in Fig. 3.9(c). For qubit operation, the $|T_+\rangle$ and $|T_-\rangle$ states can mostly be ignored, since they are only weakly coupled to $|T_0\rangle$ and $|S\rangle$ due to orbital momentum conservation. Hence, $|S\rangle$ and $|T_0\rangle$ are an adequate basis of the computational subspace. Initialization into the subspace can be provided by starting with $S(0, 2)$ as the ground state at large positive ϵ before adiabatically tuning to negative ϵ.

The qubit can be controlled by modulating ϵ and thus $J(\epsilon)$. Full control further requires a mechanism to drive transitions between the two computational states T_0 and S. This can be enabled by a magnetic field difference ΔB between the two quantum dots. In GaAs, the difference can conveniently be provided by the different distribution of nuclear spin polarization within the two dots. The resulting ΔB splits the states $|\uparrow\downarrow\rangle$ and $|\downarrow\uparrow\rangle$, which are the eigenstates for $J \ll g\mu_B\Delta B$. Hence, the eigenstates at negative ϵ are superposition states of $S(1, 1)$ and $T(1, 1)$. This allows one to control the qubit electrically via ϵ. A detailed account of the qubit dynamics and control will be given in Section 3.4.2.3, after completing our survey of solid-state qubits by considering superconducting qubits.

3.3.3 Macroscopic quantum coherence in superconductors

Before coming to superconducting qubit devices in the next section, we will discuss the concept of macroscopic quantum coherence, which is the basis for a description of the quantum dynamics of superconducting qubits.

The states of superconducting qubits are defined in terms of electromagnetic variables such as charge, current or voltage, which correspond to the collective behavior of many electrons. This might appear unusual, since most other areas for which quantum mechanics has been tested deal with a few electrons only. In this case, the quantum dynamics is usually described by treating position and momentum as conjugate variables.

One should thus ask whether the same quantization procedure also works for the quantum observables of many particles. Does it hold, e.g., for the total charge on a conductor or for a magnetic flux? This question is strongly related to the question if there is a fundamental boundary between quantum and classical behavior. To date, no definite answer to this question is known. However, the hypothesis that nothing extraordinary happens when considering increasingly macroscopic systems has not been disproved so far.

In the following, we discuss a general recipe for quantizing arbitrary dynamic systems that we will later apply to superconducting qubits. We start with the simple harmonic $L_I C$ resonator as shown in Fig. 3.10(a) where L_I describes the inductance of the circuit. The charge on its capacitor, with capacitance C, is denoted by Q. Conservation of charge implies that the flux in the inductor is given by $\Phi = L_I \dot{Q}$. The table below outlines the steps for deriving a quantum mechanical description of an arbitrary system, and applies them to our example.

General quantization procedure	Application to $L_I C$-resonator
Pick a suitable variable ("position") q.	$q = Q$
Write down the classical Hamilton function H, i.e., the energy as a function of q and $\dot{q}$.	$H = \frac{L_I \dot{Q}^2}{2} + \frac{Q^2}{2C}$
Determine the conjugate variable (canonical momentum) $p = \frac{\partial H}{\partial \dot{q}}$.	$p = L_I \dot{Q} = \Phi$
Assume a standard commutation relation $[\widehat{p}, \widehat{q}] = i\hbar$.	$[\widehat{\Phi}, \widehat{Q}] = i\hbar$
Analogous to the position representation known from mechanics, one can work with a wave function $\Psi(q)$ depending on the generalized coordinate q and a canonical momentum operator $\widehat{p} = i\hbar \frac{\partial}{\partial q}$.	$\Psi(Q)$, $\widehat{\Phi} = i\hbar \frac{\partial}{\partial Q}$

From the above derivations and, in particular, from the form of the Hamiltonian, we see that the $L_I C$-resonator forms a harmonic oscillator, not only according to a classical consideration, but also with respect to its quantum dynamics. This implies by

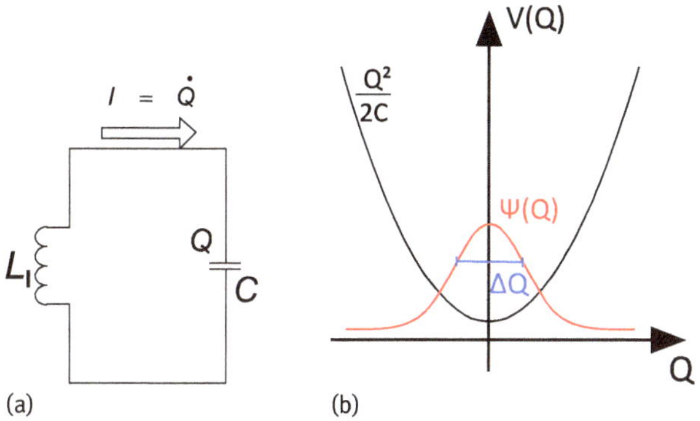

Fig. 3.10: (a) Circuit diagram of a simple $L_I C$ resonator with charge Q on the capacitor and current I through the inductor. (b) Harmonic potential of the resonator in terms of Q and the corresponding ground state wave function $\Psi(Q)$. The kinetic energy associated with the inductance leads to a quantum-uncertainty ΔQ of Q.

analogy that it has energy levels given by $E_n = (n + 1/2)\hbar\omega$ with $\omega = (L_I C)^{-1/2}$ and that there is a flux-charge uncertainty relation (Fig. 3.10(b)). Starting from a different variable, e.g., the number of electrons on the capacitor instead of its charge, or the flux in the inductor, would yield the same result, but with p and q exchanged and/or rescaled. Hence, the arbitrary choice of what variable is used as a starting point of the quantization does not affect the result.

It is interesting to note that while there is no general proof of the above procedure from first principles, it seems to be adequate to describe current experiments. This applies up to the point where decoherence becomes relevant and more advanced considerations are required as touched upon in Section 3.5. We note in passing that some researchers suggest that "new physics" might govern the transition from the quantum to the classical world beyond the concepts of Section 3.5 [41]. The proposed mechanisms include the idea that a wave function might collapse and, hence, a transition to classical behavior occurs, once the different possibilities cause a substantial difference in the curvature of space-time according to general relativity [41]. Another conjecture is that some form of universal "background noise" destroys the phase coherence of macroscopic objects, but is too weak to have any observable effect on microscopic systems [41].

Experiments aiming to demonstrate quantum behavior have shown that objects as large as C_{60} molecules (bucky balls) can interfere as expected from standard quantum mechanics in double slit experiments. Moreover, micron-scale mechanical resonators consisting of more than 10^{10} atoms also behave quantum mechanically, if cooled to their ground state, i.e., they exhibit a zero point motion [41]. These and other observations, including considerations from astrophysics, severely constrain the possible strength of the conjectured "background noise", so that the region of parameter space that cannot be ruled out is rather small.

3.3.4 Josephson relations

The central elements of virtually all types of superconducting qubits are Josephson junctions, which are tunnel junctions between two superconductors. Josephson junctions are also used in other applications, e.g., in superconducting electronics or as voltage standards. The discovery of the Josephson effect was the subject of the Nobel prize in 1973.

In most cases, Josephson junctions are formed by a thin insulating layer, called the dielectric, through which Cooper pairs can tunnel (Fig. 3.11(a)). The circuit symbol for a Josephson junction is a cross, as shown in Fig. 3.11(b). In the following, we will heuristically derive the equations of motion for such Josephson junctions, which will later form the basis for superconducting qubits.

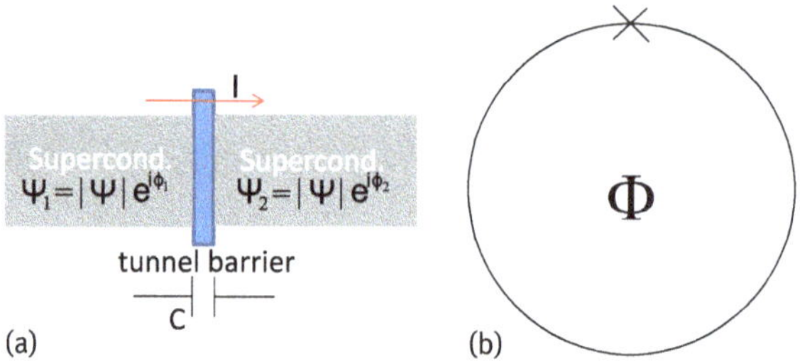

Fig. 3.11: (a) Schematic of a Josephson junction. In addition to allowing Cooper pair tunneling leading to the current I, the tunnel barrier also leads to a geometric capacitance C. (b) Circuit symbol representing a Josephson junction (cross) in a loop, which contains the magnetic flux Φ.

In the superconducting electrodes, all Cooper pairs share a common wave function Ψ (also referred to as order parameter or gap function). Its magnitude $|\Psi|^2$ reflects the so called superfluid density, which can be thought of as the density of Cooper pairs. This density is a constant material property for the conditions we are interested in. Hence, we only need to consider the phase ϕ of $\Psi = |\Psi| \cdot \exp(i\phi)$.

We denote by Ψ_i $(i = 1, 2)$ the wave function on each side of the tunnel barrier and start by applying the Schrödinger equation while identifying the energy of Cooper pairs with the electrostatic potential V_i on each side:

$$i\hbar\partial_t\Psi_i = \widehat{H}\Psi_i = -2eV_i\Psi_i\,. \tag{3.6}$$

The ac Josephson effect can be derived by rewriting the equation in terms of the phase exploiting $\partial_t|\Psi| = 0$ and taking the differences $V = V_1 - V_2$ and $\phi = \phi_1 - \phi_2$ resulting in

$$\dot{\phi} = \frac{2e}{\hbar}V\,. \tag{3.7}$$

Hence, an applied voltage across the tunnel barrier leads to a continuous change of the phase difference between the two superconductors. It is often useful to identify the phase difference ϕ with a magnetic flux Φ, which can be deliberately tuned. Therefore, one can employ a superconducting loop with negligible, geometric inductance, which is interrupted by a Josephson junction (Fig. 3.11(b)). Integrating eq. (3.7) and, according to the third Maxwell equation (eqs. (2.41)–(2.44)), using $\Phi = \int\int \vec{B} \cdot d\vec{A} = \int V(t)\, \mathrm{d}t$ for the flux Φ threading the loop gives

$$\phi = \frac{2e}{\hbar} \int V(t)\, \mathrm{d}t = \frac{2e}{\hbar} \Phi = 2\pi \frac{\Phi}{\Phi_0} , \tag{3.8}$$

with $\Phi_0 = \frac{h}{2e} = 2.07 \cdot 10^{-15}\ \mathrm{Tm}^2 (= 20.7\ \mathrm{G\,mm}^2)$ being the magnetic flux quantum (Section 1.4.3.5). Hence, the phase difference between the superconductors on both sides of the tunnel junction can be tuned by the magnetic flux threading the loop.

To understand the dynamics of a Josephson junction, one also needs a relation between the phase ϕ and the current I through the junction. In general, this current-phase relation $I(\phi)$ depends on the physical properties of the junction. However, the most widespread form, which is valid for weakly coupled junctions, can be obtained from the following heuristic argument. We start by guessing an expression for the energy of the Josephson junction, which we interpret as a potential energy E_{pot}. This energy should depend on the macroscopic wave functions Ψ_i in such a way that it does not change under an overall phase change, which should have no physical consequences. Since the tunnel coupling counteracts changes of Ψ across the junction, one may expect the energy to depend on $\Psi_1 - \Psi_2$. In addition, we look for a quadratic expression for the energy, which is usually the most simple form. All these requirements are met by the expression

$$E_{\mathrm{pot}}(\phi) = \frac{\alpha_{\mathrm{c}}}{2} |\Psi_1 - \Psi_2|^2 \tag{3.9}$$

$$= E_{\mathrm{J}}(1 - \cos\phi) , \tag{3.10}$$

where α_{c} parametrizes the coupling strength and we have defined $E_{\mathrm{J}} = \frac{\alpha_{\mathrm{c}}}{2} |\Psi_1|^2$. We have assumed that $|\Psi_1| = |\Psi_2|$, which is the case for identical superconductors on each side. However, the same result as in eq. (3.9) would also be obtained for different electrode materials. Moreover, we assumed that $|\Psi_i|$ is independent of ϕ, which holds in the limit of weak coupling, where the energy cost of a non-zero phase difference is too small to appreciably suppress superconductivity. While the above argument is heuristic, a microscopic derivation leads to the same result in the weak coupling limit, which is valid in all cases discussed here [42].

Having an expression for the potential energy, we can now obtain the current using

$$I = \frac{\text{Power}}{\text{Voltage}} = \frac{\mathrm{d}E_{\mathrm{pot}}/\mathrm{d}t}{\mathrm{d}\Phi/\mathrm{d}t} = \frac{\mathrm{d}E/\mathrm{d}\Phi \cdot \mathrm{d}\Phi/\mathrm{d}t}{\mathrm{d}\Phi/\mathrm{d}t} = \frac{\mathrm{d}E}{\mathrm{d}\Phi} . \tag{3.11}$$

which leads to

$$I(\phi) = \frac{\mathrm{d}E_{\mathrm{pot}}}{\mathrm{d}\Phi} = -\frac{2e}{h}\frac{\mathrm{d}}{\mathrm{d}\phi} E_{\mathrm{J}} \cos\phi = I_{\mathrm{c}} \sin\phi . \tag{3.12}$$

Here, $I_c = \frac{2e}{\hbar}E_J$ is the so called critical current, i.e., the largest supercurrent that can flow through the junction without causing a voltage drop.

Inserting the integrated form of eq. (3.7) into eq. (3.12), we obtain

$$I(t) = I_c \cdot \sin\left(\frac{2e}{\hbar}Vt + \varphi_0\right) \tag{3.13}$$

with φ_0 being a phase factor related to the starting conditions. This result describes the AC Josephson effect, i.e., an applied DC voltage V leads to an oscillating current with a frequency proportional to V.

Besides allowing the tunneling of Cooper pairs that gives rise to the Josephson current, the thin insulating layer also forms a capacitor with capacitance C (Fig. 3.11(a)). This creates a parallel path for AC currents that is important for superconducting qubits. To understand the resulting quantum dynamics, it is convenient to treat the charging energy of this capacitor as a kinetic energy of the phase difference ϕ reading

$$E_{\text{kin}} = \frac{C}{2}V^2 = \frac{C}{2}\left(\frac{\hbar}{2e}\right)^2 \dot{\phi}^2 = \frac{(2e\Delta N_{\text{CP}})^2}{2C} = E_c \Delta N_{\text{CP}}^2\,, \tag{3.14}$$

where V denotes the voltage across the junction, ΔN_{CP} is the number of excess or deficient Cooper pairs residing on the two electrodes, and we have used $CV = Q = 2e\Delta N_{\text{CP}}$. The term $E_c = 2e^2/C$ describes the charging energy of the junction capacitor (Section 3.3.1.2). Within this picture, the phase ϕ can be thought of as the position of a fictitious particle, whose total energy is given by $E_{\text{kin}} + E_{\text{pot}}$ according to eqs. (3.10) and (3.14). The equations of motion of ϕ are given by the Josephson relations (eqs. (3.7) and (3.12)).

Note that, as already pointed out in Section 3.3.3, the distinction between kinetic and potential energy is arbitrary. In fact, in our $L_I C$ resonator example in that section, the interpretation of magnetic flux and electrical, capacitive energy as kinetic and potential energy was interchanged. For superconducting junctions, it is more intuitive to think of the terms depending on the phase ϕ as potential energy and to think of the terms depending on $\dot{\phi}$ as kinetic energy.

It is worth pointing out that a Josephson junction can also be described as a nonlinear inductor. For small AC signals with current variations much smaller than I_c, its inductance, the so called Josephson inductances L_J, is given by (eqs. (3.8) and (3.12))

$$L_J = \frac{d\Phi}{dI} = \frac{1}{dI/d\Phi} = \frac{\Phi_0}{2\pi I_c \cos\phi}\,. \tag{3.15}$$

Here, L_J depends on I via the phase difference ϕ (eq. (3.12)). The concept of a nonlinear Josephson inductance is extremely important for understanding the behavior of superconducting electronic circuits, where Josephson junctions play a role analogous to transistors in semiconductor electronics.

Before moving on to superconducting qubits that employ Josephson junctions, we shortly discuss how these junctions can be fabricated. One widespread method is to

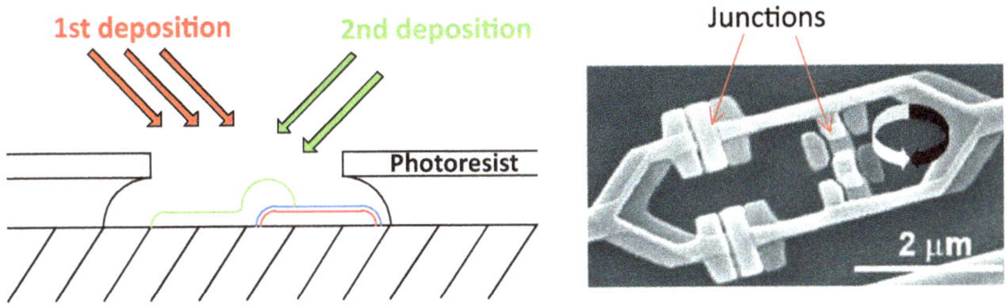

(a) (b)

Fig. 3.12: (a) Schematic of the fabrication of Josephson junctions via shadow evaporation. The substrate on which the junction is to be fabricated is coated with a relatively thick photoresist, which is patterned such that a large undercut forms. The first electrode is then evaporated, say, from the left, forming the part shown in red. Controlled exposure to oxygen forms the insulating oxide layer (amorphous Al_2O_3, blue), on top of which the second electrode is deposited from the right (green). The junction forms in the region where the two electrodes overlap. (b) Scanning electron microscopy (SEM) image of a superconducting flux qubit fabricated in this way. Two of the three Josephson junctions are marked. The "double vision effect" arises from the deposition of the whole structure through the patterned mask in the photoresist at the two different angles. (b) [34].

deposit a few nm of Al on top of a thin film of Nb, oxidize the aluminum to form the tunnel barrier, and to deposit another layer of Nb. This method is used predominantly for classical superconducting circuits designed to operate at 4.2 K (liquid helium temperature). For qubit fabrication, it is more common to use Al – Aluminum oxide – Al junctions, which are often fabricated via angle evaporation, as outlined in Fig. 3.12.

3.3.5 Superconducting qubits

Superconducting qubits are the most advanced type of solid state qubit realizations. At the time of writing (2018), cricuits with up to 20 qubits have been demonstrated, and 72 qubits in a circuit were announced. On the way to these achievements, several types of superconducting qubits have been implemented and studied, before largely converging on the so called transmon, which is currently considered most promising. We will survey four qubit types that are representative for most superconducting qubit devices and convey the most important concepts.

3.3.5.1 Phase qubit

While the phase qubit has largely been superseded by other implementations with superior coherence, it is of conceptual interest as it nicely illustrates the concept of a ficticious particle whose position is given by the phase across a Josephson junction. In the phase qubit device shown in Fig. 3.13(b), this qubit junction is the large junction on the left marked "qubit junction." The two computational states $|0\rangle$ and $|1\rangle$ are the two lowest eigenstates of the phase "particle" in the Josephson potential as shown

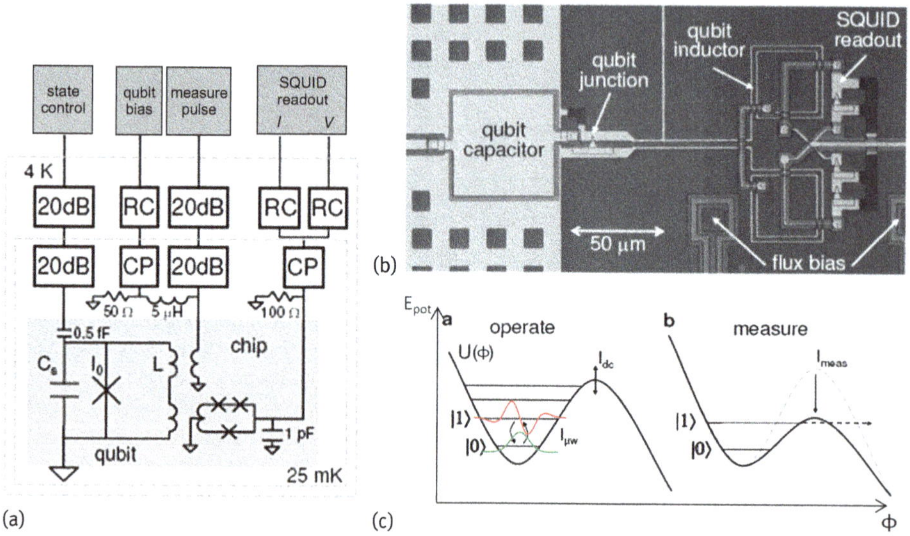

Fig. 3.13: (a) Circuit diagram of a Josephson phase qubit. (b) Image of a Josephson phase qubit. (c) Potential energy of the fictitious phase particle prior (left) and during (right) the readout process. The wave functions of two lower states $|0\rangle$ and $|1\rangle$ are drawn in colors. They span the computational subspace and can be superposed by a microwave current pulse $I_{\mu w}$, which is applied via the qubit capacitor marked in (a), (b). A current I_{meas} through the loop, called flux bias in (b), changes the potential energy for the phase ϕ (eq. (3.10)). As the barrier is lowered, the $|1\rangle$ state can tunnel out of the potential, whereas the $|0\rangle$ state remains trapped. The difference between the two resulting states can be detected using the SQUID readout. (a)–(c) [43].

in Fig. 3.13(c). Its kinetic energy arises from the shunting capacitance (eq. (3.14)). The closest physical intuition of these states is that of the ground and first excited state of an $L_I C$-oscillator. In contrast to the harmonic oscillator considered in Section 3.3.3, a strong nonlinearity of the potential makes only two states accessible.

The qubit is manipulated by applying a microwave pulse that shakes the particle basically via coupling to the magnetic flux (eq. (3.7)). This leads to a superposition state of $|0\rangle$ and $|1\rangle$, in particular if the shaking frequency is resonant with the energy distance between $|0\rangle$ and $|1\rangle$. Readout of the qubit is achieved as follows. By integrating the qubit junction in a superconducting wire loop with geometric inductance L_I and threading a flux through this loop, the cosine potential of the phase particle is tilted (Fig. 3.13(c), right). Increasing the tilt of the potential by changing the external flux causes the barrier on the right side to be lowered, so that the phase particle quickly tunnels out of the well if it is in the $|1\rangle$ state, but remains trapped if it is in the $|0\rangle$ state (Fig. 3.13(c), right). Whether the particle tunneled out of the potential well or not can be discriminated via the additional SQUID (loop with two junctions on the right of Fig. 3.13(b)). The SQUID operates as a magnetic flux sensor sensing the magnetic moment of the qubit loop and thus its current. The current is directly related to ϕ via the first Josephson relation (eq. (3.12)).

3.3.5.2 Flux qubit

The basic idea of the flux qubit (also called persistent current qubit) is to use different amounts of magnetic flux trapped in a superconducting loop to represent its computational states (Fig. 3.14(a)). It is commonly formed by placing three Josephson junctions inside a superconducting loop (Fig. 3.14(b), (c)). A second loop and a fourth junction can be added to allow the effective coupling strength of the smaller junction to be tunable by an external flux (Fig. 3.12(b)).

Figure 3.14(d) shows the energy landscape as a function of the outer junctions' phases arising from this geometry. The phase of the third junction cannot vary independently as the sum of all phase differences and the dimensionless applied flux, $2\pi\Phi/\Phi_0$, must add to a multiple of 2π due to fluxoid quantization. Since the Josephson energy E_{pot} is periodic in the phase across the junction (eq. (3.10)), so must be the total potential in both left and right phase. Hence, there are necessarily maxima and minima. According to Fig. 3.14(d), the device thus implements a two-dimensional fictitious phase particle, analogous to the one-dimensional particle of the phase qubit discussed in the previous section. The potential locally is a double well potential, e.g., along the blue or the orange line, and the two qubit states correspond to the phase particle (phases of the two Josephson junctions) being localized in the left or right well. These states can correspond to a clockwise or anticlockwise circulating current $\pm I$ in the loop (Fig. 3.14(a)). Due to the self inductance of the loop, these circulating currents also induce a flux, hence the name of the qubit. Threading an additional external flux $\Delta\Phi$ through the loop, e.g., by sending a control current I_{co} through the red wire in Fig. 3.14(b), changes the energy difference of the two states by an amount

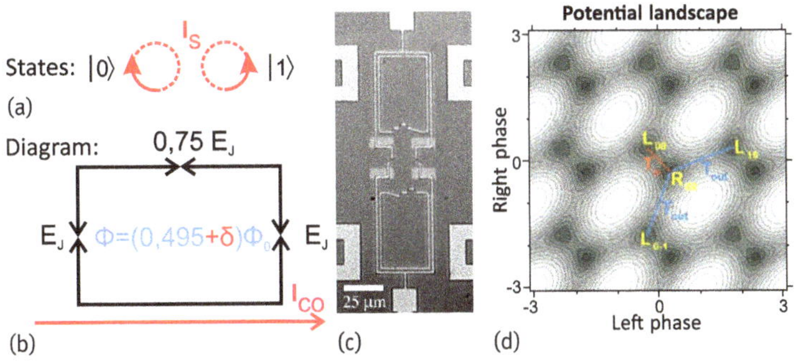

Fig. 3.14: (a) The superconducting current I_S of the two possible states of a flux qubit, represented by the opposite chirality of the current. (b) Schematic diagram of a flux qubit containing three Josephson junctions with different Josephson energies. The current I_{co} is used to change the energy difference between the two qubit states. (c) Electron microscopy image of two coupled flux qubits (inner loops) with readout circuit(outer loop). (d) The potential landscape of a single flux qubit as a function of two of the phases across two of the junctions. Darker areas correspond to a lower energy. An effective double well potential is formed along the colored lines with indicated tunnel couplings T_{out} and qubit position minima L_{nm} and R_{nm}. (a), (b), (d) [44], (c) [45].

$\epsilon = 2I\Delta\Phi$, which is formally identical to the detuning of a charge qubit discussed in Section 3.3.1.5. Just as for the charge qubit, the flux particle can also tunnel between the two wells as described by a term $\frac{\Delta}{2}\hat{\sigma}_X$. The qubit Hamiltonian is thus identical to that of a charge qubit:

$$\hat{H} = \frac{\epsilon}{2}\hat{\sigma}_Z + \frac{\Delta}{2}\hat{\sigma}_X \,. \tag{3.16}$$

The two qubit states can again be distinguished by measuring their magnetic moment with an additional SQUID surrounding the qubit (Fig. 3.14(c)). Note that if only two junctions were included in the loop, there would be only one independent phase variable, and the Josephson energy would be a sinusoidal function of this variable with a period of 2π. Since phases differing by a multiple of 2π are equivalent, there would only be a single energy minimum, and no localized states with different values of the phase variables would exist.

One drawback of the flux qubit is that it is quite susceptible to noise in the magnetic flux that sets the detuning. This flux noise typically arises from fluctuating spins on the device, which are difficult to avoid. However, flux qubits are still of high practical relevance for so-called annealing devices. A commercial annealing device is offered by D-Wave Systems Inc., but does not reveal a clear quantum speedup so far.

3.3.5.3 Cooper pair box

This type of superconducting qubit, shown in Fig. 3.15(a) and (b), is essentially a superconducting charge qubit consisting of a superconducting island, which is called a box. Contrary to the semiconductor charge qubit discussed in Section 3.3.1, there is only one small box, while the other one is replaced by a single large reservoir. As we will discuss below, this simplification is possible because of superconductivity. Another difference to the charge qubit is that there are many electrons on the metallic island. The two computational states correspond to N_{CP} and $N_{\mathrm{CP}} + 1$ Cooper pairs on it, with N_{CP} being some large number. The Josephson junction(s) between the box and the reservoir create a tunnel matrix element between the $|N_{\mathrm{CP}}\rangle$ and $|N_{\mathrm{CP}} + 1\rangle$ states by allowing Cooper pairs to hop on and off the island. One often uses two junctions enclosing a small loop, which allows the coupling energy to be tuned by threading an external flux through the loop via the Josephson relation of eq. (3.12).

To understand the energy associated with changing the number of Cooper pairs N_{CP} and thus how the qubit can be controlled, we need to consider the electrostatic energy of the qubit. In brief, it behaves like a single electron transistor (Section 3.3.1.2) with a single lead. A mathematical model can be obtained by drawing an effective circuit diagram as shown in Fig. 3.15(b). The capacitors reflect the geometric capacitances between the box and the gate, C_{Gate}, and between the qubit and the rest of the world, in particular the reservoir, $C_{\mathrm{Reservoir}}$. The latter tends to be dominated by the junction capacitance because of the very thin tunnel barrier. The relevant part of the

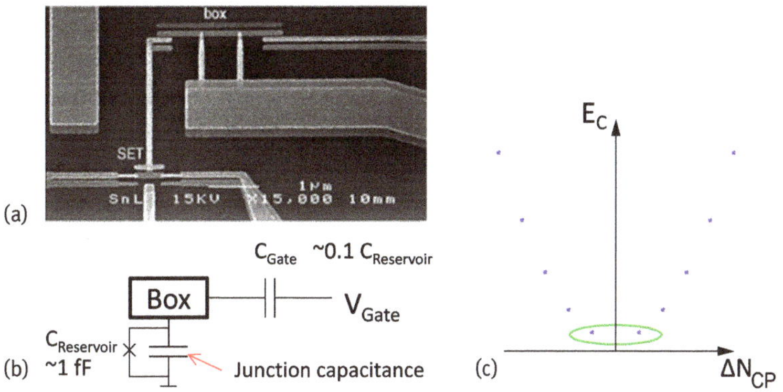

Fig. 3.15: (a) SEM micrograph of a Cooper pair box and its readout SET. (b) Circuit representation of a Cooper pair box, which is a superconducting island coupled to a reservoir. (c) Electrostatic energy of a Cooper pair box as a function of the number of Cooper pairs ΔN_{CP} added or subtracted from the Cooper pair number at equilibrium. The qubit states inside the green ellipse are those with the lowest energy, which form the computational subspace. (a) [46].

energy of the system as a function of N and the gate voltage V_{Gate} is given by

$$E(\Delta N_{\mathrm{CP}}, V_{\mathrm{Gate}}) = \frac{(2eN_{\mathrm{CP}})^2}{2C_{\parallel}} + \frac{C_{\mathrm{Gate}}}{C_{\mathrm{Reservoir}} + C_{\mathrm{Gate}}}(-2e)V_{\mathrm{Gate}}\Delta N_{\mathrm{CP}} , \tag{3.17}$$

where $\Delta N_{\mathrm{CP}} = N_{\mathrm{CP}} - N_{\mathrm{CP},0}$ and $N_{\mathrm{CP},0}$ is the equilibrium number of Cooper pairs on the box at $V_{\mathrm{Gate}} = 0$. $C_{\parallel} = C_{\mathrm{Gate}}C_{\mathrm{Reservoir}}/(C_{\mathrm{Gate}} + C_{\mathrm{Reservoir}})$ is the total capacitance of the Cooper pair box to its environment. This result can be derived formally by considering the charging energy of the capacitors and the work done by the voltage source providing V_{Gate}. Its terms can be understood as follows. The first is simply the charging energy $Q^2/(2C_{\parallel})$ of the island at $V_{\mathrm{Gate}} = 0$, which is obtained by thinking of the two capacitors connected in parallel as a single one with capacitance $C_{\parallel}$. The second term is the excess charge on the box times the potential change due to V_{Gate}, which is determined by capacitive voltage division. For a given V_{Gate}, the qubit states $|N_{\mathrm{CP}}\rangle$ and $|N_{\mathrm{CP}}+1\rangle$ are the two states with the lowest energies. We see that their energy difference is given by

$$\epsilon = \frac{C_{\mathrm{Gate}}}{C_{\mathrm{Reservoir}} + C_{\mathrm{Gate}}} 2e(V_{\mathrm{Gate}} - V_{0,\mathrm{Gate}}) ,$$

where $V_{0,\mathrm{Gate}}$ is some offset voltage (Fig. 3.15(c)). The energy difference corresponds to a detuning energy ϵ in analogy to the charge and spin qubits discussed previously. Considering that the Josephson coupling to the reservoir can induce transitions between $|N_{\mathrm{CP}}\rangle$ and $|N_{\mathrm{CP}} + 1\rangle$, we again find the familiar Hamiltonian

$$\widehat{H} = \frac{\epsilon}{2}\widehat{\sigma}_z + \frac{\Delta}{2}\widehat{\sigma}_x . \tag{3.18}$$

The drawback of the Cooper pair box is that it is sensitive to charge noise, similar to semiconductor charge qubits. This problem has been fixed with the invention of the transmon, to be discussed in Section 3.3.5.4.

It is interesting to consider why superconductivity is important for the Cooper pair box, and superconducting qubits in general. First of all, if the metal making up the qubit were in its normal state, its resistance would lead to energy dissipation causing the qubit to relax to its ground state. Secondly, the resistance would lead to noise causing loss of phase information. Finally, a normal metal island would have a quasi-continuum of states near the Fermi energy (Fig. 3.16(a)). For a 100 nm large cubic box and a lattice constant of 0.1 nm, the level spacing is of the order of (Fermi energy)/(number of electrons) $\sim 10^5\,\mathrm{K}/(100\,\mathrm{nm}/0.1\,\mathrm{nm})^3 = 0.1\,\mathrm{mK}$. This energy scale is much smaller than the few tens of mK that are conveniently reached in the lab. Consequently, thermal fluctuations would always lead to the occupation of excited states, which makes it extremely difficult to initialize the qubit in a well-defined initial state.

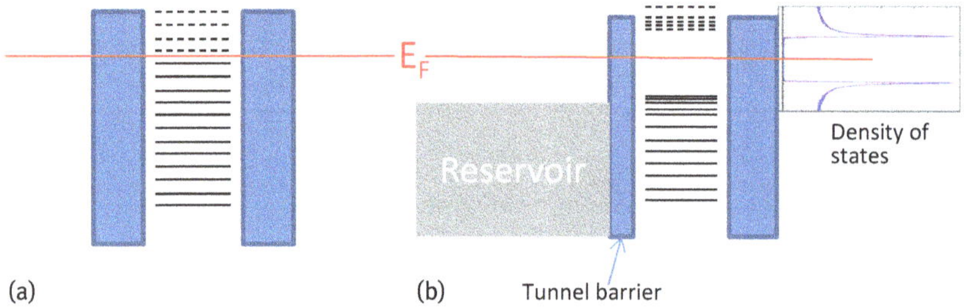

Fig. 3.16: Energy level diagram of (a) a metallic island and (b) a Cooper pair box. The superconducting gap in the latter leads to a ground state which is energetically well-separated from the excited state with a broken Cooper pair.

A superconducting island on the other hand has a gap in the density of states near the Fermi-level (Fig. 3.16(b)). In the ground state, all states below the gap are occupied and those above are empty. Hence, one needs to pay at least the gap energy to create an excitation. This energy is given by $3.5\,k_B T_C$ (in the BCS limit), where T_C ($\approx 1.2\,\mathrm{K}$ for Al) is the superconducting transition temperature. A very rough upper bound on the probability of creating an excitation is thus (Number of electrons) $\times \exp(-5\,\mathrm{K}/0.1\,\mathrm{K})$ $\sim 10^{-13}$ $\times$ (Number of electrons). In thermal equilibrium, a small island with about 10^9 electrons would thus nearly always be in its ground state. Interestingly, nonequilibrium effects, for example due to infrared radiation from 4 K, tend to be a problem in practice if one is not careful.

3.3.5.4 Transmon-qubits

The most successful version of superconducting qubits at the time of writing (2018) is the so called transmon. It is essentially a Cooper pair box with a large shunting capacitance between the box and the reservoir formed by an interdigitated capacitor (Fig. 3.17). Due to this large capacitance, the charging energy is small compared to the Josephson energy and the number of Cooper pairs is subject to large quantum fluc-

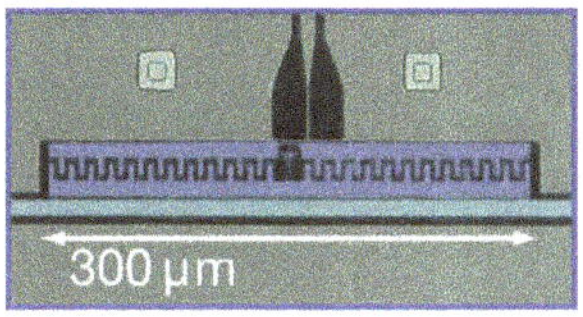

Fig. 3.17: A transmon qubit inside a transmission line. The comb structure is an interdigitated capacitor, while the qubit junctions are placed in the small loop in the middle [33].

tuations (Section 3.3.5.5). As a result, the eigenstates of the qubit do not have a well-defined charge anymore. Instead, they are close to harmonic oscillator states. As potential fluctuations only laterally translate the harmonic oscillator potential, which is now a function of charge on the island rather than flux or phase, the qubits level splitting is nearly insensitive to charge noise from its environment, which is the reason for its success. Furthermore, thanks to the nonlinearity introduced by the Josephson coupling, the qubit is not exactly harmonic and its energy levels are not equidistant. This anharmonicity is important as it allows individual transitions to be addressed by driving the qubit at the corresponding frequency, and hence to restrict excitations to the computational subspace of these two levels.

3.3.5.5 Energy scales

This non-exhaustive overview of different types of superconducting qubits already features a large variety of them. A useful way to classify them is via the ratio E_J/E_C. This basic classification can be extended by including additional energy scales, for example those associated with additional inductive elements. As we will see in the following, this ratio determines whether charge or phase are more localized. A reasonably generic Hamiltonian is given by

$$\begin{aligned}\widehat{H} &= \widehat{E}_{\text{kin}} + \widehat{E}_{\text{pot}} \\ &= E_C \widehat{N}_{CP}^2 + E_J(1 - \cos\widehat{\phi}) ,\end{aligned}$$

as derived for a single Josephson junction in Section 3.3.4 (eqs. (3.9) and (3.14)). Since we aim to treat the system quantum-mechanically, we have written the number of Cooper pairs $\widehat{N}_{CP}$ and the phase difference $\widehat{\phi}$ as canonically conjugate operators, with commutation relation $[\widehat{\phi}, \widehat{N}_{CP}] = i$ (Section 3.3.3). Note that this Hamiltonian describes the complete dynamics of the system with all degrees of freedom, whereas the two-level Hamiltonians derived in the previous sections are effective Hamiltonians describing the dynamics within the qubit subspace only. The latter are valid when no other states need to be considered. Since the above commutation relation imposes an uncertainty relation between $\widehat{N}_{CP}$ and $\widehat{\phi}$, the more precisely one quantity is defined, the more the other is smeared out.

Since the ground state of a system always minimizes the total energy, $\widehat{E}_{\text{pot}}$ tries to localize the phase by pushing it towards a potential minimum. On the other hand, the kinetic term $\widehat{E}_{\text{kin}}$ is proportional to $\widehat{N}_{CP}^2 \propto |\frac{\partial \Psi(\phi)}{\partial \phi}|^2$, where $\Psi(\phi)$ is the wave function of the phase particle. Hence, it tries to spread out the wave function by minimizing the

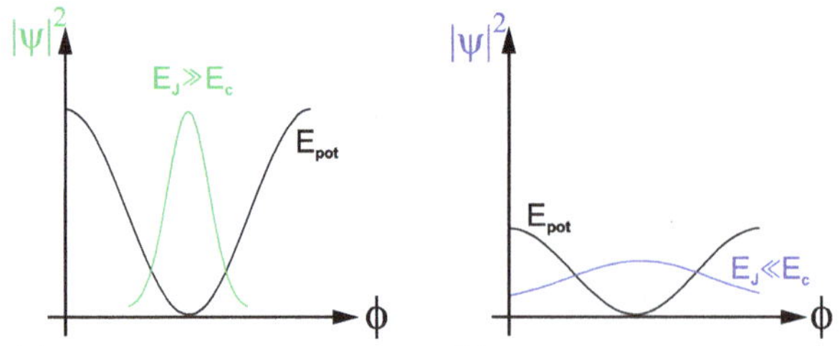

Fig. 3.18: Localization and delocalization of the wave function $\Psi(\phi)$ of the phase variable ϕ of a superconducting qubit in the Josephson potential $E_{pot}(\phi)$. (a) For $E_J \gg E_c$, the strong potential energy leads to a relatively narrow ground state wave function. (b) For $E_J \ll E_c$, the kinetic (charging) energy dominates and the wave function is broad.

gradient energy. With the prefactors in E_{kin} and E_{pot} being E_c and E_J, we see that $E_c \ll E_J$ leads to a smeared out charge and a narrowly localized phase (i.e., $\langle \Delta N_{CP}^2 \rangle \gg 1$ and $\langle \Delta \phi^2 \rangle \ll 1$), whereas the opposite is the case for $E_c \gg E_J$ (Fig. 3.18).

3.4 Dynamics and control

3.4.1 Bloch sphere

Having seen a number of different qubits with their effective Hamiltonians, which mostly can be reduced to

$$\widehat{H} = \frac{\Delta}{2}\widehat{\sigma}_x + \frac{\epsilon}{2}\widehat{\sigma}_z \tag{3.19}$$

and having identified physical parameters by which they can be manipulated, it is now time to obtain a detailed understanding of the qubit dynamics. An extremely useful tool to geometrically represent single qubit states and to gain intuition for their evolution under a Hamiltonian is the Bloch sphere.

3.4.1.1 Pure states

Any pure single qubit state can be written as $|\psi\rangle = c_0|0\rangle + c_1|1\rangle$ with complex prefactors c_0 and c_1. Given that normalization requires that $|c_0|^2 + |c_1|^2 = 1$, we can also write it as

$$|\psi\rangle = e^{i\phi_0} \sin\frac{\theta}{2}|0\rangle + e^{i\phi_1}\cos\frac{\theta}{2}|1\rangle = e^{i\phi_0}\left(\sin\frac{\theta}{2}|0\rangle + e^{i(\phi_1-\phi_0)}\cos\frac{\theta}{2}|1\rangle\right). \tag{3.20}$$

We can omit the irrelevant global phase ϕ_0, define $\phi = \phi_1 - \phi_0$ and write

$$|\psi\rangle = \sin\frac{\theta}{2}|0\rangle + e^{i\phi}\cos\frac{\theta}{2}|1\rangle. \tag{3.21}$$

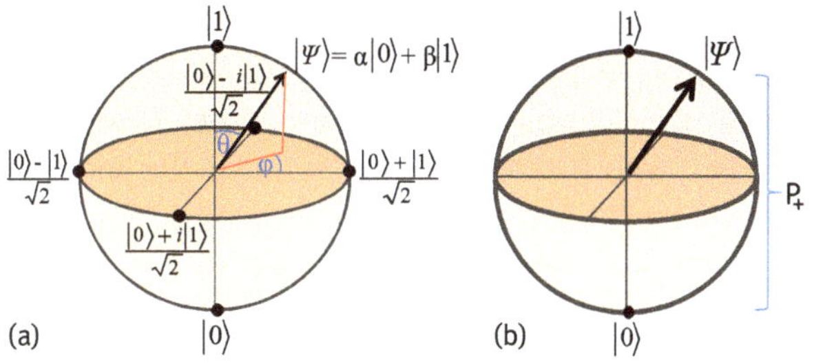

Fig. 3.19: (a) Bloch sphere representation of a quantum state $|\psi\rangle$ with its two angles θ and φ marked. The black points on the sphere are labeled with the states ending at these points. (b) The probability P_+ to measure a 1 in the $|0\rangle$, $|1\rangle$ basis is given by the z-coordinate of the state, scaled so that 0 and 1 correspond to the poles $|0\rangle$ and $|1\rangle$, respectively.

The angles ϕ and θ are uniquely defined for $0 \leq \phi < 2\pi$ and $0 < \theta < \pi$.[9] These angles look like spherical coordinates. In the following, we will show that this interpretation is meaningful. Figure 3.19(a) shows how quantum states can thus be identified with points on the surface of a sphere via eq. (3.21).

To determine the physical meaning of this identification, we define the unit vector

$$\vec{n} = (n_x, n_y, n_z) = (\sin\theta\cos\phi, \sin\theta\sin\phi, \cos\theta)\ . \tag{3.22}$$

An elementary calculation then shows that $|\psi\rangle$ is an eigenstate of the operator $\vec{n}\cdot\vec{\hat{\sigma}} = \sum_{v=x,y,z} n_v\hat{\sigma}_v$, i.e., that

$$\vec{n}\cdot\vec{\hat{\sigma}}|\psi\rangle = |\psi\rangle\ , \tag{3.23}$$

where $\vec{\hat{\sigma}}$ denotes the vector of the three Pauli matrices defined on page 204, which act on the subspace of the qubit. If our qubit is a spin 1/2, then $\vec{n}\cdot\vec{\hat{\sigma}}$ is the operator measuring the spin along the direction $\vec{n}$. The vector $\vec{n}$ is generally called the Bloch vector of $|\psi\rangle$. Hence, our state $|\psi\rangle$ represents a spin pointing along $\vec{n}$, in the sense that it is an eigenstate of the corresponding operator.[10] We thus see that the geometric interpretation is very natural. For two-level systems other than a spin, the direct physical meaning of the axis $\vec{n}$ is lost, but one can still use the same representation and think of $\vec{n}$ as a fictitious axis. It is then common to speak of a pseudospin.

The Bloch sphere picture is further established by the fact that the expectation values of the three spin operators $\hat{\sigma}_x$, $\hat{\sigma}_y$ and $\hat{\sigma}_z$ in the state $|\psi\rangle$ are given by

$$\langle\vec{\hat{\sigma}}\rangle = \langle\psi|\vec{\hat{\sigma}}|\psi\rangle = \vec{n}\ . \tag{3.24}$$

This relation also implies a simple yet very useful geometric meaning of the measurement probabilities. When measuring one of the $\hat{\sigma}_v$-operators ($v = x, y, z$), one can only

9 We omit the discussion of the poles, since it does not provide additional insight.

10 Recall that, contrary to a classical spin, spin measurements along a direction orthogonal to $\vec{n}$ would give a random result, being either 1 or −1. Hence, the measurement projects the spin to $\vec{n}$.

obtain eigenvalues ± 1, whose probabilities we denote by P_+ and $P_- = 1 - P_+$. The expectation value is thus given by $\langle \hat{\sigma}_\nu \rangle = n_\nu = 1 \cdot P_+ + (-1) \cdot P_- = P_+ - P_- = 2P_+ - 1$. Hence, the probabilities associated with measuring $\hat{\sigma}_\nu$ with input state $|\psi\rangle$ depend linearly on its Bloch-vector coordinate n_ν, with $P_\pm = 0, 1$ corresponding to the points at $n_\nu = \pm 1$ on the surface of the Bloch sphere (Fig. 3.19(b)). The same holds for any other measurement axis. For a measurement of $\hat{\sigma}_z$, whose eigenstates are $|0\rangle$ and $|1\rangle$, the result can be read off directly from eq. (3.21), noting that $P_+ = \cos^2 \frac{\theta}{2} = (1 + \cos\theta)/2$.

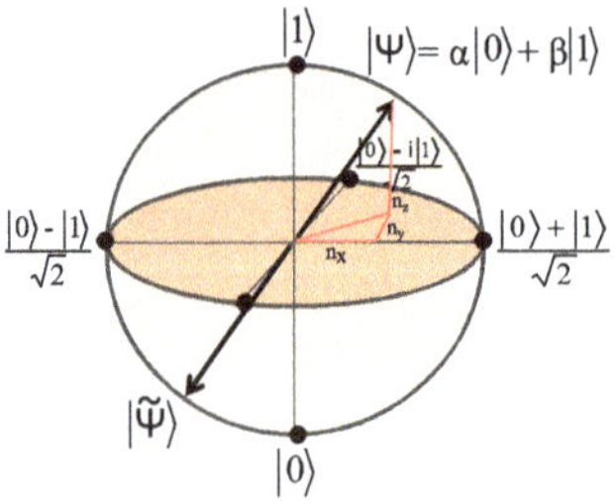

Fig. 3.20: The orthogonal states $|\psi\rangle$ and $|\widetilde{\psi}\rangle$ lie opposite to each other on the Bloch sphere and can be identified as eigenstates of a measurement operator along the axis on which they lie. The coordinates of $|\psi\rangle$ in the basis system of the principal axis of the Bloch sphere are marked in red as n_x, n_y, and n_z.

We have seen that any state $|\psi\rangle$ can be associated with a Bloch vector $\vec{n}$ and a measurement operator $\vec{n} \cdot \vec{\hat{\sigma}} |\psi\rangle$. Conversely, any vector $\vec{n}$ can be used to define two states $|\psi\rangle$ and $|\widetilde{\psi}\rangle$. These states are eigenstates of the operator $\vec{n} \cdot \vec{\hat{\sigma}} |\psi\rangle$ with different eigenvalues and hence orthogonal, i.e., $\langle \widetilde{\psi} | \psi \rangle = 0$. Furthermore, the fact that $\vec{n} \cdot \vec{\hat{\sigma}} |\widetilde{\psi}\rangle = -|\widetilde{\psi}\rangle = (-\vec{n}) \cdot \vec{\hat{\sigma}} |\psi\rangle$ implies that $|\widetilde{\psi}\rangle$ lies opposite to $|\psi\rangle$ on the Bloch sphere (Fig. 3.20).

3.4.1.2 Density matrices

We conclude our discussion of the geometric representation of states by extending it to density matrices. Remember that any pure state $|\psi\rangle$ can also be written as a density matrix $\hat{\rho} = |\psi\rangle\langle\psi|$. In addition, density matrices can also encode arbitrary statistical mixtures of states, in which case they take the form $\hat{\rho} = P_1 |\psi\rangle\langle\psi| + (1 - P_1) |\widetilde{\psi}\rangle\langle\widetilde{\psi}|$, where P_1 and $1 - P_1$ are the probabilities of the qubit being in each of the states. In general, $|\psi\rangle$ and $|\widetilde{\psi}\rangle$ can be arbitrary orthonormal states, and additional states can be added as well.

Since density matrices have to be Hermitian, they are required to have unity trace, and because $\{\hat{1}, \hat{\sigma}_y, \hat{\sigma}_y, \hat{\sigma}_z\}$ is a basis of the space of Hermitian 2×2 operators, any single qubit density matrix can be written in the form

$$\hat{\rho} = \frac{1}{2}\hat{1} + \sum_{\nu = x,y,z} n_\nu \hat{\sigma}_\nu \tag{3.25}$$

with real coefficients $\{n_\nu\}$. A simple calculation shows that

$$\langle \vec{\hat{\sigma}} \rangle = \mathrm{Tr}(\vec{\hat{\sigma}} \hat{\rho}) = \vec{n} \,, \tag{3.26}$$

so that our notation is consistent with the results found for pure states.

If $\widehat{\rho}$ is a statistical mixture of two (or more) other density matrices of the form $\widehat{\rho} = P_1\widehat{\rho}_1 + (1 - P_1)\widehat{\rho}_2$, we furthermore see from the linearity of eq. (3.25) that the Bloch vector corresponding to $\widehat{\rho}$ is a linear combination of the individual ones: $\vec{n} = P_1\vec{n}_1 + (1 - P_1)\vec{n}_2$. Hence, density matrices can correspond to points inside the Bloch sphere, and the probabilities of their constituents $\widehat{\rho}_1$ and $\widehat{\rho}_2$ correspond to coordinates along the directions of the Bloch vector constituents.

Note that there are many possible pure state mixtures that lead to the same non-pure density matrix, so that it is not possible to determine the ensemble of states that lead to a given density matrix without further information. Nevertheless, independent of its physical origin, any single-qubit density matrix can be uniquely[11] written as $\widehat{\rho} = P_1|\psi\rangle\langle\psi| + (1 - P_1)|\widetilde{\psi}\rangle\langle\widetilde{\psi}|$, with $\langle\widetilde{\psi}|\psi\rangle = 0$ so that $|\widetilde{\psi}\rangle$ and $|\psi\rangle$ lie opposite to each other on the Bloch sphere and p is the coordinate along the connecting line traversing the diameter of the sphere.

3.4.2 Single-qubit control

3.4.2.1 Visualization of evolution under a Hamiltonian

The Bloch sphere is not only useful for representing states, but also for visualizing the qubit dynamics. This enables an intuitive understanding how the qubit can be controlled with external parameters. To find out how the state of a qubit evolves, we need to solve the Schrödinger equation:

$$i\hbar\frac{\partial}{\partial t}|\psi(t)\rangle = \widehat{H}|\psi(t)\rangle\ . \tag{3.27}$$

For $\widehat{H} = \frac{\hbar\omega}{2}\widehat{\sigma}_z$, one finds

$$|\psi(t)\rangle = \widehat{U}(t)|\psi(0)\rangle = \begin{pmatrix} e^{-i\omega t/2} & 0 \\ 0 & e^{i\omega t/2} \end{pmatrix}|\psi(0)\rangle\ . \tag{3.28}$$

For $|\psi(0)\rangle$ given by our generic form (3.21), we thus have

$$|\psi(t)\rangle = e^{i\omega t/2}\sin\frac{\theta}{2}|0\rangle + e^{i(\phi-\omega t/2)}\cos\frac{\theta}{2}|1\rangle = e^{i\omega t/2}\left(\sin\frac{\theta}{2}|0\rangle + e^{i(\phi-\omega t)}\cos\frac{\theta}{2}|1\rangle\right). \tag{3.29}$$

Comparing this expression with the Bloch sphere picture (eq. (3.21), Fig. 3.19), we see that our Hamiltonian generates rotations with angular velocity $-\omega$ around the z-axis.

For a spin 1/2 qubit, this result has a simple physical interpretation in terms of the Larmor precession of the spin. In this case, the above Hamiltonian arises from the Zeeman Hamiltonian $\widehat{H} = -\frac{g\mu_B}{2}\vec{B}\cdot\vec{\widehat{\sigma}}$ if the external magnetic field $\vec{B}$ lies along the z-axis. The torque $\vec{\widehat{\tau}} = \vec{\widehat{M}}\times\vec{B}$ between the electron's magnetic moment $\vec{\widehat{M}} = \frac{g\mu_B}{2}\vec{\widehat{\sigma}}$ and $\vec{B}$ then

11 With the exception that the decomposition is not unique for the fully mixed state $\widehat{\rho} = 1/2\widehat{\sigma}_0$.

causes a precession according to $\frac{\hbar}{2}\dot{\vec{\sigma}} = \vec{\tau}$. Since this result should hold independently of our choice of z-axis, it is no surprise that any Hamiltonian

$$\widehat{H} = \frac{\omega}{2}\vec{n}\cdot\vec{\sigma} = \frac{\omega}{2}\begin{pmatrix} n_z & n_x + in_y \\ n_x - in_y & -n_z \end{pmatrix} \tag{3.30}$$

induces a rotation with angular velocity ω around the axis defined by $\vec{n}$. The proof can be carried out by explicit verification using the Schrödinger equation. Since any two-state Hamiltonian can be written as in eq. (3.30) (plus a physically unimportant term proportional to the identity), our result is general and independent of what qubit system we are considering. We thus see that the dynamics arising from any single-qubit Hamiltonian can be identified with the precession of an equivalent spin 1/2 in a fictitious magnetic field whose value can be read off from eq. (3.30). The relationships between rotation axes, states and Hamiltonians are summarized in Fig. 3.21. Operations on single qubits are often speficied by their axis and angle, e.g., as "π-pulse around the x-axis."

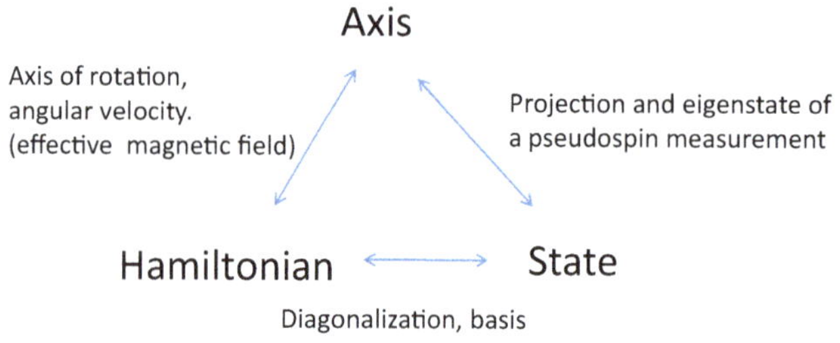

Fig. 3.21: Summary of the relations between axes on the Bloch sphere, Hamiltonians and pure qubit states.

3.4.2.2 Universal single-qubit control

Realizing quantum algorithms requires the ability to generate any desired multi-qubit unitary transformation. One can show that these can be created if any single qubit unitary operation and one particular two-qubit entangling gate, such as a controlled phase gate (CPHASE) or a controlled not gate (CNOT), can be realized.[12] Hence, it is firstly important to be able to create any desired single qubit unitary operation $\widehat{U}$. Since $\widehat{U}$ can be diagonalized and written as a diagonal matrix with phase factors on the diagonal in the eigenbasis, it can be generated by a Hamiltonian that rotates around the axis defined by the eigenstates of $\widehat{U}$. Hence, the ability of switching-on the Hamiltonian given in equation 3.30 with any given $\vec{n}$ for a variable amount of time t is sufficient to gain universal single qubit control. However, by composing a gate operation

12 Matrices describing these gates are given in eqs. (3.42) and (3.41).

of several stages with different axis $\vec{n}$, it is also possible to obtain them without implementing arbitrary values of $\vec{n}$. For example, full control can be achieved by switching between only two different evolution axis $\vec{n}$ for arbitrary durations. In Section 3.4.3, we will present a very effective and widely used specific method that is based on applying an AC signal to one of the control parameters.

3.4.2.3 Application to two-electron spin qubit

As an example, we present the operation of two-electron spin qubits in double-quantum dots in the Bloch sphere picture. As discussed in Section 3.3.2.2, the computational space of these qubits is spanned by the the states $\{|\uparrow\downarrow\rangle, |\downarrow\uparrow\rangle\}$ or equivalently by $|T_0\rangle = \frac{1}{\sqrt{2}}(|\uparrow\downarrow\rangle + |\downarrow\uparrow\rangle)$ and $|S\rangle = \frac{1}{\sqrt{2}}(|\uparrow\downarrow\rangle - |\downarrow\uparrow\rangle)$. Different conventions have been used to place these on the Bloch sphere. Here, we use that shown in Fig. 3.22, with $\{|\uparrow\downarrow\rangle, |\downarrow\uparrow\rangle\}$ on the north and south poles, respectively. $|S\rangle$ and $|T_0\rangle$ then lie on the x-axis. In Section 3.3.2.2, we have seen that the effective exchange interaction described by $J(\epsilon)$ leads to an energy splitting between $|S\rangle$ and $|T_0\rangle$. Hence, the associated Hamiltonian is $\widehat{H} = \frac{J(\epsilon)}{2}\widehat{\sigma}_x$ and generates a rotation around the x-axis. If on the other hand the two separated electrons experience a difference in the local magnetic field (ΔB), the corresponding eigenstates at $J(\epsilon) = 0$ will be $\{|\uparrow\downarrow\rangle, |\downarrow\uparrow\rangle\}$, and their energy splitting is $g\mu_B\Delta B$. Hence, the total Hamiltonian is given by

$$\widehat{H} = \frac{g\mu_B\Delta B}{2}\widehat{\sigma}_z + \frac{J(\epsilon)}{2}\widehat{\sigma}_x\,. \tag{3.31}$$

It is geometrically obvious from Fig. 3.22, that $\widehat{\sigma}_x$ and $\widehat{\sigma}_z$ are sufficient to reach any point on the Bloch sphere such that a $\widehat{\sigma}_y$ term is not required.

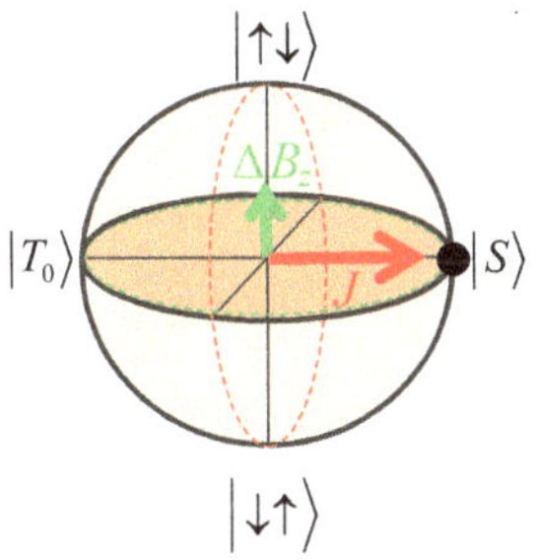

Fig. 3.22: Bloch sphere representation for a S–T_0 qubit.

Since it is difficult to realize sufficiently strong magnetic field gradients that can be varied on the nanosecond time-scale relevant for qubit operation, ΔB is practically fixed. $J(\epsilon)$ on the other hand can be varied very quickly via gate voltage pulses affecting ϵ, which is sufficient to gain universal control. For example, rotations around z are obtained at $J(\epsilon) = 0$, while x-rotations are obtained approximately for $J(\epsilon) \gg \Delta B$. Basic qubit operation can be realized by switching between those two regimes. For example, a y-rotation of $\pi/2$ is obtained from a $\pi/2$ x-rotation followed by a $\pi/2$ rotation around

z. Since this construction is subject to various systematic errors, e.g., limitations on $J(\epsilon)$ and finite pulse rise-times, more sophisticated pulse shaping is typically needed in practice for accurate control.

As we have seen in Section 3.3, several other types of qubits are described by the same Hamiltonian. Hence, the same control procedures could be applied to those other qubit systems, which indeed has been realized in some cases.

3.4.3 Rabi flopping

As we have argued in Section 3.4.2.1 based on geometric intuition, switching between any two distinct rotation axis can be used to gain universal single qubit control. For large angles between these axis, only a few rotations need to be applied in sequence. However, generating such large excursions of the control parameters with the required accuracy is often not possible or inconvenient. We thus consider the very common case where the Hamiltonian only has a small time dependent part $\widehat{H}_1(t)$, i.e.,

$$\widehat{H}(t) = \widehat{H}_0 + \widehat{H}_1(t)\,. \tag{3.32}$$

If our goal is, for example, to implement a π-pulse around the x-axis that exchanges the $|0\rangle$ and $|1\rangle$ states, we clearly cannot achieve this with a few changes of $\widehat{H}_1$ because the instantaneous rotation axis will always remain close to the z-axis (Fig. 3.23(a)). However, we will show that many oscillations of $\widehat{H}_1$ can have the desired effect.

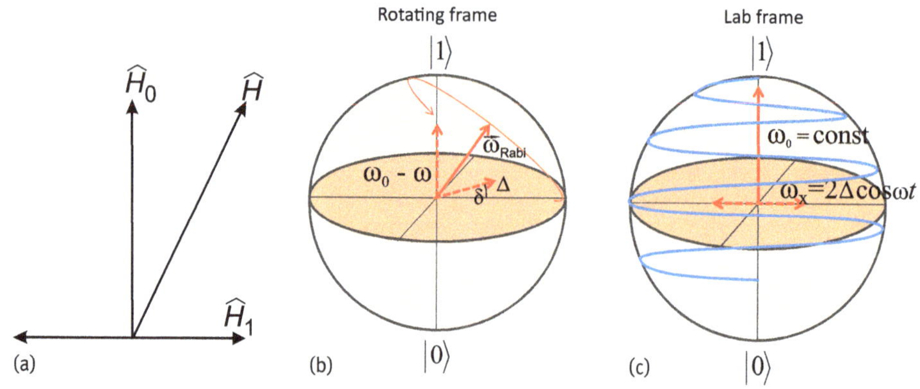

Fig. 3.23: (a) Decomposition of the Rabi Hamiltonian $\widehat{H}$ into a static part $\widehat{H}_0$ and a time-dependent part $\widehat{H}_1$. (b) Evolutionary trace in the rotating frame, where the axis of the vector $\vec{\omega}_{\text{Rabi}}$ can be tuned externally via δ, Δ_1 and $\omega - \omega_0$. (c) Corresponding evolution in the lab frame.

Since high-frequency signals are usually most easily available in sinusoidal form, we assume that

$$\widehat{H}_0 = \frac{\hbar\omega_0}{2}\widehat{\sigma}_z$$
$$\widehat{H}_1(t) = \hbar\Delta_1 \sin(\omega t + \delta)\widehat{\sigma}_x\,, \tag{3.33}$$

with $\Delta_1 \ll \omega \approx \omega_0$. Note that this Hamiltonian as well as a closely related solution was already discussed in Section 2.2.3.1 (eq. (2.75)) in the context of light-driven transitions between atomic states.

Solving the Schrödinger equation is now less straightforward because of the time dependence of $\widehat{H}(t)$. To deal with the problem, we transform into a rotating frame by defining

$$|\widetilde{\psi}(t)\rangle = \widehat{U}^\dagger(t)|\psi(t)\rangle \quad \text{with} \quad \widehat{U}(t) = \begin{pmatrix} e^{-i\omega t/2} & 0 \\ 0 & e^{i\omega t/2} \end{pmatrix}. \tag{3.34}$$

$|\psi(t)\rangle$ is the qubit state in the lab frame and $|\widetilde{\psi}(t)\rangle$ is the same state in the frame rotating around the z-axis with the same angular frequency as the drive signal. Straightforward algebra using the matrix representation of $\widehat{\sigma}_x$ and $\widehat{\sigma}_z$ (page 204) shows that the Schrödinger equation is now given by

$$i\hbar\frac{\partial}{\partial t}|\widetilde{\psi}(t)\rangle = (\widehat{H}_0 + \widehat{U}^\dagger(t)\widehat{H}_1(t)\widehat{U}(t))|\widetilde{\psi}(t)\rangle \tag{3.35}$$

$$= \left[\frac{\hbar(\omega_0 - \omega)}{2}\widehat{\sigma}_z + \frac{\hbar\Delta_1}{2}\begin{pmatrix} 0 & e^{-i\omega t} \\ e^{i\omega t} & 0 \end{pmatrix} \cdot \left(e^{i\omega t+\delta} + e^{-i\omega t-\delta}\right)\right]|\widetilde{\psi}(t)\rangle \tag{3.36}$$

$$= \left[\frac{\hbar(\omega_0 - \omega)}{2}\widehat{\sigma}_z + \frac{\hbar\Delta_1}{2}\cos\delta\,\widehat{\sigma}_x + \frac{\hbar\Delta_1}{2}\sin\delta\,\widehat{\sigma}_y + \frac{\hbar\Delta_1}{2}\begin{pmatrix} 0 & e^{-i2\omega t-\delta} \\ e^{i2\omega t+\delta} & 0 \end{pmatrix}\right]|\widetilde{\psi}(t)\rangle \tag{3.37}$$

$$\approx \left[\frac{\hbar(\omega_0 - \omega)}{2}\widehat{\sigma}_z + \frac{\hbar\Delta_1}{2}\cos\delta\,\widehat{\sigma}_x + \frac{\hbar\Delta_1}{2}\sin\delta\,\widehat{\sigma}_y\right]|\widetilde{\psi}(t)\rangle\,. \tag{3.38}$$

The last term in equation (3.37) has no effect, since $2\omega \gg |\omega_0 - \omega|, \Delta_1$ and it hence oscillates much faster than the qubit can react and averages to zero over several cycles. This very widespread approximation is called the rotating wave approximation (RWA) – it would be exact if an out-of-phase $\widehat{\sigma}_y$ term were included in the Hamiltonian $\widehat{H}_1$ of eq. (3.33), which would correspond to rotating the $\widehat{H}_1$ axis around the z-axis. The remaining terms (eq. (3.38)) have become time independent and describe a rotation around the axis $\omega_{\text{Rabi}}\vec{n} = (\Delta_1\cos\delta, \Delta_1\sin\delta, \omega_0 - \omega)$ with angular velocity $\omega_{\text{Rabi}} = \sqrt{\Delta_1^2 + (\omega_0 - \omega)^2}$. We see that by varying Δ_1, ω, and δ, which are all properties of the AC signal described by $\widehat{H}_1$, we can generate arbitrary rotations in the rotating frame. The direction of the rotation axis in the xy-plane can be tuned elegantly via the phase

δ of the excitation along $\hat{\sigma}_x$. In principle, we still have to transform back into the lab frame. However, since this transformation is the identity (up to a global phase factor of -1) whenever the fast angular rotation is by a multiple of 2π, it has no important effect. Fig. 3.23 depicts the evolution of a state starting at $|1\rangle$ in both the rotating and the stationary lab frame.

In the lab frame, this result can be understood intuitively as follows. If $\omega = \omega_0$ and $\delta = 0$, the AC excitation remains synchronized with the precession of the Bloch vector under $\hat{H}_0$. Whenever the pseudospin points to the right (left), it sees an additional effective magnetic field pointing to the right (left) that results from $\hat{H}_1(t)$. Hence, the oscillating magnetic field always rotates the (originally upwards-pointing) Bloch vector in a downwards direction until it reaches the south pole after many periods. At that point the phases are reversed and the Bloch vector is rotated back up again. In the rotating frame, this evolution is simply a rotation around the x-axis. This process is called Rabi flopping or Rabi oscillations. The time it takes to complete one rotation is inversely proportionally the excitation amplitude Δ_1. A rotation is possible as long as the phases of the precession and the excitation remain synchronized, i.e., for $|\omega - \omega_0| \ll \Delta_1$. On the other hand, synchronization is quickly lost for $|\omega - \omega_0| \gg \Delta_1$ and no Rabi flopping occurs. This resonance condition for ω with respect to ω_0 reflects the conservation of energy quantum mechanically. If the photon energy of the drive signal is not matched to the qubit level splitting, no transition occurs.

3.4.3.1 Examples

An intuitive example is again a spin ½ electron in a constant magnetic field B_0 along the z-direction driven by an AC-field along x. The corresponding Hamiltonian is

$$\hat{H} = \frac{g\mu_B B_0}{2}\hat{\sigma}_z + \frac{g\mu_B}{2}B_1(t)\hat{\sigma}_x \tag{3.39}$$

with $B_1(t) = B_1 \sin \omega t$. This control technique has been employed for decades in electron spin resonance (ESR), also called electron paramagnetic resonance (EPR), and in nuclear magnetic resonance (NMR) experiments.[13] It was demonstrated that it can also be used to manipulate an individual spin in a quantum dot. Figure 3.24a shows the experimental device. A short wire connected to an impedance-matched coplanar slot line (CPS) is placed above the quantum dot and a GHz AC electric current I_{CPS} is driven through it in order to generate $B_1(t)$. This oscillates the spin between the state $|\uparrow\rangle$ and $|\downarrow\rangle$ at a frequency $f_{Rabi} = g\mu_B B_1/h$.

After interchanging the axis of $\hat{\sigma}_x$ and $\hat{\sigma}_z$, without loss of generality, our Hamiltonian reads:

$$H = \frac{\Delta}{2}\hat{\sigma}_x + \frac{\epsilon(t)}{2}\hat{\sigma}_z\,. \tag{3.40}$$

13 For NMR experiments the Bohr magneton μ_B has to be replaced by the smaller nuclear magneton μ_N.

This is exactly the same equation as already derived for charge qubits (eq. (3.5)) and superconducting flux qubits (eq. (3.18)). Hence, Rabi oscillations are also induced by resonant AC-modulation of the detuning $\epsilon(t) \propto \sin \omega t$ (Fig. 3.24b, bottom) and can thus be used to drive charge qubits or superconducting qubits. Remember that for flux qubits, this requires an oscillating magnetic flux, i.e., an AC current, while for charge qubits, ϵ is controlled via an oscillating gate voltage. Both can be changed rather rapidly, such that the resonance condition $\omega = \omega_0$ can be fulfilled.

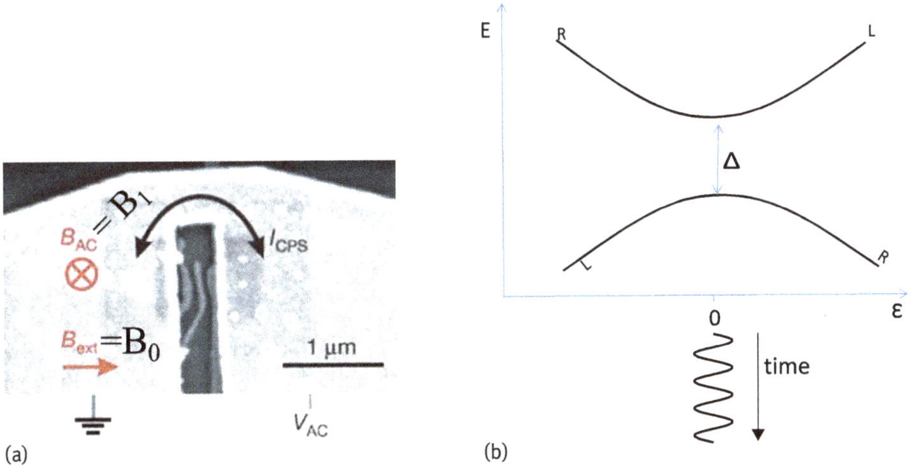

Fig. 3.24: (a) An electron spin in a quantum dot can be manipulated by an AC magnetic field $B_{AC} = B_1$ perpendicular to the external magnetic field $B_{ext} = B_0$, which sets the spin splitting energy (eq. (3.39)). The AC field is generated by sending a high frequency current I_{CPS} through a wire (white large structure encircling three visible gate fingers of the quantum dot). (b) Energy diagram of a two level qubit, e.g., of a the spin qubit in (a), a charge qubit, or a flux qubit. The driving Rabi oscillation is marked below, which corresponds to a sinusoidal variation of the detuning $\epsilon(t)$ around the degeneracy point $\epsilon = 0$. (a) [47].

3.4.4 Two-qubit operations

So far, we have discussed how to manipulate individual qubits. While important, this is not sufficient to carry out computations, which necessarily involve operations that combine the information stored in several qubits. As already mentioned in Section 3.4.2.2, it is sufficient and necessary to have single and two-qubit operations at one's disposal. Since two-qubit gates act on the four-dimensional Hilbert space of two qubits, they are represented by unitary 4×4 matrices. Common and convenient two-qubit gates are CNOT, CPHASE, $\sqrt{\text{SWAP}}$ and SWAP, whose matrices in the basis

$|00\rangle, |01\rangle, |10\rangle, |11\rangle$ are:

$$\widehat{U}_{\mathrm{CPHASE}} = \begin{pmatrix} 1 & 0 & 0 & 0 \\ 0 & 1 & 0 & 0 \\ 0 & 0 & 1 & 0 \\ 0 & 0 & 0 & -1 \end{pmatrix} \tag{3.41}$$

$$\widehat{U}_{\mathrm{CNOT}} = \begin{pmatrix} 1 & 0 & 0 & 0 \\ 0 & 1 & 0 & 0 \\ 0 & 0 & 0 & 1 \\ 0 & 0 & 1 & 0 \end{pmatrix} \tag{3.42}$$

$$\widehat{U}_{\sqrt{\mathrm{SWAP}}} = \begin{pmatrix} 1 & 0 & 0 & 0 \\ 0 & \frac{1+i}{2} & \frac{1-i}{2} & 0 \\ 0 & \frac{1-i}{2} & \frac{1+i}{2} & 0 \\ 0 & 0 & 0 & 1 \end{pmatrix} \tag{3.43}$$

$$\widehat{U}_{\mathrm{SWAP}} = \begin{pmatrix} 1 & 0 & 0 & 0 \\ 0 & 0 & 1 & 0 \\ 0 & 1 & 0 & 0 \\ 0 & 0 & 0 & 1 \end{pmatrix} . \tag{3.44}$$

For understanding CPHASE and CNOT, it is useful to consider the first qubit as a control qubit and the second qubit as a target qubit. CPHASE imprints a phase factor of -1 on the $|1\rangle$ state of the target qubit if the control qubit is in the $|1\rangle$ state. If the control qubit is in the $|0\rangle$ state, the target qubit is unchanged. The CNOT-gate works analogously, except that the target qubit is inverted instead of experiencing a phase flip. Since phase and bit flips correspond to π-rotations around different axis of the Bloch sphere, CNOT and CPHASE gates can be mapped onto each other with single qubit pulses and are thus equivalent. The SWAP gate exchanges the states of the two qubits, and $\sqrt{\mathrm{SWAP}}$ half-way exchanges them, i.e., $\widehat{U}^2_{\sqrt{\mathrm{SWAP}}} = \widehat{U}_{\mathrm{SWAP}}$. Calculations analogous to those to be presented in Section 3.5.1.1 show that CPHASE, CNOT and $\sqrt{\mathrm{SWAP}}$ can generate maximally entangled states. Furthermore, a combination of several $\sqrt{\mathrm{SWAP}}$ gates can produce a CNOT gate and vice versa. The SWAP gate in contrast cannot entangle, but is useful for moving quantum states through a multi-qubit circuit.

As an example for how to generate a two-qubit gate, consider the interaction Hamiltonian

$$\widehat{H}_{zz} = \hbar\lambda_{\mathrm{I}}(\widehat{\sigma}_z \otimes \widehat{\sigma}_z) , \tag{3.45}$$

where λ_{I} is a coupling constant. The corresponding matrix form is

$$\widehat{H}_{zz} = \hbar\lambda_{\mathrm{I}} \begin{pmatrix} 1 & 0 & 0 & 0 \\ 0 & -1 & 0 & 0 \\ 0 & 0 & -1 & 0 \\ 0 & 0 & 0 & 1 \end{pmatrix} . \tag{3.46}$$

As it is already diagonal, the exponentiation to obtain the resulting unitary matrix is trivial:

$$\widehat{U}_{zz} = e^{-\frac{i}{\hbar}\widehat{H}_{zz}t} = \begin{pmatrix} e^{-i\lambda_I t} & 0 & 0 & 0 \\ 0 & e^{i\lambda_I t} & 0 & 0 \\ 0 & 0 & e^{i\lambda_I t} & 0 \\ 0 & 0 & 0 & e^{-i\lambda_I t} \end{pmatrix} . \tag{3.47}$$

To turn this matrix into a CPHASE gate via the Hamiltonian $\widehat{H}_{\mathrm{CPHASE}}$, one needs to add single qubit phase gates and a global phase factor. Straightforward algebra shows that the desired result (eq. (3.41)) is obtained by

$$\widehat{H}_{\mathrm{CPHASE}} = \widehat{H}_{zz} + \hbar\lambda_I \left[(\hat{1} \otimes \hat{1}) - \widehat{\sigma}_z\right] \otimes \hat{1} - \hat{1} \otimes \widehat{\sigma}_z$$

and choosing $\lambda_I t = \pi/4$. Since all terms in the Hamiltonian commute, the corresponding operations can be applied sequentially. The total Hamiltonian then reads

$$\widehat{H}_{\mathrm{CPHASE}} = \hbar\lambda_I(\widehat{\sigma}_z - \hat{1}) \otimes (\widehat{\sigma}_z - \hat{1}) \tag{3.48}$$

$$= \hbar\lambda_I \begin{pmatrix} 0 & 0 \\ 0 & 2 \end{pmatrix} \otimes \begin{pmatrix} 0 & 0 \\ 0 & 2 \end{pmatrix} \tag{3.49}$$

$$= \hbar\lambda_I \begin{pmatrix} 0 & 0 & 0 & 0 \\ 0 & 0 & 0 & 0 \\ 0 & 0 & 0 & 0 \\ 0 & 0 & 0 & 4 \end{pmatrix} , \tag{3.50}$$

such that we obtain the desired $\widehat{U}_{\mathrm{CPHASE}} = e^{-\frac{i}{\hbar}\widehat{H}_{\mathrm{CPHASE}}t}$. It has a simple physical interpretation: a phase change of $4\lambda_I t = \pi$ is accumulated only by the state with both qubits in the $|1\rangle$ state.

A physical implementation of this Hamiltonian can be achieved, for example, for two electrostatically coupled charge or singlet-triplet qubits (Fig. 3.25(a)), or for two inductively coupled flux qubits (Fig. 3.14(c)). Although the configuration is symmetric with respect to the roles of the two qubits, it is useful to think of one qubit as a control and the other as a target qubit. For singlet-triplet qubits, the charge distribution of the control qubit depends on its state, i.e., it is different for the singlet and the triplet state. Electric coupling to the target qubit leads to a change of its effective detuning by an amount $\pm\Delta\epsilon$, with the sign depending on the state of the control qubit.[14] This detuning leads to a change in the level splitting of the target qubit E_{01} that is given, to linear order, by $\pm\frac{dE_{01}}{d\epsilon}\Delta\epsilon$. This mechanism is illustrated in Fig. 3.25(b) using the energy diagram of a singlet-triplet qubit (Fig. 3.9(c)). Consequently, the $\widehat{\sigma}_z$ term for the control qubit in eq. (3.45) reflects the dependence of the detuning change on its state, while the $\widehat{\sigma}_z$ term of the target qubit represents the resulting change in level splitting at the target qubit. The combination of both represents the Hamiltonian $\widehat{H}_{zz}$.

14 For flux qubits, the same considerations apply after replacing the charge distribution by a circulating current and the electric coupling between the qubits by a magnetic coupling.

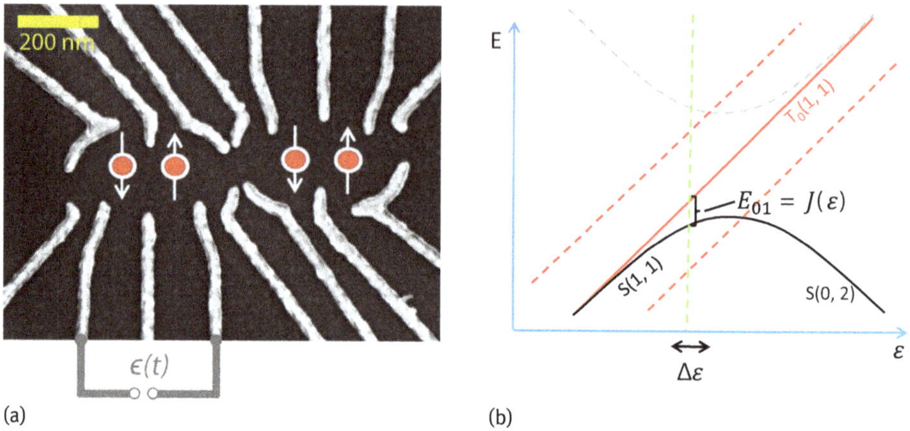

Fig. 3.25: (a) SEM image of two adjacent double quantum dots realizing two charge-coupled singlet-triplet qubits. (b) Illustration of the two qubit coupling mechanism using an energy diagram of a singlet-triplet qubit. The different slopes of the S and T-branch at the operating point (vertical dashed line) indicate that they have a different charge density (Section 3.3.1.5). The resulting state-dependent electric field of the control qubit induces a shift $\Delta\epsilon$ on the detuning of the target qubit, such that the energy splitting of the computational states of the target qubit depends on the state of the control qubit.

A similar calculation shows that a SWAP and $\sqrt{\text{SWAP}}$ gate can be generated by the exchange interaction between two spins. The physics is the same as that of the two-electron spin qubit discussed in Section 3.3.2.2, except that now each spin represents a single qubit.

A maximally entangled state can be generated via $\widehat{U}_{\text{CPHASE}}$ if both qubits are initially in a superposition state:

$$\begin{aligned}
\widehat{U}_{\text{CPHASE}}\frac{1}{2}(|0\rangle + |1\rangle) \otimes (|0\rangle + |1\rangle) \\
&= \widehat{U}_{\text{CPHASE}}\frac{1}{2}(|00\rangle + |10\rangle + |01\rangle + |11\rangle) \\
&= \frac{1}{2}(|00\rangle + |10\rangle + |01\rangle - |11\rangle) \\
&= \frac{1}{2}(|0\rangle \otimes (|0\rangle + |1\rangle) + |1\rangle \otimes (|0\rangle - |1\rangle)) \,. \qquad (3.51)
\end{aligned}$$

While such two-qubit gates and many other implementations have been demonstrated, it is generally harder to achieve a high control accuracy for two-qubit operations than for single-qubit operations. In the latter case, externally generated fields are used to control a qubit. These fields can be made relatively large, which leads to fast operations that can be completed before substantial decoherence that would cause errors. For two-qubit gates on the other hand, the dynamics arise from the often weak coupling between two qubits, e.g., via electric or magnetic stray fields of one qubit at the position of the other qubit. This makes the gates slow and thus sensitive to noise.

If the coupling is enhanced, the individual qubits also tend to become more sensitive to noise from their environment, so that little is gained.

3.5 Decoherence

3.5.1 Basic concepts of decoherence

3.5.1.1 Quantum mechanical treatment of decoherence

Probably the biggest challenge of quantum information processing is maintaining the phase information of a quantum system, which is a central ingredient to the overall concept. It is thus of great practical relevance to understand how phase information is lost, and how we can describe this so called decoherence process. Addressing this question is also of fundamental interest, as it can teach us how our classical experience of phase-free objects emerges from the underlying laws of quantum mechanics.

While a quantum mechanical measurement process inevitably introduces some randomness, the Schrödinger equation tells us that a closed system (e.g., an isolated qubit) evolves deterministically. Since the unitary time evolution operator can, in principle, always be inverted, no information contained in the initial state is lost. In other words, no decoherence occurs. One may thus correctly guess that decoherence emerges from the interaction of the system with its environment, which is often referred to as a bath. A specific example would be an electron spin coupled to a bath of nuclear spins forming the immediate environment, that is most important for decoherence.

To obtain a complete picture of the decoherence process, one should treat the full system consisting of qubit and bath quantum mechanically. The corresponding Hamiltonian can be written in the form

$$\widehat{H} = \widehat{H}_{\mathrm{Q}} + \widehat{H}_{\mathrm{B}} + \widehat{H}_{\mathrm{QB}} , \tag{3.52}$$

where $\widehat{H}_{\mathrm{Q}}$ and $\widehat{H}_{\mathrm{B}}$ describe the independent dynamics of the qubit and the bath, respectively, and $\widehat{H}_{\mathrm{QB}}$ accounts for the interaction between them. Although in general not much can be said about the form of $\widehat{H}_{\mathrm{QB}}$, it is common (and usually justified) to assume a simple form, for example, some product of a bath and a qubit operator. It is instructive and common to consider an uncorrelated initial state $|\psi\rangle = |\psi_{\mathrm{Q}}\rangle \otimes |\psi_{\mathrm{B}}\rangle$ which is a product state of a bath state $|\psi_{\mathrm{B}}\rangle$ and a qubit state $|\psi_{\mathrm{Q}}\rangle$. After a sufficiently long evolution time, such a Hamiltonian will entangle the two systems. To understand this process and how it leads to decoherence when the degrees of freedom of the bath are removed from the quantum mechanical description (traced out), we consider a toy model in which the bath consists only of a single qubit and the coupling Hamiltonian is given by

$$\widehat{H}_{\mathrm{QB}} = \frac{\hbar\lambda_{\mathrm{I}}}{4}\widehat{\sigma}_x^{\mathrm{Q}} \otimes (\widehat{\sigma}_z^{\mathrm{B}} + \widehat{1}) . \tag{3.53}$$

For simplicity, we assume $\widehat{H}_\mathrm{Q} = \widehat{H}_\mathrm{B} = 0$. The effect of this two-qubit Hamiltonian, which is very similar to the one considered in Section 3.4.4, can be understood as follows. If $|\psi_\mathrm{B}\rangle = |1\rangle$, i.e., the +1 eigenstate of $\widehat{\sigma}_z^\mathrm{B}$, then $(\widehat{\sigma}_z^\mathrm{B} + \widehat{1})|\psi_\mathrm{B}\rangle = 2|\psi_\mathrm{B}\rangle$. The bath is thus stationary under $\widehat{H}_\mathrm{QB}$ while the qubit evolves under $\frac{\hbar\lambda_\mathrm{I}}{2}\widehat{\sigma}_x^\mathrm{Q}$. If on the other hand $|\psi_\mathrm{B}\rangle = |0\rangle$, then $(\widehat{\sigma}_z^\mathrm{B} + \widehat{1})|\psi_\mathrm{B}\rangle = 0$ and both parts remain stationary. If we choose the total interaction time, at which $\widehat{H}_\mathrm{QB}$ is active, as τ_int such that $\lambda_\mathrm{I}\tau_\mathrm{int} = \pi$, we see that the qubit is inverted if and only if the bath is in the $|1\rangle$ state. This operation is just the CNOT gate. Formally, we have

$$e^{-\frac{i}{\hbar}\widehat{H}_\mathrm{QB}\frac{\pi}{\lambda_\mathrm{I}}} = \widehat{U}_\mathrm{CNOT}\,.$$

The factor π/λ_I stems from the evolution time required to complete the gate. Note that our ability to determine the action of this gate mostly by inspection and without algebra nicely demonstrates the intuitive power gained by the Bloch sphere representation that we have developed in Section 3.4.

So far, we have considered the basis states in which our gate is diagonal. On these states, its action is indistinguishable from a classical operation. The result becomes more interesting if we consider a superposition state as the initial state of the bath, e.g., $|\psi_\mathrm{B}\rangle = \frac{1}{\sqrt{2}}(|0\rangle + |1\rangle)$. One then finds that, still using $|\psi_\mathrm{Q}\rangle = |0\rangle$,

$$\begin{aligned}\widehat{U}_\mathrm{CNOT}(|\psi_\mathrm{Q}\rangle \otimes |\psi_\mathrm{B}\rangle) &= \frac{1}{\sqrt{2}}(\widehat{U}_\mathrm{CNOT}|0\rangle \otimes |0\rangle + \widehat{U}_\mathrm{CNOT}|0\rangle \otimes |1\rangle)\\ &= \frac{1}{\sqrt{2}}(|0\rangle \otimes |0\rangle + |1\rangle \otimes |1\rangle)\\ &= |\psi_\mathrm{E}\rangle\,,\end{aligned}$$

where $|\psi_\mathrm{E}\rangle$ is simply an abbreviation for the final state, which turns out to be maximally entangled. Each product of single qubit eigenstates has the same weight and factorization into a product state is not possible. The initial superposition state leads to a final entangled state via the CNOT operation. Each of its components corresponds to the results of applying the gate to the two input basis states separately. As a result, the initially uncorrelated qubits are now coherently correlated or entangled.

While for our single-qubit toy bath, it might be possible to preserve its state and disentangle it from the qubit at a later point via another interaction, this option is usually not available for a real environment with many uncontrollable degrees of freedom. Hence, it is natural to disregard the environment after the end of the interaction and ask about the state of the qubit. Formally, this means that we have to trace out the degrees of freedom of the bath. The resulting qubit density matrix is

$$\begin{aligned}\widehat{\rho}_\mathrm{Q} = \mathrm{Tr}_\mathrm{B}(\widehat{\rho}_\mathrm{QB}) &= \sum_{n=0,1} {}_\mathrm{B}\langle n|\psi_\mathrm{E}\rangle\langle\psi_\mathrm{E}|n\rangle_\mathrm{B}\\ &= \frac{1}{2}(|0\rangle\langle 0|_\mathrm{Q} + |1\rangle\langle 1|_\mathrm{Q})\end{aligned} \tag{3.54}$$

with $|n\rangle_B$ being the eigenstates of the bath. Hence, the interaction with the bath has turned the initially pure qubit state into a fully mixed state, implying that all information originally contained in the qubit has been lost.

Based on this model calculation, we can now propose an answer to why we never encounter cats in a superposition of states. Clearly, the dead-or-alive degree of freedom strongly interacts with many other degrees of freedom both of the cat itself and its environment. This uncontrolled interaction would very quickly turn a superposition of a living and a dead cat into a mixed state, which is equivalent to a description in terms of classical statistics. Speaking of a cat in a box being in a mixed state is equivalent to describing it with a probability distribution reflecting that we simply do not yet know if it has died.

This explanation of the quantum-to-classical transition is consistent with all current experimental findings. Nevertheless, as mentioned in Section 3.3.3, some researchers speculate that some new physics beyond quantum mechanics may be found when probing larger objects. Hence, one should examine increasingly large systems in detail in order to test whether a violation of the dynamics guided by the Schrödinger equation can be found.

3.5.1.2 Classical picture

Although the quantum mechanical picture discussed in the previous section can be applied to practically relevant baths such as phonons, photons or nuclear spins, it is simpler and often sufficient to treat the bath as a classical noise process. For our toy model of a single qubit bath, we can study this correspondence by assuming that the bath qubit starts out in a mixed initial state such as $\frac{1}{2}(|0\rangle\langle 0|_B + |1\rangle\langle 1|_B)$ instead of a coherent superposition state. The bath then behaves like a classical random variable controlling whether the qubit is flipped or not. A straightforward calculation, analogous to the one from the previous section, yields the same final state as eq. (3.54).

We now generalize this classical picture to the more realistic scenario of a qubit whose level splitting depends on a classical noise bath described by a time-dependent random variable $\beta(t)$. The corresponding Hamiltonian is the so called pure dephasing Hamiltonian

$$\widehat{H} = \frac{\hbar}{2}(\omega_0 + \beta(t))\widehat{\sigma}_z\,, \tag{3.55}$$

where ω_0 is a fixed level splitting in the absence of noise. The noise thus randomly changes the energy splitting between the two qubit states. The name of the Hamiltonian arises from the fact that it only affects the phase of the qubit, but not the occupation probabilities. The corresponding time evolution operator is given by

$$\widehat{U}_\beta(t) = \begin{pmatrix} e^{-i\phi(t)/2} & 0 \\ 0 & e^{i\phi(t)/2} \end{pmatrix}, \tag{3.56}$$

where the random rotation angle $\phi(t)$ is related to $\beta(t)$ via

$$\phi(t) = \omega_0 t + \int_0^t \mathrm{d}t' \, \beta(t') \,. \tag{3.57}$$

This unitary evolution is only meaningful if we consider a specific realization of $\beta(t)$. For example, one might imagine that $\beta(t)$ is measured while the qubit evolves and is thus known. Although such a measurement will not be possible in most practical cases for lack of a suitable sensor, there is nothing wrong with this notion as a thought experiment, since we assume $\beta(t)$ to be a classical variable, which always has a well defined value. For such a particular realization of $\beta(t)$, indicated by the subscript β below, an intial density matrix

$$\hat{\rho}(t=0) = \begin{pmatrix} \hat{\rho}_{11} & \hat{\rho}_{10} \\ \hat{\rho}_{01} & \hat{\rho}_{00} \end{pmatrix} \tag{3.58}$$

evolves into

$$\hat{\rho}_\beta(t) = \hat{U}_\beta(t)\hat{\rho}(t=0)\hat{U}^\dagger_\beta(t) = \begin{pmatrix} \hat{\rho}_{11} & \hat{\rho}_{10}\mathrm{e}^{-\mathrm{i}\phi(t)} \\ \hat{\rho}_{01}\mathrm{e}^{\mathrm{i}\phi(t)} & \hat{\rho}_{00} \end{pmatrix} . \tag{3.59}$$

In practice, we will not have access to each trajectory of $\beta(t)$ and thus have to treat it as a random variable. To account for this randomness, we need to average over all possible traces of $\beta(t)$. We thus obtain an expectation value with respect to the statistical averaging of β, denoted by $\langle\cdot\rangle_\beta$.

Note that there are several equivalent interpretations of the statistical distribution of $\beta(t)$. Mathematically, we may interpret them as a way to express our incomplete knowledge about β. In experiments, this concept is only meaningful if we can average over many realizations of β. Such averaging can occur by probing many identical systems (ensemble averaging), such as an ensemble of a certain spin species measured in an NMR experiment, which typically involves a macroscopic number of molecules. Alternatively, one can repeat a measurement of a single qubit many times (time averaging). The latter point of view is generally more relevant for solid state qubits, which are not easily measured in large numbers.

Carrying out the average, we obtain

$$\hat{\rho}(t) = \langle\hat{\rho}_\beta(t)\rangle_\beta = \begin{pmatrix} \hat{\rho}_{11} & \hat{\rho}_{10}\langle\mathrm{e}^{-\mathrm{i}\phi(t)}\rangle_\beta \\ \hat{\rho}_{01}\langle\mathrm{e}^{\mathrm{i}\phi(t)}\rangle_\beta & \hat{\rho}_{00} \end{pmatrix} . \tag{3.60}$$

For Gaussian distributions of $\phi(t)$, and thus of $\beta(t)$, which exhibits a variance $\langle\delta\phi(t)^2\rangle$ and a mean $\langle\phi(t)\rangle$, one can show that[15]

$$\langle\mathrm{e}^{\mathrm{i}\phi(t)}\rangle = e^{-\langle\delta\phi(t)^2\rangle/2} \cdot \mathrm{e}^{\mathrm{i}\langle\phi(t)\rangle} \,. \tag{3.61}$$

15 For example, one can use the power series of the exponential and compute the expectation value of each term, or directly evaluate the integral over $\phi(t)$ for the expectation value.

For $\phi(t)$ as given by eq. (3.57) and assuming, without loss of generality, that $\langle\beta\rangle = 0$, we have

$$\langle\delta\phi^2(t)\rangle_\beta = \left\langle\left(\int_0^t \mathrm{d}t'\ \beta(t')\right)^2\right\rangle_\beta \tag{3.62}$$

and

$$\langle\phi(t)\rangle_\beta = \omega_0 t \tag{3.63}$$

so that

$$\widehat{\rho}(t) = \begin{pmatrix} \widehat{\rho}_{11} & \widehat{\rho}_{10}\mathrm{e}^{-\mathrm{i}\omega_0 t}\mathrm{e}^{-\langle\delta\phi(t)^2\rangle_\beta/2} \\ \widehat{\rho}_{01}\mathrm{e}^{\mathrm{i}\omega_0 t}\mathrm{e}^{-\langle\delta\phi(t)^2\rangle_\beta/2} & \widehat{\rho}_{00} \end{pmatrix} \tag{3.64}$$

with $\langle\delta\phi(t)^2\rangle_\beta$ given by eq. (3.62). The factor $\mathrm{e}^{-\mathrm{i}\omega_0 t}$ represents the unitary evolution arising from ω_0, while the factor $\mathrm{e}^{-\langle\delta\phi(t)^2\rangle/2}$ describes the dephasing arising from the smearing of the phase because of the fluctuating $\beta(t)$. Since $\langle\delta\phi(t)^2\rangle_\beta$ will usually increase with t, $\mathrm{e}^{-\langle\delta\phi(t)^2\rangle_\beta/2}$ will tend to zero for large t so that $\widehat{\rho}(t)$ will turn into a diagonal from, i.e., into a mixed state with $\widehat{\rho}_{01} = \widehat{\rho}_{10} = 0$.

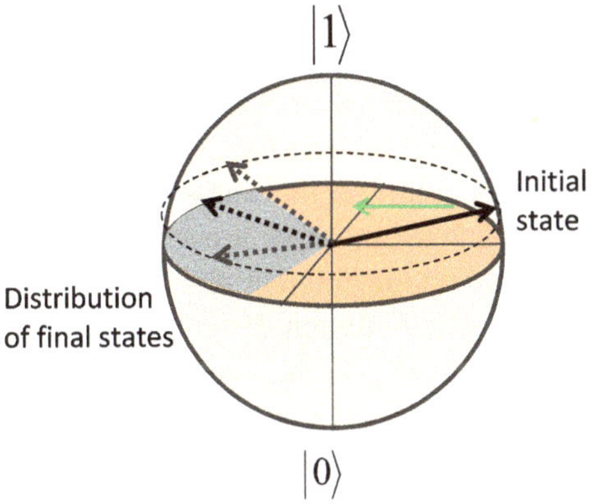

Fig. 3.26: Spin dephasing on the Bloch sphere. The initial state (full black arrow) precesses with different rates for subsequent measurements leading to different phases on the Bloch sphere (dotted arrows). Averaging over these phases leads to a mixed state inside the Bloch sphere with fixed projection to the z axis (amplitude). The average state moves continuously towards the z axis as marked by the green arrow, since the phases (dotted arrows) become more spread out.

To visualize this process, we invoke the Bloch sphere picture of Fig. 3.26. If the initial density matrix $\widehat{\rho}(t = 0)$ is a pure state, it lies somewhere on the surface of the sphere (full black arrow). The Hamiltonian (3.55) generates a rotation by $\phi(t)$ around the z-axis. The randomness of $\phi(t)$ leads to a cloud of possible final states, which are increasingly distributed over the dashed ring on the surface of the Bloch sphere (dotted arrows) at increasing t. Averaging over this distribution, one obtains a density matrix, which corresponds to a point inside the Bloch sphere. For increasing evolution time t, this point moves towards the z-axis as indicated by the green arrow. Once the uncertainty of $\phi(t)$ is large enough to evenly spread ϕ (taken modulo 2π) over the interval $[0, 2\pi)$, the isotropic distribution of final states leads to an average density matrix that is diagonal. The values of the two diagonal entries specify the position on the z axis.

3.5.1.3 T_2^* and T_2

To obtain a more concrete understanding of the two most generic types of dephasing, it is useful to treat them in a simplified, intuitive way before developing a general formalism to determine $\langle \delta\phi(t)^2 \rangle$ in Section 3.5.3.

For the sake of brevity, we will drop from now on the β-subscript in $\langle \cdot \rangle_\beta$. We first consider β to represent so called quasistatic noise, which corresponds to β being approximately constant during each evolution period of the qubit. Nevertheless, β still varies when averaging over many repetitions, but since it is approximately stationary in each run, we can disregard its explicit time dependence. For a single run, we obtain $\phi(t) = \omega_0 t + \beta t$ and thus $\langle \delta\phi(t)^2 \rangle = \langle \beta^2 \rangle t^2$ (eq. (3.57)). The decay of the off-diagonal density matrix elements is thus given by

$$\widehat{\rho}_{01}(t) = \widehat{\rho}_{10}^*(t) = \widehat{\rho}_{01}(t=0)\mathrm{e}^{\mathrm{i}\omega_0 t}\mathrm{e}^{-\frac{1}{2}\langle \beta^2 \rangle t^2} = \widehat{\rho}_{01}(t=0)\mathrm{e}^{\mathrm{i}\omega_0 t}\mathrm{e}^{-\left(\frac{t}{T_2^*}\right)^2} . \tag{3.65}$$

Here, we have defined the time scale $T_2^* = \sqrt{2/\langle \beta^2 \rangle}$, which is sometimes called the inhomogeneously broadened dephasing time. The term "inhomogeneous broadening" comes from NMR, where averaging over many spins in a slightly inhomogeneous magnetic field leads to a broadening of the resonance spectrum. Note that dephasing was studied in NMR long before the notion of a qubit was conceived.

Figure 3.27 shows a graphical representation of this result. Each colored curve shows the time dependence of $\langle \widehat{\sigma}_x(t) \rangle = \Re(\widehat{\rho}_{01}(t))$ as a function of evolution time t for a different β, with the amplitude reflecting the probability for that value of β. Averaging over all curves results in the thicker black curve. We see that while each realization of β gives an undamped oscillation, averaging over many different realizations with different frequencies leads to a decay of the averaged oscillation. Note that the Gaussian decay law of eq. (3.65) can also be understood in terms of this averaging process. Averaging over the distribution of oscillations with different frequencies is formally equivalent to a Fourier transform of the frequency spectrum. A Gaussian distribution of β is thus transformed into a Gaussian decay law.

The opposite extreme for the form of $\beta(t)$ is that of fast, uncorrelated noise, i.e., $\beta(t)$ changes much faster than the evolution time t of a single run. In this case, we can divide t into $N \gg 1$ independent intervals of duration $\Delta t = t/N$, and define β_n as the average value of β during the nth interval (Fig. 3.28). Using the absence of correlations in $\beta(t)$, which implies that $\langle \beta_i \beta_j \rangle = \delta_{ij} \langle \beta_i^2 \rangle$, we find

$$\langle \delta\phi(t)^2 \rangle = \left\langle \left(\sum_{i=1}^{N} \beta_i \Delta t \right)^2 \right\rangle = \sum_{i,j=1}^{N} \langle \beta_i \beta_j \rangle \Delta t^2 = \sum_{i=1}^{N} \langle \beta_i^2 \rangle \Delta t \frac{t}{N} = \langle \beta_i^2 \rangle \Delta t \cdot t . \tag{3.66}$$

Hence, eq. (3.64) leads to

$$\widehat{\rho}_{01}(t) = \widehat{\rho}_{10}^*(t) = \widehat{\rho}_{01}(t=0)\mathrm{e}^{\mathrm{i}\omega_0 t}\mathrm{e}^{-t/T_2} \tag{3.67}$$

with $T_2 = 2/(\langle \beta_i^2 \rangle \Delta t)$. T_2 is often called the decoherence time of the qubit. At first sight, the value of T_2 appears to depend on the choice of Δt. However, since the vari-

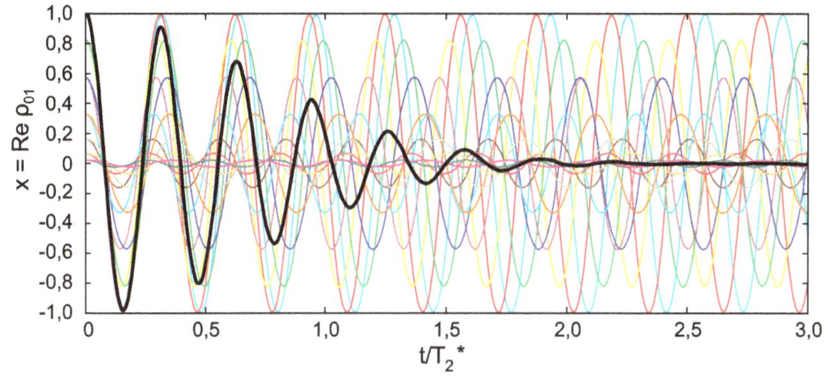

Fig. 3.27: The Gaussian decay of an ensemble of coherent oscillations of spins in a Gaussian-distributed, quasistatically fluctuating effective magnetic field corresponds to the statistical average over many sine-curves with different frequencies (colored curves) resulting in the decaying black curve after averaging. The varying amplitude of the colored curves reflects their statistical weight, hence each curve represents multiple runs of the experiment.

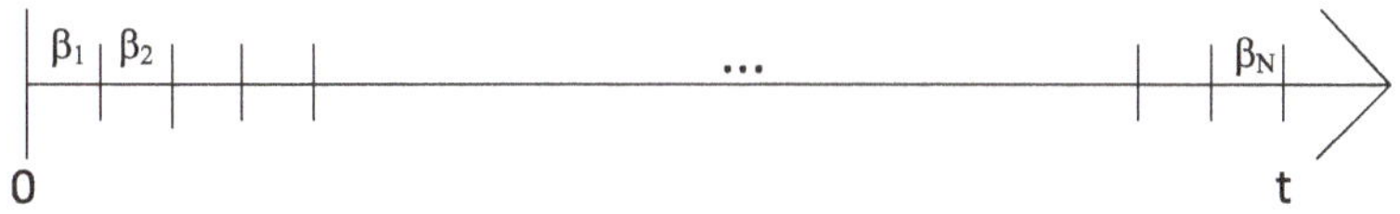

Fig. 3.28: To understand how uncorrelated (white) noise leads to an exponential decay of the phase information, it is helpful to divide the evolution time t of a single run of the experiment into many short sections with uncorrelated, constant noise value β_i.

ance of uncorrelated processes such as $\langle\beta_i^2\rangle$ decreases inversely proportional to Δt, $\langle\beta_i^2\rangle\Delta t$ is independent of Δt, as long as Δt is longer than the correlation time of $\beta(t)$. This shows that our result is meaningful.

3.5.2 Pulse sequences: Rabi, Ramsey, Hahn-echo

To characterize the qubit decoherence T_2 and to distinguish it from the dephasing time T_2^*, a number of standard pulse sequences have been developed, originally mostly for NMR experiments. We will discuss the most important ones here. They commonly involve many repetitions in order to extract probabilities with a good signal-to-noise ratio through averaging.

3.5.2.1 Rabi oscillations

As discussed in Section 3.4.3, a qubit that is driven by a resonant AC excitation, e.g., a spin qubit exposed to a magnetic field $B(t) = B_1 \cdot \cos(2\pi f_{\text{res}} t)$ that is oriented perpendicular to the external field $\vec{B}_{\text{ext}}$ and has a frequency $f_{\text{res}} = g\mu_B B_{\text{ext}}/h$. In this case, the qubit exhibits so-called Rabi flopping, i.e., the AC magnetic field generates a rotation around a horizontal axis of the Bloch sphere with a frequency proportional to the drive strength B_1.

These Rabi oscillations are typically measured by initializing the qubit in the ground state, switching on the AC drive for a variable time t and then reading out the final state along the z-axis (Fig. 3.29(a)). The oscillatory dynamics become visible by plotting the readout signal averaged over several measurement runs as a function of t. Two examples for electron spins driven by an AC magnetic field are shown in Fig. 3.29(c) and (d). One of them is obtained from a device made of isotopically purified Si (Fig. 3.29(d)). It shows a nearly perfect sinusoidal oscillation without notable

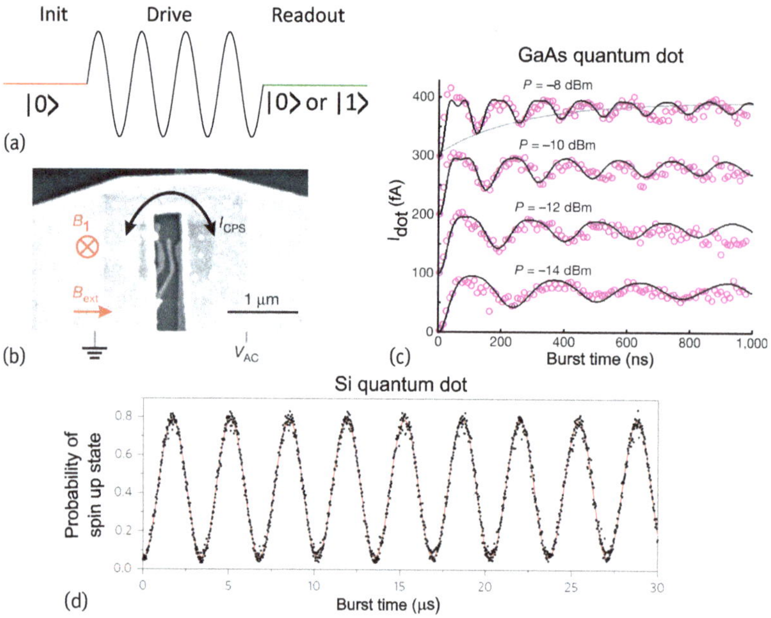

Fig. 3.29: (a) Measurement sequence for a pulse experiment to measure Rabi oscillations. The pulse of variable length with a frequency f, corresponding to the energy splitting of the two spin levels ΔE according to $f = \Delta E/h$, is applied after initializing the qubit (Init), followed by readout. The readout can only distinguish the computational states $|0\rangle$ or $|1\rangle$. (b) Electron microscopy image of the microstructured wire for the AC current (white area) (same as Fig. 3.24(a)). The wire surrounds a quantum dot, visible by part of its gate fingers (grey stripes in the center). The ac current I_{CPS} (CPS: coplanar slotline) and the resulting ac magnetic field with amplitude B_1 are marked. The additional in-plane static magnetic field B_{ext} leads to an energy splitting of the spin levels according to $\Delta E = g\mu_B B_{ext}$. (c) Rabi oscillations of an electron spin in a GaAs quantum dot at different power P of the electric current in the CPS. The power P determines the strength of B_{ac} and, hence, the frequency of the Rabi oscillation (eq. (3.38)). The decay of the oscillations arises from averaging over nuclear field fluctuations. Red dots: experimental data, black lines: fit curves. (d) Same measurement in a Si quantum dot, without any visible decay on a thirty times larger time scale than in (b). (c) [47], (d) [48]).

decay. The other one, obtained from a GaAs qubit, reveals a decay of the oscillation amplitude on the 100 ns time scale (Fig. 3.29(c)). This decay arises from the interaction with the nuclear spin bath. The nuclear spins contribute to the effective magnetic field via hyperfine interaction. Since the orientation of the nuclear spins slowly changes with time, a fluctuating effective magnetic field and hence a fluctuating f_{res} results. Consequently, the Rabi field does not match the resonance frequency perfectly, which changes the rotation axis in the rotating frame (eq. (3.38)) as well as the Rabi frequency f_{Rabi}. Hence, one averages over different frequencies and amplitudes as in Fig. 3.27 and a decaying sine-curve results.

The details of the decay curve depend on several additional factors including the driving power P. The pulse sequence of Fig. 3.29(a) is also often used to calibrate the qubit manipulation.

3.5.2.2 Measurement of the relaxation time T_1

In addition to the time scales T_2 and T_2^*, over which phase coherence is lost, an important characteristic of a qubit is the lifetime of its energy eigenstates, typically referred to as T_1. It can be measured by letting the qubit relax to the ground state and afterwards applying a so-called π-pulse, which is a Rabi oscillation of half a period, that transfers the ground state of the qubit $|0\rangle$ into the excited eigenstate $|1\rangle$. Then, one waits for a variable time t_{w}, before measuring the state of the qubit. Repeating the procedure several times each for several different waiting times t_{w} provides the probability of the qubit remaining in the excited state as a function of the waiting time t_{w}. This normally exhibits an exponential decay $P_{|1\rangle} \propto e^{-t_{\mathrm{w}}/T_1}$ with the time constant T_1 being the relaxation time. The factors determining T_1 will be discussed in Section 3.5.4.

3.5.2.3 Free induction decay (FID)

Once Rabi-oscillations or other control procedures have been calibrated, it is possible to use them in more complex pulse sequences, for example, to measure the qubit's dephasing time T_2^* directly. The most simple example is the Ramsey sequence, sometimes also referred to as free induction decay (FID, Fig. 3.30). Starting from the ground state $|0\rangle$, a $\pi/2$ Rabi pulse around a horizontal axis puts the qubit into a superposition state on the equator of the Bloch sphere. The state then precesses around the static (effective) magnetic field without an externally applied pulse, i.e., freely, for a time t. At the end of this period, the phase of the superposition state is mapped onto the amplitudes of $|0\rangle$ and $|1\rangle$ with another $\pi/2$ pulse along the same axis as the preparation pulse and the qubit is read out. If the qubit has precessed by a phase of $n \cdot 2\pi$ ($n \in \mathbb{Z}$), the two $\pi/2$ pulses add up to a π pulse such that $|0\rangle$ is transferred to $|1\rangle$. If the two $\pi/2$ pulses are separated by a free precession of angle $n \cdot 2\pi + \pi$, the second pulse rotates the state back to $|0\rangle$. For any other angle on the equator of the Bloch sphere, the second $\pi/2$ pulse leads to a superposition of $|0\rangle$ and $|1\rangle$, since the rotation is always by $\pi/2$

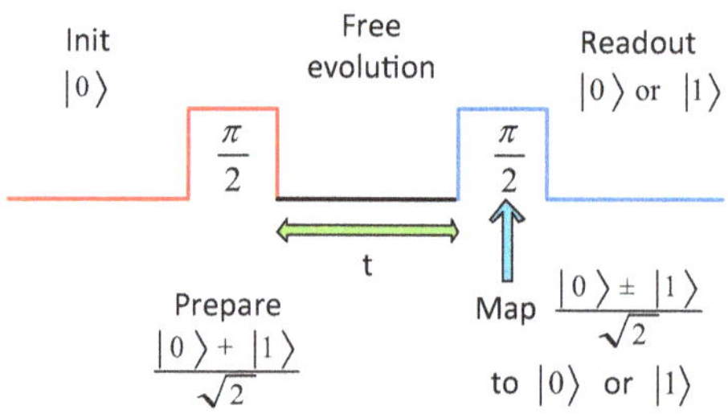

Fig. 3.30: Pulse sequence for a Ramsey experiment. The initialized ground state $|0\rangle$ is rotated onto the equator of the Bloch sphere (Fig. 3.19) using a $\pi/2$-Rabi pulse. Then, the qubit evolves freely for the time t. Another $\pi/2$ pulse maps the phase of the state on the equator to the z axis for readout, e.g., to determine the probability to have the $|0\rangle$ state $P_{|0\rangle}$. Multiple repetitions of the experiment with different t show the phase evolution of the qubit as a sinusoidal $P_{|0\rangle}(t)$ including its dephasing, which results in a damped oscillation of $P_{|0\rangle}(t)$ with damping time T_2^* (eq. (3.67)).

around the chosen axis, which mostly does not end at the poles of the Bloch sphere (Fig. 3.32). Thus, the phase acquired during the evolution time determines the readout signal. In other words, the second $\pi/2$ pulse projects the acquired phase on the equator onto the z-axis of the Bloch sphere. For coherent precession with fixed frequency ω, one obtains a sinusoidal oscillation of the probability to measure $|0\rangle$ after the second $\pi/2$ pulse. This type of oscillation is typically called Ramsey-fringes. In the presence of dephasing, the measured oscillation decays, as discussed in Section 3.5.1.2. Often, the noise sources responsible for decay have both slow and fast components. The FID measurement is typically dominated by the slow components, to which it is most sensitive as will be derived in Section 3.5.3. The decay of the FID signal is then Gaussian and reveals the dephasing time T_2^*, as discussed in Section 3.5.1.2 (eq. (3.67)).

3.5.2.4 Hahn echo

The most widespread technique to reduce the effect of slow noise and thus to extend dephasing beyond T_2^* is the Hahn echo or spin echo technique, which is a FID sequence with an additional π-pulse along a horizontal axis (Fig. 3.31). The π pulse inverts the state of the qubit half way through the experiment at time $t/2$ (Fig. 3.32).

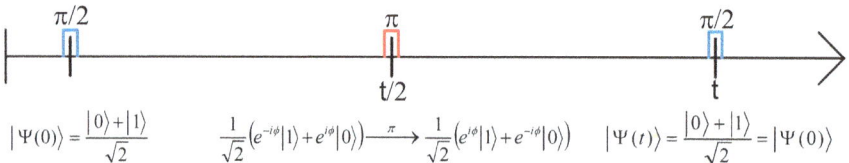

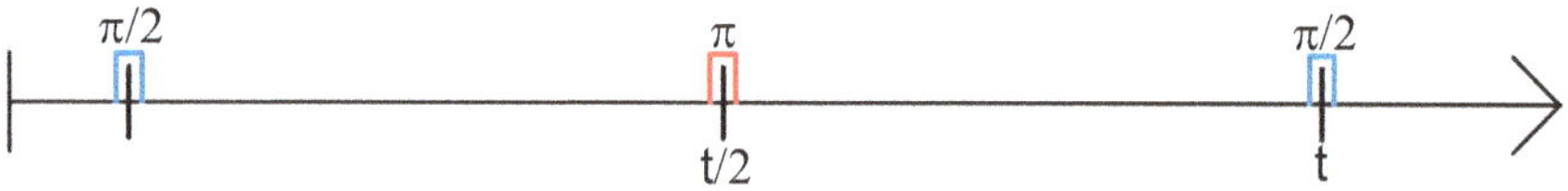

$$|\Psi(0)\rangle = \frac{|0\rangle+|1\rangle}{\sqrt{2}} \qquad \frac{1}{\sqrt{2}}\left(e^{-i\phi}|1\rangle+e^{i\phi}|0\rangle\right) \xrightarrow{\pi} \frac{1}{\sqrt{2}}\left(e^{i\phi}|1\rangle+e^{-i\phi}|0\rangle\right) \qquad |\Psi(t)\rangle = \frac{|0\rangle+|1\rangle}{\sqrt{2}} = |\Psi(0)\rangle$$

Fig. 3.31: Removal of slow noise dominating T_2^* with spin echo techniques. The phase ϕ arising from precession including any static shift in the precession frequency changes sign because of the π-pulse in the middle of the experiment. Due to the same evolution time prior and after the π pulse, the total phase at time t is zero independent of t as long as the noise is static.

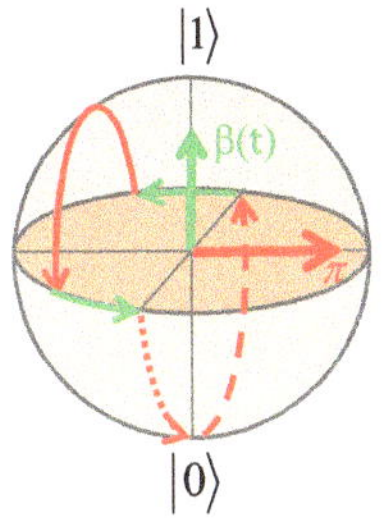

Fig. 3.32: Bloch sphere trajectory corresponding to Fig. 3.31. Starting from the $|0\rangle$-state, a first $\pi/2$ pulse (dashed red) rotates the qubit state to the equator, on which it evolves under the influence of the fluctuating level splitting $\beta(t)$ (green arrow). A π pulse (red) then flips the state to the other side of the Bloch sphere, where it evolves for the same time as before (green arrows). Finally, another $\pi/2$ pulse (red dotted) is applied. If $\beta(t)$ has not changed during this sequence, the two free evolution sections correspond to equal angles and the original state is recovered independent of t.

We will show that this inversion removes the effect of any contribution to β that is constant over the time t of a single run of the experiment. At $t/2$, the initial state $|\psi(0)\rangle = \frac{1}{\sqrt{2}}(|0\rangle + |1\rangle)$ has evolved into $\frac{1}{\sqrt{2}}(e^{-i\phi(t/2)}|0\rangle + e^{i\phi(t/2)}|1\rangle)$. The π-pulse exchanges $|0\rangle$ and $|1\rangle$, so that the state turns into $\frac{1}{\sqrt{2}}(e^{i\phi(t/2)}|0\rangle + e^{-i\phi(t/2)}|1\rangle)$, which means that the signs of the phase factors have been interchanged. For constant β, the subsequently accumulated phase from $t/2$ until t is equal to that from 0 until $t/2$, which we have denoted by $\phi(t/2)$. Hence, the total phase accumulated before and after the π-pulse vanishes and the final state $|\psi(t)\rangle$ is equal to the initial state $|\psi(0)\rangle$.

More intuitively, the length of the path that has been traveled on the equator of the Bloch sphere prior to the π pulse is the same as that of the path traveled after the π pulse (Fig. 3.32). The π pulse, being along the axis of the initial state on the equator, flips the state to a position with the same distance to the initial state, but on the opposite side. Hence, the remaining time after the π pulses naturally moves the state back to its initial position.

Since this cancellation works for any value of β that is constant over the evolution period t, the effect of quasistatic noise or inhomogeneous broadening is completely removed. For uncorrelated fast noise on the other hand, the sign change of the phase has no effect. For slowly varying β, dephasing will become noticeable once the evolution time of an individual run of the experiment is long enough for β to change significantly. If the total noise has a slow and a fast, uncorrelated component, the Hahn echo can reveal an exponential decay due to the fast component only (eq. (3.67)), while being insensitive to slow noise. This decay time constant is called the decoherence time T_2. This argument rationalizes that for truly uncorrelated noise, it is not possible to further extend the measured lifetime of the qubit's state. The absence of correlations prohibits cancellation of the accumulated phase by the Hahn echo sequence. In this case, T_2 is therefore an intrinsic property of the system including the bath of the qubit. However, not all noise processes of an experiment can unambiguously be broken down in this way. In general, many different frequency components of the noise contribute with different weight. Then, the observed coherence time can be extended further by more frequently flipping the qubit during a measurement run. Consequently, the definition of T_2 becomes non-universal and needs to be interpreted in the context of how it was measured. The association of the FID with a Gaussian decay on the scale of T_2^* and the Hahn echo with T_2 is therefore not a strict law, but rather a rule of thumb. One thus often speaks of the FID time constant or the Hahn echo time constant to accurately describe the experimental findings. In the next two sections, we will develop a formalism to compute the effect of arbitrary Gaussian noise whose correlations can have arbitrary time scales.

3.5.3 Dephasing: general case

This section provides an advanced treatment of dephasing that generalizes and formalizes the above considerations. It reflects the standard method for interpreting and modeling experiments. The results can quantitatively predict the decay of the phase coherence for arbitrary noise correlations and for several π-pulse sequences applied. In Section 3.5.3.1, we will introduce the spectral density as a mathematical tool to describe noise with arbitrary correlations. It generalizes the notions of white, fully uncorrelated and quasistatic, perfectly correlated noise as already discussed. This concept is of much broader relevance than for qubit dephasing. For example, it can also be used for a detailed description of the sensitivity of measurement instruments. Section 3.5.3.2 will then introduce a formalism that allows one to derive the decay of the qubit coherence from a given noise spectrum. Going beyond the basic concepts, it is mostly aimed at students who are mathematically inclined or are considering working in the field.

3.5.3.1 Noise spectral density

To generalize from the limits of short and long correlation times, we need to characterize a noise process $\beta(t)$, which can fluctuate on arbitrary timescales. We still assume that the noise process is Gaussian and, without loss of generality, that the time averaged noise $\langle\beta(t)\rangle = 0$. Such a noise process can be described by a correlation function

$$K_\beta(\Delta t, t) = \langle\beta(t)\beta(t + \Delta t)\rangle \, . \tag{3.68}$$

Generally, this correlation function can change with time, i.e., each measurement run or each set of consecutive measurement runs delivers another type of correlation function. This is indicated by the additional t in the argument of K_β. Here, we will restrict our considerations to stationary noise, whose statistical properties are independent of this time t. In this case, one can think of taking the expectation value $\langle\cdot\rangle_t$ as a complete averaging over infinite times t, i.e., $\langle\beta(t)\beta(t+\Delta t)\rangle = \langle\beta(t)\beta(t+\Delta t)\rangle_t$. For Gaussian noise, all other properties, including in particular higher order correlation functions, can be derived from $K_\beta(\Delta t)$. According to the central limit theorem, a noise process that is the sum of many different contributions is Gaussian even if the individual constituents are non-Gaussian as long as some additional assumptions that ensure convergence and are not very restrictive hold.[16] Hence, our model is quite general. The typical behavior

16 Wikipedia: Central limit theorem.

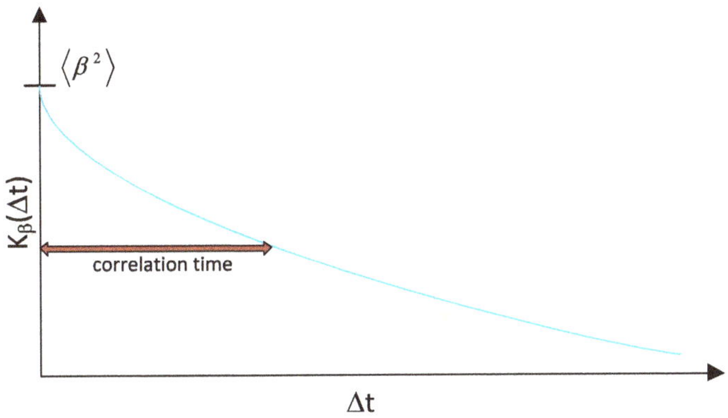

Fig. 3.33: Schematic of a noise correlator: the time over which the noise correlator $K_\beta(\Delta t)$ decays can be identified with the correlation time of the fluctuating variable β.

of $K_\beta(\Delta t)$ is a monotonous decay over some time scale t_{noise}, which corresponds to the correlation time of $\beta(t)$ (Fig. 3.33). $K_\beta(0) = \langle\beta^2\rangle$ is the instantaneous variance of $\beta(t)$, which however may be singular for some noise models.

A very convenient quantity is the noise spectral density, also called noise spectrum, $S_\beta(\omega)$ of β. It is defined as the Fourier transform of $K_\beta(\Delta t)$:

$$S_\beta(\omega) = \int \mathrm{d}\Delta t\, \mathrm{e}^{\mathrm{i}\omega\Delta t} K_\beta(\Delta t)\,. \tag{3.69}$$

The inverse relation is thus

$$K_\beta(\Delta t) = \int \frac{\mathrm{d}\omega}{2\pi} \mathrm{e}^{-\mathrm{i}\omega\Delta t} S_\beta(\omega)\,. \tag{3.70}$$

The noise spectrum is not only important for qubit dephasing, but also for other applications requiring a statistical description of space or time dependent variables, including measurement noise. Two important relations that help to interpret the meaning of the spectrum are

$$\langle\beta^2\rangle = K_\beta(0) = \int \frac{\mathrm{d}\omega}{2\pi} S_\beta(\omega) \tag{3.71}$$

and the Wiener–Khinchin theorem, according to which

$$S_\beta(\omega) = \lim_{t\to\infty} \frac{\left\langle \left| \int_0^t \mathrm{d}\tilde{t}\beta(\tilde{t})\mathrm{e}^{\mathrm{i}\omega\tilde{t}} \right|^2 \right\rangle_\beta}{t}\,. \tag{3.72}$$

The limit $t \to \infty$ is approached when t is much longer than the correlation time of the noise, so that many independent samples of β are taken. Both of these relations show that the spectrum $S_\beta(\omega)$ indicates to what extent each frequency ω is present in the noise process. According to eq. (3.71), $S_\beta(\omega)$ is the contribution of fluctuations at frequency ω to the total variance of β. Similarly, eq. (3.72) says that $S_\beta(\omega)$ is the variance of the Fourier transforms at frequency ω of an ensemble of time traces of $\beta(t)$.

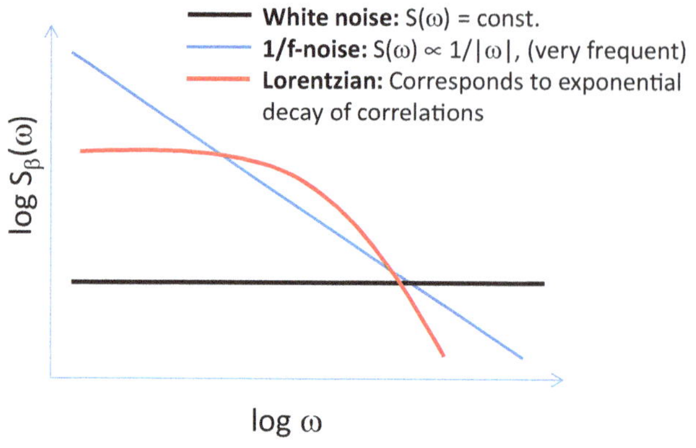

Fig. 3.34: Examples of typical noise spectra shown on a log–log plot.

Figure 3.34 shows a few important examples for $S_\beta(\omega)$. For white noise, $S_\beta(\omega)$ is constant and hence $K_\beta(\Delta t) \propto \delta(\Delta t)$, implying uncorrelated noise. For a Lorentzian or a similar spectrum strongly decaying with frequency above a cutoff, the cutoff frequency specifies the inverse correlation time.

3.5.3.2 Filter-function formalism

We are now in a position to derive the decay of qubit coherence for a rather general case by assuming that $\beta(t)$ is Gaussian and stationary and has an arbitrary spectrum S_β. Furthermore, we allow π-pulses to be applied at times $\tau_1, \ldots, \tau_m$ in order to generalize from the Hahn echo sequence. Extending our considerations in Section 3.5.2.4, we can describe their effect in terms of a sign change of the noise operator, which acts on the phase of the qubit by definition, whenever a π-pulse is applied. The qubit, subject to the π-pulses and noise $\beta(t)$, behaves like a qubit without pulses but subject to a noise $c(t)\beta(t)$, with $c(t) = \pm 1$ changing sign at every pulse. This result can be formally derived by switching into a time dependent reference frame that follows each π-pulse, making use of the fact that $\hat{\sigma}_{x(y)}\hat{\sigma}_z\hat{\sigma}_{x(y)} = -\hat{\sigma}_z$. With these definitions, we can compute

$$\begin{aligned}\langle \phi^2(t)\rangle &= \left\langle \left(\int_0^t \mathrm{d}t'\; c(t')\beta(t') \right)^2 \right\rangle \\ &= \int_0^t \mathrm{d}t' \int_0^t \mathrm{d}t''\; c(t')c(t'')\langle \beta(t')\beta(t'')\rangle \\ &= \int \frac{\mathrm{d}\omega}{\pi} S_\beta(\omega) \frac{F(\omega t)}{\omega^2}\,. \end{aligned} \tag{3.73}$$

To obtain the last line, we have substituted

$$\langle \beta(t')\beta(t'')\rangle = K_\beta(t''-t') = \int \frac{\mathrm{d}\omega}{2\pi} \mathrm{e}^{-\mathrm{i}\omega(t''-t')} S_\beta(\omega) \tag{3.74}$$

and defined the so called filter function

$$F(\omega t) = \frac{\omega^2}{2}\int_0^t \mathrm{d}t' \int_0^t \mathrm{d}t''\; c(t')c(t'')\mathrm{e}^{\mathrm{i}\omega(t''-t')}$$
$$= \frac{\omega^2}{2}\left|\int_0^t \mathrm{d}t'\; c(t')\mathrm{e}^{\mathrm{i}\omega t'}\right|^2$$
$$= \frac{1}{2}\left|\sum_{m=0}^{N}(-1)^m(\mathrm{e}^{\mathrm{i}\omega\tau_{m+1}} - \mathrm{e}^{\mathrm{i}\omega\tau_m})\right|^2 .$$

The definition of the π-pulse times τ_m was extended by $\tau_0 = 0$ and $\tau_{m+1} = t$. The filter function $F(\omega t)$ is a property of the pulse sequence. According to eq. (3.73), it describes how much each frequency contributes to dephasing. As specific examples, we compute the results for FID with

$$c(t) = 1$$
$$\Longrightarrow F(\omega t) = 2\sin^2\left(\frac{\omega t}{2}\right)$$

and the Hahn echo, where

$$c(t) = \begin{cases} +1 & (t' < t/2) \\ -1 & (t' > t/2) \end{cases}$$
$$\Longrightarrow F(\omega t) = 8\sin^4\left(\frac{\omega t}{4}\right) .$$

These filter functions are plotted in Fig. 3.35(a). In addition, the filter function for a so called 2-pulse CPMG sequence[17] and the Lorentzian spectrum of Fig. 3.34 (exponential decay of the correlations) are shown. The sequence $c(t')$ for the CPMG pulses, which has a π-pulse at $t/4$ and $3t/4$, is plotted in Fig. 3.35(c). Note that much more sophisticated sequences have been devised based on the same ideas in order to minimize dephasing.

Based on these filter functions, we can understand why the Hahn echo is insensitive to low frequency noise. For the Hahn echo, $F(\omega t)/\omega^2 \approx \omega^2 t^4/32$ vanishes for $\omega \to 0$ so that very little weight is given to low frequency noise and, as expected, zero frequency contributions, which are constant in time, are entirely irrelevant. For the FID on the other hand, $F(\omega t)/\omega^2 \approx t^2/2$ remains nonzero down to $\omega = 0$, so that low frequency noise can contribute significantly to dephasing. For either sequence, one can also recover the exponential decay for white noise found earlier by inserting $S_\beta(\omega) = \text{const.}$ into the integral of eq. (3.73).

17 It is named after its inventors Carr, Purcell, Meiboom and Gill.

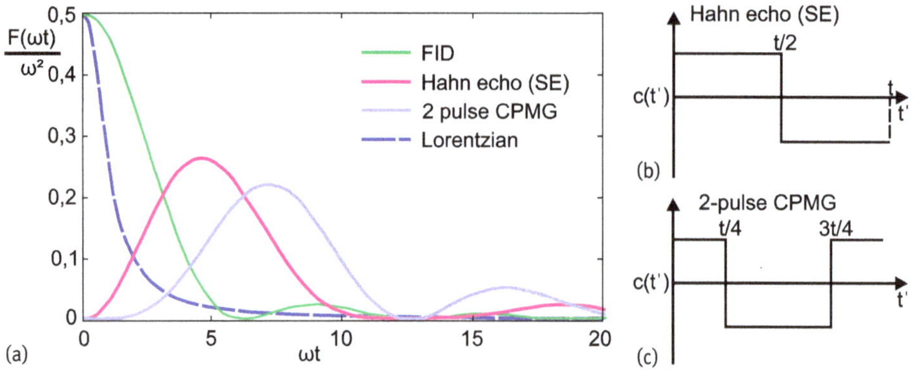

Fig. 3.35: (a) Comparison of filter functions $F(\omega, t)/\omega^2$ for different pulse sequences. (b) Pulse sequence for a spin echo (SE) measurement dubbed Hahn echo. (c) CPMG pulse sequence for only two pulses.

3.5.4 Energy relaxation

So far, we have only considered pure dephasing from a Hamiltonian due to the noise function $\beta(t)$ (eq. (3.55)). This noise cannot drive transitions between the $|0\rangle$ and $|1\rangle$ eigenstates, because it contains no suitable matrix elements contributing to the $\widehat{\sigma}_x$ or the $\widehat{\sigma}_y$ term of the qubit Hamiltonian. As a result, only the phase of superposition states will be randomized, while the population probability of the basis states corresponding to the z-coordinate of the qubit state is conserved. However, noise terms coupling to $\widehat{\sigma}_x$ and/or $\widehat{\sigma}_y$ can appear, which thus induce transitions between the eigenstates of the qubit. The associated transition rate can be characterized experimentally with the scheme described in Section 3.5.2.2. In our study of Rabi oscillations (Section 3.4.3), we have seen that in order to drive such transitions with a coherent signal, it needs to be approximately resonant with the qubit's level splitting ω_0. By switching into the rotating frame, one can similarly show that only noise near $\omega = \omega_0$ is effective at causing transitions between the eigenstates. Depending on the temperature T_{noise} of the noise, the qubit will either relax to the ground state ($k_B T_{\text{noise}} \ll \hbar\omega_0$) or towards a fully mixed state ($k_B T_{\text{noise}} \gg \hbar\omega_0$). A formal analysis of such energy relaxation and heating requires a quantum model for the noise, which is beyond the scope of this course. Having said that, since such transitions between eigenstates can be identified with classical bit flips, they are less quantum-mechanical than pure dephasing. Since noise spectral densities usually have much less structure at high frequencies compared to the low frequency range that is more relevant for dephasing, the noise can normally be approximated as white in the relevant high frequency range, so that energy relaxation is in most cases exponential in time with decay time constant T_1. One can show that T_2 has an upper bound of $2T_1$, but one often finds much shorter T_2 because of excess low frequency noise with respect to the white noise limiting T_1.

3.5.5 Physical sources of noise and decoherence

3.5.5.1 Overview

Having developed a framework for the description of decoherence, we now turn to physical mechanisms that can cause a noise process as described by $\beta(t)$. We first give a qualitative overview of the most common effects and then treat two specific examples in more detail.

Many qubits work with energy splittings of order 10 GHz $\geq 5k_\mathrm{B}T/h$ for $T \leq$ 100 mK, so that in thermal equilibrium only the ground state is populated. Energy relaxation is thus associated with relatively fast noise processes in the same frequency range. The most relevant ones are phonons and photons, i.e., mechanical and electrical noise. In some cases they couple directly to a qubit (e.g., electric fields to a charge qubit), in some other cases more intricate coupling mechanisms such as spin-orbit coupling or a change of the band energies in response to lattice deformations are involved. Defect states that absorb energy at the qubit frequency can also play a role. A detailed general treatment [49] shows that the asymmetry between excitation and relaxation, required to reach thermal equilibrium with different occupation probabilities of the ground and the excited state, is due to quantum fluctuation of the noise field. The zero-point fluctuations of the noise field (e.g., electric or magnetic field for photons) can relax the qubit, but not excite it. The underlying physics is closely related to the Einstein coefficients, which describe absorption as well as spontaneous and stimulated emission of photons within a two-level system, e.g., of an atom (Section 2.2.3.2). The two-level system considered there corresponds to the qubit, and the photons to a noise field. The zero-point fluctuations lead to spontaneous emission, but not to absorption as they cannot provide energy to the qubit. For dephasing, which is more sensitive to low-frequency noise, an even broader range of relevant noise processes is encountered. In brief, any kind of randomly fluctuating field coupled to the qubit can be a source of decoherence, independent of whether its dynamics have a thermal, a quantum mechanical or an instrumental origin. Prominent examples include charge noise from charge traps at defects or dopants in the material, spins from impurities or dangling bonds causing magnetic field or flux noise, Johnson noise from resistors, and the hyperfine interaction between an electron spin and fluctuating nuclear spins. The latter two will be discussed in more detail in the next two sections. Last but not least, instrumental noise sources, e.g., from measurement and control instruments connected to an experimental setup, must be considered carefully to avoid dephasing.

3.5.5.2 Johnson noise

According to the fluctuation-dissipation theorem, any dissipating physical system also produces noise. In thermal equilibrium, the noise intensity is directly related to the temperature and the degree of dissipation. The probably most common manifestation is electrical noise from resistors. The resulting voltage noise spectral density is given by $S_V(\omega) = 2R\hbar\omega \coth(\hbar\omega/k_B T))$, where R is the value of the resistor and T its temperature. For $\hbar|\omega| \ll k_B T$, one recovers $S_V = 2Rk_B T$ representing classical, thermal Johnson noise.[18] Evidently, it is an example for white noise. For $\hbar\omega \gg k_B T$, the result $S_V = 2R\hbar\omega$ is independent of temperature and reflects quantum mechanical zero-point fluctuations. The asymmetry between positive and negative frequencies in this regime reflects the difference between the ability of the resistor to absorb ($\omega > 0$, zero-point fluctuations important) and emit ($\omega < 0$, zero-point fluctuations play no role) energy.

As solid state qubits are usually controlled by some kind of electrical signal, they are by definition exposed to Johnson noise and care must be taken to avoid too much decoherence via this noise. The resistor can be the output impedance of a control instrument, or an attenuator, i.e., an impedance matched voltage divider, that is placed at low temperature to reduce the noise originating from room temperature equipment. An example of such a setup is shown in Fig. 3.13(a).

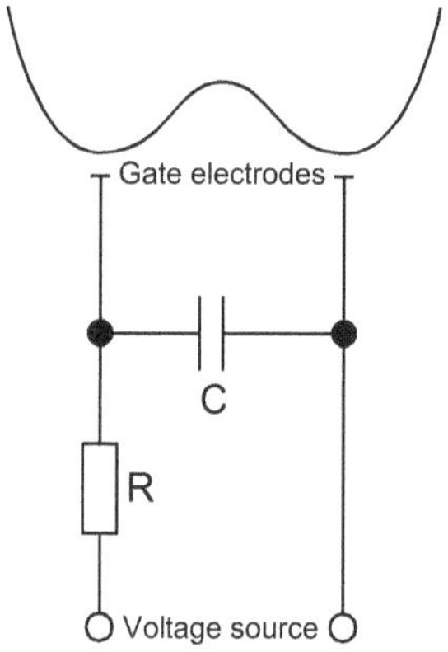

Fig. 3.36: Example of a qubit control circuit in which the capacitor filters the Johnson noise emitted by the resistor, leading to a Lorentzian noise spectrum $S_\beta(\omega)$ (Fig. 3.34). The double well potential of, e.g., a charge qubit (Fig. 3.2) is sketched on top.

Although thermal Johnson noise is intrinsically white, an example for low frequency noise is obtained if it is filtered with a capacitor to ground, as shown in Fig. 3.36. The combination of the resistor and the capacitor forms a RC-filter that attenuates the amplitude of the Fourier components of any signal by a factor of $1/(1 + i\omega RC)$. As the noise spectral density reflects the square of the Fourier components according to the Wiener–Khinchin theorem (eq. (3.72)), the noise is attenuated by the frequency-dependent factor $1/|1 + i\omega RC|^2 = 1/(1 + (\omega RC)^2)$. The resulting spectrum is Lorentzian, as shown in Fig. 3.34. Depending on the time scale considered, it can be approximated

18 Compared to the more common form $S_V = 4Rk_B T$, a factor of 2 is missing, because we use the convention that one has to integrate over both positive and negative frequencies to obtain the variance of the noise signal.

as white noise ($\omega \ll 1/RC$) or quasistatic noise ($\omega \gg 1/RC$). For intermediate regimes, the filter-function formalism of Section 3.5.3.2 must be applied to compute the quantitative decay of the phase coherence of the qubit from the spectrum of the noise.

3.5.5.3 Hyperfine interaction of electron and nuclear spins

As a second example of physical decoherence processes, this section discusses the hyperfine interaction between fluctuating nuclear spins and an electron spin qubit. The intention is to convey a flavor of the intricate physics that can be associated with the dephasing of qubits. The topic is well suited for this purpose as the underlying interactions and material parameters are well-understood. Many other decoherence mechanisms are either rather trivial or very difficult to understand, because the nature and coupling of material imperfections responsible for them are poorly known. For technological purposes, the hyperfine interaction can be avoided by using materials which are free of nuclear spins. The use of isotopically purified Si that contains nearly exclusively the spin-free ^{28}Si isotope is considered most promising in this respect.

However, the hyperfine interaction has been studied intensely in the context of GaAs spin qubits. Hence, a detailed understanding has been obtained, including possibilities to manipulate the nuclear bath such that it is less detrimental for the qubits. We shortly describe the main results at the end of this chapter of the book, since it brings the reader close to current research topics. Some techniques and concepts are specific examples of the general principles already discussed. Others are more specific and go beyond the generic considerations so far. We will adopt a style that is close to that of a scientific review paper, i.e., some arguments are less explicitly described, either because they are assumed to be known or the interested reader is expected to refer to the original literature.[19]

Readers who are not interested in this detailed flavor of research can proceed directly to Section 3.6.

3.5.5.3.1 Hyperfine interaction

The spin of a confined electron in a gate-defined GaAs quantum dot is coupled to about $N_\mathrm{I} = 10^6$ nuclear spins via the hyperfine interaction. In natural Si, the number is about two orders of magnitude smaller due to the smaller wave function of the electron and because the only spin-carrying ^{29}Si isotope has a natural abundance of only 5%. The hyperfine interaction is usually quantified in terms of an effective Overhauser magnetic field, B_nuc, that acts like a real magnetic field, but arises from direct local coupling of the electron spin to the nuclear spins. Full polarization of the nuclear spins, i.e., all nuclear spins pointing in the same direction, would correspond to an Over-

19 The material in this section is adapted from a review article [50] and from lecture notes for a summer school [51].

hauser field of about 4 T for the electron spins in GaAs. Statistical fluctuations are a factor $\sqrt{N_I}$ smaller and thus amount to a few mT. The fluctuations of the Overhauser field thus have a significant impact on the coherence of the electron spin.

The dominant contribution to the electron-nuclear hyperfine interaction originates from the contact Fermi interaction. In first order perturbation theory, it can be described by the Hamiltonian [52]:

$$\widehat{H}_{\mathrm{hf}} = V_{\mathrm{uc}} \sum_j \widetilde{A}_j |\psi(\vec{R}_j)|^2 \vec{\widehat{S}} \cdot \vec{\widehat{I}}_j \tag{3.75}$$

where V_{uc} is the volume of the unit cell per atom, $\vec{\widehat{S}}$ is the electron spin operator, $\vec{\widehat{I}}_j$ is the spin operator of a nucleus located at position $\vec{R}_j$, $\psi(\vec{R}_j)$ is the envelope of the electron wave-function at that position, and summation goes over all nuclei in the crystal. The coupling constants $\widetilde{A}_j$ are about 50 μeV for all isotopes in GaAs.

Using the identity $\vec{\widehat{S}} \cdot \vec{\widehat{I}}_j = 1/2(\widehat{I}_+^j \widehat{S}_- + \widehat{I}_-^j \widehat{S}_+) + \widehat{I}_z^j \widehat{S}_z$ where $\widehat{I}_\pm^j$ and $\widehat{S}_\pm$ are the nuclear and electron spin raising and lowering operators, respectively, eq. 3.75 can be rewritten. This emphasizes two parts: a dynamical part $\propto (\widehat{I}_+^j \widehat{S}_- + \widehat{I}_-^j \widehat{S}_+)$ responsible for the transfer of the angular momentum between the two spin systems, and a static part $\propto \widehat{I}_z^j \widehat{S}_z$, contributing to the energy splitting of the two spin system and thus to its precession frequency. The effect of the static part can be described in terms of an effective magnetic field acting on the electron spin (the Overhauser field) and on individual nuclear spins (the Knight field). The Overhauser field, a collective effect of a large number of nuclei, is described using the electron g-factor, g_e, as:

$$\widehat{B}_{\mathrm{nuc}} = \frac{V_{\mathrm{uc}}}{g_e \mu_B} \sum_j \widetilde{A}_j |\psi(\vec{R}_j)|^2 \widehat{I}_z^j . \tag{3.76}$$

Using the mean spin of the different nuclei of isotope η, $\langle \widehat{I}_\eta \rangle$, the Overhauser (or hyperfine) field reads:

$$B_{\mathrm{nuc}} = \frac{\sum_\eta \widetilde{A}_\eta \langle \widehat{I}_\eta \rangle}{g_e \mu_B} . \tag{3.77}$$

with $\widetilde{A}_\eta$ being the coupling constant for nuclear isotope η. The total electron Zeeman splitting $E_{Z,\mathrm{tot}}$, in the presence of both B_{nuc} and the external magnetic field B_{ext}, is thus determined by the total magnetic field $B_{\mathrm{ext}} \pm B_{\mathrm{nuc}}$, where both fields are assumed to be directed along the same axis, for the sake of simplicity:

$$E_{Z,\mathrm{tot}} = g_e \mu_B (B_{\mathrm{ext}} \pm B_{\mathrm{nuc}}) = g_e \mu_B B_{\mathrm{ext}} + \Delta E_Z \tag{3.78}$$

where $\Delta E_Z = \pm g_e \mu_B B_{\mathrm{nuc}}$ is the Overhauser shift.

The nuclear field B_{nuc} statistically fluctuates around its average as a result of the slow redistribution of nuclear spin polarization due to nuclear spin flips via dipolar coupling between distant nuclear spins or via virtual excitations of the electron spin. In the limit of large N_I, where N_I is the effective number of nuclei, this can be described by a Gaussian distribution with the standard deviation $\sigma_{B_{\mathrm{nuc}}} \sim \widetilde{A}/\sqrt{N_I}$ [53, 54]

with $\widetilde{A}$ being the weighted average of the $\widetilde{A}_\eta$. For a typical number of $N_I = 10^6$ nuclei interacting with an electron confined in a gated GaAs quantum dot, this results in $\sigma_{B_{nuc}} \sim 3\,\text{mT}$.

In a FID experiment (Section 3.5.2.3) with a large number of identical measurements, electron spins that are all initialized in the same superposition state will exhibit different dynamics as they will evolve in different effective magnetic fields. This is the case of quasistatic noise (eq. (3.65)). When averaging over many measurements, one observes spin dephasing with a dephasing time

$$T_2^* \simeq \frac{g_e \mu_B \sigma_{B_{nuc}}}{\hbar} \tag{3.79}$$

of the order of 10 ns due to $\sigma_{B_{nuc}} \approx 3$ mT [53–56]. Much of the dephasing due to the slowly fluctuating random nuclear field can be counteracted by spin-echo techniques (Section 3.5.2.4). The remaining decay of the electron spin coherence, with characteristic timescale $T_{2,\text{echo}}$, provides information on the timescale of the nuclear field fluctuations [55–59].

3.5.5.3.2 Probing nuclear spins

The nuclear polarization along the external magnetic field can be probed by measuring the shift of the electron Zeeman energy E_Z due to the nuclear field B_{nuc}. According to eq. (3.78)), the determination of $E_{Z,\text{tot}}$ in a known B_{ext} reveals B_{nuc}. $E_{Z,\text{tot}}$ can be measured, e.g., by finding the drive frequency that is resonant with $E_{Z,\text{tot}}$ and thus most effective for driving Rabi oscillations (Section 3.4.3.1 and 3.5.2.1). These Rabi oscillations can be induced through electron spin resonance (ESR) [47, 61] or the so-called electric-dipole spin resonance (EDSR) using electric rather than magnetic fields [62, 63]. Typically, the resonance frequency, $\omega = E_{Z,\text{tot}}/\hbar$, is in the microwave regime. The width of the resonance peak gives a bound for $\sigma_{B_{nuc}}$.

An alternative to spectroscopic measurements are time-resolved measurements as FID measurements (Section 3.5.2.3), in which one probes how the time evolution of the electron spin is affected by the nuclear field. For instance, a single spin precessing about a static magnetic field sees its Larmor precession frequency modified by $g_e \mu_B B_{nuc}/h$. Thus, a measurement of the precession rate reveals the value of B_{nuc}. If B_{nuc} fluctuates over time, a time-averaged measurement of the electron spin precession will contain a spread of precession rates, leading to decay of the envelope with time constant $T_2^* = \hbar\sqrt{2}/(g_e \mu_B \sigma_{B_{nuc}})$[56].

This principle has been applied in particular to two-electron spin qubits (Section 3.3.2.2, Fig. 3.37(a)). Starting from the singlet ground state at large positive detuning ϵ, a fast pulse to large negative detuning initiates an oscillation between the singlet state S and the $(m = 0)$-triplet state T_0 with frequency $f = |g_e \mu_B \Delta B_{nuc}|/h$ due to the splitting of $S(2, 0)$ and T_0 (Fig. 3.37(a), left inset). The splitting is caused by the imbalance in Overhauser fields of the two quantum dots, which mixes the singlet and the triplet state, leading to new eigenstates $|\uparrow\downarrow\rangle$ and $|\downarrow\uparrow\rangle$. Consequently, the strength

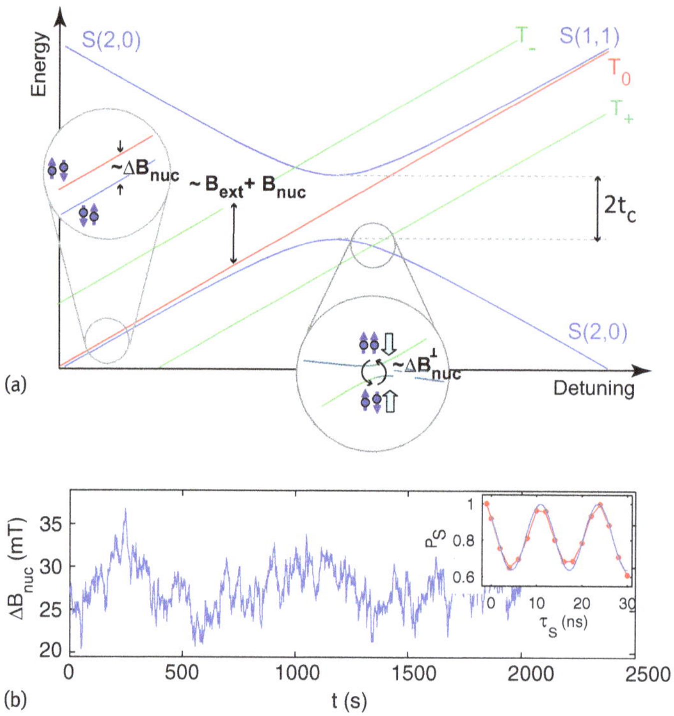

Fig. 3.37: Measurements of the nuclear spin bath evolution via probing of the singlet-triplet electron spin qubit. (a) Effects of the hyperfine interaction illustrated for the energy diagram of a singlet-triplet qubit (Fig. 3.9(c)). At the far left where $J \approx 0$, the eigenstate space of $S(1,1)$ and T_0 leads to two energetically separated states only due to a different Overhauser field in the two quantum dots. The corresponding states $|\uparrow\downarrow\rangle$ and $|\downarrow\uparrow\rangle$ are depicted in the left inset. The degeneracy point of S and T_+ (other inset) is split by the difference in the transversal hyperfine field of the two quantum dots. It can be used for polarizing the nuclear spins via spin transitions of the electron spins between the spin-polarized triplet state and the singlet state [50]. (b) Time trace of the difference between the z-components of the Overhauser field in the two quantum dots, ΔB_{nuc}. Each data point reflects a measurement of the precession of the qubit in the Overhauser field via its probability P_S to be in the singlet state, as shown exemplary in the inset. The measured precession frequency is given by $g_e \mu_B \Delta B_{\mathrm{nuc}}/h$ [60]. After [50].

of the splitting, which determines the oscillation frequency, is proportional to ΔB_{nuc}. Using single shot readout of the electron positions, i.e., distinguishing the state $S(0,2)$ from the state $T_0(1,1)$, via RF-reflectometry of an adjacent QPC or SET, such a measurement of ΔB_{nuc} involving thousands of measurement runs can be completed within less than 10 ms [64, 65]. Even faster measurements are possible with more advanced statistical methods [66], so that a detailed picture of the dynamics can be obtained. Importantly, the time resolution is sufficient to probe the relatively slow Overhauser field fluctuations in real-time (Fig. 3.37(b)).

Measuring the precession frequency of an $S - T_0$ qubit only gives access to the Overhauser field gradient ΔB_{nuc}. The average Overhauser field in a double quantum dot can additionally be measured from the inter-dot detuning ϵ^*, for which $T_+(1,1)$, the $m = 1$ triplet, crosses $S(0,2)$, the singlet branch (Fig. 3.37(a), middle inset) [67]. At this degeneracy point, the S and T_+ state are subject to rapid mixing mediated by the nuclei, since the Zeeman splitting of the nuclei matches the energy difference between S and T_+. This can be detected via spin to charge conversion as in the case of $S-T_0$ mixing (Section 3.3.2.2). Given the measured dependence of ϵ^* on $B_{\mathrm{ext}} + B_{\mathrm{nuc}}$, B_{nuc} can be extracted from ϵ^* obtained at a given B_{ext}. The underlying mechanism is that electron spins can be flipped whenever two states with different electron spins are nearly degenerate. The spin-angular momentum of the electrons can then be transferred to the nuclear spins through the processes described by the terms $\hat{I}^j_+\hat{S}_-$ and $\hat{I}^j_-\hat{S}_+$, which are called electron-nuclear flip-flops. In contrast, for electron spin states that are well separated in energy, the energy mismatch between electron and nuclear Zeeman splitting suppresses such flip-flop processes. By determining the detuning of maximal electron spin flip rate, one can hence deduce ϵ^* and thus B_{nuc}.

3.5.5.3.3 Nuclear spin dynamics

In order to understand how nuclear spin fluctuations affect electron spin coherence in detail, which helps to implement remedies, it is useful to consider the dynamics of the nuclear spins themselves. Two different pictures are useful. At short time scales, the fluctuations of the Overhauser field are given by a varying hyperfine interaction due to the nuclear Larmor precession and a few isolated flip-flop events between different nuclear spins driven by the term $J_{\mathrm{Dipole,n}} \cdot \hat{I}^{j_1}_+ \hat{I}^{j_2}_-$ within the Hamiltonian, which is caused, e.g., by a dipolar coupling of strength $J_{\mathrm{Dipole,n}}$. At longer time scales, series of nuclear flip-flops lead to a diffusion-like redistribution of the local nuclear polarization, which gets laterally shifted between the areas inside and outside of a quantum dot. Hence, the Overhauser field in the quantum dot changes with time. The cross-over between these two different regimes is roughly given by the coupling strength between the nuclear spins ($J_{\mathrm{Dipole,n}}/h \sim 10\,\mathrm{kHz}$).

The diffusive long-time behavior of the nuclear polarization has been probed by directly measuring the fluctuations of the Overhauser field [68] using methods discussed in Section 3.5.5.3.2. Characterization up to 100 Hz shows a Lorentzian-like spectrum of the nuclear hyperfine field with roll-off frequencies (reduction of $S_\beta(\omega)$ by a factor of 1/2) in the range of 0.1 to 10 Hz (Fig. 3.38(a)). The relaxation dynamics of a dynamically induced nuclear polarization [69], which occurs on a time scale of tens of seconds to minutes, draws a similar picture. At low magnetic fields ($\lesssim 20\,\mathrm{mT}$), one finds an about tenfold speedup of spin diffusion, i.e., dephasing via the Overhauser field [68] that likely reflects the activation of additional nuclear spin diffusion channels by the reduced Zeeman energy mismatch between electrons and nuclei, such that electron mediated spin transfer between nuclear spins can take place. Such electron mediated nuclear spin diffusion also leads to a dependence of the decay rate of an induced nuclear spin polarization on the occupancy of the dot with electrons, since the later flip-flop process requires the presence of an electron [69].

The short-time nuclear spin dynamics have been probed via electron spin dephasing under a Hahn-echo sequence (Section 3.5.2.4), i.e., inversion of the electronic spin state halfway through an interval of free evolution. This technique is only sensitive to the changes of the Overhauser field during the evolution, whereas the effects of slower fluctuations as well as of the static magnetic field are canceled. It thus gives detailed insight into the nuclear spin dynamics on the microsecond time scale. Figure 3.38(b) shows such measurements on double quantum dots [57]. The monotonic decay of the Hahn-echo signal with total evolution time, τ, at high fields is a result of the diffusive dynamics of B^z_{nuc} due to the dipolar coupling $J_{\mathrm{Dipole,n}}$ [71, 72]. The oscillations found at lower fields, which eventually turn into full collapses and revivals, were first predicted based on a fully quantum mechanical treatment of the interaction between nuclear spins and electron spins[73, 74], but can also be understood intuitively with a semiclassical model [70]. The semiclassical model, as outlined below, is in excellent agreement with the data. Hence, we have another example where the full complexity of the quantum mechanical description is not required to obtain an accurate picture for the relevant mechanisms.

For each electron spin, the Zeeman energy is proportional to the total magnetic field $B_{\mathrm{tot}} = \sqrt{(B_{\mathrm{ext}} + B^z_{\mathrm{nuc}})^2 + {B^{\perp}_{\mathrm{nuc}}}^2} \approx B_{\mathrm{ext}} + B^z_{\mathrm{nuc}} + {B^{\perp}_{\mathrm{nuc}}}^2/2B_{\mathrm{ext}}$ (Fig. 3.38(c), top left). Dephasing is caused by fluctuations of this level splitting and is thus related to the time-dependence of both the parallel and the transverse nuclear spin leading to components, B^z_{nuc} and ${B^{\perp}_{\mathrm{nuc}}}^2$, respectively. The time-dependence of B^z_{nuc} is mainly caused by nuclear spin diffusion (SD) via $\hat{I}^{j_1}_+\hat{I}^{j_2}_-$ as discussed above and is predicted to cause a $\exp(-(\tau/T_{\mathrm{SD}})^4)$ decay of the echo signal of the spin qubit [71, 72]. This decay law is generic for noise with low weight at high frequencies. The upper cutoff in the fluctuation spectrum originates from the required time for individual flip-flops to occur, i.e., the nuclear spin flip-flop rate.

The origin of the collapses and revivals in the echo signal at intermediate fields (0.07–0.2 T) provides a rather intuitive understanding of the nuclear dynamics causing the dephasing. The collapses arise from an oscillation associated with ${B^{\perp}_{\mathrm{nuc}}}^2$. The transverse nuclear field, $\vec{B}^{\perp}_{\mathrm{nuc}}$, is a vector sum of contributions from the three nuclear species ^{69}Ga, ^{71}Ga and ^{75}As (Fig. 3.38(c), top right), present in GaAs. Due to the different precession rates of these species, one gets three oscillation frequencies of $B^{\perp}_{\mathrm{nuc},i}$, which add up to a beating pattern (Fig. 3.38(c), bottom). While the Larmor precession frequencies of the different species are fixed, the amplitude and phase fluctuates over the course of many repetitions. Hence, the resulting phase is randomized and the echo signal is suppressed. However, if the precession period of the electron is a multiple of all three relative Larmor periods of the nuclei, which is approximately possible due to a fortuitous equidistant spacing of these frequencies, the resulting oscillation of the Overhauser field imprints no fluctuating net phase on the precessing electron spin

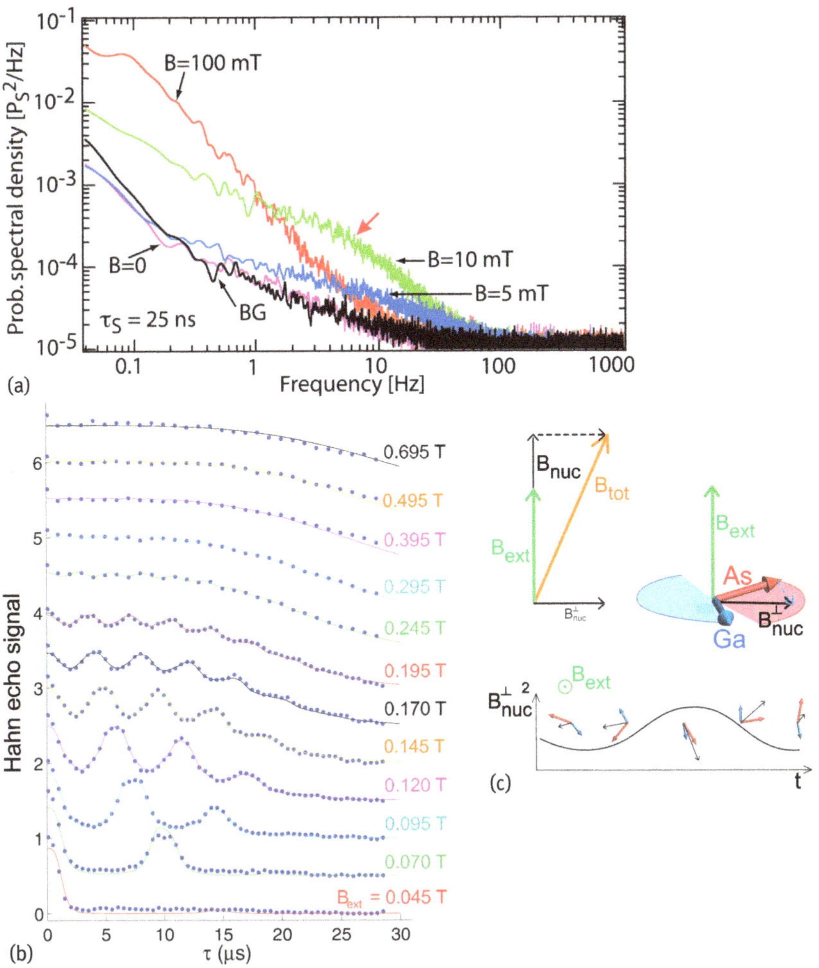

Fig. 3.38: Dynamics of nuclear spins in a double quantum dot. (a) Frequency spectra of the fluctuations of the nuclear hyperfine field, obtained from time traces of the singlet probability of the qubit after precession in the Overhauser field over a fixed evolution time τ_S = 25 ns. A significant speedup of the dynamics is observed at low magnetic fields (< 10 mT), where the high frequency signal around 10 Hz is larger than at higher B fields (red arrow at the 10 mT curve). The mostly Lorentzian shape (Fig. 3.34) of the spectra can be explained in terms of nuclear spin diffusion. (b) Hahn echo signal from a S–T_0 qubit as a function of the total evolution time τ, for different values of magnetic field. Exchanging the two electrons at time $\tau/2$ via a gate voltage pulse causes them to see the same static hyperfine field, so that only fluctuations during τ reduce the probability of the electrons to return to their initial state (Fig. 3.32). The maxima in the oscillating signal reflect the fact that the fluctuations of the Overhauser field cancel at particular evolution times of the qubit. (c) Illustration of the semiclassical model [70] used for the fits (solid lines) in (b). The top left image shows the contributions to the local effective field. The upper right image visualizes the different phases of the different nuclear spins due to different precession rates in the external field. This leads to different contributions to the transversal hyperfine field $B^{\perp}_{nuc}$, which evolve with time and can partly cancel. The resulting total transversal hyperfine field as a function of time is shown in the lower image with the constituting components of two nuclear fields marked. (a) after [68], (b), (c) after [57].

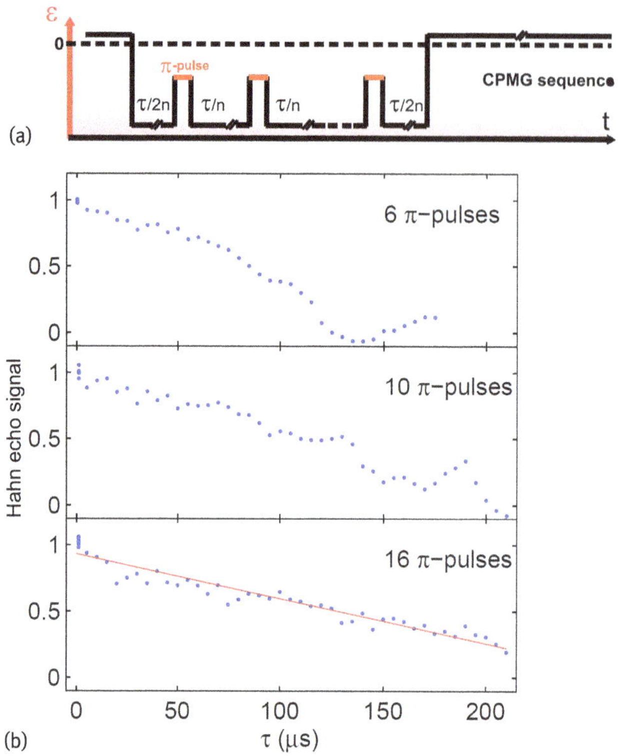

Fig. 3.39: Extension of the coherence time of a $S-T_0$ qubit via dynamical decoupling. (a) CPMG pulse sequence consisting of a string of n Hahn echo pulses. (b) Hahn echo signal under different CPMG sequences. The linear fit to the 16-pulse data (red line) intersects the x axis at $\tau = 276$ μs, which can be taken as a rough estimate or lower bound of the coherence time T_2. (a), (b) [57].

and the echo amplitude revives. This intuitive picture can be turned into a quantitative model by treating the three components of $\vec{B}^{\perp}_{\text{nuc}}(t)$ as classical random variables and averaging over the Gaussian distribution of their initial values. The model also explains the faster decay of the echo envelope at low fields (Fig. 3.38(b)) in terms of dephasing of the nuclear spin precessions themselves.

While the relatively simple Hahn echo sequence is convenient for detailed modeling of the complex dephasing processes arising from the Larmor precession of the nuclei, it is not ideal for optimizing the lifetime of a superposition state of the electron spins. Indeed, substantially longer coherence times were obtained with a CPMG sequence [60], which consists of a n-fold repetition of the Hahn-echo, thus requiring n π pulses as shown in Fig. 3.39(a). As shown in Fig. 3.35(a), the filter function of this sequence increases more slowly from $\omega = 0/\text{s}$ with increasing frequency than for the Hahn echo sequence. Thus, noise components at small but nonzero frequencies, up to about $\omega t = 5$ in Fig. 3.35(a), contribute much less to the total phase variance $\langle \phi^2(t) \rangle$

according to eq. (3.73). For the system at hand, this protection can be expected to be particularly effective at fields above ~ 0.2 T, where oscillations of the Hahn echo signal are not observable and a smooth decay due to low-frequency fluctuations occurs (Fig. 3.38(b)). This expectation is confirmed by the data (Fig. 3.39(b)), which showcase the results of CPMG-experiments for $n = 6$, 10 and 16 pulses at $B_{\text{ext}} = 0.4$ T. For $n = 16$, the echo signal persists for more than 200 µs. Experiments with up to 1600 pulses have extended the decay time to about 1 ms. It is likely that an improved pulse sequence, based on a better quantitative understanding of the influence of the hyperfine induced noise and different pulse sequences, can lead to even larger coherence times.

3.6 Outlook

This chapter of the book describes quantum computing as a relatively concrete prospective application building on many concepts that emerged from fundamental quantum mechanical considerations. Realizing solid state qubit devices requires, moreover, a detailed understanding of solid state concepts like band structure, superconductivity, noise processes, nanofabrication and many others.

Since the largely curiosity-driven first proof-of-principle experiments, the field has advanced tremendously. For both, superconducting and semiconductor qubits, the coherence time, which is a key figure of merit, has improved by orders of magnitude. It exhibits a growth similar to Moore's law of conventional electronics. Researchers have learned how to initialize, read and couple qubits with an accuracy that approaches and in some cases fulfills the requirements for error correction. Based on this impressive success, many workers now believe that it is just a matter of time until useful quantum computers will become reality. This optimism has already triggered considerable yet largely exploratory investment by IT companies and motivated focused efforts to realize multi-qubit quantum processors.

Nevertheless, a closer examination reveals that many challenges have yet to be overcome. For transmons, the so far most successful superconducting qubits, it appears at the time of writing (2018) that the relatively low anharmonicity makes it difficult to avoid the occupation of higher excited states outside the computational subspace when operating multi-qubit devices. This severely impacts the gate fidelity. The first efforts to realize quantum circuits with up to tens of qubits resulted in a substantial degradation of the fidelity. For semiconductor qubits, accurate two-qubit operations are a matter of active research, and the step to larger qubit numbers is yet to be made. A remaining qualitative challenge is to realize inter-qubit coupling over distances of a few micrometers. This seems necessary in devices with more than a handful of qubits in order to provide the required space to connect wires to all qubits in an on-chip architecture.

Both superconducting and semiconductor qubits still leave room for reducing decoherence. For example, material imperfections can give rise to charge and flux noise, which limits the coherence time of the qubits. Some of these effects are still poorly understood and warrant more basic condensed matter research. Moreover, it is not clear that the qubit types that are most advanced today will prevail and turn out to be truly scalable. Besides the improvement of well-established qubits, new ideas still emerge. A particularly hot topic is the quest to realize so-called Majorana qubits, which might provide superior coherence properties and gate operations by leveraging topological effects to protect against errors. One suggested approach builds on topological materials (Sections 5.4.6 and 5.4.7). Another one is based on a combination of proximity-induced superconductivity and spin-orbit coupling in semiconductor nanowires.

Beyond quantum computation which is currently the biggest challenge in quantum technology research, many other applications arise from the detailed understanding of the dynamical quantum behavior of physical objects. They range from quantum communication employing photons for the secure generation of keys for message encryption [27] to different types of improved sensing by exploiting quantum degrees freedom. For example, one can employ a spin-polarized defect center in diamond, the so-called nitrogen-vacancy center, which provides, e.g., very sensitive temperature and magnetic field sensing at nm spatial resolution under ambient conditions [75]. This offers major perspectives for analytic tools both in biology and chemistry. Additionally, this kind of research will help to obtain a better intuition for the unusual consequences of quantum physics including the non fully established nature of the transition from quantum objects to macroscopic objects.

Appendix: Definitions

$\hat{\sigma}$-matrices

The $\hat{\sigma}$-matrices, also called Pauli matrices, are given by

$$\hat{\sigma}_x = \begin{pmatrix} 0 & 1 \\ 1 & 0 \end{pmatrix}, \quad \hat{\sigma}_y = \begin{pmatrix} 0 & -i \\ i & 0 \end{pmatrix}, \quad \hat{\sigma}_z = \begin{pmatrix} 1 & 0 \\ 0 & -1 \end{pmatrix}. \tag{3.80}$$

For our purposes, it is convenient to work in the basis $\{|1\rangle, |0\rangle\}$, so that $\hat{\sigma}_z|1\rangle = |1\rangle$ and $\hat{\sigma}_z|0\rangle = -|0\rangle$.

Thomas Brückel

4 Correlated electrons in complex transition metal oxides

4.1 Introduction

In the previous chapters of this book, we have discussed the consequences of quantum mechanics on the properties of solid state electrons mostly on the single or two particle level. This revealed the important influence of the phase of the electronic wave functions in mesoscopic electronic transport and a general understanding of the optical properties of solids, where the interaction between the electrons led only to relatively simple modifications such as the excitonic binding energy $E_{\mathrm{Ryd,X}}$ or the dielectric constant of the material ε. In addition, we have learned how to gain an unprecedented control of the quantum mechanical properties, including the dynamics, for single-electron and two-electron systems in spin qubits or in many-particle states in superconducting qubits. There, the electron-electron interaction was mostly used as an exchange coupling or as a classical repulsive energy for read-out.

In the following chapters of the book, we will deal with more complex electron systems, where the Coulomb repulsion between many electrons is decisive for the resulting electronic structure. These materials are said to exhibit strong electronic correlations, i.e., the movement of one electron depends on the positions and movements of all other electrons due to the long-range Coulomb interaction. With this definition, one could naively think that all materials should show such strong electronic correlations. However, in purely ionic systems, the electrons are confined to the immediate neighborhood of the respective atomic nucleus. Because of the strong ionic bond to the nucleus, electronic excitations in the eV range are needed to remove an electron from its nucleus. This suppresses the influence of the slightly smaller electron-electron correlation energies. On the other hand, in ideal metallic systems, the conduction electrons screen the long-range Coulomb interaction between two chosen electrons. Electrons in the vicinity of any given conduction electron are being pushed away, leaving a surplus of positively charged atomic cores thus largely compensating for the negative charge of the electron. Therefore, while electronic correlations are also present in these systems and lead, for example, to magnetism, the main properties can be explained in simple models, where electronic correlations are either entirely neglected (e.g., the free electron Fermi gas as a model for simple metals as Na) or taken into account only in low order approximations (Fermi liquid, exchange interaction). In highly correlated electron systems, such simple approximations can qualitatively fail and entirely new phenomena appear, implying also related novel functionalities. These so-called emergent phenomena are not predictable by well-controlled material-specific

https://doi.org/10.1515/9783110438321-004

model descriptions, i.e., they cannot be anticipated from the static, local interactions among the electrons and between the electrons and the lattice [76]. This is a typical example of complexity, where the laws that describe the behavior of a complex system are qualitatively different from those that govern its units [77]. It makes highly correlated electron systems a challenging research field at the forefront of condensed matter research. The central challenge is that one cannot reliably predict the properties of these materials starting from first principles, i.e., there is no theory, which can handle the huge number of interacting degrees of freedom. While the underlying fundamental principles of quantum mechanics (Schrödinger equation, relativistic Dirac equation) and statistical mechanics (maximization of entropy under given boundary conditions) are well known, there is currently no way to solve the many-body problem for the $\sim 10^{23}$ particles of a solid.

Exemplary properties of strongly correlated electron systems providing emergent phenomena and/or novel functionalities are:

High temperature superconductivity

This phenomenon was first reported in 1986 by Bednorz and Müller [78] (Fig. 4.1), who received the Nobel Prize for this discovery. There is still no commonly accepted mechanism for the coupling of electrons into Cooper pairs, let alone a theory which can predict high temperature superconductivity or its transition temperatures quantitatively. This lack of understanding is the more surprising, when one considers the huge

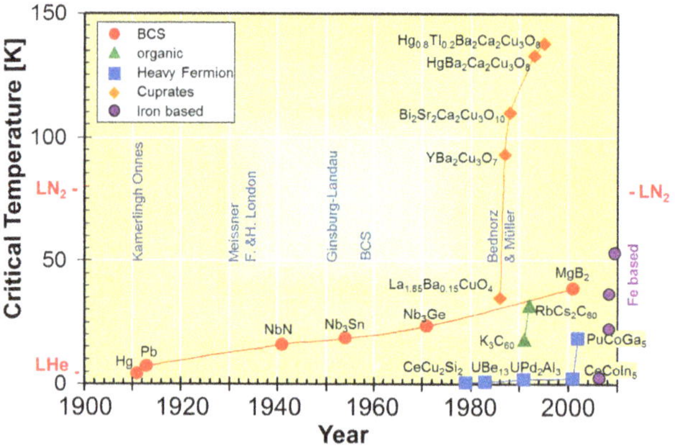

Fig. 4.1: History of superconductivity up to 2010: Besides some breakthrough discoveries or theories (vertically written names), the development of the critical temperature for superconductivity versus the year of its discovery is plotted. The compounds marked with red dots follow the famous BCS theory, while for the high temperature superconductors (diamonds) no generally accepted theory exists. Note, that we only show values for the critical temperature under ambient pressure. In 2015, it has been demonstrated that superconductivity in H_2S sets in already at 203 K, but at an extremely high pressure around 150 GPa.

number of solid-state physicists worldwide, who are trying to solve this problem. High temperature superconductivity has, nevertheless, already some applications like for so-called SQUID (superconducting quantum interference device) magnetic field sensors (Section 3.3.5), superconducting generators or motors, and high field magnets, but might in the future have even economically more relevant applications, e.g., for loss-free energy storage or energy transport.

Colossal magnetoresistance effect (CMR)
The CMR was discovered in transition metal oxide manganites. It describes a large change of the electrical resistance in an applied magnetic field [79], if compared to the field free situation. The effect can be used in magnetic field sensors and is related to the giant magnetoresistance effect[1], which is employed, e.g., in the read heads of magnetic hard discs.

The magnetocaloric effect [81]
This describes a temperature change of a material in an applied magnetic field and can, for example, be used for magnetic refrigeration. Magnetic refrigeration in, e.g., air conditioning systems could save up to 30% of electric energy compared to the conventional vapor cycle cooling and also employs no environmentally hazardous refrigerant liquids. This effect occurs also in materials without strong electronic correlations, if a strong spin-lattice coupling is present.

The multiferroic effect [82]
This describes the simultaneous occurrence of various ferroic orders in one material. This could be ferromagnetism, ferroelectricity or ferroelasticity. If the respective degrees of freedom are strongly coupled, one can switch one of the orders by applying the conjugate field of the other order parameter. In certain multiferroic materials, the application of a magnetic field can switch the ferroelectric polarization or the application of an electric field can switch the magnetization of the material. Future applications of multiferroic materials in computer storage elements are apparent. One could either imagine elements, which store several bits in form of magnetic and electric polarization, or one could apply the multiferroic properties for an easy switching of the memory element by one order (e.g., ferroelectricity) and an easy read-out by another order (e.g., ferromagnetism).

1 The giant magnetoresistance effect [80] is an effect that occurs in artificial magnetic thin film multilayers. It was discovered independently by P. Grünberg and A. Fert, who received the Nobel Prize in 2007 for their discovery.

Metal-insulator-transitions

Such a transition is observed e.g., in magnetite (Verwey transition [83], Fig. 4.2) or certain vanadites. It is due to strong electronic correlations and could be employed as electronic switches, where the resistivity of the material changes by many orders of magnitude upon change of an external thermodynamic parameter (e.g., temperature, pressure, field).

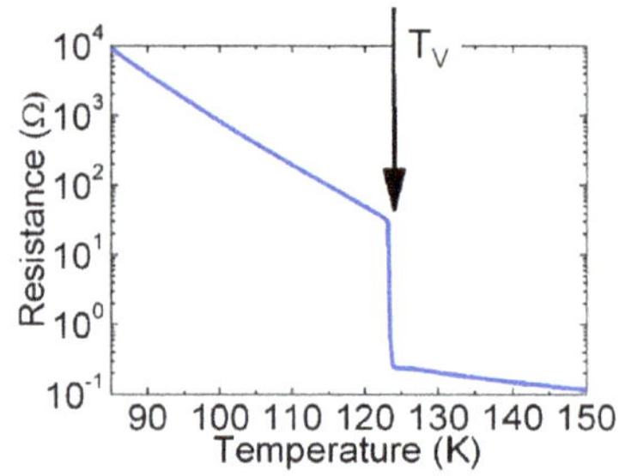

Fig. 4.2: Electrical resistance as a function of temperature for a magnetite Fe_3O_4 sample. The Verwey-transition is visible from the step in the resistance at $T_V \approx 123$ K.

Negative thermal expansion [84]

This describes the decrease in volume with increasing temperature and is another example of the novel properties of these materials. Such negative thermal expansion can originate from a coupling of different phonon modes such as in graphene, but can also have an electronic origin. Most obvious, phase transitions as discussed in this chapter of the book lead to a different occupation of orbitals in a crystal, which can lead to an expansion of the material with decreasing temperature. Changes in the electronic structure due to correlations can also appear continuously with temperature such that the size of the unit cell changes continuously via electronic effects. Negative thermal expansion coefficients ranging down to $-4 \cdot 10^{-5}$/K have been observed, e.g., for $CeAl_3$ at temperatures below $T = 1$ K [84].

It is likely that many more of such emergent phenomena will be discovered in the near future, which makes research on highly correlated electron systems an exciting topic. At the same time, the research is quite challenging due to the missing unified theory. This research area is, hence, located at the intersection between fundamental science, striving for basic understanding of the electronic correlations, and possible technological applications, connected to the new functionalities [85].

Here, we will focus on a fundamental description of a few well understood phenomena, which are selected to serve as a base to tackle more complex subjects in correlation physics.

4.2 Electronic structure of solids

In order to be able to tackle the effects of strong electronic correlations, we will first recapitulate the textbook knowledge of the electronic structure of solids [86]. The description usually starts with the adiabatic or Born-Oppenheimer approximation. The argument is made, that the electrons are moving so much faster than the nuclei, that they can instantaneously follow the movement of the much heavier nuclei. Thus, they feel a static nuclear potential. This approximation serves to separate the lattice and electronic degrees of freedom. Often, one makes a further approximation and considers the nuclei to be at rest in their equilibrium positions.

The potential energy as seen by a single electron in the averaged field of all other electrons and the atomic core potential is depicted schematically (for a one dimensional system) in Fig. 4.3. The following simple models are typically used to describe the electrons in such a crystalline solid:

Free electron Fermi gas: Here, a single electron moves in a three dimensional potential well with infinitely high walls corresponding to the crystal surfaces. All electrons move completely independent, i.e., the Coulomb interaction between the electrons is not considered explicitly. Only the Pauli exclusion principle is being taken into account in order to eventually occupy the resulting single-particle states via the Fermi–Dirac distribution function.

Fermi liquid: Here the electron-electron interaction is accounted for in a first approximation by introducing quasiparticles, so-called dressed electrons, which have a charge e, and a spin $\frac{1}{2}$ like the free electron, but an effective mass m^*, which can differ from the free electron mass m_e (Section 1.2.2). This model has been mostly used in Chapter 1 of the book[2]

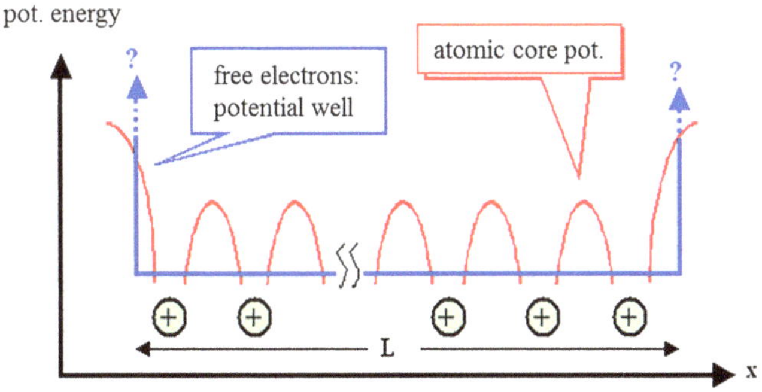

Fig. 4.3: Schematic plot of the potential energy of an electron in a one-dimensional solid, once in the case of free electrons, where the electron is described as moving in a potential well with infinitely high walls (blue) and once taking into account the potential arising from the Coulomb interaction with the atomic cores (turquoise circles marked by +), which is periodic in the infinite solid (red).

2 Note, however, that in Section 1.2.2 the effective mass has been introduced as being due to an electron–nucleus interaction, while here it is due to electron–electron interactions.

Band structure model: This model takes into account the periodic potential of the atomic cores at rest, i.e., the electron moves in the static potential of the atomic cores and the average potential from the other electrons. The model results in Bloch waves as the single-particle wave functions for the quasi-particles (eq. (2.90)). This model was the base for Chapter 2 and 3 of the book.

Considering the strength of the long-range Coulomb interaction, it is surprising that these simple models describe basic properties of many materials rather adequately. The band structure model is even successful in describing multiple properties quantitatively accurate, if the electron-electron interaction is approximated as a fine-tuned static repulsive potential acting on a particular electron (Section 5.1.2). All three models have in common that the electron is eventually described by a single particle wave function. Electronic correlations are typically only crudely approximated, e.g., via the exchange interaction[3], which accounts for phenomena such as ferromagnetism, or via the weak attractive interaction mediated through lattice vibrations, which explains superconductivity in the BCS model[4] employing the resulting pairing of electrons to Cooper pairs, which itself can form a Bose–Einstein condensate.

So far, we have sketched the typical description of introductory textbooks for solid state physics. Of course, there exist more advanced theoretical concepts, which try to take into account the many-particle electronic correlations. The strong Coulomb interaction between the electrons is considered, e.g., in density functional theory by the so-called "LDA+U" approximation,[5] in so-called dynamical mean field theory (DMFT) or by a combination of the two in various degrees of sophistication [87]. Still, all these extremely powerful and complex theories often fail to predict even the most simple physical properties, such as whether a material is a conductor or an insulator.

Let us come back to the band structure of solids, describing it from another viewpoint. In the so-called tight binding model, one conceptually starts from isolated atoms, where the energy levels of the electrons in the Coulomb potential of the corresponding nucleus is firstly calculated (Fig. 4.4(a)). If such atoms are brought together, the wave functions of the electrons from different sites start to overlap, leading to a broadening of the atomic energy levels via different linear combinations of levels from different atoms (Fig. 4.4(b)). This eventually leads to the electronic bands in solids. The closer the atoms are brought together, the more the wave functions overlap, the more the electrons will be delocalized, and the broader are the corresponding bands. This relationship between constituting orbitals and bandwidths is depicted graphically as a function of atomic number in Fig. 4.4(c).

---

3 This term, a specific quantum mechanical part of the Coulomb interaction, is described in detail in Section 4.4.1.

4 J. Bardeen, L. N. Cooper, and J. R. Schrieffer, Nobel Prize in Physics 1972.

5 LDA: local density approximation.

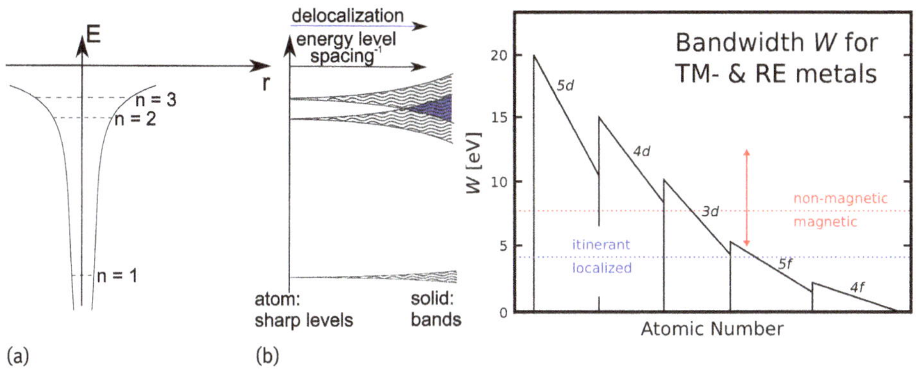

Fig. 4.4: Development of a band structure starting from isolated atoms such as in a simple tight binding model (schematic). (a) Potential energy for an electron interacting with an atomic core (solid lines) and the corresponding sharp single particle energy levels (dashed lines). (b) Atomic energy levels broaden into bands, the more the wave functions of neighboring atoms overlap. (c) Schematic of band width as a function of atomic number for the rare-earth and transition metals. Below a certain band width W, the electrons remain localized (dotted blue line). For partially filled shells such electrons can be magnetically ordered. But even itinerant electrons can remain magnetic up to a certain band width (dotted red line). Above a band width of typically 8 eV, the electrons will be itinerant, hence, the material will be metallic and non-magnetic.

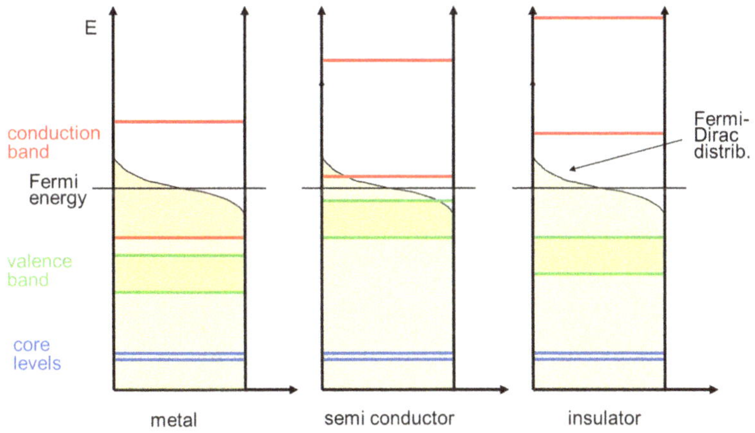

Fig. 4.5: Schematic band structure for a simple metal, a semiconductor and an insulator. Bands are the areas marked blue, green, and red, which are plotted together with the Fermi–Dirac distribution function (yellow area).

If electronic correlations are not too strong, the electronic properties can be described by such a band structure, which allows one to predict, whether a material is a metal, a semiconductor or an insulator. This is shown in Fig. 4.5. At $T = 0\,\mathrm{K}$, all electronic states are being filled up to the Fermi energy E_F. At finite T, the Fermi–Dirac distribution $f(E, T)$ (eq. (1.39)) describes the occupancy of the energy levels. If the Fermi energy lies somewhere in the middle of the conduction band, the material will be metallic, at least, at low disorder (Section 1.6). If it lies in the middle between valence band and conduction band, and these two are separated by a large energy gap, the material will show insulating behavior. Finally, if the gap is small enough to allow thermal excitations of electrons from the valence band to the conduction band, we have semiconducting behavior. We emphasize again that this band structure model describes the electrons by single particle wave functions. Thus, one might ask: Where are the electronic correlations?

4.3 Electronic correlations

It turns out that electronic correlations are particularly important in materials, which have some very narrow bands. This occurs for example in transition metal oxides or transition metal chalcogenides[6] as well as in some light rare earth intermetallics (heavy fermion systems). Let us chose CoO as a typical and simple example of a transition metal oxide. CoO has a rock salt structure as depicted in Fig. 4.6.

The most symmetric unit cell as depicted in Fig. 4.6 is face centered cubic (fcc) and contains four formula units. The primitive unit cell of the fcc lattice is, however, smaller. It is spanned by the basis vectors

$$\vec{a} = \frac{1}{2}a(\vec{e}_x + \vec{e}_y)\,, \quad \vec{b} = \frac{1}{2}a(\vec{e}_y + \vec{e}_z)\,, \quad \vec{c} = \frac{1}{2}a(\vec{e}_z + \vec{e}_x)\,. \tag{4.1}$$

Here a is the lattice constant, and $\vec{e}_x$, $\vec{e}_y$, $\vec{e}_z$ are the unit basis vectors of the original fcc unit cell. Therefore, the primitive unit cell contains exactly one cobalt and one oxygen atom. The electronic configurations of the neutral atoms are: Co: $[\mathrm{Ar}]3\mathrm{d}^7 4\mathrm{s}^2$; O: $[\mathrm{He}]2\mathrm{s}^2 2\mathrm{p}^4$. In the solid, the atomic cores of Co and O have, hence, the electronic configuration of Ar and He, respectively. These core electrons are very strongly bound

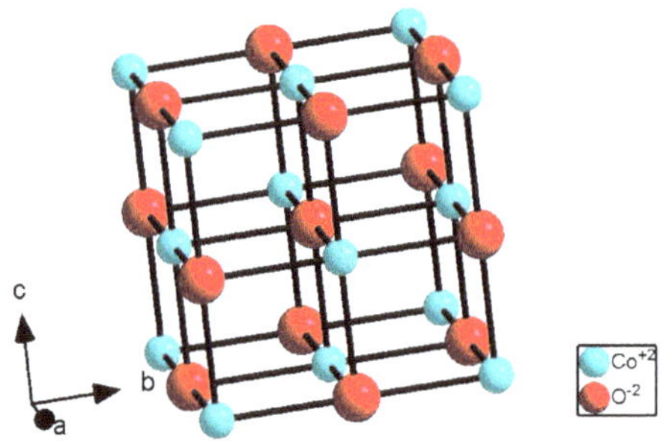

Fig. 4.6: Unit cell of CoO (rock salt structure), which exhibits the face centered cubic (fcc) structure for each sublattice implying four formula units per unit cell.

6 Chalcogenides are compounds of the heavier chalcogens (group VI elements of the periodic table, particularly sulfides, selenides, tellurides). Albeit oxygen is in the same group of the periodic table, oxides are usually not called chalcogenides.

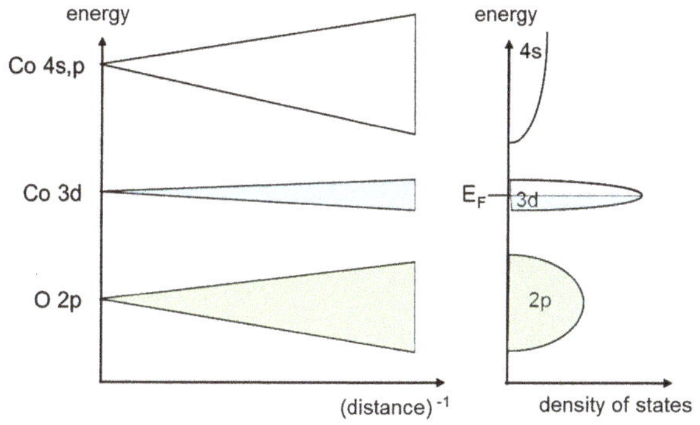

Fig. 4.7: Expected schematic band structure of a simple 3d-metal oxide, such as CoO, according to the tight binding model. The oxygen 2p-orbitals are fully occupied with six electrons per primitive unit cell, i.e., they lie below the Fermi level. Also the oxygen 2s-orbitals are filled completely, by two electrons per unit cell (not shown). According to the arguments given in the text, the 3d band must then be partially occupied by seven electrons per unit cell, such that we get a metal.

to the nucleus and need not to be considered for the usual energy scales for excitations in a solid. We are left with nine outer electrons for the Co and six outer electrons for the O atom, so that the total number of electrons per primitive unit cell is $9+6 = 15$. Therefore, we have an uneven number of electrons in the primitive unit cell. According to the Pauli principle, each electronic single-particle state can be occupied by two electrons, one with spin up and one with spin down. Therefore, with an uneven number of electrons, we must have at least one partially filled band and CoO must be a metal. Via the tight binding approximation of band theory, the naively expected electronic band structure of CoO can be, thus, graphically constructed as plotted in Fig. 4.7.

What does the experiment tell us? In fact, CoO is a very good insulator as the resistivity at room temperature amounts to $\rho \sim 10^8$ Ωcm. This value can be compared to a good conductor like iron, which has a resistivity of about 10^{-7} Ωcm. The resistivity of CoO corresponds to activation energies of about 0.6 eV or a temperature equivalent of $0.6\,\text{eV}/k_B \simeq 7000\,\text{K}$, which means there is a large band gap making CoO a good insulator. Hence, the band theory, predicting metalic behavior, already breaks down for a very simple oxide, consisting of only one transition metal and one oxygen atom.

In order to understand the reason for this dramatic failure of band theory, let us consider an even simpler example: the alkali metal sodium. It has the electronic configuration: $[\text{Ne}]3s^1 = 1s^2 2s^2 2p^6 3s^1$. Following our argumentation for CoO, sodium obviously has a half-filled 3s band and, therefore, is a metal. This time, our prediction is correct, i.e., the electrical resistivity at room temperature is about 5×10^{-6} Ωcm. However, what happens, if we pull the atoms further apart and increase the lattice constant continuously? The simple band theory argument predicts that sodium remains a metal for all distances, since the 3s band will always be half-filled. This contradicts our intuition. At a certain critical separation of the sodium atoms, there must be a transi-

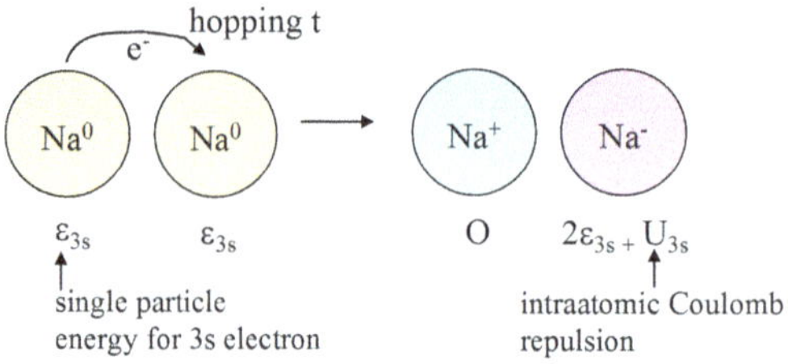

Fig. 4.8: Illustration of the hopping process of an electron between two sodium atoms leading to charge fluctuations. The energies of the two different charge configurations are marked below the sketches.

tion from a metal to an insulator. It was Sir Nevill Mott (Nobel Laureate in physics of 1977), who predicted this metal-to-insulator transition, which is therefore called the Mott-transition [88] (see also Section 1.6.2). The physical principle can be made clear with the illustration in Fig. 4.8.

On the left of Fig. 4.8, two neutral sodium atoms are depicted. The atomic energy levels of the outer electrons correspond to an energy ϵ_{3s}. The wave functions of the 3s electrons overlap giving rise to a finite probability that an electron can hop from one sodium atom to the other one. Such a delocalization of the electrons is favored according to the Heisenberg uncertainty principle

$$\Delta p \Delta x \geq \frac{\hbar}{2} . \tag{4.2}$$

A less localized electron wave function (larger Δx) requires less plane waves with different $p = \hbar k$ (smaller Δp) to be constructed. Since the kinetic energy E_{kin} of an electron plane wave is proportional to p^2, the electron can gain kinetic energy, if it becomes more delocalized. Figure 4.8 on the right shows the situation after the electron transfer. Instead of neutral atoms, there is one Na^+ and one Na^- ion. Thus, we have to pay a price for the double occupation of the 3s states on the Na^- ion, namely the intra-atomic Coulomb repulsion between the two electrons denoted as U_{3s}. This is a very simplistic picture, where we assume that the electron is either located on one or the other Na atom. It describes the two main energy terms by only two parameters, namely the hopping matrix element t, connected to the kinetic energy gain via the overlap of two distinct, atomically centered electronic wave functions (eq. (4.21)), and the intra-atomic Coulomb repulsion U, connected to the potential energy due to the Coulomb interaction between the two electrons at one site (eq. (4.20)). Most importantly, we have replaced the long range Coulomb potential, proportional to $\frac{1}{r}$, with its leading term, the onsite Coulomb repulsion U only. More realistic models would have to take higher order terms into account, but already the simple model employing U and t leads to very rich physics in many body systems. Figure 4.8 also shows that the electronic conductivity requires an energy penalty. It is connected with charge fluctuations on the different sites, where the corresponding charge transfer costs energy U, typically in the order of 1 eV to 10 eV. Only if the gain in kinetic energy due to the hopping t is larger than this penalty in potential energy U, we get metallic behavior for the system. If the sodium atoms are now being separated more and more, the intra-atomic Coulomb repulsion U maintains its value, while the hopping matrix element t, which depends on the overlap of the wave functions, will diminish. At a certain critical value of the lattice parameter a, the penalty in potential energy is larger than the gain in kinetic energy and conductivity is suppressed. This is the physical principle of the Mott-type metal-insulator transition. In the following section, we will formulate some of these arguments more quantitatively, taking into account the spin degree of the electron for the most simple case of a two atom molecule.

4.4 The spin of the electron: exchange interaction

Up to now, we have considered the kinetic energy of the electrons in form of the hopping matrix element t and the leading term of the Coulomb interaction in the form of the intra-atomic Coulomb repulsion U. We have not yet considered the spin degree of freedom of the electrons, which leads to additional quantum mechanical correlation effects. We will study these for the case of a molecule consisting of two hydrogen atoms, as might have been treated previously in a quantum mechanics course. The situation is depicted schematically in Fig. 4.9.

The ground state wave function of the electron at the hydrogen atom A located at $\vec{R}_A$ is the eigenfunction of the Hamiltonian of this atom with the ground state energy as its eigenvalue:

$$\widehat{H}_{\text{Atom}}(\vec{r}-\vec{R}_A)\psi_{1s}(\vec{r}-\vec{R}_A) = E_{1s}\psi_{1s}(\vec{r}-\vec{R}_A) = E_{\text{Atom}}\psi_{1s}(\vec{r}-\vec{R}_A)\,. \tag{4.3}$$

The analoguous expression is valid for atom B. The Hamiltonian for the hydrogen molecule consisting of two hydrogen atoms reads:

$$\begin{aligned}\widehat{H} &= \widehat{H}_{\text{Atom}}(\vec{r}_1-\vec{R}_A) + \widehat{H}_{\text{Atom}}(\vec{r}_2-\vec{R}_B)\\ &\quad + \frac{e^2}{4\pi\varepsilon_0}\left(-\frac{1}{|\vec{r}_1-\vec{R}_B|} - \frac{1}{|\vec{r}_2-\vec{R}_A|} + \frac{1}{|\vec{r}_1-\vec{r}_2|} + \frac{1}{|\vec{R}_A-\vec{R}_B|}\right)\,.\end{aligned} \tag{4.4}$$

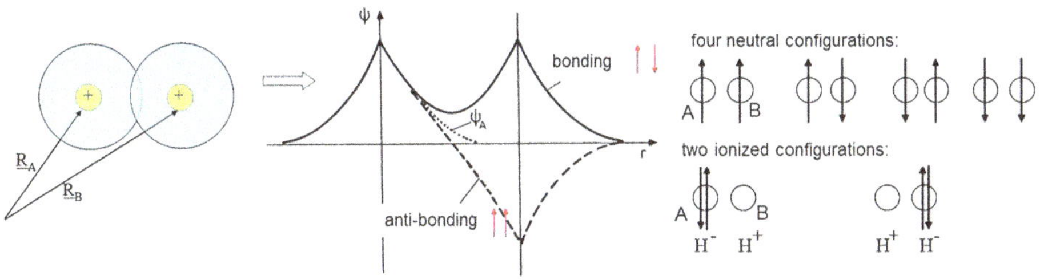

Fig. 4.9: The hydrogen-molecule. Left: schematic plot of the molecule in real space with the positions $\vec{R}_A$ and $\vec{R}_B$ of the nuclei (yellow) and the surrounding 1s electron orbitals (blue). Middle: The molecular orbitals as linear combinations of the 1s hydrogen wave functions in Heitler-London approximation. Right: Semiclassically possible spin configurations of the two electrons (arrows) being located at the different atom cores (top) or at one of the two atom cores (bottom).

The vectors $\vec{r}_1$ and $\vec{r}_2$ denote the positions of the two electrons. The first line denotes the Hamiltonians of the single atoms. The second line consecutively denotes the two attractive Coulomb interactions of the electrons with the respective other nucleus, the Coulomb repulsion between the two electrons, and the Coulomb repulsion between the two nuclei.

We will use perturbation theory to tackle this Hamiltonian, using the abbreviations $\psi_{1s}(\vec{r}-\vec{R}_A) := \psi_A(\vec{r})$ and $\psi_{1s}(\vec{r}-\vec{R}_B) := \psi_B(\vec{r})$ for the single-particle wave functions. Since our approximation assumes that the electrons are always in atomic 1s orbitals, it is sufficient to represent the wave functions by spin arrows at the two atom sites. The six possible configurations are depicted in Fig. 4.9 on the right.

4.4.1 Direct exchange in non-orthogonal orbitals

The overlap integral of the two atomic wave functions is defined as:

$$\int \psi_A^*(\vec{r})\psi_B(\vec{r})\mathrm{d}^3\vec{r} = l_{\mathrm{OV}} \,. \tag{4.5}$$

If the distance between the two protons is large enough and, thus, the overlap integral l_{OV} is small, we can assume that the low lying energy levels mostly belong to the neutral configurations of Fig. 4.9.

Then, we can construct a ground state wave function in the four dimensional Hilbert space, which excludes the two ionized configurations. This is the well-known Heitler-London approximation. Since the Hamiltonian is isotropic in spin space, the total spin S and its component S_z along quantization axis z are good quantum numbers. We therefore construct the eigenstates of the Hamiltonian (4.4) as linear combinations of the four neutral configurations of Fig. 4.9, which are eigenstates of $\hat{\vec{S}}$ and $\hat{S}_z$ with corresponding eigenvalues S and S_z. For fermions, the total antisymmetric basis states of the two-particle system can be constructed by Slater-determinants:

$$\begin{aligned} S=1\,,\quad S_z=1:\ \Psi_1 &= \frac{1}{\sqrt{2(1-l_{\mathrm{OV}}^2)}}\begin{vmatrix}\psi_A(\vec{r}_1)|\uparrow\rangle_1 & \psi_A(\vec{r}_2)|\uparrow\rangle_2\\ \psi_B(\vec{r}_1)|\uparrow\rangle_1 & \psi_B(\vec{r}_2)|\uparrow\rangle_2\end{vmatrix}\\ &= \frac{1}{\sqrt{2(1-l_{\mathrm{OV}}^2)}}|\uparrow,\uparrow\rangle[\psi_A(\vec{r}_1)\psi_B(\vec{r}_2)-\psi_A(\vec{r}_2)\psi_B(\vec{r}_1)] \end{aligned} \tag{4.6}$$

$$\begin{aligned} S=1\,,\quad S_z=0:\ \Psi_2 &= \frac{1}{2\cdot\sqrt{(1-l_{\mathrm{OV}}^2)}}[\psi_A(\vec{r}_1)\psi_B(\vec{r}_2)-\psi_A(\vec{r}_2)\psi_B(\vec{r}_1)]\\ &\quad [|\uparrow,\downarrow\rangle+|\downarrow,\uparrow\rangle] \end{aligned} \tag{4.7}$$

$$S=1\,,\quad S_z=-1:\ \Psi_3 = \frac{1}{\sqrt{2(1-l_{\mathrm{OV}}^2)}}[\psi_A(\vec{r}_1)\psi_B(\vec{r}_2)-\psi_A(\vec{r}_2)\psi_B(\vec{r}_1)]|\downarrow,\downarrow\rangle \tag{4.8}$$

$$S = 0\,, \qquad S_z = 0:\ \Psi_4 = \frac{1}{2\cdot\sqrt{(1+l_{\mathrm{OV}}^2)}}[\psi_A(\vec r_1)\psi_B(\vec r_2) + \psi_A(\vec r_2)\psi_B(\vec r_1)]$$
$$[|\uparrow,\downarrow\rangle - |\downarrow,\uparrow\rangle]\,. \tag{4.9}$$

In first order perturbation theory, the energy eigenvalues are obtained as expectation values of the Hamiltonian (eq. (4.4)) for triplet (eqs. (4.6)–(4.8)) and singlet states (eq. (4.9)):

$$E_S = \langle\Psi_4|\widehat{H}|\Psi_4\rangle = 2E_{\mathrm{Atom}} + \frac{\widetilde{C}+\widetilde{J}}{1+l_{\mathrm{OV}}^2} \tag{4.10}$$

$$E_T = \langle\Psi_1|\widehat{H}|\Psi_1\rangle = 2E_{\mathrm{Atom}} + \frac{\widetilde{C}-\widetilde{J}}{1-l_{\mathrm{OV}}^2} \tag{4.11}$$

with the Coulomb-term

$$\widetilde{C} = \int d^3\vec r_1 \int d^3\vec r_2 |\psi_A(\vec r_1)|^2 \frac{e^2}{4\pi\varepsilon_0|\vec r_1 - \vec r_2|}|\psi_B(\vec r_2)|^2$$
$$- \int d^3\vec r_1 \frac{e^2}{4\pi\varepsilon_0|\vec r_1 - \vec R_B|}|\psi_A(\vec r_1)|^2 - \int d^3\vec r_2 \frac{e^2}{4\pi\varepsilon_0|\vec r_2 - \vec R_A|}|\psi_B(\vec r_2)|^2 \tag{4.12}$$

and the exchange term

$$\widetilde{J} = \int d^3\vec r_1 \int d^3\vec r_2 \frac{e^2}{4\pi\varepsilon_0|\vec r_1 - \vec r_2|}\psi_A^*(\vec r_1)\psi_B(\vec r_1)\psi_B^*(\vec r_2)\psi_A(\vec r_2)$$
$$- l_{\mathrm{OV}} \int d^3\vec r_1 \frac{e^2}{4\pi\varepsilon_0|\vec r_1 - \vec R_B|}\psi_A^*(\vec r_1)\psi_B(\vec r_1)$$
$$- l_{\mathrm{OV}} \int d^3\vec r_2 \frac{e^2}{4\pi\varepsilon_0|\vec r_2 - \vec R_A|}\psi_B^*(\vec r_2)\psi_A(\vec r_2)\,. \tag{4.13}$$

Both terms, $\widetilde{C}$ and $\widetilde{J}$, contain the interaction between the electrons and the interaction of an electron with the respective other nucleus. The splitting between triplet and singlet states amounts to:

$$\Delta E_{TS} = E_T - E_S = 2\frac{l_{\mathrm{OV}}^2\widetilde{C} - \widetilde{J}}{1 - l_{\mathrm{OV}}^4}\,. \tag{4.14}$$

Within the four-dimensional Hilbert space considered here, i.e., for the low lying neutral configurations, the effect of the full Hamiltonian (4.4) can, hence, be described by a simpler effective Hamiltonian, which acts on the spin operators only. Using the relation for the Eigenvalue of the square of the spin operator $\widehat{\vec S}^2 = \hbar^2 S(S+1)$ for electron spins with spin quantum number $S = 1/2$, one obtains:

$$(\widehat{\vec S}_1 + \widehat{\vec S}_2)^2 = \widehat{\vec S}_1^2 + \widehat{\vec S}_2^2 + 2\widehat{\vec S}_1\cdot\widehat{\vec S}_2 = 2\frac{1}{2}\frac{3}{2}\hbar^2 + 2\widehat{\vec S}_1\cdot\widehat{\vec S}_2 = \frac{3}{2}\hbar^2 + 2(\widehat{\vec S}_1\cdot\widehat{\vec S}_2) \tag{4.15}$$

and

$$(\hat{\vec{S}}_1+\hat{\vec{S}}_2)^2=\hbar^2\cdot\begin{cases}2 \text{ triplet}\\ 0 \text{ singlet}\end{cases}\quad\Rightarrow\quad 2\hat{\vec{S}}_1\cdot\hat{\vec{S}}_2+\frac{1}{2}\hbar^2=\hbar^2\begin{cases}1, \text{ triplet}\\ -1, \text{ singlet}\,.\end{cases}\tag{4.16}$$

Consequently, the effective Hamiltonian takes the form of the well-known Heisenberg Hamiltonian:

$$H_{\text{eff}}=J\hat{\vec{S}}_1\cdot\hat{\vec{S}}_2\quad\text{with}\quad\hbar^2J=\Delta E_{TS}=2\frac{l_{\text{OV}}^2\widetilde{C}-\widetilde{J}}{1-l_{\text{OV}}^4}\,.\tag{4.17}$$

Here we have suppressed all spin independent energy terms, since we are only interested in the energy difference between the different spin configurations, which involve the two 1s states of the H atoms. It is important to note that the exchange parameter J can take positive or negative values, depending on the values of $\widetilde{C}$, $\widetilde{J}$ and l_{OV}. Let's consider two limiting cases:

1. Vanishing overlap integral $l_{\text{OV}}=0$, i.e., orthogonal wave functions: In this case, the exchange parameter J is negative and the ground state is a triplet state. One speaks of a ferromagnetic exchange interaction, since $J<0$ leads to a parallel alignment of the spins in the ground state. This case is realized, for example, for orthogonal atomic orbitals and is the justification for Hund's first rule.
2. Large overlap with $l_{\text{OV}}^2\widetilde{C}>\widetilde{J}$: In this case, the exchange parameter J is positive and the ground state is a singlet. One speaks of an antiferromagnetic exchange interaction, since $J>0$ leads to an antiparallel alignment of the spins in the ground state. This is the case for the hydrogen molecule as depicted schematically in Fig. 4.9. The wave functions of eqs. (4.6)–(4.9) are only large between the two nuclei, if $S=0$ (singlet state). Otherwise, the minus sign between the two terms leads to a vanishing wave function amplitude between the nuclei. In case of antiparallel spins (singlet state), the enhanced negative charge between the positive nuclei leads to binding through the attractive Coulomb interaction of these parts of the electrons to the two protons, as illustrated in Fig. 4.9, middle.

4.4.2 Kinetic exchange

The Heitler-London approximation assumes that the four neutral configurations depicted in Fig. 4.9 (right) are sufficient to describe the H_2 molecule and, thus, ignores the two ionized configurations. However, for short distances, i.e., large overlap integrals, these ionized states have to be taken into account. They are described by two-particle wave functions Ψ_{IA} and Ψ_{IB}, where both electrons are localized in the same orbital, i.e., either both at atom A or both at atom B, respectively:

$$\Psi_{IA}(\vec{r}_1,\vec{r}_2)=\frac{1}{\sqrt{2}}\psi_A(\vec{r}_1)\psi_A(\vec{r}_2)[|\uparrow,\downarrow\rangle-|\downarrow,\uparrow\rangle]\tag{4.18}$$

$$\Psi_{IB}(\vec{r}_1,\vec{r}_2)=\frac{1}{\sqrt{2}}\psi_B(\vec{r}_1)\psi_B(\vec{r}_2)[|\uparrow,\downarrow\rangle-|\downarrow,\uparrow\rangle]\,.\tag{4.19}$$

Only the singlet configuration is possible due to the Pauli principle. Because of the intra-atomic Coulomb interaction

$$U = \left\langle \Psi_{IA} \left| \frac{e^2}{4\pi\varepsilon_0 |\vec{r}_1 - \vec{r}_2|} \right| \Psi_{IA} \right\rangle$$
$$= \int d^3\vec{r}_1 \int d^3\vec{r}_2 |\psi_A(\vec{r}_1)|^2 \frac{e^2}{4\pi\varepsilon_0 |\vec{r}_1 - \vec{r}_2|} |\psi_A(\vec{r}_2)|^2 , \tag{4.20}$$

these ionized states are higher in energy than the neutral states. The parameter U will appear again, when we introduce the Hubbard model. It is commonly called the "Hubbard-U" or the on-site Coulomb term. Since for the calculation of U both electrons are located at the same atom, while they are at different atoms for the calculation of $\tilde{C}$ and $\tilde{J}$, the relation holds: $U > \tilde{C} > \tilde{J}$. Neutral states and ionized states are connected through the matrix element

$$-t = -\int d^3\vec{r}\psi_A^*(\vec{r}) \frac{e^2}{4\pi\varepsilon_0} \frac{1}{|\vec{r} - \vec{R}_B|} \psi_B(\vec{r}) , \tag{4.21}$$

where $-t$ denotes the amplitude for hopping of an electron from atom A to atom B. Basically, the electron from atom A gains energy by exploring the attractive potential of the atom core B as well.

Note that the spin quantum numbers must be maintained during the hopping process, if hopping does not involve spin-orbit terms. The hopping processes are, hence, forbidden for the triplet states of the neutral configuration due to the Pauli-principle. Hereby we neglect the simultaneous hopping of two electrons, which is a second order hopping process, dubbed a co-tunneling process, and much less likely than the first order hopping process described by t. Thus, the most relevant matrix element for hopping is:

$$\langle \Psi_4 | \widehat{H} | \Psi_5 \rangle = -2t \quad \text{with} \quad \Psi_5 = \frac{1}{\sqrt{2}} (\Psi_{IA} + \Psi_{IB}) . \tag{4.22}$$

Due to the inversion symmetry of the H_2-molecule, only the wave function with even parity $\Psi_{IA} + \Psi_{IB}$ plays a role. In order to demonstrate the effect of the ionic states on the energy levels, we consider a simplified toy model, where we neglect the Coulomb matrix elements $\tilde{C}$ and $\tilde{J}$, as well as the overlap-integral l_{OV}. Then the Hamilton-matrix reads:

$$\widehat{H} = \begin{pmatrix} 0 & 0 & -t & -t \\ 0 & 0 & +t & +t \\ -t & +t & U & 0 \\ -t & +t & 0 & U \end{pmatrix} \begin{matrix} |\uparrow, \downarrow\rangle \\ |\downarrow, \uparrow\rangle \\ |\uparrow\downarrow, \cdot\rangle \\ |\cdot, \uparrow\downarrow\rangle . \end{matrix} \tag{4.23}$$

The change of sign between column 1 and column 2 results from the antisymmetry of the wave functions for fermions (Pauli principle), i.e., a sign change if the particles are exchanged. Diagonalization of this Hamilton-matrix gives the energy eigenvalues of the remaining singlet states:

$$E^{\pm} = \frac{1}{2}(U \pm \sqrt{U^2 + 16t^2}) = \frac{1}{2}\left(U \pm U\sqrt{1 + \frac{16t^2}{U^2}}\right) \approx \begin{cases} U + \frac{4t^2}{U} \\ -\frac{4t^2}{U} \,. \end{cases} \tag{4.24}$$

Here, we have used the approximation for the relevant case of correlation physics $U \gg t$ via the expansion $\sqrt{1+x} \approx 1 + \frac{1}{2}x$. The additional lowering of the energy levels due to the possible ionic configurations leads to a further contribution to the exchange interaction in the Heisenberg-Hamiltonian for singlet states. Since this contribution stems from the hopping of the electrons between the atoms, it is denoted as kinetic exchange. In a good approximation, the exchange parameter of the Heisenberg–Hamiltonian $\widehat{H}_{\text{eff}} = J\vec{\widehat{S}}_1 \cdot \vec{\widehat{S}}_2$ is given as a sum of direct and kinetic exchange:

$$\hbar^2 J \approx \hbar^2 \left(J_{\text{dir}} + J_{\text{kin}}\right) = 2\frac{l_{\text{OV}}^2 \widetilde{C} - \widetilde{J}}{1 - l_{\text{OV}}^4} + \frac{4t^2}{U} \,. \tag{4.25}$$

The effect of the direct and kinetic exchange is summarized in the energy level diagram of Fig. 4.10, which shows the effect of the various parameters U (intra-atomic Coulomb interaction), $\widetilde{C}$ and $\widetilde{J}$ (inter-atomic Coulomb interaction and exchange term) and t (hopping-matrix element).

Figure 4.10 illustrates how kinetic exchange leads to a further lowering of the singlet state and thus enhances the antiferromagnetic exchange interaction. The reason

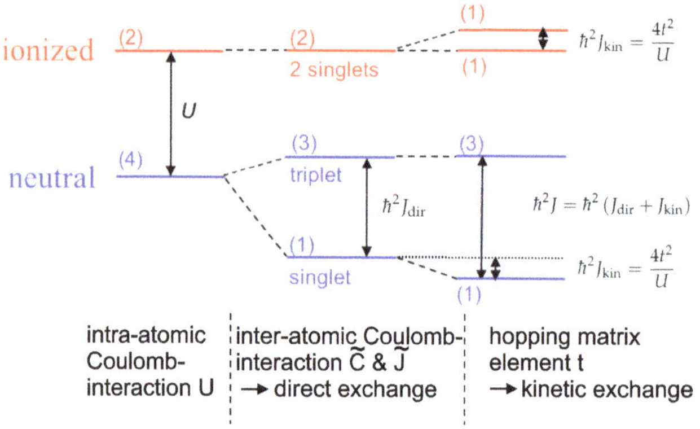

Fig. 4.10: Schematic energy level diagram (not to scale). The numbers in parenthesis denote the level degeneracies. Left: Splitting between neutral and ionized configurations due to the intra-atomic Coulomb interaction. Middle: Inter-atomic Coulomb interaction with its exchange term taken into account. This leads to a splitting of the low lying neutral states into a singlet and a triplet state. The Heisenberg-Hamiltonian describes this splitting by the exchange parameter of the direct exchange interaction J_{dir}. Right: Effect of the hopping processes for non-vanishing hopping-matrix-element t. This leads to a mixing of the excited ionized states into the low lying neutral states and thus to an additional lowering of the singlet state. Within the Heisenberg-model, this effect is described by the kinetic exchange term $\hbar^2 J_{\text{kin}} = \frac{4t^2}{U}$.

is that hopping processes are only permitted if neighbouring electrons have antiparallel spins. Due to these hopping processes, the electrons can lower their energy.

The hydrogen molecule, discussed here, is the most simple system for which correlation effects due to ionized states matter. Hopping processes are only possible, if neighbouring spins are antiparallel. This leads to an additional shift of the energy levels. We will recognize such terms again, if we discuss the so-called Hubbard-model, which allows the description of correlated electron systems in a solid.

4.4.3 Superexchange interaction

So far, we have discussed direct and kinetic exchange, where we assumed that the wave functions of the involved atoms overlap. In general, this is not the case for transition metal oxides. Let us look at the example of CoO (Fig. 4.6). Neighbouring cobalt atoms which feature unpaired electrons are separated by an intermediate oxygen atom. Nevertheless, cobalt oxide is an antiferromagnet with a Néel temperature of about 290 K. This means that the magnetic exchange interaction between the cobalt-3d-electrons has to be mediated by the oxygen-2p-orbitals, which are responsible for the chemical bonds. The situation is depicted in Fig. 4.11. The resulting effective spin-spin interaction is called superexchange interaction.

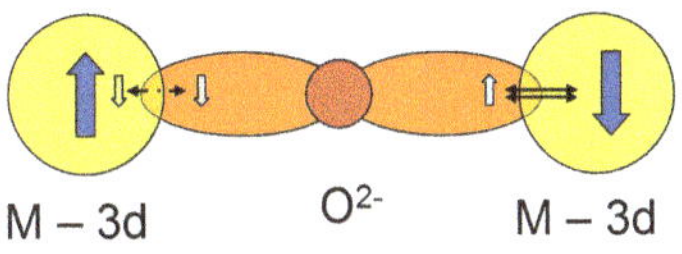

Fig. 4.11: Schematic plot of the superexchange interaction between two 3d-metal ions (M-3d) mediated via an oxygen-2p-orbital (O^{2-}). While the oxygen-p-orbital is shown explicitly, the 3d-orbitals are only shown schematically indicating the total spin at the metal ion (dark blue arrows). Through the direct and kinetic exchange between the metal-3d-electrons and the oxygen-2p-electrons at both sides, an effective exchange interaction between the metal-3d-orbitals is achieved.

The superexchange interaction can be quantitatively described in analogy to the direct and kinetic exchange, see e.g. [89]. The so-called Goodenough–Kanamori–Anderson rules can be applied to estimate the sign of the superexchange interaction. Magnetic superexchange depends critically on the occupation of the metal orbitals as well as on the angle between the metal atoms and the mediating oxygen atom. For a 180° superexchange interaction, i.e., the two metal atoms and the oxygen atom (M – O – M) lie on one line, one obtains an antiferromagnetic exchange between the metal orbitals which are half-filled. Qualitatively, this is rationalized by bonds between the two half-filled metal orbitals and one oxygen p-orbital occupied by two electrons (Fig. 4.11). Each p-electron binds to one of the neighboring metal atoms with a total spin of the pair as given by the sign of the M-O exchange interaction. Since the two electrons from the same oxygen orbital must have opposite spins, the two metal orbitals must also have opposite spins. This leads to an antiferromagnetic coupling of the two metal atoms. A ferromagnetic exchange is instead found between half-filled and empty or full- and half-filled metal orbitals. In the latter case, one can argue that the additional electron in the full orbital can only gain kinetic energy by delocalization, if the surrounding half-filled orbitals all provide the same empty spin level, such that they order ferromagnetically. This leads to a ferromagnetic coupling between them. The 90° superexchange (M–O–M angle of 90°) for two half-filled metal orbitals is also ferromagnetic. Here, the two metal atoms bind to different oxygen p-orbitals. Orienting the spins of the binding p-electrons parallel reduces their Coulomb repulsion due to the corresponding antisymmetric spatial two-particle wave function (eqs. (4.6)–(4.9)). This is analogous to the reduced Coulomb repulsion of parallel spin orientations of electrons within atoms, which leads to Hund's first rule.

The theoretical derivation of the Goodenough–Kanamori–Anderson rules goes beyond the scope of this lecture, but can be found in the literature, e.g. [90]. Note that the Goodenough–Kanamori–Anderson rules are semi-empirical for predicting the sign and strength of the superexchange interaction and are being used differently in the literature. Besides the superexchange interaction, there is the so-called double-exchange interaction in systems, where the metal-ions have different valences. We will discuss the double-exchange interaction in more detail in the context of the properties of manganites.

4.5 Hubbard model

For the case of the extended system of a solid, the processes discussed in Sections 4.3 and 4.4 can be cast most conveniently into a model Hamiltonian, if we choose a representation in second quantization of quantum field theory for the many body wave functions. This is the so-called Hubbard model [91]. If we consider only one relevant band, one can write down the so-called single band Hubbard Hamiltonian:

$$\widehat{H} = -t \sum_{j,l \in \mathrm{N.N}} \sum_{\sigma} (\widehat{c}^{+}_{j\sigma} \widehat{c}_{l\sigma} + \widehat{c}^{+}_{l\sigma} \widehat{c}_{j\sigma}) + U \sum_{j} \widehat{n}_{j\uparrow} \widehat{n}_{j\downarrow} \,. \tag{4.26}$$

The operator $\widehat{c}^{+}_{j\sigma}$ creates an electron in the tight binding (Wannier)-state $\psi(\vec{r} - \vec{R}_j)|\sigma\rangle$ with spin σ at the atom with position $\vec{R}_j$. Such Wannier states can be constructed from Bloch states by overlapping the corresponding plane waves of different wave vector $\vec{k}$, all with the same phase at atom site $\vec{R}_j$. The Wannier states are, hence, localized at the corresponding atom j. Wannier states with different j are quasi-orthogonal, as described, e.g., in introductory textbooks on quantum mechanics. They are used as an adequate basis for tight binding type Hamiltonians. The sum index $j, l \in \mathrm{N.N}$ means that only nearest neighbor sites are considered for hopping. The term $\widehat{n}_{j\sigma}$ is the occupation operator $\widehat{c}^{+}_{j\sigma} \widehat{c}_{j\sigma}$ of the corresponding Wannier state. The Hubbard U is the

Coulomb repulsion in one orbital at one site:

$$U = \int d^3\vec{r}_1 \int d^3\vec{r}_2 \frac{e^2}{4\pi\varepsilon_0} \frac{|\psi(\vec{r}_1 - \vec{R}_j)|^2 |\psi(\vec{r}_2 - \vec{R}_j)|^2}{|\vec{r}_1 - \vec{r}_2|} . \tag{4.27}$$

The parameter t is the hopping amplitude between neighboring sites $\vec{R}_1$ and $\vec{R}_2$:

$$t = \int d^3\vec{r}\psi(\vec{r} - \vec{R}_1) \frac{e^2}{4\pi\varepsilon_0} \frac{1}{|\vec{r} - \vec{R}_2|} \psi(\vec{r} - \vec{R}_2) . \tag{4.28}$$

The Hubbard model is a so-called "lattice Fermion model", since only discrete lattice sites are being considered. It is the most simple way to incorporate correlations due to the Coulomb interaction, since it takes into account only the strongest contribution, the onsite Coulomb term U. Nevertheless, it still contains rich physics related to experimentally observed phenomena. This includes the electronic phases of ferromagnetic or antiferromagnetic metals and insulators or charge and spin density waves [91]. A realistic Hamiltonian should contain many more intersite terms due to the long range Coulomb interaction and it is quite likely that additional, so far unknown electronic phases are contained in such more realistic models.

The most direct consequence of the onsite Coulomb interaction U is that additional so-called Hubbard bands are created due to possible hopping processes. For a single band, this is illustrated in Fig. 4.12.

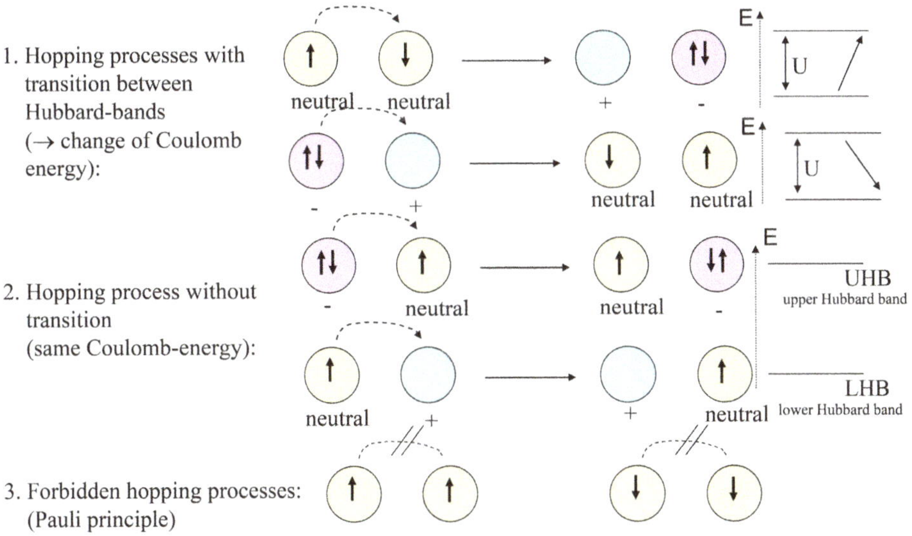

Fig. 4.12: Hopping within a one-band Hubbard model. Left column: name of the hopping process. Middle column: illustration of electron hopping processes between neighboring atoms. Right column: corresponding energy level schemes with required band transitions indicated by diagonal arrows.

The first row in Fig. 4.12 shows hopping processes, which are connected with a change of the total Coulomb energy. The second row shows hopping processes which do not involve a change in Coulomb energy. The last row shows hopping processes which are forbidden in a single band due to the Pauli principle. From Fig. 4.12, we identify two different energy states. Configurations, for which both electrons are located on the same atom are higher by the onsite Coulomb repulsion U as compared to configurations, where the electrons are not on the same atom. In a solid, we expect that these two energy levels will broaden into bands due to the delocalization of the electrons on many atoms as driven by the hopping matrix element t. These two bands are called the lower Hubbard band (LHB) and the upper Hubbard band (UHB). In the following, we will discuss some limiting cases in order to understand the different terms in the Hubbard Hamiltonian and to make the connection to the Hubbard band model and the Heisenberg model.

4.5.1 Band dispersion in the Hubbard model

The Hubbard-Hamiltonian consists of two terms: one with the hopping matrix element t as a prefactor and one with the on-site Coulomb repulsion U as a prefactor. Let us first consider the case where the Hubbard U becomes very small and can be neglected, which reveals the so-called band Hamiltonian:

$$\widehat{H}_{\text{Band}} = -t \sum_{i,j\in \text{N.N}} \sum_{\sigma} (\widehat{c}^{+}_{j\sigma}\widehat{c}_{i\sigma} + \widehat{c}^{+}_{i\sigma}\widehat{c}_{j\sigma}) \,. \tag{4.29}$$

In a crystalline solid, one usually employs translational symmetry, which helps to diagonalize the band Hamiltonian by expanding the creation and annihilation operators into a Fourier series:

$$\widehat{c}_{j\sigma} = \frac{1}{\sqrt{N}} \sum_{\vec{k}} \widehat{c}_{\vec{k}\sigma} e^{i\vec{R}_j\vec{k}} \,, \tag{4.30}$$

where N is the number of atoms and $\vec{k}$ a wavevector. In the most simple case, we assume an infinite one-dimensional lattice of equally spaced atoms, i.e.,

$$\vec{R}_j = j \cdot a \,, \quad \frac{1}{N}\sum_j e^{i(k-k')ja} = \delta_{kk'} \,. \tag{4.31}$$

Inserting eqs. (4.30) and (4.31) in eq. (4.29) yields:

$$\begin{aligned}
\widehat{H} &= -t \sum_{j,\sigma} \sum_{k,k'} \frac{1}{N} \widehat{c}^{+}_{k\sigma}\widehat{c}_{k'\sigma} \\
&\quad \cdot \left[e^{-i(ja)\cdot k} \left(e^{i(j+1)a\cdot k'} + e^{i(j-1)a\cdot k'} \right) \cdot \frac{1}{2} + e^{i(ja)k} \left(e^{-i(j+1)a\cdot k'} + e^{-i(j-1)a\cdot k'} \right) \cdot \frac{1}{2} \right] \\
&= -t \sum_{\sigma k} \widehat{c}^{+}_{k\sigma}\widehat{c}_{k\sigma} \left(e^{iak} + e^{-iak} \right) \\
&= -2t \sum_{\sigma k} \widehat{n}_{k\sigma} \cos(a \cdot k) \,.
\end{aligned} \tag{4.32}$$

This result can be generalized for a three-dimensional cubic lattice ($j = 1, 2, 3$) as follows:

$$\widehat{H} = \sum_{\vec{k}\sigma} \epsilon_{\vec{k}} \widehat{n}_{\vec{k}\sigma} \,, \quad \epsilon_{\vec{k}} = -2t \sum_{j=1}^{3} \cos(k_j \cdot a) \,. \tag{4.33}$$

Now we have decomposed the Hamiltonian into a sum of independent modes characterized by wavevector $\vec{k}$ and spin quantum number σ with occupation operator $\widehat{n}_{\vec{k}\sigma}$ having the energy eigenvalue $\epsilon_{\vec{k}}$. The energy of these modes is proportional to the hopping matrix element t and varies with the components of the wave vector as a cosine function. This can be interpreted as the band dispersion of the electron gas without interactions ($U = 0$) and within the Hubbard model. The bandwidth W of this model is proportional to the number of neighbors z_{NN} ($z_{\mathrm{NN}} = 2$ for a one-dimensional chain, $z_{\mathrm{NN}} = 6$ for a three-dimensional simple cubic lattice) and the hopping matrix element t: $W = 2z_{\mathrm{NN}} \cdot t$.

4.5.2 Heisenberg model and Hubbard model

Next, we look at the opposite limit of vanishing hopping matrix element t. The effect of the second term in the Hubbard-Hamiltonian can best be described in a local basis since the eigenvalues depend on the occupation of each single site with electrons. This is illustrated in Table 4.1, where we pick out the site j, which can be occupied with 0, 1 or 2 electrons. We will calculate the corresponding local Coulomb term for this site (last row). This term has a non-vanishing value of U, only if site j has a double occupancy, obviously with electrons of opposite spin due to the Pauli principle.

Tab. 4.1: On-site Coulomb energy for different occupancies at site j.

	$j-1$	j	$j+1$	$j+2$	$U\widehat{n}_{j\uparrow}\widehat{n}_{j\downarrow}$
$\lvert 0\rangle_j$	•	∘	•	•	0
$\lvert\uparrow\rangle_j$	•	↑	•	•	0
$\lvert\downarrow\rangle_j$	•	↓	•	•	0
$\lvert d\rangle_j$	•	↓↑	•	•	U

We will take this as the starting point for a calculation within first order perturbation theory using the small parameter t as the perturbation. It is expected that the low-lying excitations in the lower Hubbard band correspond to magnetic fluctuations, i.e., electrons change their spin orientation without hopping to another site. In contrast, excitations with higher energy represent transitions into the upper Hubbard band corresponding to charge fluctuations, i.e., electrons change their site via hoppping and, hence, induce double occupations of sites. We will now attempt to reproduce the separation into low-lying magnetic and higher-energy charge fluctuations within the Hubbard model and derive a Hamiltonian for the low-lying magnetic fluctuations. As in Section 4.4, we choose a simplified model of a pair of atoms only. We consider the following four states, which correspond to the two electrons being located on the two sites 1 and 2, but with all possible, different spin quantum numbers.

$$\begin{aligned} |\uparrow\uparrow\rangle &= \hat{c}^{+}_{1\uparrow}\hat{c}^{+}_{2\uparrow}|00\rangle \\ |\uparrow\downarrow\rangle &= \hat{c}^{+}_{1\uparrow}\hat{c}^{+}_{2\downarrow}|00\rangle \\ |\downarrow\uparrow\rangle &= \hat{c}^{+}_{1\downarrow}\hat{c}^{+}_{2\uparrow}|00\rangle \\ |\downarrow\downarrow\rangle &= \hat{c}^{+}_{1\downarrow}\hat{c}^{+}_{2\downarrow}|00\rangle \,. \end{aligned} \tag{4.34}$$

Here, $|00\rangle$ is the vacuum state having zero electrons on both sites. In first order perturbation theory, the higher lying excited states with double occupancy at one site ($|0d\rangle$, $|d0\rangle$) are mixed into the low-lying states through the hopping matrix element t:[7]

$$|\uparrow\uparrow\rangle_1 = |\uparrow\uparrow\rangle\,; \quad |\uparrow\downarrow\rangle_1 = |\uparrow\downarrow\rangle + \frac{t}{U}\left(|0d\rangle + |d0\rangle\right)\,. \tag{4.35}$$

First order perturbation theory is justified as long as the on-site Coulomb repulsion U is sufficiently large compared to the hopping matrix element t. We can now calculate the expectation value of the band Hamiltonian (eq. (4.29)) in order to obtain the resulting shift in energy:

$$\begin{aligned} {}_1\langle\uparrow\downarrow|\hat{H}_{\mathrm{Band}}|\uparrow\downarrow\rangle_1 &= -t\cdot\left(\langle\uparrow\downarrow| + \frac{t}{U}(\langle 0d| + \langle d0|)\right) \\ &\quad \left[\hat{c}^{+}_{1\uparrow}\hat{c}_{2\uparrow} + \hat{c}^{+}_{2\uparrow}\hat{c}_{1\uparrow} + \hat{c}^{+}_{1\downarrow}\hat{c}_{2\downarrow} + \hat{c}^{+}_{2\downarrow}\hat{c}_{1\downarrow}\right] \\ &\quad \left(|\uparrow\downarrow\rangle + \frac{t}{U}(|0d\rangle + |d0\rangle)\right) \\ &= -\frac{4t^2}{U}\,. \end{aligned} \tag{4.36}$$

The expectation value of the Coulomb term (second term in eq. 4.26) can be calculated in an analogous manner:

$${}_1\langle\uparrow\downarrow|\hat{H}_{\mathrm{Coulomb}}|\uparrow\downarrow\rangle_1 = \frac{2t^2}{U}\,. \tag{4.37}$$

For the total Hubbard Hamiltonian we obtain as a sum of band and Coulomb contribution:

$${}_1\langle\uparrow\downarrow|\hat{H}_{\mathrm{Hubbard}}|\uparrow\downarrow\rangle_1 = -\frac{2t^2}{U}\,, \tag{4.38}$$

analogously

$${}_1\langle\downarrow\uparrow|\hat{H}_{\mathrm{Hubbard}}|\downarrow\uparrow\rangle_1 = -\frac{2t^2}{U}\,. \tag{4.39}$$

7 Identical states are found for the two lower starting configurations of the list $|\downarrow\downarrow\rangle$ and $|\downarrow\uparrow\rangle$.

For states where the electron spins are pointing up on both sites or down on both sites, there is no energy shift since hopping between the sites is forbidden. Therefore, the diagonal elements of the Hubbard matrix in this subspace as described by first order perturbation theory are given by:

$$\widehat{H}_{\text{eff}}^{\text{Diagonal}} = -\frac{2t^2}{U}\left(\frac{1}{2} - \frac{2}{\hbar^2}\widehat{S}_1^z\widehat{S}_2^z\right) . \tag{4.40}$$

The non-diagonal matrix elements can be calculated analoguously:

$$_1\langle\uparrow\downarrow|\widehat{H}_{\text{Hubbard}}|\downarrow\uparrow\rangle_1 = -\frac{2t^2}{U} . \tag{4.41}$$

Combining these results, we finally obtain an effective Hamiltonian for the low-lying states with magnetic fluctuations in the following form:

$$\widehat{H}_{\text{eff}} = \frac{1}{\hbar^2}\frac{4t^2}{U}\hat{\vec{S}}_1 \cdot \hat{\vec{S}}_2 . \tag{4.42}$$

Thus, we have derived an effective Hamiltonian for the low-lying states in the limit of large U for half-filling of a single band. It takes exactly the form of the Heisenberg-Hamiltonian (eq. (4.17)) with antiferromagnetic exchange parameter $J = \frac{1}{\hbar^2}\frac{4t^2}{U}$. Decisive for our argument is the band term of the Hubbard-Hamiltonian which leads to a weak delocalization of the electrons ($\propto \frac{t}{U}$) so that the lowering in energy due to delocalization is larger than the increase in Coulomb repulsion energy, when electrons probe the occupation of the neighboring sites through virtual hopping processes. This is nothing else than the kinetic exchange interaction discussed in Section 4.4.2.

4.5.3 Hubbard band model

From Sections 4.5.1 and 4.5.2, we can now understand, how the Hubbard bands develop in the solid (Fig. 4.13). The energy spectrum of the single site is dependent on the occupation. A single electron occupies the lowest available atomic energy level ϵ_{at}. Due to the intra-atomic Coulomb interaction, a second electron has to occupy an energy level which is higher by U. If neighboring sites exist, the electron can hop from site to site due to a non-vanishing hopping amplitude t. In a solid, the sharp atomic energy levels broaden into energy bands (Section 4.5.1). The upper and lower Hubbard bands are separated by the Hubbard U (from band center to band center) and have a band width of $W = 2z_{NN} \cdot t$, where z_{NN} denotes the number of nearest neighbors. Note that the two bands are not caused by the interaction of electrons with the periodic potential of the lattice, but are solely due to correlations in the electron system, i.e., due to the electron-electron interaction. As a result, the existence of the Hubbard bands depends on the electronic occupation. Figure 4.12 illustrates, how the energy terms for simple hopping processes depend on the occupation of neighboring sites and how

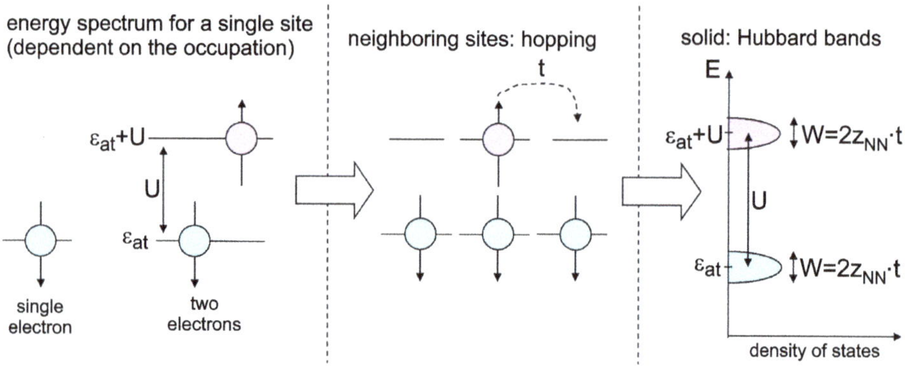

Fig. 4.13: Illustration of the development of the energy spectrum at large U from a single site (left) via several neighboring sites (middle) to the solid (right).

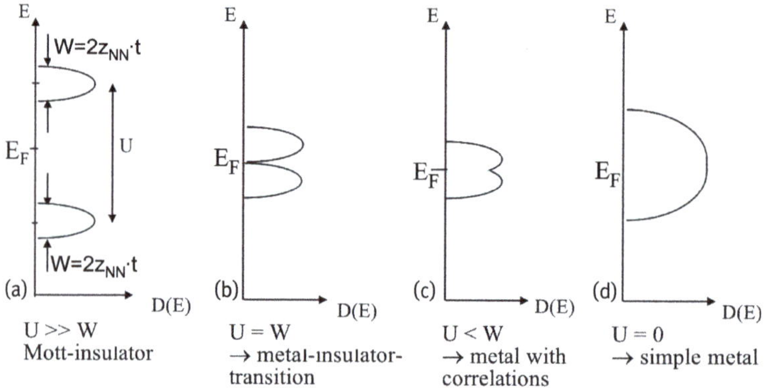

Fig. 4.14: Density of states $\mathcal{D}(E)$ for a Mott–Hubbard-transition for a half-filled band (compare e.g., CoO), (a) The intra-atomic Coulomb interaction U is significantly larger than the band width $W = 2z_{NN} \cdot t$. (b), (c) U decreases relative to W. (d) Limiting case $U = 0$ (see text).

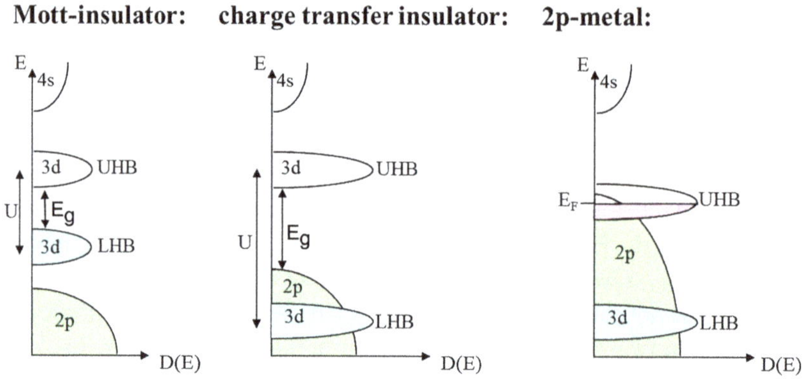

Fig. 4.15: Illustration of the band structure with multiple bands. Left: Mott-insulator. Middle: Charge transfer insulator. Right: 2p-metal.

the hopping can transport spin information. The apparently simple single electron operator t gets complex by many body aspects due to the presence of U.

If the two Hubbard bands are well separated as in Fig. 4.13, i.e., the Coulomb repulsion U dominates over the hopping term t, we have an insulating state for half filling, since only the lower Hubbard band is occupied with electrons. With the Hubbard bands, we can nicely illustrate the Mott–Hubbard transition (Fig. 4.14). In the limiting case $U \gg W = 2z_{\mathrm{NN}} \cdot t$, upper and lower Hubbard band are well separated. For half filling, the Fermi level lies between both bands. Only the lower band is filled. This situation describes a Mott insulator. Charge transport is suppressed by the strong on-site Coulomb repulsion U. If U is decreased relative to W we obtain a special case at $U = W$. Here, the two bands touch and an insulator-metal transition occurs. If U is only a little smaller than the bandwidth W, we obtain a metal with strong correlations.[8] Finally for $U \ll W$, the correlations can be neglected and we have the case of a simple metal. Note, that it is typically easier to change W than to change U. For example, one can apply pressure to change the distance between neighboring sites and, thus, the overlap integral between neighboring orbitals, which determines t (eq. (4.21)).

So far, we have discussed the single band Hubbard Hamiltonian for 3d electrons, where the single band splits into the two Hubbard bands. In oxides, the 2p bands of oxygen and the 4s bands of the transition metal atoms also play a role. These bands are wider and, thus, exhibit a higher mobility of the charge carriers compared to the partly localized 3d electrons. Depending on the position of the additional bands relative to the d band, we obtain the different cases illustrated in Fig. 4.15. Most simply, the Hubbard bands are clearly separated from the oxygen 2p and metal 4s bands, such that the system remains a simple Mott insulator (left). In case of the so-called charge transfer insulator (middle), the lower Hubbard band lies within the oxygen 2p bands, such that the activation energy for electrical conductivity gets reduced. For the Mott insulator, this activation energy is the distance between the upper edge of the LHB and the lower edge of the UHB, while for the charge transfer insulator, it is the distance between the upper edge of the oxygen 2p band and the lower edge of the UHB. Charge transfer can occur between the oxygen 2p and the metal 3d electrons. Using the energy gap E_{g} (Fig. 4.15, left and middle), the temperature dependence of the conductivity can be described by the Arrhenius-law (see also Section 1.6.3):

$$\sigma \sim \mathrm{e}^{-\frac{E_{\mathrm{g}}}{k_{\mathrm{B}}T}} . \tag{4.43}$$

For the case of the 2p-metal shown on the right of Fig. 4.15, the UHB and the Fermi-level, both, lie within the oxygen 2p states implying a metal. The 2p holes have a higher mobility than the 3d holes and thus determine the conductivity.

8 A more elaborate analysis shows that, in contrast to the simplified illustration of Fig. 4.14, an additional peak appears at E_{F} due to quasiparticle excitations.

4.5.4 Spin-orbit interaction

So far, we have considered 3d metal oxides, since correlation phenomena can be prominent in these compounds. What about the 4d and 5d metal oxides? While 3d electrons are rather well localized and thus electronic correlations play an important role, 4d and 5d electrons are significantly more delocalized, i.e., the bandwidth W is generally larger than the Hubbard U: $W > U$.

As an example antiferromagnetic Sr_2IrO_4 is an insulator, which is unexpected for a system with a half- filled outermost electronic 5d shell and spatially extended orbitals (^{77}Ir: $[Xe]4f^{14}5d^7 6s^2 \rightarrow Ir^{4+}$: $[Xe]4f^{14}5d^5 \Rightarrow$ half-filled 5d shell). Since electronic correlations are small, the conventional Mott mechanism cannot be invoked to explain the insulating behavior. However, besides the on-site Coulomb interaction U, two other energy terms become important for 5d electron systems, namely the spin-orbit interaction in combination with the crystalline field interaction. The corresponding additional terms in the Hamiltonian read:

$$\widehat{H}_{+} = \zeta \widehat{L} \cdot \widehat{S} - \Delta_{\mathrm{CF}} \widehat{L}_z^2 , \quad \zeta \propto Z^4 . \tag{4.44}$$

Here, ζ is the spin-orbit parameter and Δ_{CF} is the crystal field splitting parameter, where ζ scales with the atomic number Z to the fourth power: $\zeta \propto Z^4$. Therefore, the spin-orbit interaction becomes relevant for 5d systems, while it is mostly negligible for 3d electrons. The effect of the various energy terms can be illustrated within the schematic band structures for the $5d^5$ irridates as shown in Fig. 4.16, where the label t_{2g} refers to a specific crystalline field state (Section 4.6).

Figure 4.16(a) shows the band structure in the absence of perturbations. The 5d band is partly filled, the bandwidth W is large. For a realistic $U < W$ as for 5d bands, U is too small to open a gap at the Fermi energy. The band structure without electronic correlations U, but with spin orbit interaction ζ is shown in Fig. 4.16(b). The spin orbit coupling splits the otherwise degenerate 5d-t_{2g} states into $J_{\mathrm{eff}} = \frac{1}{2}$ and $J_{\mathrm{eff}} = \frac{3}{2}$ states, where J_{eff} denotes the total angular momentum. The upper $J_{\mathrm{eff}} = \frac{1}{2}$ state with a bandwidth $w < W$ is, consequently, half filled. Finally, the band structure with electronic correlations U and with spin orbit interaction ζ is shown in Fig. 4.16(c), where the narrow $J_{\mathrm{eff}} = \frac{1}{2}$ band splits into a lower and upper Hubbard band, such that the system turns into an insulator.

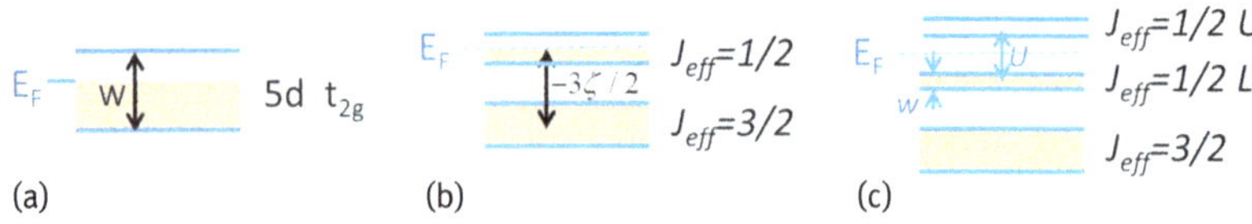

Fig. 4.16: Schematic band structures partly including spin orbit interaction and correlations: (a) Single particle band structure without perturbations. (b) Band structure with spin orbit interaction of strength ζ, but without correlations. (c) Same as (b), but with correlation the term U.

4.6 Crystal field effects, Jahn–Teller distortion and orbital ordering

4.6.1 Crystal field splitting

As mentioned in the introductory Section 4.1, correlated electron systems are systems where the electrons are in between the limiting cases of being localized and fully itinerant. In the last section, we have discussed the fundamental effects of electronic correlations between neighboring metal ions, also taking the spin degree of freedom into account. In this section, we want to study the effect of the local environment of these metal ions. These interactions with the surrounding atoms and ions play an important role in the highly correlated transition metal oxides. In general, the metal ion is surrounded by oxygen atoms. Due to the strong electronegativity of the oxygen atoms, charge is pulled away from the metal atoms towards the oxygen atoms and the metal-oxygen bond is partly covalent and partly ionic, but leaning more towards the latter. Therefore, we will often talk about metal and oxygen ions keeping in mind that this is a crude approximation. The anisotropic charge density distribution around the metal ions can be represented in the most simple approximation by point charges leading to the so-called crystal field theory. A more precise description is provided by the ligand field theory, which takes the additional covalent contribution to the bond into account. A quantitative description of the resulting effects requires symmetry considerations in the framework of group theory. Here, we will give simple and hand waving arguments to describe the crystal field effects. The Stark-effect known from atomic physics tells us that a homogeneous electric field changes the energy levels of an atom. The crystal field theory can be considered as a generalization of the Stark effect for anisotropic charge distributions.

For atoms with closed shells, which do not have electric multipole moments, we do not expect an anisotropic interaction with an external electrical field. This implies that crystal field effects are only important for atoms with open shells, for which the degeneracy of the $2J + 1$ ground state levels (according to Hund's rules) are being lifted by the interaction with the anisotropic charge distribution. This leads to the so-called crystal field multiplets. Through the spin-orbit interaction, this splitting will lead to an anisotropy for the local spin orientation, the so-called crystal field anisotropy.

The crystal field effect for an atom with 3d wave functions in an octahedral environment of oxygen atoms is illustrated in Fig. 4.17. The 3d orbitals $d_{3z^2-r^2}$ and $d_{x^2-y^2}$ are pointing in the direction of the negative point charges, while the orbitals d_{zx}, d_{yz}, and d_{xy} point in between the negative point charges. Thus, a 3d electron in one of the three latter orbitals feels a weaker Coulomb repulsion from the oxygen atoms. This leads to a splitting of the 3d energy levels as shown in Fig. 4.18.

In crystal field theory, the energy levels are labelled with symmetry quantum numbers. Due to the anisotropic charge distribution of the environment, the full three di-

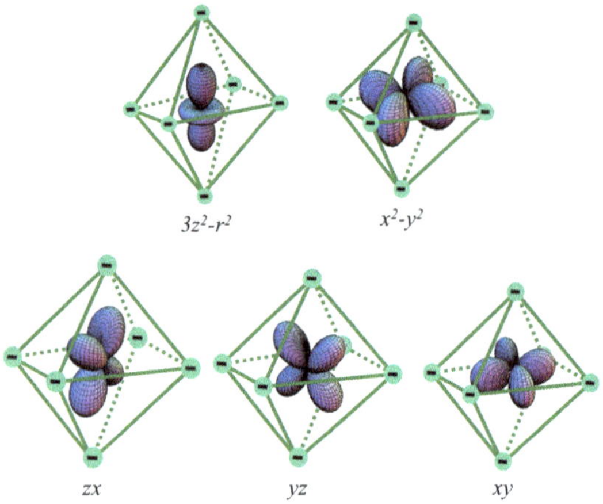

Fig. 4.17: Illustration of the crystal field effect for a 3d metal ion in an oxygen octahedron. Within the crystal field approximation, the effect of the anisotropic charge distribution around the metal ion is described by replacing the oxygen atoms with negative point charges at the corners of the octahedron. For the different 3d orbitals, the Coulomb interaction with these point charges is different (see text).

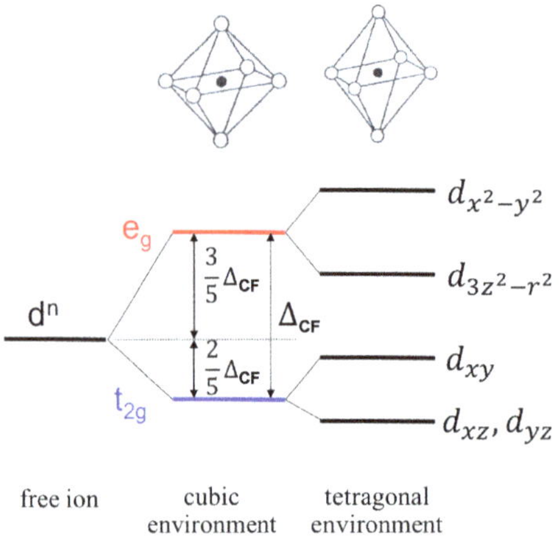

Fig. 4.18: (a) Crystal field splitting in a cubic crystal with octahedral symmetry. The 3d states split into three lower lying so-called t_{2g}-states and two energetically higher e_g-states. With decreasing symmetry of the surrounding, further splitting occurs. The latter is illustrated as an octahedron elongated along the z direction corresponding to the tetragonal point symmetry. Due to the larger distance between the $d_{3z^2-r^2}$-orbital and the neighboring point charges along z compared to the $d_{x^2-y^2}$-orbital with respect to its neighboring point charges, the energy level of the $d_{3z^2-r^2}$-orbital is lower than the one of the $d_{x^2-y^2}$-orbital. The same argument applies for the lowering of the d_{xz} and d_{yz} orbitals with respect to the d_{xy} orbital.

Tab. 4.2: Typical strength of different contributions to the Hamilton operator for 3d-electrons.

3d ion:	$\hat{H}$	=	$\hat{H}_{kin}$	+	$\hat{H}_{Coul}$	+	$\hat{H}_{CF}$	+	$\hat{H}_{LS}$
strength					strong		medium		weak
					1–10 eV		≈ 1.5 eV		≈ 0.1 eV
orbital					$\hat{\vec{L}} = \sum_i \hat{\vec{l}}_i$		$\langle \hat{\vec{L}} \rangle \approx \vec{0}$		
spin					$\hat{\vec{S}} = \sum_i \hat{\vec{s}}_i$				weak anisotropy

mensional rotational symmetry corresponding to the symmetry group SO(3) is broken and the magnetic quantum number m is not a good quantum number anymore. Therefore, the labelling of the energy levels employs the so-called irreducible representations of the point group symmetry of the metal ion. One, two and three dimensional representations (corresponding to the degree of degeneracy of the energy level) are labelled with the letters A, E and T, respectively. The influence of the crystal field leads to a (partial) suppression of orbital angular momentum $\vec{L}$, which is dubbed the quenching of $\vec{L}$. Orbitals corresponding to A and E representations have an orbital angular momentum of $\langle \vec{L} \rangle \approx \vec{0}$, while orbitals belonging to T-representations can have an orbital angular momentum quantum number of $l = 1$.

Finally, one should note that the symmetry of the ionic environment of the metal ion determines the number and type of different levels, while the charge states of the ions determine the size of the splitting between the levels (parameter Δ_{CF} in Fig. 4.18) and the sequence of the levels. With similar arguments as used to explain the splitting in Fig. 4.18, one comes to the conclusion that for a tetrahedral environment, the e_g-levels are lower than the t_{2g}-levels. While there is cubic point symmetry for both configurations (octahedral and tetrahedral), the positions of the e_g- and t_{2g}-levels are inverted. The size of the crystal field effects for the 3d-transition metal ions is very different as for the 4f rare-earth ions. The 3d electrons are in the outer shells and, hence, are influenced strongly by the surrounding ions, while the electric field of the nucleus is partially shielded by the inner electrons. Thus, the spin-orbit interaction, induced by the electric field of the nucleus, is smaller for 3d-electrons than the crystal field effects from the surrounding ions. The various contributions to the Hamilton operator, as kinetic energy $\widehat{H}_{\mathrm{kin}}$, Coulomb interaction of the 3d-electrons with the nucleus $\widehat{H}_{\mathrm{Coul}}$, crystal field effect $\widehat{H}_{\mathrm{CF}}$, and spin-orbit interaction $\widehat{H}_{LS}$, exhibit the order of strengths as shown in Table 4.2.

The influence of the different terms on the orbital momentum $\vec{L}$ and the spin momentum $\vec{S}$ is also given, i.e., $\vec{L}$ is quenched by $\widehat{H}_{\mathrm{CF}}$ and $\widehat{H}_{LS}$ gives rise to a small magnetic anisotropy.

For the 4f-ions, the sequence of the energy contributions is different, since the 4f-electrons are located within the $5s^2p^6$-shell and, hence, relatively well shielded from the surrounding atoms, but feel a relatively large effective field from the nucleus. Thus, the spin-orbit coupling is larger than the crystal field effect. As for the 3d-ions, the term $\widehat{H}_{\mathrm{Kin}} + \widehat{H}_{\mathrm{Coul}}$ commutates with the orbital and spin angular momentum operators, so that the eigenstates can be labelled with the corresponding quantum numbers. In the next order of perturbation theory, spin and orbital angular momentum combine to the total angular momentum $\widehat{\vec{J}}$ via the spin-orbit interaction $\widehat{H}_{LS}$. Finally, the $2J + 1$-times degenerate energy level splits up under the influence of the crystal field $\widehat{H}_{\mathrm{CF}}$. For the rare-earth ions, the crystal field splitting is only some 10 meV, while for the 3d-transition metals it is ~ 1 eV. This leads to the following sequence of energies for the 4f ions (Table 4.3).

Tab. 4.3: Typical strength of different contributions to the Hamilton operator for 4f-electrons.

4f ion:	$\hat{H}$	=	$\hat{H}_{\mathrm{kin}}$	+	$\hat{H}_{\mathrm{Coul}}$	+	$+\hat{H}_{LS}$	+	$\hat{H}_{\mathrm{CF}}$
strength					strong 1–10 eV		medium ≈ 0.25 eV		weak ≈ 0.01 eV
orbital					$\hat{\vec{L}} = \sum_j \hat{\vec{l}}_i$		$\hat{\vec{J}} = \hat{\vec{L}} + \hat{\vec{S}}$		$2J + 1$-degeneracy partly lifted
spin					$\hat{\vec{S}} = \sum_j \hat{\vec{s}}_i$				

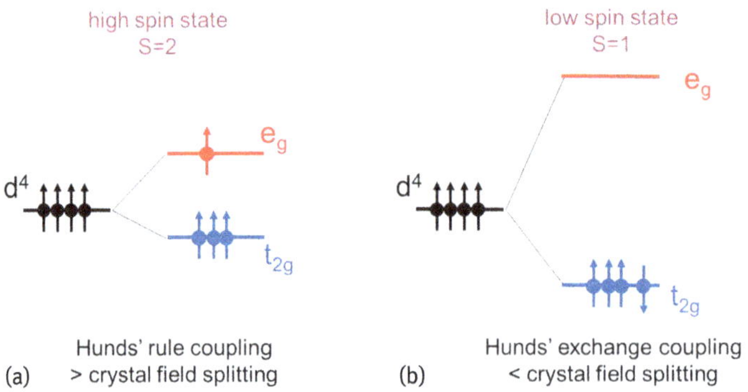

Fig. 4.19: Occupation of the crystal field levels of a Mn^{3+}-ion (four electrons in the 3d-shell) in an octahedral environment. Left: Hund's intra-atomic exchange interaction is larger than the crystal field splitting. This results in a high spin state with $S = 2$. Right: Hund's intra-atomic exchange interaction is smaller than the crystal field splitting. This results in a low spin state with $S = 1$.

So far, we have only discussed the splitting of the energy levels in the crystalline electric field. Next, we will study how the electrons occupy these levels. The competition between the intra-atomic exchange interaction (Hund's rules) and the crystal-field-splitting can lead to interesting effects. We will discuss this for a particular example, namely a Mn^{3+}-ion in an octahedral environment. The electron d-state configuration of the free Mn^{3+}-ion is $3d^4$. According to Hund's rule, all electron spins in the 3d shell of the free Mn^{3+}-ion will be parallel, hence, maximizing the total spin angular momentum S. This is a consequence of the intra-atomic exchange between orthogonal orbitals (page 218). In Fig. 4.19, the four electrons are distributed on the crystal field levels, once for the case that the intra-atomic exchange is larger than the crystal field splitting and once for the opposite case.

For the free Mn^{3+}-ion, following the first and second Hund's rule (maximized S and L), the four orbitals with quantum numbers $m = +2, +1, 0$ and -1 of $\hat{L}_z$ are all occupied with one electron, having the same spin direction. In the octahedral crystal field, there are, however, only three orbitals for the ground state multiplet t_{2g}. These will be occupied with electrons with parallel spins (spin-up in Fig. 4.19) due to the intra-atomic ferromagnetic exchange interaction. For the 4th electron, there are two possibilities:

1. Hund's exchange coupling J_H is larger than the crystal field splitting Δ_{CF}, and, hence, S will be maximized. The electron has to occupy one orbital of the energetically higher e_g-crystal field-level resulting in the high spin state with $S = 2$ as depicted on the left of Fig. 4.19.
2. The crystal field splitting is larger than Hund's exchange interaction. Then, it is energetically favorable to occupy one of the t_{2g}-orbitals with two electrons. According to the Pauli-principle, this is only possible with antiparallel spins resulting in the low spin state with $S = 1$ as shown on the right of Fig. 4.19.

4.6.2 Jahn–Teller effect

The example of a Mn^{3+}-ion in an octahedral environment can also be used to illustrate another crucial effect relevant for the complex transition metal oxides, which is called the Jahn–Teller effect. In the high spin state, there is one electron in the e_g-orbitals. Since there are two orbitals for this electron, we deal with an orbital degeneracy. The electron can be either in the $d_{x^2-y^2}$-orbital or in the $d_{3z^2-r^2}$-orbital. The Jahn–Teller theorem states: If the symmetry of the crystal field is so high that the ground state of an ion is orbitally degenerate, it is energetically favorable to distort the crystal lattice, such that the orbital degeneracy is lifted. This lowers the energy for one of the two orbitals, which then can be occupied by the electron. The other orbital gets higher in energy, but it does not contribute to the total energy of the system, since it is empty. Such an energy gain by lifting of the orbital degeneracy through the Jahn–Teller distortion is illustrated in the energy-level-diagram of Fig. 4.20.

Jahn–Teller distortions can only occur, if the original degeneracy is higher than the so-called Kramers minimum. Time reversal symmetry implies, at least, a two-fold degeneracy of each energy level. Since the Jahn–Teller distortion does not break time-reversal symmetry, the minimum degeneracy remains the same after the Jahn–Teller

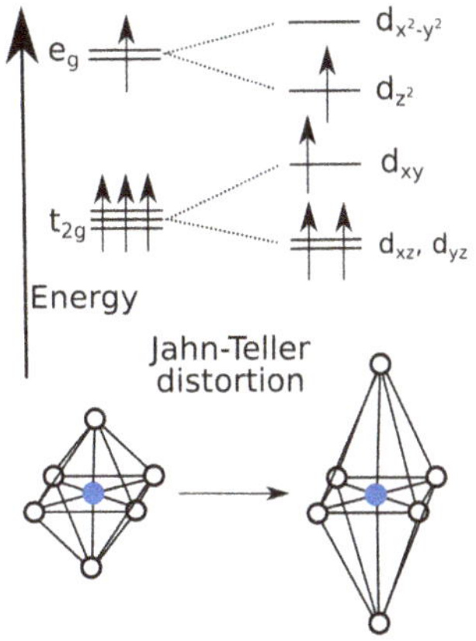

Fig. 4.20: A $3d^4$-ion in an octahedral environment has an orbital degeneracy in the e_g-crystal field level: the one electron can occupy either of the two e_g orbitals. The Jahn–Teller effect lifts this orbital degeneracy by a distortion of the crystal lattice. Here, d_{z^2} is used as an abbrevation for $d_{3z^2-r^2}$.

distortion. Hence, the original degeneracy must be larger than two (the Kramers minimum) in order to enable a Jahn–Teller effect.

The distortion of the lattice via the Jahn–Teller effect can be derived by a simple, phenomenological energy consideration. The increase in elastic energy E_{elastic} due to a distortion of the octahedron is

$$E_{\text{elastic}} \sim \alpha_{\text{el}}(\delta z)^2 \,, \tag{4.45}$$

assuming the usual parabolic potential minimum for the interatomic distance, here between the oxygen atom and the metal atom within the octahedron (α_{el}: elastic constant, δz: deviation of interatomic M-O distance from equilibrium). On the other hand, the lowering in electronic energy by the reduced Coulomb repulsion from the surrounding oxygen atoms can be approximated via Taylor expansion of the Coulomb potential by:

$$E_{\text{electronic}} \sim -\beta_{\text{el}}\delta z \,. \tag{4.46}$$

The total energy change due to the distortion by δz is, thus, given by:

$$E_{\text{total}} \approx \alpha_{\text{el}}(\delta z)^2 - \beta_{\text{el}}\delta z \,. \tag{4.47}$$

The ground state is determined by the energy minimum, which leads to:

$$0 = \frac{\partial E_{\text{total}}}{\partial z} \approx 2\alpha_{\text{el}}\delta z - \beta_{\text{el}} \Rightarrow \delta z \approx \frac{\beta_{\text{el}}}{2\alpha_{\text{el}}} \,. \tag{4.48}$$

For a transition metal ion in an octahedral oxygen environment, the x-, y- and z-axis are equivalent due to the cubic symmetry. Therefore, a Jahn–Teller distortion can occur along either of these axes. This leads to the so-called dynamic Jahn–Teller effect, where the whole complex can tunnel between these energetically equivalent states. In the crystal lattice, the so-called cooperative Jahn–Teller effect can eventually emerge. While at high temperatures, the Jahn–Teller effect will be dynamic, at lower temperatures, the distortions at different lattice sites will depend on each other due to the elastic interactions between neighboring octahedra.

4.6.3 Orbital ordering

The cooperative Jahn–Teller effect leads to so-called orbital order, which is characterized by a regular pattern of Jahn–Teller distortions throughout the entire crystal or, at least, through an extended part of it, corresponding to a domain of ordered distortions. Therefore, at a certain temperature, a phase transition appears which leads to a lowering of the space-group-symmetry of the crystal. The electron orbitals are then occupied in a regular spatial order throughout the crystal.

Besides via the cooperative Jahn–Teller effect, orbital order can also be induced by an electron exchange process, which involves spin and orbital degrees of freedom.

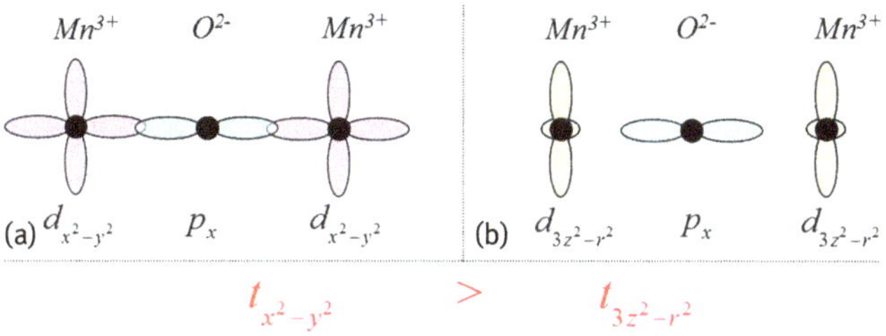

Fig. 4.21: Superexchange interaction between two Mn^{3+}-ions mediated through a p_x orbital of an O_2^--ion. The hopping matrix element is different for hopping between the $d_{x^2-y^2}$ wave functions (left) and for hopping between the $d_{3z^2-r^2}$-wave functions (right), due to the different overlap of adjacent wave functions. For the sake of simplicity, we display only hopping processes between the same type of wave function on both sites, albeit interorbital hopping is also possible.

Since two orbitals are involved, we have to generalize the one-band-Hubbard-model and consider two degenerate orbitals at each site, for example the e_g-orbitals $\psi_a = d_{x^2-y^2}$ and $\psi_b = d_{3z^2-r^2}$. With two orbitals, there is a possibility of having a double-occupation at one site with electrons of the same spin-direction. In such a case, we have to consider the intra-atomic exchange coupling as well. As illustrated in Fig. 4.21, for the case of a superexchange between two Mn^{3+}-ions mediated by a O_2^--ion, the hopping matrix elements can be different for the two different orbitals, since the overlap of the wave functions is different.

Having this in mind, the possible hopping processes for the two-orbital model are plotted in Fig. 4.22 for the case that interorbital hopping is weak and can be, hence, neglected. Process number one is forbidden according to the Pauli-principle. Processes number two and three describe the kinetic exchange (Section 4.4.2), in the example being an antiferromagnetic superexchange (Section 4.4.3), with the Hamiltonian (4.42)

$$\widehat{H} = \frac{4t^2}{U}\frac{1}{\hbar^2}\sum_{i,j}\widehat{\vec{S}}_i \cdot \widehat{\vec{S}}_j \,. \tag{4.49}$$

An antiferromagnetic configuration is favored by this term. However, for two degenerate orbitals, process number four is lowest in energy, since in general $U_{ab} < U_{aa}, U_{bb}$, where the two indices describe the two orbitals on the same site, which are occupied by the interacting electrons. An additional energy reduction can be achieved through Hund's intra-atomic exchange interaction J_H. This leads to the so-called Kugel–Khomskii exchange interaction [92]. In case of a two-fold orbital degeneracy and half-filled bands, the intra-atomic exchange interaction favors a ferromagnetic ordering between neighboring sites in order to allow hopping into triplet states made of different orbitals on the same atom. This is necessarily accompanied by an antiferro-orbital-ordering, i.e., neighboring metal atoms alternately occupy different orbitals, such that hopping can take place between the same type of orbitals. Note that antiferro-orbital ordering does not imply an antiferromagnetic order by the orbital magnetic momenta, which are typically quenched (Table 4.2). Instead, it denotes

1. Ψ_a triplet:
$\quad$ hopping forbidden (Pauli)
$$\to E_{a,T} = E_{b,T} = 0$$

2. Ψ_a singlet:
$\to$ kinetic exchange $\quad E_{a,\text{singlet}} = -\dfrac{4t_a^2}{U_a}$ $\quad \Psi_b$ singlet: $\quad E_{b,\text{singlet}} = -\dfrac{4t_b^2}{U_b}$

3. Ψ_a/Ψ_b-singlet:
$$E_{ab,\text{singlet}} = -\dfrac{4t_a t_b}{U_{ab}}$$

4. Ψ_a/Ψ_b-triplet:

$E_{ab,\text{triplet}} = -\dfrac{4t_a t_b}{U_{ab} - J_H}$ $\quad$ also: "a" or "b" electron can hop back $\quad E_{ab,\text{triplet}} = -\dfrac{(t_a+t_b)^2}{U_{ab} - J_H}$

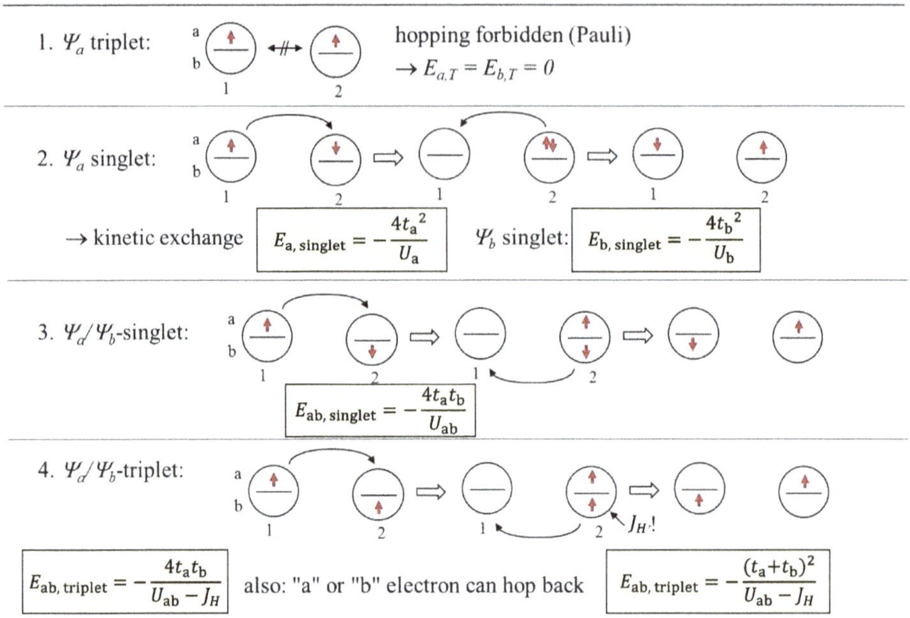

Fig. 4.22: Illustration of virtual hopping processes between neighboring sites in case of two degenerate orbitals ψ_a and ψ_b, e.g., the two e_g crystal field orbitals. Left column: Ground state spin configuration. Other columns: The up to three static configurations related to a hopping process. If hopping is possible, the corresponding energy gain is given below in a yellow box.

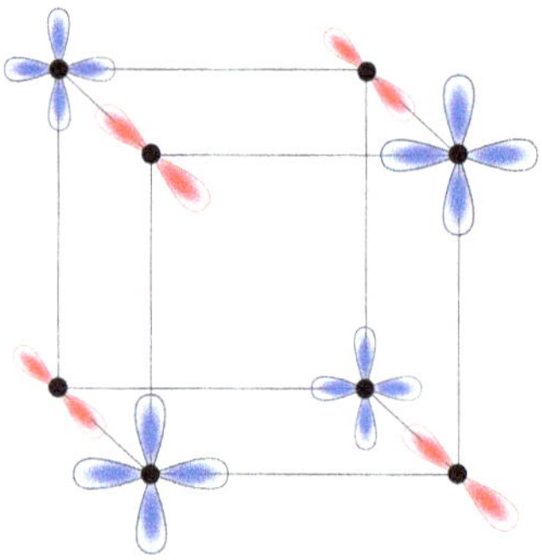

Fig. 4.23: The kinetic exchange of e_g-electrons favors a configuration, where occupied $d_{x^2-y^2}$ and $d_{3z^2-r^2}$ orbitals alternate on neighboring sites. Such an orbitally ordered state is called an antiferro-orbital phase.

only a long-range alternating occupation of two different types of Wannier-states for neighboring metallic ions, hence, a superstructure of charge distributions. An example of antiferro-orbital ordering on a cubic lattice is depicted in Fig. 4.23. Alternating, the lobe like red orbital ($d_{3z^2-r^2}$) and the cross-like blue orbital ($d_{x^2-y^2}$) are filled with an electron.

4.7 Example: doped manganites

We will now discuss one example of highly correlated electron systems in detail, the mixed valence manganites (see e.g. [93]). Their stoichiometric formula is $A_{1-x}B_xMnO_3$, where A is a trivalent cation (A = La, Pr, Nd, Sm, Eu, Gd, Tb, Dy, Ho, Er, Y, Bi) and B is a divalent cation (B = Sr, Ca, Ba, Pb). The mixture of divalent and trivalent cations leads to a mixed valence of the manganese ions. If we neglect covalent bonding[9] and describe these compounds in a purely ionic model, charge neutrality requires that manganese exists in two valence states:

$$Mn^{3+}: [Ar]3d^4 \quad \text{and} \quad Mn^{4+}: [Ar]3d^3 \,. \tag{4.50}$$

Now consider mixed valence manganites. The structure of these mixed valence manganites is related to the perovskite structure depicted in Fig. 4.24.

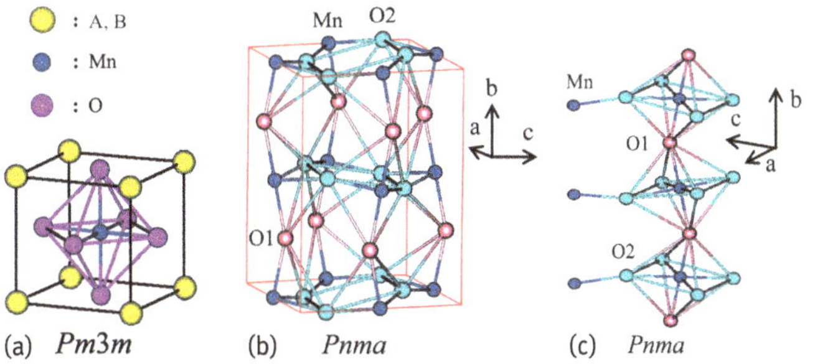

Fig. 4.24: The perovskite structure of doped manganites $A_{1-x}\,B_xMnO_3$. Left: The ideal cubic structure with space group Pm3m. Middle: The distorted orthorhombic structure with space group Pnma. Right: Illustration of the tilt of the oxygen octahedra in the distorted orthorhombic structure. (adapted from H. Li, Ph. D. thesis, RWTH Aachen University 2018).

Perovskite is the mineral $CaTiO_3$, which has a cubic crystal structure. The smaller Ca^{2+} metal cation is surrounded by six oxygen atoms forming an octahedron, which looks like the MnO_6 octahedron in Fig. 4.24, left. Such octahedra are centered on the eight corners of a simple cubic unit cell, i.e., the corners of the cube are each covered by the blue atom in Fig. 4.24, left. The larger Ti^{4+} metal cation (yellow in Fig. 4.24, left) is filling the centre of the cube in between the octahedra. Such an ideal cubic perovskite structure is extremely rare. It only occurs when the sizes of the different metal ions match, such that they fill the spaces between the oxygen atoms ideally. Usually, there is a misfit between the mean ionic radii of the A and B ions, which leads to sizeable distortions. The resulting structure is illustrated in the middle of Fig. 4.24. An important feature of this structure is the tilting of the corner shared oxygen octahedra as

9 This crude assumption is often made, albeit not completely valid (see below).

illustrated on the right of Fig. 4.24. Such an orthorhombic structure is for example realized in $LaMnO_3$. Orthorhombic structures occur, if the so-called tolerance factor T_{ion}, which measures the misfit between the ionic radii R_X, deviates significantly from one. The tolerance factor T_{ion} is defined as:

$$T_{ion} = \frac{1}{\sqrt{2}} \frac{\langle R_{A,B} \rangle + \langle R_O \rangle}{\langle R_{Mn} \rangle + \langle R_O \rangle} . \tag{4.51}$$

For the manganites, the octahedral surrounding of the Mn ions leads to a crystal field splitting, as shown in Fig. 4.18. The energy distance Δ_{CF} between the t_{2g} and the e_g levels is ~ 2 eV. If we now consider a Mn^{3+} ion with $3d^4$ configuration, the occupation of the crystal field levels depends on the ratio between the crystal field splitting and the intra-atomic exchange (Fig. 4.19). Usually, the intra-atomic exchange interaction amounts to about 4 eV and is stronger than the crystal field splitting, which favors the high spin state $S = 2$ of Fig. 4.19. Hence, four electrons with parallel spins occupy the three t_{2g} levels and one of the two e_g levels. The orbital angular momentum in the manganites is typically quenched, i.e., $\langle \hat{\vec{L}} \rangle \approx \vec{0}$. The e_g electron is prone to an orbital degeneracy, i.e., the electron can either occupy the $3d_{3z^2-r^2}$ or the $3d_{x^2-y^2}$ orbital. The resulting Jahn–Teller effect (Fig. 4.20 in Section 4.6.2), leads to a further splitting of the two e_g levels by typically ~ 0.6 eV. The Jahn–Teller effect couples the electronic degrees of freedom to the lattice degrees of freedom. This coupling occurs only for the Mn^{3+} ion (the so-called Kramer ion), since the Mn^{4+} ion with only three 3d electrons cannot lower its energy by a lattice distortion. A transfer of charge between neighboring Mn ions is, hence, accompanied by a change of the local distortion of the surrounding oxygen octahedra. This leads to a hopping process of a so-called lattice polaron, which consists of the electron and the surrounding distortion. Consequently, charge fluctuations, mediated by hopping processes of the e_g electron, are coupled to fluctuations of the lattice distortion. The Jahn–Teller effect can also lead to a long-range orbital order at lower temperature. This occurs, e.g., for the $LaMnO_3$ parent compound. Here, all Mn ions are trivalent and are expected to undergo a Jahn–Teller distortion as shown in Fig. 4.25. To minimize the elastic energy of the lattice, the Jahn–Teller distortions on neighboring sites are partly pointing in a different direction. Below the Jahn–Teller transition temperature $T_{JT} \approx 780$ K, a cooperative Jahn–Teller effect occurs, i.e., a long range superstructure of local lattice distortions. Correspondingly, the Mn $3d_{3z^2-r^2}$ orbitals exhibit an accompanying orientational order (Fig. 4.25). Again, the orbital ordering is not related to a magnetic order, but only provides a long range order of anisotropic charge distribution around the nuclei. As the temperature is further lowered, antiferromagnetic spin order sets in at the Néel temperature of 145 K. In $LaMnO_3$, the spins of the Mn^{3+} ions exhibit the so-called A-type order (black arrows in Fig. 4.25), i.e., spins within one plane are parallel, while spins of neighboring planes are coupled antiferromagnetically via a superexchange process (Section 4.4.3). The combination of A-type antiferromagnetic order and alternating occupation of different d-type orbitals results from a complex interplay be-

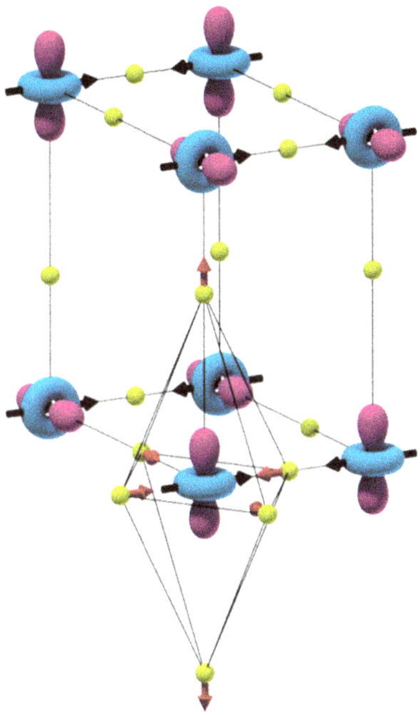

Fig. 4.25: Orbital order in $LaMnO_3$. Below the Jahn–Teller transition temperature (780 K), Jahn–Teller distortions of the oxygen octahedra occur as shown for one manganese ion by the elongated octahedron at the bottom (not to scale!). This elongation induces an orbital order of the surrounding eg orbitals (violet lobes with blue rings) of Mn^{3+} ions. An antiferromagnetic spin order (black arrows) sets in below $T_{Neel} \approx 145$ K. Oxygen: yellow spheres, La: not shown.

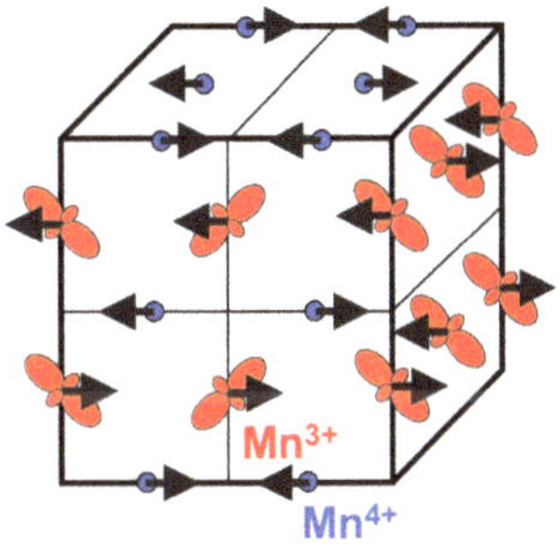

Fig. 4.26: Simultaneous order of charge (color), orbital occupation (direction of occupied d-orbital) and spin (arrow) in a half-doped manganite.

tween structural, orbital and spin degrees of freedom, which is governed by the relative strengths of the different coupling mechanisms. The situation becomes even more complex for doped manganites, where the charge on the Mn site becomes an additional degree of freedom due to the two possible valences Mn^{3+} and Mn^{4+}. In order to minimizes the Coulomb interaction between neighboring Mn sites, so-called charge order can develop. This is shown for the example of a half-doped manganite in Fig. 4.26.

These half-doped manganites show antiferromagnetic spin order, a checkerboard-type charge order with alternating Mn^{4+} and Mn^{3+} sites and a zigzag-type orbital order of the occupied e_g states on the Mn^{3+} sites. This is only one example of the complex ordering phenomena in doped mixed valence manganites. Many others exist, which leads to novel phenomena and functionalities like the colossal magnetoresistance effect as discussed in the following.

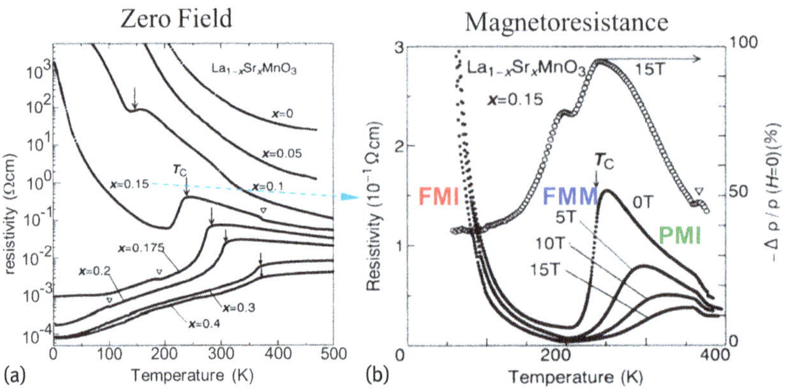

Fig. 4.27: Resistivity in the $La_{1-x}Sr_x MnO_3$ series. (a) Resistivity in zero magnetic field for various compositions from $x = 0$ to $x = 0.4$ as marked. (b) Resistivity for $x = 0.15$ in different magnetic fields. Additionally, the magnetoresistance is shown (right scale). It is defined as the magnetic field induced change in resistivity $\Delta\rho$ relative to its value at 0 T $\rho(H = 0)$. PMI: paramagnetic insulator, FMM: ferromagnetic metal, FMI: ferromagnetic insulator. After [94].

How are these ordering phenomena related to the macroscopic properties of the crystal? To answer this question, let us look at the resistivity of doped Lanthanum-Strontium-Manganites (Fig. 4.27). The zero field resistance changes dramatically with composition. The $(x = 0)$-compound shows insulating/semiconducting behavior as the resistivity increases with decreasing temperature.[10] The higher doped compounds, e.g., at $x = 0.4$, are metallic as the resistivity decreases with decreasing temperature. Note, however, that the resistivity is still about three orders of magnitude larger than for typical good metals such as Cu. At an intermediate composition $(x = 0.15)$, the samples exhibit insulating behavior at higher temperatures down to about 250 K. Then a dramatic drop of the resistivity indicates an insulator-to-metal transition, while an upturn below about 210 K reveals insulating behavior again. The metal-insulator transition occurs at a temperature, where the ferromagnetic long-range order sets in. Around this temperature, one also observes a very strong dependence of resistivity on an external magnetic field. This is the so-called colossal magnetoresistance effect. In order to appreciate the large shift in the maximum of the resistivity curve with field (Fig. 4.27, right), one should remember that the energy scales connected with the Zeeman interaction of a spin 1/2 electron are very

10 By definition, insulators and metals are distinguished by their change of resistivity ρ with temperature T, i.e., a metal shows increasing ρ with T and an insulator shows decreasing ρ with T. The initial reasoning is that for insulators/semiconductors the number of charge carriers, contributing to the electrical transport, increases with T via thermal activation. This leads to a larger conductivity with increasing T. For metals, the charge carrier density is largely independent of T. However, the dominating scattering probability by phonons increases with T, leading to a lower conductivity with increasing T (see also Section 5.3.1).

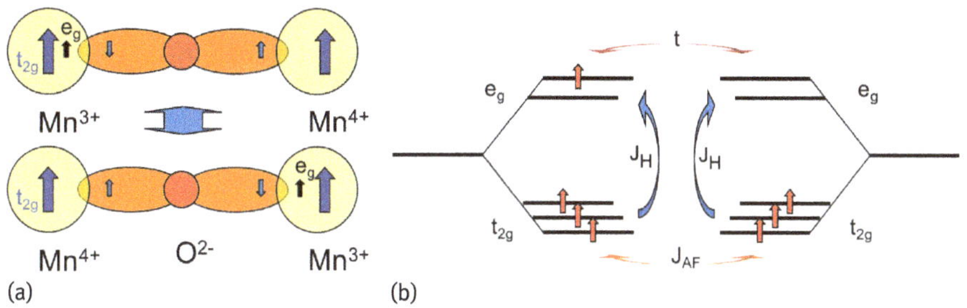

Fig. 4.28: Schematic representation of the double exchange interaction. Left: Transfer of an e_g electron via the intervening 2p orbital of the O^{2-} ion from a Mn^{3+} ion to a Mn^{4+} ion. Right: Energy level diagram for the two Mn atoms. There is an antiferromagnetic exchange interaction J_{AF} between the t_{2g} electrons of neighboring ions (superexchange). Within the Mn ions, Hund's rule coupling J_H is assumed to be larger than the crystal field splitting. The term t represents hopping between the two Mn sites. Only if the t_{2g} spins of both Mn atoms are parallel, can the e_g electron hop between the two sites. If the t_{2g} spins are antiparallel, hopping is suppressed by the strong Hund's rule coupling between e_g and t_{2g} spins on the same atom. If the double exchange is stronger than J_{AF}, the Mn spins tend to align parallel.

small. The energy splitting of the two possible spin levels within an external magnetic field of 1 T is only 0.12 meV, which corresponds to a temperature equivalent of 1.3 K.

Can we understand this strong dependence of the resistance on an external field in simple terms? Indeed, there is a mechanism for a magnetic exchange interaction, which can give rise to a ferromagnetic order and at the same time is connected with conductivity. This mechanism is called double exchange and is depicted schematically in Fig. 4.28. It can only occur between transition metal ions of different valences. In the case depicted on the left of Fig. 4.28, an e_g electron from a Mn^{3+} ion hops into the oxygen 2p orbital, while simultaneously an oxygen 2p electron hops on the Mn^{4+} site. This effectively transports an electron from a Mn^{3+} site to a Mn^{4+} site. Since e_g and t_{2g} electrons are strongly coupled through the intra-atomic exchange coupling, this transfer of an electron from Mn^{3+} to Mn^{4+} can only occur, if the spins of the t_{2g} electrons of the two neighboring metal ions are parallel. For an antiparallel orientation of the neighboring t_{2g} spins, hopping is suppressed due to the penalty of the intra-atomic exchange energy J_H on the target ion. Therefore, the double exchange between Mn^{3+} and Mn^{4+} ions is ferromagnetic. The ferromagnetic exchange process is directly connected with conductivity, which is simply realized by the hopping processes. In terms of the double exchange mechanism, we can now explain the insulator-to-metal transition occurring at the Curie temperature T_{Curie}. In the paramagnetic state, the spin directions of the t_{2g} electrons at neighboring ions fluctuate with respect to each other, thus suppressing the hopping of the e_g electron. The system behaves like an insulator. As soon as ferromagnetism sets in, hopping between neighboring Mn sites can occur and the resistivity drops. An applied magnetic field aligns the Mn spins even above

T_{Curie}. The induced magnetization permits an increased hopping of the e_g electrons and, thus, leads to a decrease of resistivity. Thus, the simple model of the double exchange interaction explains the observed magnetoresistance qualitatively. However, it has been shown that the double exchange interaction alone gives the wrong magnitude for the colossal magnetoresistance effect [95]. Other effects, such as the electron-phonon interaction, have to be taken into account. Since the entire discussion above starting from ionic states is a crude approximation, it is not exactly valid for real systems. Clarifying the situation of real systems usually requires a detailed interplay between theory and experiment. Therefore, the following section will introduce leading experimental methods for the study of the complex ordering phenomena and of the possible excitations of the charge, orbital, spin, and lattice degrees of freedom.

4.8 Experimental techniques: neutron- and X-ray scattering

Finally, we discuss how the various ordering phenomena can be studied experimentally. Obviously, we need probes with atomic resolution, which interact with the spins and with the charges of the system. It turns out, that neutron and x-ray scattering are ideal for studying the complex ordering phenomena and their excitation spectra. The lattice and spin structure can be studied with neutron diffraction using a polycrystalline or a single crystalline sample. Figure 4.29 shows the example of a powder spectrum of a $La_{7/8}Sr_{1/8}MnO_3$ material.

Preferably, the structure determination of polycrystalline materials is done by a simultaneous refinement of neutron and x-ray powder diffraction spectra, as the two probes have different contrast mechanisms. Neutrons interact with the nuclei of the atoms in contrast to x-rays, which interact with the electron cloud. Consequently, an x-ray spectrum contains less precise information on the structural parameters for the oxygen atoms, since these rather light atoms (few electrons) scatter much weaker than the heavier metal atoms (many electrons).

Neutrons have the additional advantage of a vanishing form factor for nuclear scattering, since the nucleus is point-like on the length scale of the wavelength of thermal neutrons, which is about 0.1 nm. Therefore, they give information up to large momentum transfer. This is particularly useful for the determination of the thermal parameters as described by Debye–Waller factors. Moreover, through their nuclear magnetic moment, neutrons are sensitive to the magnetic induction B in the sample, which allows one to determine the magnetic structure from a powder diffraction pattern.

Via the refinement of the fit to the experimental data, one can show, e.g., that the low temperature structure of a compound is monoclinic or even triclinic, i.e., there exist additional distortions from the Pnma structure introduced in the previous sec-

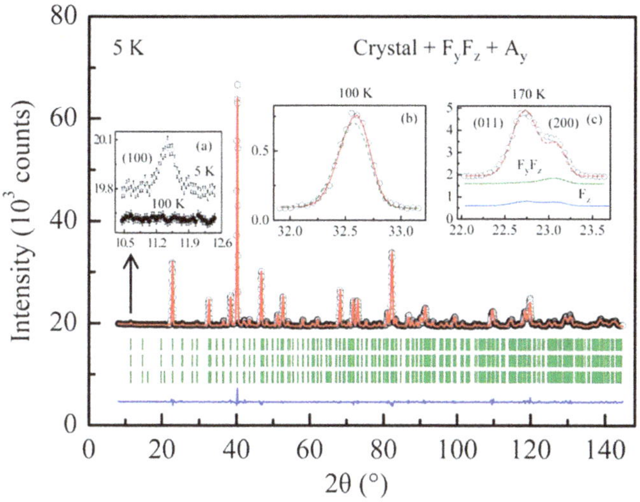

Fig. 4.29: High resolution neutron diffractogram of a powdered single crystal of $La_{7/8}Sr_{1/8}MnO_3$. Circles are the data points, the red solid line is the result of the fit via structural refinement. Structural and magnetic Bragg reflections are marked by the green vertical lines below the spectrum. The blue solid line underneath shows the difference between the observed and the simulated spectrum. Insets: Different regions of the spectrum at higher resolution partly measured at different temperatures. Moreover, fit curves or different contributions to fit curves (F_yF_z, F_z) are added as solid lines. For example, the inset marked (a) shows the appearance of a magnetic Bragg reflection at low temperature. It emerges at a low angle 2θ, which, hence, indicates a relatively large magnetic unit cell (from: Haifeng Li, PhD Thesis, RWTH Aachen University (2008)).

tion[11] (Fig. 4.24). Ferromagnetic order becomes visible by intensity on top of the structural Bragg peak within neutron diffraction patterns. Antiferromagnetic order is usually (but not always) connected with an increase in the unit cell size, which shows up by additional superstructure reflections between the main nuclear reflections (inset (a) in Fig. 4.29). It is beyond the scope of this lecture to discuss the experimental and methodological details of such a structural analysis or to present detailed results on specific model compounds. For this, we refer to the literature, e.g. [93]. We just mention that with detailed structural information, one is able to determine the lattice and spin structure as well as the charge and orbital order, which can then be related to macroscopic phenomena such as the colossal magnetoresistance (Fig. 4.27).

At first sight, it might be surprising that neutron diffraction provides information about charge order, since neutrons are neutral particles. Obviously, charge order is not determined directly by neutrons. However, in a transition metal-oxygen bond, the

11 Synchrotron-based x-ray powder diffraction, which provides the highest resolution in reciprocal space, is the best method to determine such small distortions of the low temperature structure of a compound, as peak splittings can be detected significantly better than with laboratory-based x-ray- or neutron powder diffraction methods.

Tab. 4.4: R_0 values of cation-oxygen bonds in manganese perovskites as needed for the bond valence calculation of eq. (4.52) [96].

Ions	**La^{3+}**	**Pr^{3+}**	**Nd^{3+}**	**Sm^{3+}**	**Eu^{3+}**	**Gd^{3+}**	**Tb^{3+}**	**Dy^{3+}**	**Er^{3+}**
R_0 (Å)	2.172	2.138	2.105	2.090	2.074	2.058	2.032	2.001	1.988
Ions	Tm^{3+}	Yb^{3+}	Y^{3+}	Ca^{2+}	Sr^{2+}	Ba^{2+}	Mn^{3+}	Mn^{4+}	
R_0 (Å)	1.978	1.965	2.019	1.967	2.118	2.285	1.760	1.753	

bond length depends on the charge of the transition metal ion. The higher the positive charge of the transition metal, the shorter will be the bond to the neighboring oxygen atoms due to Coulomb attraction. This qualitative argument is quantified in the so-called bond valence sum, which is an empirical correlation between the chemical bond length R_{ij} and the bond valence s_{ij}:

$$s_{ij} = \exp\left(\frac{R_0 - R_{ij}}{B}\right) . \tag{4.52}$$

Here, B is a constant ($B = 0.37$ Å according to [96]) and the values for R_0 differ for different cation-oxygen bonds as given in Table 4.4 [96]. Finally, the valence or oxidation state V_i of the cation i can be determined by the sum of the bond valences around the respective atom i according to:

$$V_i = \sum_j s_{ij} . \tag{4.53}$$

Even though this method is purely empirical, it proves to be rather precise. Most importantly, the values of the valences found with this method differ significantly from a purely ionic model. Instead of integer differences between charges on different transition metal ions, one finds differences of a few tenth of a charge of an electron.
Just like charge order, orbital order is not directly accessible to neutron diffraction, since orbital order represents an anisotropic charge distribution and neutrons do not directly interact with the charge of the electrons. However, we have seen in the discussion of the Jahn–Teller effect that an orbital order is connected with a distortion of the bond lengths within the anion complex surrounding the metal cation. In this way, by a precise determination of the structural parameters from a combined neutron and x-ray powder diffraction experiment, one can determine, in favorable cases, the ordering pattern of all four degrees of freedom: lattice, spin, charge and orbitals.

One can ask, whether there is a more direct way to determine charge and orbital order. The scattering cross section of x-rays contains the atomic form factors, which are Fourier transforms of the charge density distribution of the electrons within an atom. Hence, one might think that charge and orbital order can be determined by x-ray scattering directly. However, the effect is usually too weak. As discussed in the last paragraphs, it is only a fraction of an elementary charge, which contributes to charge or orbital ordering. In the following, we will estimate the resulting change in scattering intensities. For the case of a Mn atom, the atomic core has the [Ar] electron configuration, i.e., 18 electrons are in closed shells with spherical charge distributions. For the Mn^{4+} ion, three further electrons are in the t_{2g} levels. These 21 electrons contribute to the scattered intensity, besides the small charge difference between the neighboring Mn ions, which amounts to $\sim 0.2\,e$ only. Since a scattering experiment measures intensities and not amplitudes, we get the relative contribution from the charge ordering as: $\frac{0.2^2}{21^2} \approx 1\times10^{-4} = 0.01\%$. In this estimate, we have even ignored the scattering from all the other atoms, so that detection of charge or orbital order is even more difficult. There is, however, a way to enhance the scattering from non-spherical charge distributions, the so-called anisotropic anomalous x-ray scattering. It was first discussed by Templeton and Templeton [97] and applied for orbital order in manganites by Murakami et al. [98]. The principle of the technique is shown in Fig. 4.30.

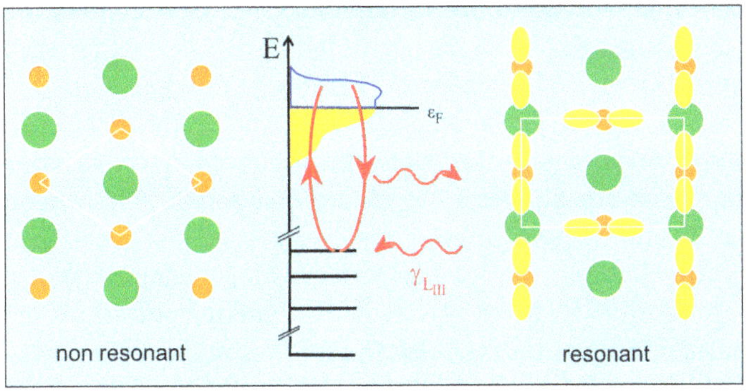

Fig. 4.30: Illustration of the principle of anisotropic anomalous x-ray scattering for a hypothetical 2-dimensional compound consisting of two atoms (orange and green circles) with different number of electrons. Left: Reconstruction of the charge distribution from a laboratory x-ray source. Non-resonant x-ray scattering is mainly sensitive to the spherical charge distribution. A unit cell as shown by the white lines is deduced. Middle: Resonant x-ray scattering process (see text). Right: Charge distribution as deduced from an anomalous x-ray scattering experiment. An orbital ordering pattern (yellow lobes) becomes apparent. The unit cell (white rectangle) is larger, which leads to aditional superstructure reflections at resonance.

It depicts the scattering result from a hypothetical two-dimensional compound consisting of two atoms with different number of electrons. Non-resonant x-ray scattering, as possible with a laboratory x-ray source, is mainly sensitive to the spherical charge distribution. A reconstruction of the charge distribution from such an experiment might look as shown on the left of Fig. 4.30. The corresponding crystal structure is described by a primitive unit cell indicated by the white lines. In order to enhance the scattering from the non-spherical part of the charge distribution, an experiment at a synchrotron radiation source can be employed, where the energy of the x-rays can be tuned to the absorption edge of a core electron level as shown in the middle of Fig. 4.30. Now, second order perturbation processes occur, where a photon induces virtual transitions of an electron from the core level to empty states above the Fermi energy and back with reemission of a photon of the same energy as the incident photon. Neglecting polarization of the x-ray beam for simplicity and focussing on the energy dependence, the form factor $\tilde{f}_i$ for dipolar scattering from the initial state i through an intermediate state j (e.g., $i = 1\,\mathrm{s}$, $j = 4\mathrm{p}_{x,y,z}$ at the K edge) takes the form:

$$\tilde{f}_i = \sum_j \frac{\langle i|\vec{p}_E|j\rangle\langle j|\vec{p}_E|i\rangle}{E_j - E_i - \hbar\omega - i\gamma/2} \,. \tag{4.54}$$

Here, $\vec{p}_E$ denotes a dipolar momentum operator (Section 2.2.3.1), $\hbar\omega$ is the photon energy, E_i and E_j are the energies of the initial and intermediate state, respectively, and γ is a damping factor due to the finite life time of the intermediate state.

Since second order perturbation processes have a resonant denominator, this scattering will be largely enhanced close to the absorption edge. If the intermediate states in this resonant scattering process are connected to the orbital ordering, scattering peaks from the orbital order will be enhanced. Thus, in the resonant scattering experiment, orbital order can become visible as depicted schematically on the right of Fig. 4.30. With the shown arrangement of orbitals, the primitive unit cell is larger than the unit cell deduced from the non-resonant scattering experiment (shown on the left). An increase of the unit cell size in real space is connected to a decrease of the distance of the reciprocal lattice points, i.e., an increase in the number of Bragg reflections. Therefore, orbital order is visible by a resonant scattering process via the appearance of additional superstructure reflections. The anisotropic anomalous x-ray scattering is, hence, sensitive to the tiny anisotropic local charge distributions around an atom. An additional advantage of the technique is that it is element specific due to the different absorption edge energies for the different elements. Hence, it combines diffraction and spectroscopy. Figure 4.31 is a sketch of the structure of $LaMnO_3$ which can be deduced from such data. From the intensity of the reflections, which exhibits a strong dependence on the x-ray photon energy due to the second order perturbation process, one deduces, e.g., the presence of an orbital polaron lattice.

So far, we have discussed powerful experimental techniques to determine the various ordering phenomena in complex transition metal oxides. Scattering can, however, provide more information. For example, quasielastic diffuse scattering reveals information on fluctuations and short range correlations. Hence, short range correlations of polarons above the phase transition temperature, magnetic correlations in a paramagnetic phase, local dynamic Jahn–Teller distortions above the Jahn–Teller transition temperature, etc. can be probed, which helps, e.g., to determine the dominating mechanism for a particular phase transition. The strength of the relevant interactions can, moreover, be determined from inelastic scattering experiments. For example, one can quantify the strength of the exchange interaction from measurements of the spin wave spectra. Subsequently, one can compare this result with models for superexchange or double exchange in order to verify or falsify the corresponding model explanations,

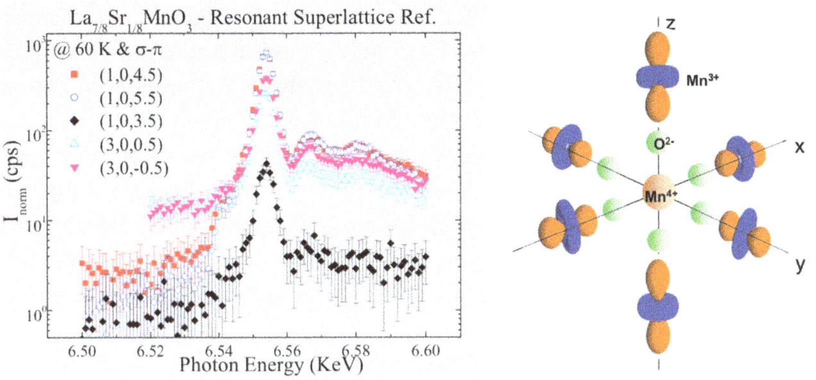

(a) (b)

Fig. 4.31: Left: Photon energy dependence of the logarithmically displayed intensity I_{norm} of some resonant superlattice reflections from $La_{7/8}Sr_{1/8}MnO_3$. For the displayed reflections with half indices along the *c* direction, a strong resonant enhancement at the K-absorption edge of Mn is visible at 6.55 keV. Right: Interpretation of a experiment in real space by an alternating arrangement of Mn^{3+} and Mn^{4+} ions. The additional electron of the Mn^{3+} ion occupies an e_g orbital, which points towards the Mn^{4+} ion. This arrangement is called an orbital polaron. In the ferromagnetic, insulating phase of $La_{7/8}Sr_{1/8}MnO_3$ below 155 K, the orbital polarons arrange into an orbital polaron lattice with long range order.

e.g., for the colossal magnetoresistance effect as discussed in Section 4.7 (Fig. 4.27). However, this goes beyond the scope of these introductory sections.

4.9 Summary

This chapter of the book has introduced the physics of highly correlated electron systems, where the electron-electron interaction strongly drives the behavior. We have found novel types of exchange interactions, such as kinetic exchange, superexchange and double exchange, and novel ordering phenomena of orbits, charges, spins, and lattice distortions. We have restricted ourselves to the paradigmatic case of complex transition metal oxides and chalcogenides.

The complexity of the correlated electron systems arises from the competing degrees of freedom: charge, lattice, orbit, and spin. The ground state is a result of a detailed balance between different energy scales influencing the different degrees of freedom. This balance can be disturbed by external fields or other thermodynamic parameters, giving rise to new ground states or complex collective behavior. Examples are the ordering phenomena discussed in this chapter, but also Cooper pairing in superconductors, so-called spin Peierls transitions in one dimensional systems (Section 5.2.1), etc. The sensitivity to external parameters as well as the novel types of ground states of the systems enable novel functionalities, such as the colossal magnetoresistance, which can be exploited for magnetic sensing, high temperature superconductivity, or multiferroic behavior. A theoretical description of these complex systems starting from first principles, as from the Schrödinger equation for quantum mechanics or from the maximization of entropy for statistical physics, fails due to the large number of strongly interacting particles. Hence, toy models have been developed to describe the emergent phenomena of these complex systems, which often capture the essential properties. The ongoing search for more adequate model descriptions and for novel phenomena, which surprisingly emerge due to the incomplete theoretical description so far, places highly correlated electron systems to the forefront in modern condensed matter physics.

We have additionally described that neutron and x-ray scattering are central experimental tools to disentangle the complexity of these systems experimentally. They are employed to determine the various ordering phenomena as well as the fluctuations and excitations of the relevant degrees of freedom.

Apart from these fundamental aspects, the novel properties led also to novel applications as already discussed in the introduction of this chapter (Section 4.1). Most obviously, high-temperature superconductors, which are not understood on a fundamental level so far in terms of their coupling process, are applied, e.g., for resistance-free cables in power plants, very sensitive magnetometers or high-field magnets. Other effects might be used as well, e.g., the colossal magnetoresistance (Fig. 4.27) for magnetic sensing or multiferroic effects, where different orders are tightly linked to each other in a single material, for memories. There, one order is employed for reading while the other order is employed for writing such that both processes can be optimized independently.

In the following chapter, we will discuss other types of electron phases, which are guided by various interactions such as electron-disorder, electron-phonon and electron-electron interaction. Very similarly, we will find that first-principle methods largely fail to describe the relevant properties. Typically, a concerted approach of theory and experiment develops an intuitive, but nevertheless scientifically stringent understanding of the key properties by employing adequate, partly ad-hoc approximations.[12] This can lead to novel classification schemes of different electron phases as the one by topological indices, where, e.g., an abstract winding property of the band structure determines the phase of the electron system completely. The most prominent example of the latter is the topological insulator phase, which via such a winding number provides conducting surfaces independent of the detailed atomic structure of the surfaces. Hence, the approach to develop novel types of simplified toy models beyond the established first-principle approaches proves to be extremely succesful for the understanding of solids.[13]

12 An example is the largely incomplete approximation of ionic binding in the manganites, which, however, led to decisive insights into the hopping properties of the electrons.

13 The textbooks [91, 99, 100] are recommended for further reading on correlation physics.

Markus Morgenstern

5 Interactions and topology for itinerant electrons

5.1 Introduction

5.1.1 Outline

As outlined in the previous chapter of this book, the single-particle band structure description for electrons in solids can break down completely due to the electron-electron interactions. For example, fully occupied Hubbard bands develop from a half-filled single-particle band, such that the system turns from a metal into an insulator (Section 4.5). Intriguingly, one cannot capture this behavior via a parameter-free description based on first principles. Instead, one has to apply instructive toy models as the Hubbard Hamiltonian (eq. (4.26)), that can explain the qualitative behavior of the electrons in the solid straightforwardly. The Hubbard model can then act as a starting point to implement further relevant interactions such as the spin-orbit interaction (eq. (4.44)) or electron-lattice couplings (Section 4.6). This eventually leads to the explanation of other observed effects such as polaron hopping or orbital ordering.

These type of model approaches proved to be extremely successful to get a systematic understanding of a multitude of effects observed in the correlated oxides and chalcogenides. Nevertheless, the approach is incomplete by construction, e.g., ignoring the long-range part of the Coulomb interaction. It is believed that this incompleteness hides additional interesting properties such as the still not understood origin of high temperature superconductivity (Section 4.1). In other words, it is believed that the incompleteness of the developed toy models explains most of the missing understanding of solid state properties. Hence, the partial failure in quantitative understanding does not indicate any failure of the first principle assumptions as given by the Schrödinger or Dirac equation, but points to inappropriate approximations during the solution of these equations.

Therefore, a central task is to develop adequate toy models for the approximate description of the multiple degrees of freedom in a solid. These toy models must be combined with the knowledge, that can be gained by first-principle methods mostly based on the density functional theory (DFT) approach. DFT can be employed to narrow down the adequate parameter regimes. This way, the toy models get a predictive character to be confirmed by experiments.

In the following final chapter of this book, additional toy models are introduced, that apply to the more intinerant s-type or p-type electrons, as present, e.g., in doped semiconductors (Figs. 2.8 and 5.1(a)). Hence, the starting configuration of the tight binding model, that is used for the oxides and basically localizes electrons at single atomic sites, has to be replaced by a starting configuration operating with Bloch states

https://doi.org/10.1515/9783110438321-005

or, even more simplified, with plane wave states $\psi(\vec{x}) \propto e^{i\vec{k}\vec{x}}$ as for free electrons. Consequently, the phase of the electron wave function can be decisive for the behavior of the system as discussed for some simple systems already in Section 1.4 and 1.5.

In first principle methods, electrons in crystalline solids are described as Bloch waves, i.e., each quasiparticle is mapped to a single particle wave function $\psi_j(\vec{x}) = u_{n\vec{k}}(\vec{x}) \cdot e^{i\vec{k}\cdot\vec{x}}$ (eq. (2.90)). Thereby, n is the band index, $\vec{k}$ the wave vector, $u_{n\vec{k}}(\vec{x})$ a function that exhibits the periodicity of the lattice vectors, and $j \in \mathbb{N}$ a counting index. A typical band structure (Fig. 5.1(a)) displays the energy of these single particle states as a function of the wave vector $\vec{k}$, that is the primary good quantum number in a periodic crystal. A rather successful model to treat the resulting quasiparticle states is to fill pairs of electrons with opposite spin into the states of lowest energy up to the Fermi level E_F at $T = 0$ K. At larger T, one can use the Fermi–Dirac distribution function (eq. (1.39)) for filling. An excitation of the solid, e.g., by absorbing a photon, is then described as a transition of one electron from an occupied to an unoccupied single-particle state (Section 2.3.1). The calculated short-range atomically periodic parts $u_{n\vec{k}}(\vec{x})$ as well as the long-range wave parts $e^{i\vec{k}\cdot\vec{x}}$ of the Bloch state are, hereby, very realistic descriptions. They can be imaged, e.g., by scanning tunneling spectroscopy (STS) (Fig. 5.1(b)–(f)), where the visibility of the long-range part typically requires scattering at defects in order to fix the phase of the wave in real space. The band structure model also describes a large part of other solid state properties adequately such as, e.g., conductivity σ. However, several predictions of the band structure model are qualitatively wrong. This was outlined in the previous chapter of the book (Chapter 4) and will be the focus of this chapter of the book, too.

We will again go beyond the band structure description, but in contrast to the previous chapter, we will start from rather simple band structures as derived from s-type and p-type electrons. We will systematically discuss the influence of different types of interactions beyond the band structure description. These interactions partly change the behavior of the electron system completely. They are electron-phonon, electron-disorder and electron-electron interaction. Some examples have already been discussed in previous sections. For instance, weak localization (Section 1.4.3.7) is a consequence of the interplay of crystalline disorder with the phase of a single particle wave function at the Fermi level, eventually leading to localization of electrons via constructive interference. This effect, not present in the single-particle band structure description, drives the system insulating, which would be a metal according to its single-particle band structure. Another example is the Mott–Hubbard transition (Section 4.5.3), where the electron-electron interaction (described by the Hubbard U) drives the single-particle metal into an insulator.

In this chapter, we will extend the learned approaches used to describe such phase transitions by analyzing several other phase transitions of the electron system, that are driven by the strength of a particular interaction. Thereby, we will also employ a novel classification scheme of electron phases, which is topology. It boils down the properties of an electron phase, e.g., conductivity, to a single integer number deduced from the Schrödinger or Dirac equation. Such a number can also be found from a single-particle band structure, such that the single particle states determine the collective properties as, e.g., surface conductivity directly. Importantly, the number called the topological index makes the found behavior robust against small continuous changes of the Hamiltonian as, e.g, the interaction of the electrons with disorder. Thus, it allows distilling of robust properties already from the incomplete descriptions as via first-principle methods or adequate toy models without including small disturbing interaction terms explicitly. More intriguingly, from a practical point of view, the novel classification scheme discovered useful properties of solids, that were simply overlooked for decades, albeit the corresponding solids were available and experimental hints to these particular properties were present in measured data.

5.1.2 Approximations of the band structure model

Before starting the description of the interaction effects, we recall, once more, the approximations leading to the description of electrons by Bloch waves (Section 4.2). A solid consists of N_A nuclei and $b \cdot N_\mathrm{A}$ electrons with b being the average number of electrons per atom. Quantum-mechanically, this ensemble is described by the many-particle wave function $\Psi_{\mathrm{total}}(\vec{x}_1, \vec{x}_2, \ldots \vec{x}_{N_\mathrm{A}}, \vec{x}_{N_\mathrm{A}+1}, \ldots \vec{x}_{N_\mathrm{A}+bN_\mathrm{A}}, t)$.[1] The function Ψ_{total} must be a solution of the Schrödinger equation or the Dirac equation including all energy terms of the system. Additionally, it must obey the exchange statistics of fermions and bosons.

The first approximation separates the electron system from the nuclei (Born–Oppenheimer-approximation) (Section 4.2). The electronic wave function, that reads $\Psi_{\mathrm{el}}(\vec{x}_{N_\mathrm{A}+1} \ldots \vec{x}_{N_\mathrm{A}+bN_\mathrm{A}})$, is calculated for fixed positions of the nuclei. The interaction potential between the nuclei is then reconstructed from the electron energy at different fixed positions of the nuclei, such that the solutions of the resulting Schrödinger equation for the nuclei reveal the excitation energies of the nuclear system (phonons). The central argument in favor of the separation is, that the electrons are much lighter than the nuclei, such that they can follow the motion of the nuclei rather instantaneously. One can estimate the error in energy made by the separation $\Psi_{\mathrm{total}} = \Psi_{\mathrm{nuclei}} \cdot \Psi_{\mathrm{el}}$ as $\Delta E = 1/b \cdot (m_\mathrm{e}/M_\mathrm{n})^{0.75} \cdot E_{\mathrm{electrons}} = 1/b \cdot (m_\mathrm{e}/M_\mathrm{n})^{0.25} \cdot E_{\mathrm{nuclei}}$ with m_e and M_n being the mass of electron and nucleus, respectively, and $E_{\mathrm{electrons}}$ and E_{nuclei} the energies of the electron and the nuclear system, respectively. Thus, ΔE is small with respect to E_{nuclei} and $E_{\mathrm{electrons}}$. Nevertheless, the properties of the system can be dramatically changed by the small contribution ΔE of the electron-phonon-interaction. Prominent examples are superconductivity as discussed in more basic textbooks of condensed matter physics and charge density waves as described in Section 5.2.1.

1 Here, we ignore the spin degree of freedom of each electron and nucleus, for the sake of simplicity.

The second approximation is the separation of the electron wave function in single particle states:

$$\Psi_{\text{el}}(\vec{x}_1,\ldots\vec{x}_{bN_A}) = P_{\leftrightarrow}\left(\prod_{i=1}^{bN_A}\psi_i(\vec{x}_i)\right), \tag{5.1}$$

$$E_{\text{electrons}} = \sum_{i=1}^{bN_A} E_i \,. \tag{5.2}$$

Hereby, $P_{\leftrightarrow}$ makes the required permutations in order to get an antisymmetric wave function Ψ_{el} as necessary for fermions. The twofold occupation of each single particle state by electrons with opposite spin (Pauli principle) is a direct consequence of this antisymmetry. The separation into single particle states works, if the individual electrons are independent, i.e., if the Coulomb interaction between the electrons gets negligible. This is not the case, since typical Coulomb interaction energies between electrons in a solid are as large as the kinetic energies of the electrons (1–10 eV) (Section 4.3). However, using an iterative process within the so-called local density approximation, one can approximate these interactions by an additional, effective electrostatic potential for a particular electron. In the most simple, quasi classical approximation, the potential $\Phi^j_{\text{electrons}}(\vec{x})$ for the electron in state ψ_j reads:

$$\Phi^j_{\text{electrons}}(\vec{x}) = \frac{1}{4\pi\varepsilon_0}\cdot\iiint\frac{n_{\text{e},j}(\vec{y})}{|\vec{y}-\vec{x}|}\,\text{d}^3\vec{y}\,. \tag{5.3}$$

Here, $n_{\text{e},j}(\vec{y}) = \sum_{i\neq j}|\psi_i(\vec{y})|^2$ is the electron density given by all the other electrons. The potential energy $\Delta E_{\text{pot},j}$ of the electron j in the electric field of all other electrons is then: $\Delta E_{\text{pot},j} = \langle\psi_j(\vec{x})|\Phi^j_{\text{electrons}}(\vec{x})|\psi_j(\vec{x})\rangle$. The iterative process to determine $\Delta E_{\text{pot},j}$ starts with a set of single particle wave functions, calculates the electron density $n_{\text{e},j}(\vec{y})$ and the corresponding $\Phi^j_{\text{electrons}}(\vec{x})$ from these wave functions, solves the Schrödinger equation with $\Phi^j_{\text{electrons}}(\vec{x})$ via matrix diagonalisation, and uses the novel wave functions from the diagonalisation to repeat the whole loop. The loop repetition stops, if the changes in energy per loop are small. Notice, that the diagonalisation is an approximation, too, since it requires an infinitely large $n\times n$ matrix to be exact, but uses only a finite number of matrix elements in practice.

Moreover, the loop neglects the required antisymmetrisation of Ψ_{el}, that would lead, in second quantisation, to terms of the form:

$$\widehat{H}_{\text{ee}} = \sum_{i,j,k,l} u_{ijkl}\;\hat{c}_i^\dagger\hat{c}_j^\dagger\hat{c}_k\hat{c}_l \tag{5.4}$$

with

$$u_{ijkl} = \iiint\left(\iiint\psi_i^\star(\vec{x})\cdot\psi_j^\star(\vec{y})\cdot\frac{e^2}{4\pi\varepsilon_0\cdot|\vec{x}-\vec{y}|}\cdot\psi_k(\vec{x})\cdot\psi_l(\vec{y})\,\text{d}^3\vec{y}\right)\text{d}^3\vec{x}$$

with $\hat{c}_i$ ($\hat{c}_i^\dagger$) being annihilation (creation) operators of electrons. Notice that, in eq. (5.3), only the terms u_{ijij} contribute. The terms u_{ijji} are the exchange terms (Sec-

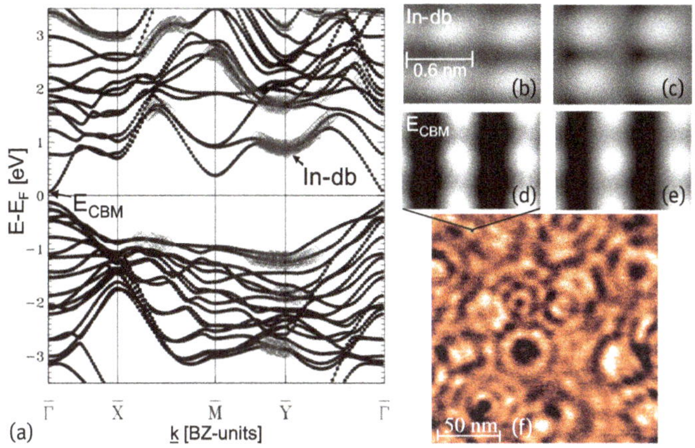

Fig. 5.1: (a) Band structure of a slab of an InAs crystal. The bands are projected to the (110) surface, i.e., only their k_x and k_y value are displayed. Each point is a single Bloch state as calculated by density functional theory (DFT) within the local density approximation (LDA). The Bloch states are equidistant in $\vec{k}$ space. The conduction band minimum E_{CBM} and a surface state (In-db) are marked. (b), (d) Measured $|u_{n\vec{k}}(\vec{x})|^2$ using scanning tunneling spectroscopy (STS) for the states marked in (a) as In-db (db = dangling bond) (b) and E_{CBM} (d). More precisely, the image displays the sum of all $|u_{n\vec{k}}(\vec{x})|^2$ for all states at the chosen energy. (c), (e) Calculated sum of all $|u_{n\vec{k}}(\vec{x})|^2$ at the energies marked in (a). (f) Large-scale STS image at energy $E - E_F = 0.05$ eV. The concentric rings are a 2D-cut through the 3D standing electron waves, that form spherical standing waves around defects $|\psi_i(|\vec{x}|)|^2 \propto 1/|\vec{x}|^2 \cdot \cos^2(|\vec{k}| \cdot |\vec{x}| + \varphi)$ ($|\vec{x}|$: distance to defect, φ: phase shift). The wave length $\lambda = 2\pi/|\vec{k}|$ is ~ 40 nm; (d) is a zoom of (f) as marked. (a)–(e) [101], (f) [102].

tion 4.4.1). All other terms are called correlation terms. Within the local density approximation, these exchange and correlation terms are approximated as an additional electrostatic potential $\Phi_{\mathrm{xc}}(\vec{x})$, called the exchange-correlation potential. $\Phi_{\mathrm{xc}}(\vec{x})$ can be approximated analytically for a free electron gas (no nuclei) as a functional of the electron density $n_{\mathrm{e}}(\vec{x})$. This offers a reasonable starting point for the development of adequate potentials in the presence of nuclei. A multitude of approximate functionals $\Phi_{\mathrm{xc}}(n_{\mathrm{e}}(\vec{x}))$ have been derived for different solids with different potentials of nuclei in order to approximate the electron system as good as possible.[2]

Using these functionals of $\Phi_{\mathrm{xc}}(n_{\mathrm{e}}(\vec{x}))$, the loop described above can be used to determine the single particle wave functions iteratively. Formally, these single particle wave functions are variational parameters only, but they approximate the real single particle wave functions adequately for most cases.

2 This approach can be extended by combining it with the Hubbard model (Section 4.5) via including an on-site term U into the DFT calculation in a method called LDA+U (LDA: local density approximation) or by the inverse mapping of the DFT band structure to an adequate local model, that is called dynamical mean-field theory (DMFT).

The band structure in Fig. 5.1(a) and the atomically periodic parts of the squared single particle wave functions in Fig. 5.1(c) and (e), as calculated by such a method, e.g., agree excellently with the experimental data (Fig. 5.1(b) and (d)). The method gets worse, if the electron-electron interaction gets stronger with respect to the kinetic energy of the electrons. This is the case for the strongly localized d- and f-electrons in oxides, where the Hubbard model reveals better results than a DFT approach (Section 4.5). For nearly free electron systems (itinerant electron states), the electron-electron interaction dominates with respect to the kinetic energy at low electron density, in particular, if the dimension of electron motion is reduced to 2 or 1 or in strong magnetic fields. This either leads to localized electron phases (Wigner crystal, Section 5.5.2.1) or it implies, that the excitations of the system are genuinely of a many-particle type as described in Section 5.5.2.3.

A third approximation within the band structure description is the perfectly periodic lattice potential. This implies the lattice periodicity of $|\psi_i(\vec{x})|^2$ for each electronic state i, i.e., the description by extended Bloch waves. In real crystals, the periodicity is not perfect. Firstly, the crystal has boundaries. This leads to additional states, the so-called surface states (not discussed here), that are the solutions with an imaginary $\vec{k}$ component perpendicular to the surface (eq. (1.28) in Section 1.4.3). Moreover, the interior of the crystal contains defects such as vacancies, dislocations, or foreign atoms. For many properties of the solid as, e.g., the heat capacity, these defects are only a minor perturbation. For other properties as, e.g., the low temperature conductivity, they are constitutional (e.g., Section 1.4.3.7). There are regimes, where the disorder changes the properties of the system completely, as within the quantum Hall regime: the disorder in the atomic lattice leads to a Hall resistance R_{Hall} of nearly always the $1/n$-fold ($n \in \mathbb{N}$) of the natural constant h/e^2 (h: Planck's constant) with a measured precision of $\Delta R_{\text{Hall}}/R_{\text{Hall}} \simeq 10^{-10}$ (Section 5.3.2). This spectacular precision also led to the more abstract description of the whole system by the already mentioned single integer index on the base of a topological analysis. This type of analysis has afterwards been extended to other solids, where it led to the prediction of dissipationless electron transport at surfaces and edges. The corresponding edge transport was indeed subsequently found in experiments (Section 5.4).

In summary, there are three major approximations for the description of solid state electrons as Bloch waves, i.e., one neglects electron-phonon, electron-disorder, and the dynamics of electron-electron interactions. All three of them imply novel phases and phenomena of the electron system not covered by the band structure description as exemplary described in the following.[3] The most complex description results from the electron-electron interaction as partly already covered in Section 4.5–4.7. An

3 Another approximation is the separation of spin and spatial degrees of freedom of the electrons, that neglects the spin-orbit interaction (Section 4.5.4), itself originating from the relativistic description of spins within the Dirac equation. This leads to additional phenomena, e.g., the Rashba effect (e.g. [103]), that are beyond the scope of this book.

experimentally simple approach to study the consequences of such interactions are so-called artificial atoms or quantum dots (Section 5.5.1). Their investigation basically matches the few particle description of exchange effects in Section 4.4.

We will start with a consequence of the electron-phonon interaction (Section 5.2), will then describe some consequences of the electron-disorder interaction (Section 5.3) including the topological description of solids (Section 5.4). Finally, we will discuss the electron-electron interaction starting with the quantum dots (Section 5.5.1), before we proceed with higher dimensions employing itinerant s- or p-type electrons (Section 5.5.2). For the latter, we will learn how quasiparticles can emerge in solids prone to interaction effects, which are qualitatively different from the constituting particles. As an example, we discuss composite fermions, which can exhibit a charge, which is only a fraction of the electron charge e and a mass that depends on the electron density n_e.

5.2 Electron-Phonon Interaction

Since the electron-phonon interaction is small with respect to the electronic and phononic energy, it can be described via perturbation theory. In second quantisation, one finds for the electron system:

$$\widehat{H}_{\mathrm{e,ph}} = \sum_{n,n',\vec{k},\vec{q},\vec{G},j} M^{j,n,n'}_{\vec{q}+\vec{G},\vec{k}} \cdot \left(\widehat{b}^{\dagger}_{\vec{q},j} + \widehat{b}_{-\vec{q},j}\right) \cdot \widehat{c}^{\dagger}_{n',\vec{k}-\vec{q}-\vec{G}} \cdot \widehat{c}_{n,\vec{k}} \,. \tag{5.5}$$

This corresponds to a process, where an electron is initially in state n, $\vec{k}(\widehat{c}_{n,\vec{k}})$ and then scattered into state n', $\vec{k}-\vec{q}-\vec{G}$ $(\widehat{c}^{\dagger}_{n,\vec{k}-\vec{q}-\vec{G}})$, while it creates a phonon $(\widehat{b}^{\dagger}_{\vec{q},j})$ or destroys it $(\widehat{b}_{-\vec{q},j})$. Here, $\vec{k}$ and $\vec{q}$ are the wave vectors of the initial electrons and phonons, which must be summed up across the whole first Brillouin zone for both partners. The term $\vec{G}$ is a reciprocal lattice vector, describing so-called Umklapp scattering processes. Thereby, the translational symmetry of the crystal lattice implies a modified momentum conservation, that applies only modulus reciprocal lattice vector $\vec{G}$, since the translational symmetry is not continuous, but only periodic with the lattice vectors. The indices n, n' are electron band indices and j $(= 1, 2, 3)$ sums the three possible, linearly independent vibrational directions of the phonons. The creation and annihilation operators $\widehat{b}^{\dagger}_{\vec{q},j}$ and $\widehat{b}_{\vec{q},j}$ for the phonons as well as $\widehat{c}^{\dagger}_{n,\vec{k}}$ and $\widehat{c}_{n,\vec{k}}$ for the electrons are linked to each other via matrix elements $M^{j,n,n'}_{\vec{q}+\vec{G},\vec{k}}$. The sum, thus, considers all processes as shown in Fig. 5.2(a) and (b), i.e., scattering of an electron via phonon creation or via phonon annihilation.

The calculation of the matrix elements is laborious and is not shown here. For a lattice with a single atom of mass M_{n} within the unit cell, it reads:

$$M^{j,n,n'}_{\vec{q}+\vec{G},\vec{k}} = \sqrt{\frac{\hbar \cdot N_{\mathrm{A}}}{V^2_{\mathrm{uc}}}} \cdot \frac{1}{\sqrt{M_{\mathrm{n}} \cdot \omega_{\mathrm{ph},j}(\vec{q})}} \cdot \mathrm{i} \cdot (\vec{q}+\vec{G}) \cdot \vec{e}_j(\vec{q}) \cdot \widetilde{\Phi}^{\mathrm{el}}_{\vec{q}+\vec{G}} \cdot \langle n', \vec{k}-\vec{q}-\vec{G} | n, \vec{k}\rangle \tag{5.6}$$

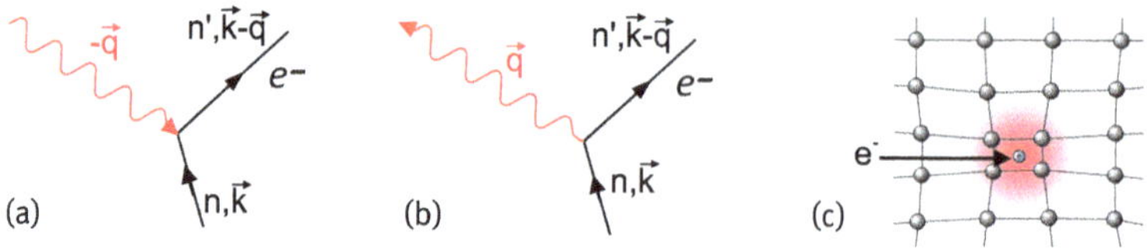

Fig. 5.2: (a) Process of phonon annihilation. (b) Process of phonon creation: straight black lines mark electronic states, curved red lines are phonon states. (c) Illustration of electron-phonon interaction.

with V_{uc} being the volume of the unit cell, N_{A} the number of atoms, $\omega_{\mathrm{ph},j}(\vec{q})$ the phonon frequency, $\vec{e}_j$ the unit vector in the direction of vibration of the phonon, $\langle n', \vec{k} - \vec{q} - \vec{G}|n, \vec{k}\rangle$ the overlap integral of the two atomically periodic parts of the contributing electron states, and $\widetilde{\Phi}^{\mathrm{el}}_{\vec{q}+\vec{G}}$ the Fourier component of the ionic potential acting on the electron and corresponding to the wave vector $\vec{q} + \vec{G}$.

Generally, the dynamical modification of the ionic potential due to the vibration leads to a mixing of similar single-particle states of the electrons. The process is particularly efficient, if the mass of the ion is small ($\omega_{\mathrm{ph},j} \propto 1/\sqrt{M_{\mathrm{n}}} \Rightarrow 1/\sqrt{M_{\mathrm{n}} \cdot \omega_{\mathrm{ph},j}} \propto M_{\mathrm{n}}^{-0.25}$) and, if the Fourier component of the ionic potential, that corresponds to the scattering vector, is large. Moreover, scattering is favored for longitudinal phonons ($\vec{e}_j \parallel \vec{q}$) except for Umklapp scattering ($\vec{G} \neq 0$). Finally, the scattering is favored for electron states of the same band, because states in the same band exhibit similar orbital wave functions increasing the overlap integral. Consequently, the overlap integral is often approximated by $\delta_{nn'}$.

Using Coulomb potentials for the ions $\Phi^{\mathrm{el}}(\vec{x}) \propto 1/|\vec{x}|$, – a good approximation in insulators-, one finds $\widetilde{\Phi}^{\mathrm{el}}_{\vec{q}} \propto 1/\vec{q}^2$ and, therefore, at small phonon energies ($\omega_{\mathrm{ph}} \propto |\vec{q}|$) $M_{\vec{q},\vec{k}} \propto |\vec{q}|^{-1.5}$ and at large phonon energies ($\omega_{\mathrm{ph}} \approx$ const.) $M_{\vec{q},\vec{k}} \propto |\vec{q}|^{-1}$, i.e., the interaction with longitudinal phonons close to the center of the Brillouin zone, called Γ, is strongest. In metals, the ionic potentials are strongly screened by the conduction electrons leading approximately to $\Phi^{\mathrm{el}}(\vec{x}) \propto \delta(\vec{x}) \Rightarrow \widetilde{\Phi}^{\mathrm{el}}_{\vec{q}} =$ const., i.e., to $M_{\vec{q},\vec{k}} \propto |\vec{q}|^{0.5}$ ($M_{\vec{q},\vec{k}} \propto |\vec{q}|$) for small (large) phonon energies, such that phonons at the Brillouin zone boundary contribute most strongly.

Often, the electron-phonon interaction is illustrated as in Fig. 5.2(c). The charge of the passing electron locally strains the atomic lattice. When the electron is gone, the strain relaxes, thereby inducing a lattice vibration. Of course, the electron can also decelerate an already existing vibration, if passing at the right vibrational phase. This corresponds to a phonon annihilation.

The most simple consequence of the electron-phonon interaction is a change of the single particle energy of the electrons. For s-type states, that can be described well by a parabolic dispersion ($E = \hbar^2\vec{k}^2/(2m^*)$) with effective mass m^*, the electron-phonon interaction changes m^*. Table 5.1 shows that the measured mass is up to 50% larger than calculated, while neglecting the electron-phonon interaction, i.e.,

Tab. 5.1: Effective mass m^*_{LDA} as calculated without electron-phonon interaction using the local density approximation and experimentally determined effective mass m^*_{Exp} for s-type states at the Fermi level within simple metals (m^* is given in units of the bare electron mass m_e).

metal	Li	Na	K	Rb	Cs	Al
m^*_{LDA} (m_e)	1.54	1.01	1.07	1.19	1.53	1.05
m^*_{Exp} (m_e)	2.22	1.24	1.21	1.37	1.80	1.49

the electrons are more difficult to move due to their interaction with phonons. This corresponds to the idea that the electron is accompanied by a lattice strain during its movement through the crystal, that has to be accelerated together with the electron. This is similar to the polaronic effect discussed in Section 4.7.

A second, intuitively simple consequence of the electron-phonon interaction is the electrical resistivity, that, at room temperature, is typically dominated by the scattering of electrons at phonons. Within the Boltzmann approximation, one firstly calculates, with the help of Fermi's Golden Rule (eq. (2.99)), the transition probabilities between the single particle states $n\vec{k}$ and $n'\vec{k}'$ of the electrons: $W_{n\vec{k}\to n'\vec{k}'} \propto |\langle n'\vec{k}'|\widehat{H}_{\mathrm{e,ph}}|n\vec{k}\rangle|$. After convoluting with the distribution functions (Fermi–Dirac for electrons, Bose–Einstein for phonons), one finds after a lengthy calculation that the temperature dependence of the resistivity $\rho(T)$ for scattering at acoustic phonons reads approximately [104]:

$$\begin{aligned} \rho(T) &\propto T^5 \quad \text{for} \quad T \ll \Theta_{\mathrm{D}} \\ \rho(T) &\propto T \quad \text{for} \quad T \gg \Theta_{\mathrm{D}} \end{aligned} \tag{5.7}$$

with Θ_{D} being the Debeye temperature of the material. Such a temperature dependence has indeed been found for simple metals such as Na and Cu. For longitudinal optical (LO) phonons, we have $\omega_{\mathrm{ph}}(\vec{q}) \simeq \omega_{\mathrm{ph,LO}} \approx \text{const}$, i.e., $M_{\vec{q},\vec{k}} \propto |\vec{q}|^{-1} \cdot \omega_{\mathrm{ph,LO}}^{-0.5}$ leading to $\rho(T) \propto T$. Acoustic phonons in piezoelectric materials are a peculiarity, that exhibit a T^4 dependence of the resistivity at low temperature.

Besides these purely quantitative consequences of the electron-phonon interaction (heavier mass and limitation of the mean free path or, more formally, modification of the real and imaginary part of the self energy), one also finds qualitative changes of the electron system driven by the electron-phonon interaction, i.e., phase transitions. The most prominent example is the superconducting phase appearing at low temperature. The explanation within BCS theory considers electron-phonon interactions in second order perturbation theory, i.e., $\widehat{H}_{\mathrm{e,ph}}$ is described by processes as shown in Fig. 5.3.

It turns out that these processes, effectively coupling two electrons, lead to a reduction of the energy of the electrons, if the energy of the contributing phonon is larger than the energy difference between the initial and final state of each electron. Due to this effective binding energy, two electrons form a stable pair with bosonic character

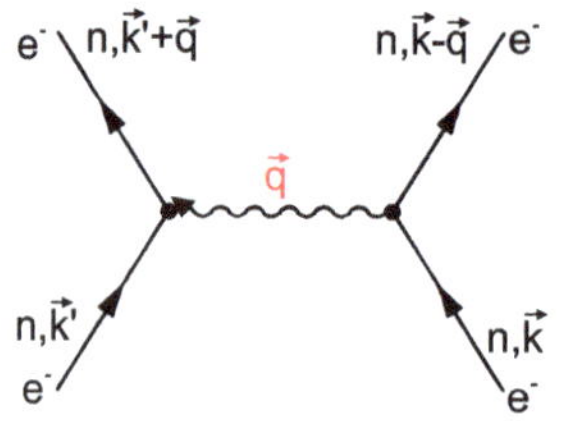

Fig. 5.3: Scattering process as derived within second order perturbation theory of the electron-phonon interaction, eventually leading to Cooper pairs.

(spin = 1 or 0) such that these bosons (Cooper pairs) can condense, within a Bose–Einstein like transition, into a phase where all bosons occupy the energetically lowest state in a phase coherent manner. This phase is the celebrated superconducting condensate with the favorable property of zero electrical resistance. Note that although the electron-phonon interaction is weak (Section 5.1.2), second order perturbation theory in this weak interaction can be decisive for the behavior of the system.

5.2.1 Peierls transition

The time period between the discovery of superconductivity (1908) and its explanation (1957) by the BCS theory amounts to five decades, which is largely due to the intuitively unexpected dominance of processes within second order perturbation theory.[4] For a long period, a phase transition that appears in first order perturbation theory was thought to be a likely explanation. This phase transition, called the Peierls transition, will be discussed in the following.

Experimentally, one observes for some materials that the electron charge distribution gets inhomogeneous below a certain temperature, i.e., it forms a static wave pattern, whose wave length is half the Fermi wave length λ_F. Figure 5.4(b) shows a scanning tunneling microscopy (STM) image of such a charge density wave. Besides the atomic periodicity, that corresponds to the known $|u_{n\vec{k}}(\vec{x})|^2$ of the Bloch waves, one recognizes a periodicity with the threefold atomic distance, which is rather precisely $\lambda_F/2$. At higher temperature, this additional periodicity disappears (Fig. 5.4(c)).[5] Simultaneously with the appearance of the charge density wave, one observes a lateral shift of the nuclei by about 1% of the lattice constant exhibiting a static wave pattern of the nuclear shifts (atomic superstructure) with wave length $\lambda_F/2$ as well. This is called a frozen phonon, since the wave pattern corresponds to a standing phonon wave with frequency $\omega_{\text{ph}} = 0/\text{s}$, i.e., with energy $E_{\text{ph}} = 0\,\text{meV}$. This phonon is, thus, already excited at temperature $T = 0\,\text{K}$, i.e., it is frozen.

---

4 The electron-phonon interaction is a weak perturbation, that might lead one to believe that it is negligible in second order perturbation theory.

5 STM approximately maps the contours of constant charge density of all states between the Fermi level of the sample and the Fermi level of the tip, that are different due to an applied tunneling voltage.

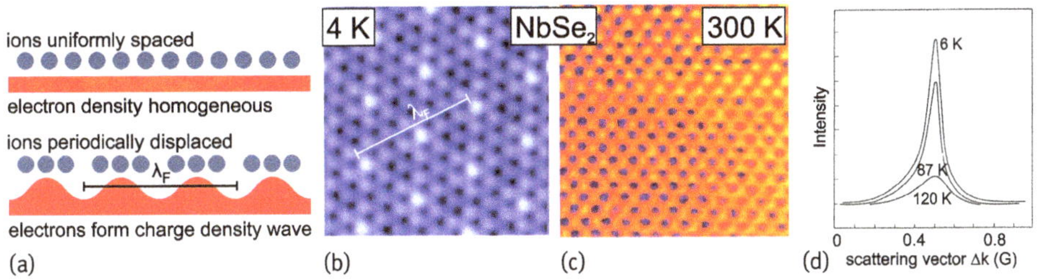

Fig. 5.4: (a) Model of atomic arrangement above (top) and below (bottom) the transition temperature of the Peierls transition. The shifts of the blue atoms (ions) are largely exaggerated. The Fermi wave length λ_F is marked. The pink area indicates the resulting distribution of electron charge. (b) Scanning tunneling microscopy (STM) image of $NbSe_2$ below the Peierls transition temperature. Besides the atomic periodicity, one observes a long-range periodicity with period $\lambda_F/2$. (c) STM image of $NbSe_2$ above the Peierls transition temperature. Only the atomic periodicity is visible. (d) Intensity of x-ray diffraction of $K_2Pt(CN_4)Br_{0.3}$ as a function of the polar scattering angle (scattering vector Δk in units of the reciprocal lattice vector G) at different temperatures as marked. The strength of the peak, that belongs to the charge density wave, disappears at the Peierls transition temperature, (d) [105].

Figure 5.4(a) schematically shows the shift of the nuclei (ions) and Fig. 5.4(d) displays x-ray diffraction data for the diffraction spot belonging to a charge density wave as a function of temperature. The transition temperature is $T \approx 100$ K, around which the peak gets weaker and broader with increasing temperature, i.e., the charge density wave continuously disappears, while the domains of phase coherent charge density wave areas get smaller. Besides these observations, that principally could also indicate an entropy driven structural phase transition, one finds that the phonon at wave length $\lambda_F/2$ (wave vector: $|\vec{q}| = 2 \cdot |\vec{k}_F|$) exhibits a phonon energy reduction as shown in Fig. 5.5(a) and that the electron system exhibits a band gap E_{Gap} at the Fermi level of order 10–100 meV (Fig. 5.5(b)). The latter can be confirmed, e.g., by measuring the temperature dependence of the electrical conductivity $\sigma(T)$ (Section 1.6.3). Figure 5.5(c) shows $\sigma(T)$ of a certain material in the inset. The Peierls transition is indicated by the maximum of the conductivity. Above 250 K, σ decreases with increasing temperature as usual for metals, since the increasing number of phonons increases the scattering probability for the electrons. Below 250 K, σ increases with increasing T as usual for insulators, since the number of excited charge carriers into the conduction band increases with T. The Arrhenius plot in the main image shows a nearly linear behavior below 250 K, such that an energy barrier for excitations of $E_{Gap} = 80$ meV can be deduced according to $\sigma(T) \propto \exp(-E_{Gap}/k_B T)$ (k_B: Boltzmann constant). Hence, a band gap developed at E_F because of the Peierls transition. With the help of tunneling spectroscopy, the band gap can also be probed directly.

Most materials showing a Peierls transition exhibit either a strongly one dimensional (1D) structure, i.e., electronic dispersion and electrical conductivity are much stronger in one direction than in the two perpendicular ones, or large parts of the Fermi

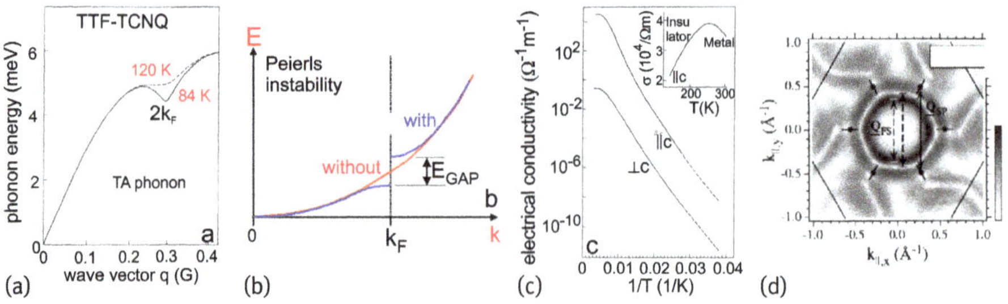

Fig. 5.5: (a) Phonon dispersion of the transversal acoustic (TA) branch of TTF-TCNQ at two temperatures slightly below and slightly above the Peierls transition temperature as determined by inelastic neutron diffraction. (b) Schematic of the electron dispersion $E(\vec{k})$ close to E_F above (red) and below (blue) the Peierls transition temperature. (c) Electrical conductivity of $K_2Pt(CN_4)Br_{0.3}$ below the Peierls transition temperature displayed as an Arrhenius plot. The linearity shows activated behavior of the type $\sigma(T) \propto \exp(-E_{Gap}/k_B T)$ with a deduced band gap of $E_{Gap} \approx 80$ meV. The two curves probe two different crystallographic directions. The inset shows the conductivity in linear $\sigma(T)$ representation. (d) A 2D section through the 3D Fermi surface for a material that shows a 2D Peierls transition (determined by photoelectron spectroscopy). The data are differentiated, i.e., the gradient of the photoelectron intensity I_{el}, $\nabla_{\vec{k}} I_{el}(E = E_F, k_x, k_y)$, is displayed in order to increase the visibility. Black lines mark the Brillouin zone boundaries. The vectors $\vec{Q}_{FS}$ and $\vec{Q}_{SP}$ are the preferential scattering vectors of the electron-phonon interaction leading to the Peierls transition. (a), (c) [105], (d) [106].

line of a largely two-dimensional (2D) material consist of parallel lines. The latter is shown in Fig. 5.5(d), where the inner Fermi line is mostly a hexagon, where the opposite edges of the hexagon run parallel.

The driving force of the Peierls transition is displayed in Fig. 5.4(a). The creation of the frozen phonon leads to periodic areas of increased positive charge by the ionic cores. Arranging the negative electronic charge density (pink area) in a way, such that the maxima are at these points gains Coulomb energy $E_{\text{e-ion}}$. However, the frozen phonon is not the most favorable arrangement of the ions such that one has to pay elastic energy E_{elastic} in order to create the frozen phonon. The decisive question, thus, is, if the energy difference $\Delta E = E_{\text{e-ion}} - E_{\text{elastic}}$ is negative, i.e., favorable to realize a charge density wave. This, interestingly, is always the case for 1D electron systems, as will be shown below.

The second question is about the favorable wave length. For the answer, one has to consider that the electron-phonon interaction, in first order perturbation theory, corresponds to scattering processes as described by eq. (5.5). This requires that the initial state of the electron is occupied while the final state is empty. Without additional energy and at $T = 0$ K, this is only possible at E_F. The corresponding process is displayed for a 1D system in Fig. 5.6(a). Obviously, the favorable wave vector of the phonon is $\vec{q} = 2 \cdot \vec{k}_F$. This fits with the experimental wave length of the frozen phonons. In addition, the resulting superposition of the two electronic states with opposite wave vector $(\vec{k}_F, -\vec{k}_F)$ creates a standing wave with a charge density $(|\psi_i(\vec{x})|^2)$ of the exper-

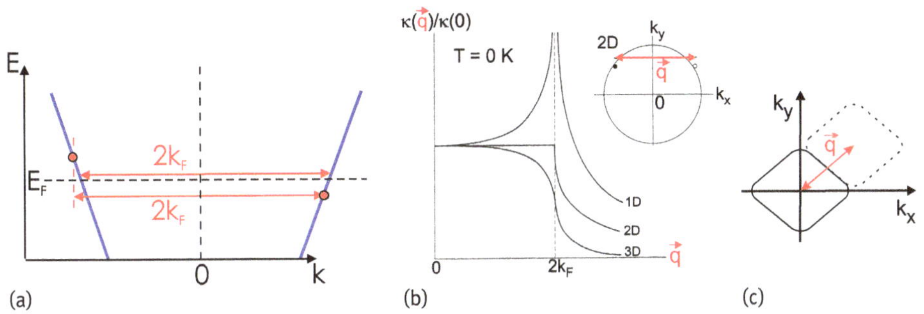

Fig. 5.6: (a) Scattering processes (red double arrows) leading to a charge density wave within a 1D system. The system is displayed by its $E(k)$ dispersion around E_F (blue lines). (b) Polarizability $\kappa(q)$ for a 1D, a 2D and a 3D electron system with parabolic, isotropic dispersion. Inset: Fermi circle for 2D electrons with one indicated phononic scattering vector $\vec{q}$. (c) Illustration of Fermi nesting: Using the indicated phononic wave vector $\vec{q}$, one can combine all states on the Fermi line, that overlap after the Fermi line is additionally shifted by $\vec{q}$ (dashed line) with respect to the original Fermi line (solid line). (b) [105].

imentally observed periodicity $\lambda_F/2$. The charge density wave at this wave length can be supported by processes as marked by the two red dots in Fig. 5.6(a). These processes also lead to a standing wave with periodicity $\lambda_F/2$, but require the energy difference between the two states to be provided, e.g., by annihilating a phonon.

Formally, we will restrict our description of the Peierls transition to 1D. We start with the electrostatic attraction $E_{\text{e-ion}} = \delta\rho_{c,q} \cdot \widetilde{\Phi}_q$, where $\delta\rho_{c,q}$ and $\widetilde{\Phi}_q$ are the amplitude of the charge density wave of the electrons and the amplitude of the electrostatic potential of the frozen phonon, respectively, both at wave vector q. The charge density wave is described as the polarization of the sea of electrons by the potential of the frozen phonons: $\delta\rho_{c,q} = \kappa(q) \cdot \widetilde{\Phi}_q$. The missing polarizability function $\kappa(q)$ is now calculated in first order perturbation theory. The corresponding novel many-particle wave function $\widetilde{\Psi}_{q,\text{el}}$ of the charge density wave with wave vector q reads:

$$\widetilde{\Psi}_{q,\text{el}} = \Psi_{0,\text{el}} + \sum_{\pm k} \frac{\left\langle \Psi^{\pm}_{k,\text{el}} \middle| \widehat{H}_{\text{e,ph}} \middle| \Psi_{0,\text{el}} \right\rangle}{\epsilon_k - \epsilon_{k\pm q}} \Psi^{\pm}_{k,\text{el}} \,. \tag{5.8}$$

Thereby, $\Psi_{0,\text{el}}$ is the ground state many-particle wave function without electron-phonon interaction, $|\Psi^{\pm}_{k,\text{el}}\rangle = \widehat{c}^{\dagger}_{k\pm q}\widehat{c}_k|\Psi_{0,\text{el}}\rangle$, and ϵ_k, $\epsilon_{k\pm q}$ the single particle energies of the unperturbed electronic states.[6] The created charge density wave $\delta\rho_c(x)$ is the difference between the charge densities of the two many-particle wave functions $\widetilde{\Psi}_{q,\text{el}}$ and $\Psi_{0,\text{el}}$. Using $\widehat{H}_{\text{e,ph}} \propto \sum_q \sqrt{q} \cdot \widetilde{\Phi}_q(\widehat{b}_{\pm q} + \widehat{b}^{\dagger}_{\mp q})\widehat{c}^{\dagger}_{k\pm q}\widehat{c}_k$ as for a material interacting via longitudinal acoustic phonons (page 257ff.), we find after replacing the electron

6 A state at k is transferred to a state at $k \pm q$.

operators $\hat{c}_k^\dagger$ und $\hat{c}_k$ by Fermi–Dirac distribution functions according to $(1 - f_k), f_k$:[7]

$$\kappa(q) \propto \sqrt{q} \cdot \left[\sum_k \frac{f_k \cdot (1 - f_{k+q})}{\epsilon_k - \epsilon_{k+q}} + \sum_k \frac{f_k(1 - f_{k-q})}{\epsilon_k - \epsilon_{k-q}} \right] . \tag{5.9}$$

Here, f_k is the value of the Fermi–Dirac distribution function for single particle states with wave vector k. In eq. (5.9), we restricted ourselves to phonon creation processes, that do not depend on the Bose–Einstein distribution function of the phonons, i.e., they are possible even at $T = 0$ K. The first (second) term describes processes, that transfer electrons from k to $k + q$ $(k - q)$.

Using a parabolic dispersion of the electrons $\epsilon_k = \frac{\hbar^2 k^2}{2m^*}$, we find $\epsilon_k - \epsilon_{k\pm q} = -\frac{\hbar^2}{2m^*}(q^2 \pm 2kq)$. For $T = 0$ K (Fermi function = step function), we can additionally select the k-space intervals where the electron is scattered from an occupied state ($f_k = 1$) to an unoccupied state $f_{k\pm q} = 0$. The initially occupied electron states can be, at most, $\pm q$ away from the closest Fermi wave vector. This leads to:

$$\begin{aligned} \kappa(q) &\propto -\frac{2m^*}{\hbar^2} \sqrt{q} \cdot \left[\sum_{k=k_F-q}^{k_F} \frac{1}{q^2 + 2kq} + \sum_{k=-k_F}^{-k_F+q} \frac{1}{q^2 - 2kq} \right] \\ &\propto -\frac{2m^*}{\hbar^2} \sqrt{q} \cdot \left[\int_{k_F-q}^{k_F} \frac{1}{q^2 + 2kq} \, \mathrm{d}k + \int_{-k_F}^{-k_F+q} \frac{1}{q^2 - 2kq} \, \mathrm{d}k \right] \end{aligned} \tag{5.10}$$

In the second line, we replaced the sum by an integral, exploiting the high density of states in k space for an extended crystal. By substituting $\tilde{k} = -k$ for the second term, one can verify straightforwardly that both terms are identical leading to:

$$\kappa(q) \propto -\frac{4m^*}{\hbar^2} \sqrt{q} \cdot \int_{k_F-q}^{k_F} \frac{1}{q^2 + 2kq} \, \mathrm{d}k . \tag{5.11}$$

In order to simplify the denominator, we substitute $\tilde{k} = q^2 + 2kq$ with $\frac{\mathrm{d}\tilde{k}}{\mathrm{d}k} = 2q$ and corresponding replacement of the integration limits resulting in:

$$\kappa(q) \propto -\frac{4m^*}{\hbar^2} \frac{1}{2\sqrt{q}} \cdot \int_{q^2+2(k_F-q)\cdot q}^{q^2+2k_F q} \frac{1}{\tilde{k}} \, \mathrm{d}\tilde{k} \tag{5.12}$$

$$= -\frac{2m^*}{\hbar^2} \frac{1}{\sqrt{q}} \cdot \ln\left(\frac{q^2 + 2k_F q}{2k_F q - q^2} \right) = -\frac{2m^*}{\hbar^2} \frac{1}{\sqrt{q}} \cdot \ln\left(\frac{q + 2k_F}{2k_F - q} \right) . \tag{5.13}$$

7 Applying the annihilation operator $\hat{c}_k$ to a many particle state returns a factor, that is the occupation probability of the corresponding single-particle state, i.e., f_k, while applying the creation operator $\hat{c}_k^\dagger$ returns a factor describing the emptiness of the state, i.e., $(1 - f_k)$. This averages over many realizations via the Fermi–Dirac distribution function, hence, it leads a mean-field description of the interaction term $\widehat{H}_{\mathrm{e,ph}}$.

Obviously, the polarizability diverges towards $-\infty$ for $q = 2 \cdot k_\mathrm{F}$, i.e., the energy gain $E_\text{e-ion} = \kappa(q)\widetilde{\Phi}_q^2$ for a charge density wave coupled to a frozen phonon of wave vector $2k_F$ diverges, too. Consequently, the charge density wave is the ground state at $T = 0\,\mathrm{K}$, independent of the required finite elastic energy E_elastic. Generally, the divergence is caused by the divergence of the perturbational terms for the two contributing electron states at k_F and $-k_\mathrm{F}$ having the same energy.[8] More descriptively, the energy gain arises, because the two states k_F and $-k_F$ can form two linearly independent standing waves with maxima either on the high density area of the frozen phonon or on the low density area of the frozen phonon (Fig. 5.4(a)) without energy penalty. Formation of one of the standing waves gains energy, while the other is loosing energy. Since only one of the two waves must be occupied at k_F (at E_F, the Fermi function is $f_k(E_\mathrm{F}) = 1/2$), the system gains energy by occupying the energetically favorable standing wave. This process is directly represented by an experimentally observed band gap at E_F.

For $T > 0\,\mathrm{K}$, the Fermi function $f_k(E)$ is not a step function anymore such that additional processes contribute. States above E_F are partly occupied and states below E_F are partly empty, such that (quasi-)elastic scattering processes are enabled at these energies, too. Importantly, the sign of the perturbational term changes at $\epsilon_k = \epsilon_{k\pm q}$ (eq. (5.8)), such that repulsive and attractive terms contribute at finite T. Hence, these extra processes do not stabilize the charge density wave additionally. Formally, one approximates the electron dispersion by straight lines around E_F, i.e., $\epsilon_k \propto k$, and neglects the influence of the thermally excited phonon bath (phonon annihilation), such that one gets after keeping the Fermi functions of eq. (5.9) and considering the phonon energy $\hbar\omega_\mathrm{ph}(q)$ explicitly as the energy difference between initial and final state:[9]

$$\kappa(2k_\mathrm{F}) \propto \sqrt{2k_\mathrm{F}\mathcal{D}(E_\mathrm{F})} \int_0^{\tilde{x}} \frac{\tanh x}{x} \mathrm{d}x \tag{5.14}$$

$$= \sqrt{2k_\mathrm{F}\mathcal{D}(E_\mathrm{F})} \ln(2.28 \cdot \tilde{x}) \tag{5.15}$$

with

$$\tilde{x} = \frac{\hbar\omega_\mathrm{ph}(2k_\mathrm{F})}{k_\mathrm{B}T} \tag{5.16}$$

with $\mathcal{D}(E_\mathrm{F})$ being the electron density of states at E_F, that is independent of energy for the approximated linear, one dimensional $E(k)$ dispersion close to E_F and determines how many states are available for scattering. The energy gain does not diverge anymore. It, moreover, decreases with increasing T. This is due to the increasing contribu-

8 Notice that perturbation theory requires that the matrix element of two degenerate states with the perturbing Hamiltonian $\widehat{H}_\mathrm{e,ph}$ must be zero, that is always possible, such that a more detailed analysis including the overlap integral in the matrix element is required. However, this does not change the general observation and will not be performed here.

9 $\tanh(x) = \frac{e^x - e^{-x}}{e^x + e^{-x}}$.

tion of scattering processes, that are not exchanging a momentum of $2\hbar k_F$, such that there is no clear selection anymore for a favorable wave vector of the frozen phonon. Additionally, the changing sign of the interaction term contributes. Hence, $E_{\text{e-ion}} = \kappa^2(2k_F)\widetilde{\Phi}_{2k_F}$ must be compared with the loss of elastic energy E_{elastic}.

Therefore, one approximates the temperature independent, elastic energy by a parabolic function around the equilibrium positions of the nuclei (linear chain with springs between the atoms) (eq. (4.45)). Hence, one takes $E_{\text{elastic}} \simeq 1/2 \cdot \widetilde{D} \cdot u_q^2$ with u_q being the amplitude of the displacement of the nuclei from the equilibrium position and $\widetilde{D}$ being an effective spring constant representing the binding forces between the atoms. Moreover, the amplitude of the electrostatic potential $\widetilde{\Phi}_q$ is assumed to be linear in the displacement amplitude $\widetilde{\Phi}_q = A_{\text{Ion}} \cdot u_q$ (eq. (4.46)). This is correct for small displacements (Section 4.6.2). The term A_{ion} is proportional to the ionic charge. Setting the temperature dependent Coulomb energy $E_{\text{e-ion}}$ equal to the elastic energy E_{elastic}, one eventually gets an estimate for the Peierls transition temperature T_C, that reads

$$T_C \simeq 2.28 \cdot \hbar\omega_{\text{ph}}(2k_F)/k_B \cdot \exp\left(-\frac{\chi\widetilde{D}}{A_{\text{ion}}^2 \cdot \mathcal{D}(E_F)}\right) \tag{5.17}$$

with χ being a material dependent constant. As expected, a large density of states at E_F ($\mathcal{D}(E_F)$) and a large ionic charge (large A_{ion}) are favorable for the Peierls transition, while a rigid lattice (large $\widetilde{D}$) is not favorable. Moreover, a large phonon energy helps, since this allows that the electron states can couple across a larger energy region by $2k_F$ scattering such that more electrons can gain from the charge density wave.

The charge density wave, in turn, is attractive for the created frozen phonon, which explains the energy gain of the phonon, as visible in Fig. 5.5(a). Moreover, the frozen phonon leads to a novel unit cell of the atomic lattice with reciprocal unit vector $2k_F$. This induces a band gap at the novel Brillouin zone boundary k_F, as usual for Brillouin zone boundaries. The electronic states close to k_F that are in phase (or 180° shifted) with respect to the frozen phonon periodicity will gain (lose) energy. This explains the band gap of Fig. 5.5(b) from a slightly different perspective and accordingly the corresponding $\sigma(T)$ in Fig. 5.5(c).

For electron systems in higher dimensions, the polarizability $\kappa(q)$ at $T = 0$ K does not diverge at $2k_F$. The result for electron systems with isotropic and parabolic dispersion, which leads to a Fermi circle (Fermi sphere) in 2D (3D), is shown in Fig. 5.6(b). The constant behavior of $\kappa(q)$ in 2D is clarified by the inset. Each phonon vector $q < 2k_F$ can couple exactly two possible pairs of electronic states such that the corresponding density of states $\mathcal{D}(E_F, \vec{k})$ is infinitesimally small, being zero in the thermodynamic limit. Consequently, the transition temperature is $T = 0$ K. However, if one increases the corresponding (joint) density of states belonging to a particular $\vec{q}$, a Peierls transition can also appear in higher dimensions. Figure 5.6(c) illustrates a 2D-Fermi line, where complete sections of the Fermi line can be coupled by a single phonon vector. The coupled density of states gets larger than zero and a Peierls transition takes place at finite T. For the system shown in Fig. 5.5(d), the Fermi line exhibits a hexagon in the Brillouin zone center, such that coupling vectors such as the one displayed as $\vec{Q}_{FS}$ can lead to three charge density waves in three different directions. This qualitatively explains the charge density wave pattern of Fig. 5.4(b), albeit the details are more complex and are still not settled completely (e.g. [107]). In particular, these details decisively include the wave vector dependence of $M^{j,n,n'}_{\vec{q}+\vec{G},\vec{k}}$ in real systems.

In closing this section, we briefly discuss two final aspects:

1. Since the electron-electron repulsion for extended states is usually weaker for parallel than for antiparallel spins due to the exchange term (Section 4.4.1), the charge density wave is often accompanied by a spin density wave, i.e., the maxima of the charge density wave are alternately filled preferentially with electrons of spin ↑ and electrons of spin ↓. This results in a spin density wave with twice the wave length of the charge density wave, respectively, with wave vector k_F. Mostly, the spin density wave appears at lower temperature than the charge density wave, such that one observes two phase transitions.
2. Using the arguments above, the existence of the charge density wave can be explained, but not its relative phase with respect to the atomic lattice. If the wave length is $\lambda_F/2 = n/m \cdot a$ with n, m being integer numbers ($n > m$) and a the lattice vector, this is called a commensurable charge density wave. Then, it is obvious, that the maximum of the charge density wave is at an ionic center where it can profit most strongly from the highest positive charge density of the ions. If, in contrast, the ratio of $\lambda_F/2$ and a is not a rational number, then there is no optimal position. The energy of the system is independent of the relative phase between the atomic lattice and the charge density wave. Such a charge density wave necessarily includes regions, where the charge density maximum sits above an ion core and regions, where the charge density maximum is located exactly between two ion cores. This situation does not change at all, if one changes the position of the charge density wave with respect to the atomic lattice. Consequently, the charge density wave can be moved through the crystal without any energy penalty. This idea has originally been pursued as an explanation for the disappearing electrical resistance of superconductors, but turned out to be wrong. In reality, each deviation from the perfect lattice (atomic defects) destroys the equivalence of all relative phases, since the defect energetically prefers either a minimum or a maximum of the charge density wave. This is called pinning of the charge density wave at the defect. It leads to the insulating behavior as shown in Fig. 5.5(c).

5.3 Consequences of potential disorder

In this section, we will discuss consequences of disorder within the atomic lattice. The most intriguing result in this respect is the quantum Hall effect. Due to the presence of the disorder and independent of the details of the disorder, one gets an extremely precise measurement of natural constants. The relation h/e^2 (e: electron charge,

h: Planck's constant) can be determined with an error of less than 10^{-10} by probing the Hall resistance of an imperfect 2D crystal. In fact, this effect defines the standard unit Ω for the resistance and defines together with the Josephson effect the SI-unit Ampére. A novel definition of the mass (kg) based on natural constants has been launched in 2018. It replaces the primary kilogram in Paris. Probably, the quantum Hall effect will play an important role in the resulting measurement protocols.

Prior to discussing the quantum Hall effect in Section 5.3.2, we will shortly introduce disorder induced localization as already discussed in Section 1.6 with respect to the resulting transport properties. Here, we will focus more on the resulting wave function properties.

5.3.1 Metal-insulator transition

One can not avoid deviations from the perfect arrangement of atoms in an ideal crystal, even if the production processes are continuously improved in order to decrease these deviations. The central reason is the relation between atomic mobility and entropy. Mostly, crystals are grown by slow cooling from the melt, i.e., the crystal gets enough time to realize the configuration of minimum free energy. But since diffusion processes depend exponentially on temperature, the waiting time gets too large at a certain minimum temperature, such that the atomic arrangement is basically frozen at that temperature during cooling. A finite temperature, however, increases the entropic contribution to the free energy, such that deviations from the ideal arrangement appear in equilibrium. At lower temperature, the minimum of free energy exhibits less deviations, but the mobility of the atoms is already too low to transfer the crystal from the current local minimum of internal energy into the global one. An ideal crystal, thus, requires either no energy barriers between local and global minima or a way to construct the crystal artificially atom by atom. The latter is possible, e.g., with a scanning tunneling microscope (STM), but only for a few hundred atoms in reasonable time. Partly, the disorder is even constitutional for the functionality of the material, e.g., if one dopes a semiconductor by foreign atoms, i.e., by donors or acceptors.

In 3D, there are basically two types of phase transitions driven by disorder. Metals become insulators (Section 1.6 and Fig. 4.27) and insulators become metallic. Recall that an insulator is a material with conductivity $\sigma = 0\,\mathrm{S/m}$ at $T = 0\,\mathrm{K}$ and a metal is a material with $\sigma > 0\,\mathrm{S/m}$ at $T = 0\,\mathrm{K}$. However, this is practically useless, since one can not measure at $T = 0\,\mathrm{K}$. In practice, one measures the temperature dependence of the conductivity or resistivity ρ. If ρ decreases with decreasing temperature, the material is called metallic, following the idea that the density of charge carriers is constant, but the scattering probability decreases with decreasing temperature, e.g., due to the decreasing number of phonons or due to the decreasing phase space for electron-electron scattering. If ρ is increasing with decreasing temperature, the material is called an insulator (or semiconductor), for which the charge carrier density gets significantly reduced with decreasing temperature (page 242, Section 1.6). This practical definition is not unambiguous, since some materials exhibit non-monotonic $\rho(T)$ dependence (Fig. 4.27), that we ignore in the following for the sake of simplicity.

Figure 5.7(a) shows a successful experiment for an alloy. The resistance $R(T)$ is recorded repeatedly, while the crystal is bombarded by α particles. This increases the disorder within the crystal continuously and induces a change in the slope of $R(T)$, that initially decreases and finally increases with decreasing temperature. The metal-insulator transition (MIT) is defined at the point of disorder, where the sign of the slope of $R(T)$ changes.

As already discussed in Section 1.6, the reason for the insulating properties at higher disorder is not the formation of a band gap at E_F, as e.g., during a Mott–Hubbard transition (Fig. 4.14), but the creation of localized states. The probability density of these states is concentrated in a certain area. Consequently, they barely contribute to the electrical transport between two macroscopically separated electrodes and its contribution to the transport decreases with decreasing T (Section 1.6.3). An example is shown in Fig. 5.7(b)–(e). For a regular electrostatic potential, as shown in

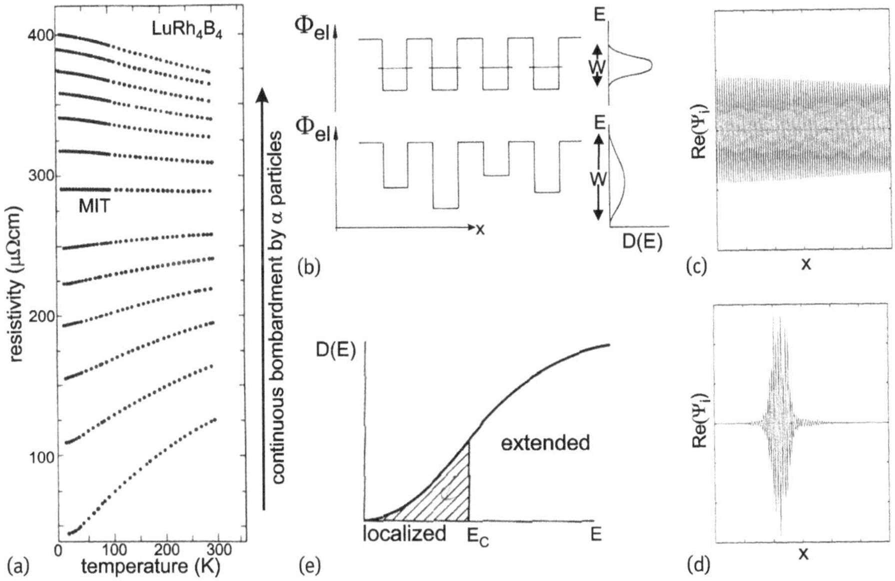

Fig. 5.7: (a) Resistivity as a function of temperature for $LuRh_4B_4$. The data are recorded continuously from the bottom to the top while the crystal is bombarded by α particles. (b) Schematic of the electrostatic potentials $\Phi_{el}(x)$ of a 1D crystal without (top) and with (bottom) potential disorder. The corresponding density of states $\mathcal{D}(E)$ is plotted on the right. (c) Real part of $\psi_i(x)$ for an extended state. (d) Real part of $\psi_i(x)$ for a localized state. (e) Density of states $\mathcal{D}(E)$ of a disordered crystal with localized states (hatched area). E_C is called the critical energy or the mobility edge of the metal-insulator transition (similar to Fig. 1.28). (a) R. C Dynes et al. in [108].

the top part of Fig. 5.7(b), one gets Bloch states and a band width W, that depends on the strength of the hopping parameter t (eq. (4.33)). Varying the atomic potentials, as in the bottom part of Fig. 5.7(b), leads to areas that are on average more attractive to the electrons than others. Localizing the probability amplitude $\psi_i(\vec{x})$ of a single particle state in these areas costs kinetic energy (stronger curvature of $\psi_i(\vec{x})$), but this can be compensated by the gain in electrostatic energy. Thus, one gets some states, that are localized in these areas (example in Fig. 5.7(d)). Their probability amplitude decreases roughly exponentially with distance from the center of mass (eq. (1.98)). However, other states are still extended similar to Bloch states. They only exhibit a slightly varying amplitude of $\psi_i(\vec{x})$ with position (Fig. 5.7(c)). Recall that different (stationary) states have to be orthogonal. This avoids states to be located in the same area.[10] It turns out that the localized states have energies at the rims of the bands as one would expect, since these states mostly explore the extrema of the potential landscape. Hence, there is a transition between localized states at lowest energies in the band and extended states towards the energetic center of the band as displayed in Fig. 5.7(e). These two types of states are separated by the critical energy E_{C}, also called the mobility edge in Section 1.6. Moreover, the bands get broader due to these localized states as shown on the right of Fig. 5.7(b). Localized states have indeed been observed in experiments. A very direct experimental observation, where a light wave instead of an electron wave is localized within a disorder potential, is shown in Fig. 5.8(b).

In order to calculate states within a disorder potential, one typically uses tight-binding models (Section 4.5) reading:

$$\hat{H} = \sum_{n,m} t_{n,m}\, \hat{c}_n^\dagger \hat{c}_m + \sum_n \Phi_{n,\mathrm{el}}\, \hat{c}_n^\dagger \hat{c}_n \tag{5.18}$$

with n, m counting the lattice positions and hopping amplitudes $t_{n,m}$, that are mostly taken to be non-zero only for neighboring atoms. $\Phi_{n,\mathrm{el}}$ are the spatially varying electrostatic potentials of the individual atoms, that in the most simple calculations are distributed randomly (Fig. 5.7(b)). Notice that the potential disorder term is distinct from the Hubbard U term in eq. (4.26). The former operates with a single occupation operator $\hat{n}_n = \hat{c}_n^\dagger \hat{c}_n$, while the later uses two occupation operators $\hat{n}_n \hat{n}_n$, hence counting double occupation of a site n.

The probability amplitudes of the different single particle states within the disorder potential are then determined by matrix diagonalization, where Wannier states localized at the atoms (Section 4.5) are used as a starting base for the diagonalization. The energy interval, where one finds localized states,[11] increases with increasing disorder. This is displayed in Fig. 5.8(a). The physical reason for the increasing

10 Additionally, the electron-electron repulsion naturally avoids that all states are localized in the areas of strongest attractive ionic potential.

11 A state is called localized, if $|\psi_i(\vec{x})|^2$ decreases with the distance from the center of mass $\vec{x}_0$ of the probability density of the state. Mostly, one can fit an exponential function $\overline{\psi_i}(|\vec{x} - \vec{x}_0|) \propto$

energy interval of localization is, that one can curve the probability density increasingly strongly, since the increasing gain in potential energy at larger disorder increasingly compensates for the loss in kinetic energy by the wave function curvature. The stronger curvature implies a better localization of states on small areas such that more localized states, orthogonal to each other, become possible. The existence of localized and extended states does not depend on the details of the chosen disorder. Within the calculation of Fig. 5.8(a), a simple cubic atomic lattice is used with a hopping rate between neighbors of $t = 1/6$. The disorder is the randomly chosen potential $\Phi_{n,\mathrm{el}} \in [-\widetilde{W}/2, +\widetilde{W}/2]$. For disorder strength $\widetilde{W} < 2.6$, both, localized and extended states exist. Hence, it depends on E_F, if the system is an insulator (localized states at E_F) or a metal (extended states at E_F). This implies a metal-insulator transition with E_F as the control parameter. Alternatively, the disorder strength $\widetilde{W}$ can be used as the control parameter as in the experiment of Fig. 5.7(a). Above $\widetilde{W} = 2.6$, the electrons at E_F are always localized independent of the position of E_F.

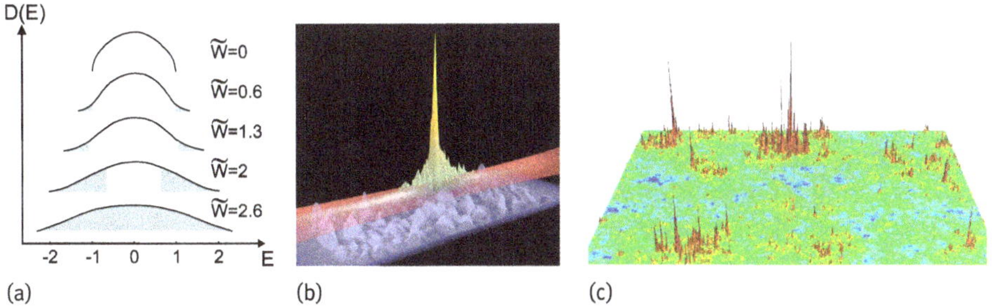

Fig. 5.8: (a) Calculated density of states $\mathcal{D}(E)$ of a cubic lattice for increasing disorder $\widetilde{W}$ of the electrostatic potential of the atoms as marked ($\Phi_{n,\mathrm{el}} \in [-\widetilde{W}/2, +\widetilde{W}/2]$). Energy intervals of localized states are filled in gray. (b) Experimentally determined $|\Psi_i(\vec{x})|^2$ of a localized state (yellow-green). The disorder potential $\Phi_\mathrm{el}(\vec{x})$ is shown below in violet. The experiment is performed with light waves instead of electron waves, which facilitates the imaging. (c) Calculated $|\psi_i(\vec{x})|^2$ of a critical state at the transition energy from localized to extended states. (b) [109], (c) courtesy of A. Mildenberger, see also [110].

Note that the order parameter of these phase transitions is not temperature, but electron density or disorder. As other phase transitions, it exhibits universal properties, that do not depend on the details of the system. An important example is the exponent of the localization length as a function of energetic distance from the mobility edge E_C. Close to E_C, one finds: $\xi(E) \sim |E - E_\mathrm{C}|^{\nu_\mathrm{C}}$ with ν_C being the critical exponent, that only depends on the general symmetries of the Hamiltonian, such as time-reversal symmetry, charge conjugation symmetry, chiral symmetry, and the spatial dimension d

$\exp(|\vec{x} - \vec{x}_0|/\xi)$ (eq. (1.98)) to the state at large $|\vec{x}|$. ξ is called the localization length, that diverges at the transition between localized and extended states.

of the electronic system. The study of such universalities of so-called quantum phase transitions is a scientific subject in itself, which is beyond the scope of this book. An interesting aspect are the fractal properties of the critical state at E_C. This state is neither a 0D state, i.e., a localized state existing only in a finite area, nor a 3D state, i.e., a state that is extended across the whole crystal, but a filamentary state with a non-integer dimension in between. This dimension can be calculated by the so called box counting method (see, e.g., Wikipedia) applied to the probability density of an electron $|\psi_i(\vec{x})|^2$. A calculated $|\psi_i(\vec{x})|^2$ of a critical state at E_C is shown in Fig. 5.8(c). It exhibits so-called multi-fractal properties, i.e., the fractal dimension (calculated by the box counting method) differs for different momenta of the probability distribution, i.e., the dimension of $|\psi_i(\vec{x})|^{2n}$ depends on the chosen $n \in \mathbb{N}$. The distribution of the fractal dimensions for different n, however, is universal, i.e., it depends only on the general symmetries of the Hamiltonian and the details are not important.

The opposite behavior to metals, where disorder drives the system insulating, is observed for semiconductors and insulators, where foreign atoms can drive the system metallic. As discussed in more basic textbooks of condensed matter physics, a doping atom can be approximated as a Coulomb potential $\Phi_{\text{el}}(\vec{x}) \propto 1/|\vec{x}|$ for one electron (or one hole), analogously to the hydrogen atom, but with reduced binding energy $E_{\text{Bind}} = m^*/m_e\varepsilon^2 \cdot 13.6\,\text{eV}$ (m^*: effective mass, ε: dielectric constant) (see also Section 2.3.3).[12] At $T = 0\,\text{K}$ and low donor density, the electrons are bound to the donor and only get mobile after thermal excitation into the conduction band. Thus, according to our definition, the material is an insulator. However, if the density of donors (or acceptors) is increased, neighboring Coulomb potentials overlap as shown in Fig. 5.9(a). The resulting bound states appear at different energies, i.e., we get a so-called impurity band of localized states, where the probability amplitudes of the states start to get distributed across several doping atoms. However, the electrons are still localized with their center of mass at the corresponding donor. This tendency is supported by the electron-electron repulsion, that avoids having several electrons at the same donor. The situation is similar to the Mott–Hubbard transition of a half filled band (Section 4.3 and 4.5). Within the localized regime, the electrical transport is governed by thermally induced hopping of the electrons between localized states as indicated by the red arrow in Fig. 5.9(a) (Section 1.6.3). The transport described by matrix elements as $\langle\psi_i|\widehat{H}_{\text{e,ph}}|\psi_j\rangle$, thus, depends strongly on temperature. One finds, e.g., $\sigma(T) \propto \exp\left(-\beta/\sqrt[4]{T}\right)$ within the model of variable range hopping as outlined in Section 1.6.3.2 (eq. (1.110)).

Increasing the donor density further, naturally leads to a decreasing potential maximum around the donors as shown in Fig. 5.9(b), such that the majority of the states belonging to the donors are not bound anymore. Alternatively, one can employ the knowledge from Section 4.5. One can argue in analogy that the gain via hopping t

12 This applies to a parabolic band, where m^* is well defined.

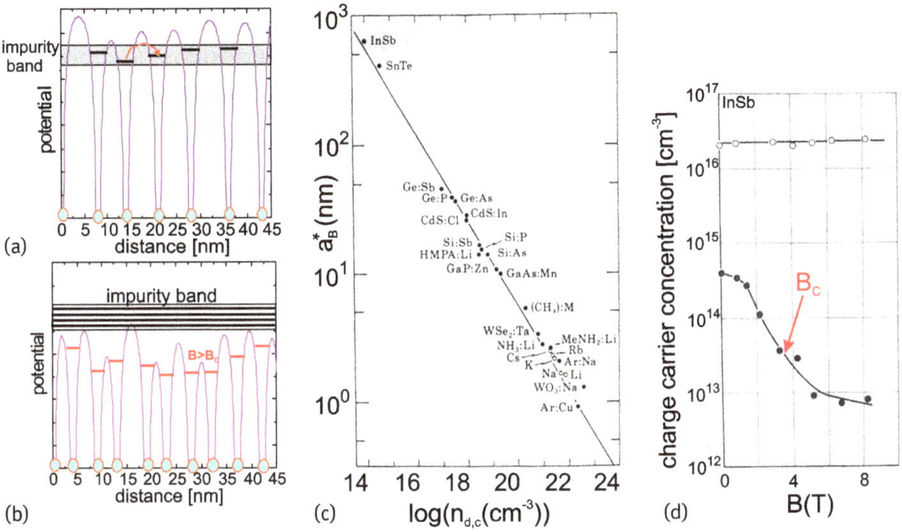

Fig. 5.9: (a) Disorder potential resulting from a row of donors (Coulomb potentials) with varying distances. Donors are marked by turquoise circles at the bottom. The thicker black, horizontal lines mark the energies of bound electron states. The greyish rectangle around these states marks the impurity band. A hopping process is indicated by a red arrow. (b) Same as (a) at larger donor density. Extended states are marked by continuous black lines. Above a critical magnetic field B_c, the states get localized as marked by red, horizontal lines. (c) Double logarithmic plot of the critical donor density $n_{d,c}$ as a function of the Bohr radius of the donor atoms a_B^*. The materials are labeled as crystal:donor or only as crystal, if little dependence on the dopant material is found. The line displays the relation $a_B^* = 0.27 \cdot n_{d,c}^{-1/3}$. (d) $\vec{B}$ field dependence of effective charge carrier density of InSb as determined by the Hall effect at two different donor concentrations (full circles: $n_d = 4 \cdot 10^{14}/\text{cm}^{-3}$, empty circles: $n_d = 2 \cdot 10^{16}/\text{cm}^{-3}$). The critical $\vec{B}$ field B_c is marked. (c) [111].

(eq. (4.28)) gets larger than the energy penalty via on-site interaction U (eq. (4.27)), while replacing the atomic potentials from Section 4.5 by the randomly distributed donor potentials. Since the bandwidth W increases with t, the impurity band additionally starts to overlap with the original conduction band. Consequently, we get occupied, extended states with E_F within the overlaping bands as required for a metal. The transition, where E_F passes from localized to extended states, such that the system turns from an insulator into a metal, is also called Mott transition in analogy to Section 4.3. The order parameter is the dopant density and the transition appears at the critical dopant density $n_{d,c}$. The value of $n_{d,c}$ turns out to be related to the Bohr radius of the donor $a_B^* = \varepsilon \cdot m_e/m^* \cdot a_B$ ($a_B = 0.053$ nm: Bohr radius of H atom):

$$a_B^* \cdot n_{d,c}^{1/3} = 0.27 \,. \qquad (5.19)$$

This so-called Mott criterion is excellently fulfilled for many dopants in different materials as shown in Fig. 5.9(c). Although a bit oversimplified, one can argue that the states get extended if the Bohr radius of neighboring, bound electrons overlap.[13] Thus, also doped semiconductors exhibit a disorder driven phase transition, now from the insulator to the metal, with the dopant density n_d as the control parameter.

At donor (or acceptor) concentrations slightly above $n_{\mathrm{d,c}}$, one can drive the metal back into an insulator by applying a magnetic field $\vec{B}$.[14] The central reason is, that the probability amplitude of the bound electrons gets laterally squeezed by the magnetic field in all directions perpendicular to the field, such that the effective hopping amplitude t gets smaller (eq. (4.28)). The squeezing can be understood semiclassically. While the Bohr radius corresponds to the classical radius of the electron paths encircling the positive ionic core, electrons in $\vec{B}$ field get additionally moved in circles with a radius called the cyclotron radius $r_\mathrm{c} = mv/eB$ (v: electron velocity, $B = |\vec{B}|$). Quantization of this radius results in the so called Landau radius (also called magnetic length) $l_B = \sqrt{\hbar/(eB)} = 29\,\mathrm{nm} \cdot B[T]^{-1/2}$ of the circular movement in the ground state, that decreases with increasing $\vec{B}$ independent of v. As soon as $l_B < a^*_\mathrm{B}$, the wave function decreases in lateral extension such that the Coulomb potential of the donor can bind the electron despite the reduced potential maximum around the donor. These magnetically confined states are marked as red lines in Fig. 5.9(b). Since the resulting reduction of free charge carriers additionally increases the screening length λ_screen and, thereby, the Coulomb potential strength $\Phi_\mathrm{Coul}(\vec{x}) \propto 1/|\vec{x}| \cdot \exp(-|\vec{x}|/\lambda_\mathrm{screen})$, respectively, the maxima around each donor, one observes a relatively sharp transition at the critical field B_C. Such a transition is called magnetic freeze-out[15] and can be probed experimentally, if a^*_B is large as, e.g., in InSb (Fig. 5.9(d)). The remaining carrier density above B_C originates from thermally excited electrons, which thus disappear at $T \to 0\,\mathrm{K}$. Consequently, the $\vec{B}$ field can also be a control parameter for a metal insulator transition (see also Section 1.5.1.2). Notice that $\vec{B}$ is an elegant parameter to tune t, which can drive the Mott–Hubbard transition (Section 4.5). However, the disorder in the system and the fact that the orbital states in a donor potential are only separated by a few meV, such that multiple bands contribute, makes the interpretation of the data more complex.

5.3.2 Quantum Hall effect

In 2D, there are many more metal-insulator transitions driven by a $\vec{B}$ field, where each insulating phase is accompanied by a quantized value of the Hall resistance R_Hall.

13 A more detailed calculation has to consider the effect of electron-electron repulsion as well, similar to Section 4.5.

14 Formally, $\vec{B}$ is the magnetic induction.

15 The magnetic-freeze-out effect is also relevant for stars and the creation of the universe, since $\vec{B}$ obviously influences the transition from a plasma made of electrons and nuclei to atoms, but at much larger $\vec{B}$, since for atoms we have $a^*_\mathrm{B} \simeq 0.05\,\mathrm{nm}$.

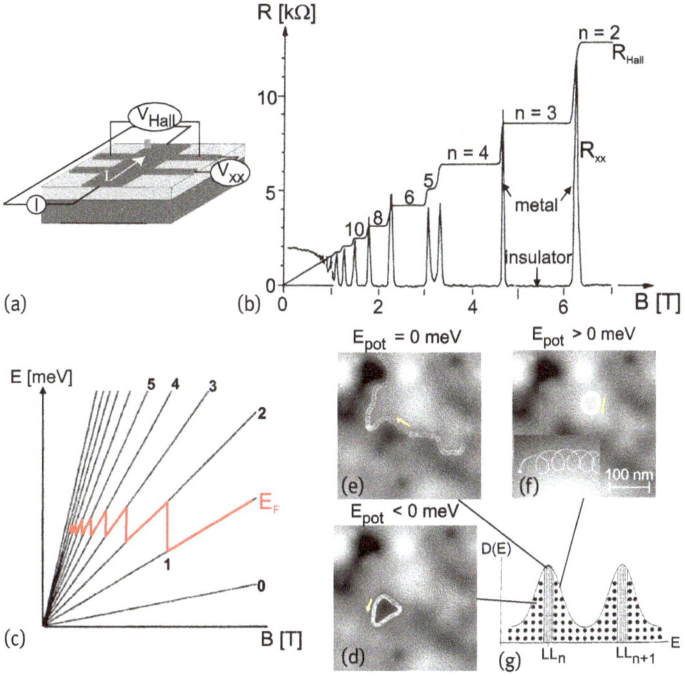

Fig. 5.10: (a) Sketch of a setup to measure the quantum Hall effect. (b) Longitudinal resistance $R_{xx} = V_{xx}/I$ and Hall resistance $R_{\mathrm{Hall}} = V_{\mathrm{Hall}}/I$ of a 2D electron system as a function of $\vec{B}$ field. (c) Landau levels LL_n as a function of $\vec{B}$ field with Fermi energy E_F marked in red. (d)–(f) Potential disorder (grey scale image) and classically resulting electron paths (white lines) within a magnetic field of $B = 6$ T at different potential energies E_{pot} as marked. The Inset in (e) shows a magnification of the cycloid path of an electron. (g) Density of states $\mathcal{D}(E)$ of a 2D electron system in $\vec{B}$ field with marked Landau levels LL_n, LL_{n+1} and assigned states from (d)–(f). (b) after https://www.nobelprize.org/nobel_prizes/physics/laureates/1998/press.html.

R_{Hall} is the ratio of the voltage V_{Hall} measured perpendicular to the direction of current I and the current I. The quantized Hall resistance $R_{\mathrm{Hall}} = V_{\mathrm{Hall}}/I$ appears to be $R_{\mathrm{Hall}} = h/e^2 \cdot 1/n = 25.8\,\mathrm{k\Omega}/n$, with I being the current through the 2D system and n being a positive integer. Figure 5.10(b) shows the measurement for a 2D system of GaAs using the setup in Fig. 5.10(a). The areas with $R_{xx} \approx 0\,\Omega$ are the insulating regions (explanation below), that exhibit plateaus of $R_{\mathrm{Hall}}(B)$ with the values $h/e^2 n$, where n gets smaller with increasing B.

In order to understand the quantum Hall effect (QHE), we firstly consider the behavior of 2D electronic states in the $\vec{B}$ field. The electrons move in circles, if a $\vec{B}$ field perpendicular to the 2D sample is applied. This leads to the well known quantization of the kinetic energy (rotational energy) described by the Landau energies E_n^{LL}. Formally, it is a quantization of the orbital momentum, which for a parabolic dispersion $E = \hbar^2 k^2/2m^*$ leads to $E_n^{\mathrm{LL}} = (n + 1/2) \cdot \hbar e B/m^*$. In 2D, we get a complete quantization, if the $\vec{B}$ field is not parallel to the 2D plane. The corresponding energy levels, equidistant by $\hbar\omega_c = \hbar eB_\perp/m^*$ ($B_\perp$: $\vec{B}$ field perpendicular to 2D electron system), are called Landau levels and are highly degenerate. Each Landau level can be occupied by one electron per area A of a flux quantum $\Phi_0 = B_\perp \cdot A = h/2e$.[16] This leads to a degeneracy density (number of states per m^2) of $n_{LL} = 1/A = 2 \cdot eB_\perp/h$. The degeneracy of a Landau level, thus, increases with $B_\perp$ such that, at constant electron density n_e, the number of occupied Landau levels decreases with increasing $B_\perp$. The Fermi level E_F resides within the mth Landau level with $m = \lfloor n_e/n_{LL} \rfloor$[17] and jumps to a lower Landau level, if the summed degeneracy density of all Landau levels at lower energy is larger than n_e. The movement of E_F within the Landau levels at constant n_e is depicted as a red line in Fig. 5.10(c). We will see that the jumps between the Landau levels are important for the QHE, but that we will need the additional ingredient of disorder to smooth these jumps such that the QHE becomes a robust feature. The electrostatic potential of the atoms is mostly irrelevant as long as the magnetic length l_B is significantly larger than the lattice constant, i.e., up to $B_\perp \simeq 1000\,\text{T}$. In the following, we will restrict ourselves to magnetic fields perpendicualr to the 2D electron system, such that $B = B_\perp$.

The required potential disorder appears naturally because of the charged dopants within doped semiconductors, where the quantum Hall effect is usually observed. This potential disorder is typically long-range (correlation length: $\sim 0.1{-}1\,\mu\text{m}$) as shown as a greyscale pattern in Fig. 5.10(d)–(f), in particular, if the dopants are placed remotely (typically by about 100 nm) from the 2D electron system (2DES). Hence the correlation length is large with respect to l_B, respectively r_c. The originally degenerate states of one Landau level are then simply distributed at different energies. This can be qualitatively understood by a classical consideration. If the cyclotron radius is much smaller than the correlation length of the potential disorder, the electron will be alternately accelerated and decelerated during its circular path by the in-plane electric field corresponding to the potential disorder. Consequently, the radius of curvature $r_c \propto v$ changes continuously. This results in a cycloid path as shown in the inset of Fig. 5.10(f). Importantly, the cycloid path moves the electron perpendicular to the direction of the electric field, i.e., along equipotential lines of the disorder potential. In fact, also quantum mechanically, one finds single particle states with $|\psi_i(\vec{x})|^2$ located along an equipotential line of the potential disorder. Perpendicular to the equipotential line, $|\psi_i(\vec{x})|^2$ exhibits a full width at half maximum of about $\sqrt{2n+1} \cdot l_B$ and n nodes for Landau level n. In particular, the potential energy of the equipotential line,

16 Recall that the flux quantum is defined for Cooper pairs leading to the factor of two in the denominator of the flux quantum. The corresponding phase preserving area encircled by a single electron contains a flux of $\Phi = h/e = 2\Phi_0$. This area can be occupied by two electrons, one electron for each spin direction. Notice that $\Phi = h/e$ is the smallest flux unit that can be introduced into a system of coherently moving electrons.

17 $\lfloor x \rfloor$ is the largest integer that is smaller than $x \in \mathbb{R}$.

belonging to a particular $|\psi_i|^2$, adds to the kinetic energy of the electron given by the Landau level energy E_n^{LL}. Thus, states with small (large) potential energy occupy the area along equipotential lines around valleys (hills) of the potential disorder. Such states are sketched in Fig. 5.10(d) and (f). They are localized since the equipotential lines are closed contours. However, in the energetic center of the potential landscape, there is one state (equipotential line) that traverses the whole crystal, i.e., it is extended (Fig. 5.10(e)). In fact, one can show rigorously, that an infinitely large 2D system with disorder has exactly one extended state per Landau level. Figure 5.11(a)–(b) maps the $|\psi_i(\vec{x})|^2$ of localized (a) and extended (b) states by scanning tunneling spectroscopy.

Within Fig. 5.10(g), the density of states $\mathcal{D}(E)$ is plotted where states of one Landau energy correspond to one peak, that, however is broadened by the different potential energies of the states, such that the peak widths reflects the strength of the potential disorder.

With increasing $\vec{B}$, E_{F} shifts within a particular Landau level from high to low energies, if n_{e} = const., since n_{LL} increases for each Landau level. Consequently, E_{F} always coincides with localized states, i.e., the 2D system is an insulator ($\sigma(T = 0\,\mathrm{K}) = 0\,\mathrm{S}$), except if E_{F} passes the extended state of the particular Landau level and, thus, the 2D system gets metallic.

Notice that conductivity σ and resistivity ρ are (2×2) tensors in a $\vec{B}$ field, since the current in one direction implies a voltage in two perpendicular directions. With $\vec{j} = \boldsymbol{\sigma} \cdot \vec{E}$ and $\boldsymbol{\rho} = \boldsymbol{\sigma}^{-1}$, one straightforwardly finds $\sigma_{xx} = 0\,\mathrm{S} \Longleftrightarrow \rho_{xx} = \frac{\sigma_{xx}}{\sigma_{xx}^2 + \sigma_{xy}^2} = 0\,\Omega$, as long as $\sigma_{xy} \neq 0\,\mathrm{S}$, i.e., counter-intuitively insulators exhibit zero resistance. Hence, the B-field regions with $R_{xx} \approx 0\,\Omega$ in Fig. 5.10(b), indeed, can be explained by E_{F} being in the area of localized states, where conductivity is suppressed. In contrast, if E_{F} is at the energy of an extended state, a peak in $R_{xx}(B)$ is observed.

The remaining conductivity at $T > 0\,\mathrm{K}$, if E_{F} is at localized states, depends on the energy distance of E_{F} to the extended state. Since this distance gets decreasingly smaller with decreasing potential disorder, the insulating behavior gets more pronounced with increasing disorder, as long as the scattering length is significantly larger than l_B such that Landau quantization appears.

Quantitatively, the distance between Landau levels is relatively small, i.e., $\hbar\omega_{\mathrm{c}} = 0.18\,\mathrm{meV/T}) \cdot Bm_{\mathrm{e}}/m^*$, such that the disorder has to be firstly reduced in order to separate the Landau levels and, moreover, systems with small m^*, as e.g., GaAs ($m^* = 0.067 \cdot m_{\mathrm{e}}$) are preferred. In metals, so far, the quantum Hall effect has not been observed, since the life time of electrons in metals is typically shorter than the time required for a full cyclotron orbit. Hence, there is no quantization of the circular orbit and, thus, no Landau levels. Here, the dynamic disorder prohibits the QHE.

An additional complication is the spin degree of freedom of the electrons, whose degeneracy is lifted in $\vec{B}$ field, such that one gets two different spin energies for the same Landau level (Zeeman effect). The energy distance is is the Zeeman energy $E_{\mathrm{Z}} = g\mu_{\mathrm{B}}B$ with g being the effective gyromagnetic factor of the electrons in the material

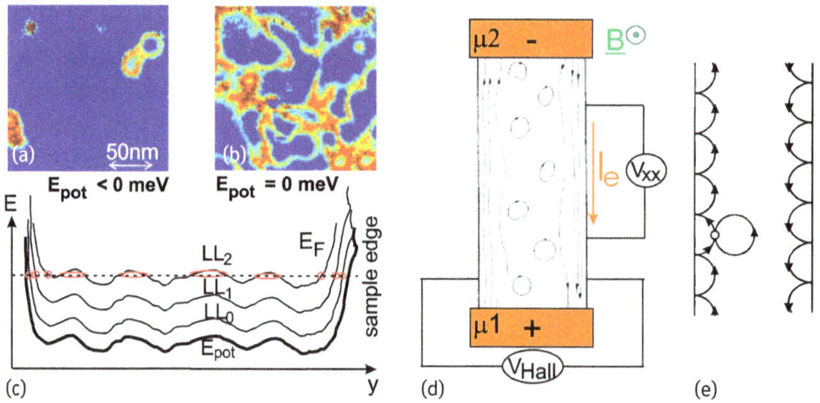

Fig. 5.11: (a) Localized states and (b) extended states of the lowest Landau level LL_0 as probed by scanning tunneling spectroscopy at $B = 12\,T$ and $T = 0.3\,K$. (c) Thick, curved line: electrostatic potential E_{pot} for a sample with edges and disorder. Thinner curved lines: potential energy is shifted by the kinetic energy of the Landau level LL_n. The resulting states at E_F belonging to the third Landau level LL_2 are shown as red ellipses. The red circles at the edges are edge states. (d) Top view of the drifting electron paths for states at E_F (lines with arrows). Electrodes are drawn as orange rectangles. The Hall voltage V_{Hall}, the longitudinal voltage V_{xx}, the potential of the two electrodes μ_1, μ_2 ($\mu_1 > \mu_2$), and the electron current $I_e = -I$ are indicated. (e) Schematic showing that scattering can not change the direction of propagation of the electron (thick lines with arrows) at the edge as long as the distance between the scattering centers (small circle) is larger than the cyclotron radius. (a), (b) [112].

and the Bohr magneton $\mu_B = 9.27 \cdot 10^{-24}$ J/T. Each spin level exhibits a degeneracy density $n_{LL}/2 = eB/h$ including one extended state. The pairs of metallic states are indeed visible in Fig. 5.10(b) left and right of the Hall Plateaus, which are marked by $n = 5$ and $n = 3$.

The quantized Hall resistance (plateaus) within the insulating regimes is explained most straightforwardly by the existence of edge states. Figure 5.11(c) displays the electrostatic potential $E_{pot}(y)$ perpendicular to the direction of current flow with disorder and the edges of the sample (thick line). The disorder potential is additionally displayed after adding the kinetic energy of the Landau levels, hence, $E_{pot}(y)+(n+1/2)\hbar\omega_c$ as marked by LL_n (thin lines). The cycloid paths at E_F are running along equipotential lines of Landau level LL_2, as shown by red lines in Fig. 5.11(c). They are localized at hills of the potential disorder and, thus, the 2D system is insulating. Towards the edge of the sample, $E_{pot}(y)$ must increase such that it matches the vacuum energy (or another type of confinement energy) somewhere at the edge. The region between the edge and interior of the sample is thereby as large as about the screening length of the material being 100–1000 nm for semiconducting 2D systems in $\vec{B}$ field. Within this region, all energy lines of Landau levels, that are in the interior of the sample below E_F, must cross E_F. The corresponding states at these crossing points (marked by red circles in Fig. 5.11(c)) perform its cycloid path along the equipotential line at the edge. Thus, the corresponding state called an edge state surrounds the whole 2D sample in one preferential direction. These paths together with the closed paths of the localized states at E_F are shown as viewed from the top in Fig. 5.11(d). Obviously, the edge states connect different electrodes and, thus, can transport electrons albeit only in one direction. For the displayed polarity of the electrodes, only the three edge states at the right edge can transport electrons as indicated by the arrow marked I_e. Within these edge channels, the electrons, moreover, can not be backscattered as shown in Fig. 5.11(e). Even if they are scattered, the Lorentz force will turn them back into the same direction, i.e., they will never move backwards on scales larger than the cyclotron radius, at least, as long as the scattering length is larger than about twice the cyclotron radius. Since one can not get a potential difference along the edge channels, if the electrons are not backscattered, one measures $V_{xx} = 0\,\text{V}$ between two electrodes along the edge channel, i.e., one finds indeed the insulating behavior, $R_{xx} = V_{xx}/I = 0\,\Omega$, as discussed above. The potential difference applied between the two orange electrodes, injecting and removing electrons, drops exclusively at the beginning of the second electrode (μ_1), where the electrons get adapted to μ_1 by inelastic scattering processes within the metal. This is indeed measurable.[18]

Since the electrons are injected only in one of the edge channels, one gets a potential difference between the edges, i.e., a Hall voltage. Formally, this Hall voltage is $\mu_2 - \mu_1$, since the left channel equilibrates with μ_1 and the right channel equilibrates with μ_2 (Fig. 5.11(d)). The relation of this Hall voltage to the current can be calculated realizing that the edge channels are one-dimensional ballistic conductors (Section 1.4.1), where all electrons between μ_1 and μ_2 contribute to the current. Each channel, thus, contributes $I_n = \int_{\mu_1}^{\mu_2} \mathcal{D}_R(E) \cdot e \cdot v(E)\,dE = e/h \cdot (\mu_2 - \mu_1) = e^2/h \cdot V_{ext}$ with velocity $v(E) = dE/dk \cdot \hbar^{-1}$, 1D density of states of spin polarized electrons moving in the same direction $\mathcal{D}_R(E) = (dE/dk \cdot 2\pi)^{-1}$ (eq. (1.13)), and external voltage $V_{ext} = (\mu_2 - \mu_1)/e$. Hence, each 1D channel (edge state) contributes with e^2/h to the conductivity at the edge.

Since $V_{ext} = V_{Hall}$, the Hall conductance is simply $G_{Hall} = I/V_{Hall} = ne^2/h$ with n being the number of edge channels and, thus, the number of occupied extended states in the interior of the sample. The Hall resistance is the inverse of G_{Hall}, if $\sigma_{xx} = 0\,\text{S}$, i.e., $R_{Hall} = h/e^2 n$ in agreement with the experimental data of Fig. 5.10(b).

Only if an extended state from the interior is at E_F, the current partly runs along this state such that G_{Hall} is not quantized anymore. Backscattering can take place within the extended state at the many crossing points of the extended state as visible in Fig. 5.11(b). Consequently, the longitudinal resistance R_{xx} is larger than $0\,\Omega$ and

---

18 To show the position of voltage drop, one covers the 2D system with a thin film of suprafluid He. The energy that gets dissipated by the electrons at the position of voltage drop heats the sample locally, such that an increased pressure within the He film appears inducing a little fountain at this point [113].

R_{Hall} is not quantized anymore. If B is further increased, E_{F} gets below the energy of the extended state, such that the sample becomes insulating again ($\sigma_{xx} = 0\,\mathrm{S}$) and the number of edge states is reduced by one. Consequently, G_{Hall} is dropped by e^2/h. This explains the steps in the Hall resistance straightforwardly by largely employing classical considerations of the electron paths. Quantum mechanics was only necessary to explain the development of Landau levels via the quantization of the kinetic energy and for the derivation of the universal conductance e^2/h of a ballistic channel (Section 1.4.1).

This explanation, however, raises a number of questions. Most importantly, there is no fundamental reason that the edge states do not get gapped at E_{F}, such that the system becomes an insulator with $\sigma = 0\,\mathrm{S}$. A mechanism to gap a 1D electron system (1DES) has been shown, e.g., in Section 5.2.1. In principle, details of the edge properties can distinguish, if the 1DES gets gapped or not. Experiments, however, show that the quantized plateaus of R_{Hall} are very robust with respect to the details of the edge and are found in many different materials. Hence, the question arises, if there is a more fundamental reason to explain the quantum Hall effect. This is the topic of the next section.

5.4 Topological analysis of electron systems in solids

There is a more general way to describe the quantum Hall effect. It is based on a so-called topological number, the Chern number n_{Ch}. The Chern number is an integer. It can be described as a geometrical feature of the Hilbert space of the wave functions in two dimensions exposed to a magnetic field. Thus, it is called topological.[19]

The approach is more difficult to illustrate. However, it secures the quantized Hall conductance, independent of the details of the edges. Most favorably, the topological approach is transferable to other systems, which has been done very successfully within the last decade. For example, it led to a totally new class of materials, that can be insulating in the bulk, but are necessarily conducting at the surface. These new type of materials are called strong 3D topological insulators (3DTIs) and several tens of 3DTIs have been verified experimentally. In principle, this is the third possible type of conductance in solids besides metals and insulators. Topologically, this class is distinct from all other materials by a single number ν_0. This number is $\nu_0 = 1$ for 3DTIs, i.e., for the insulators, that necessarily have surface states on all surfaces, and it is $\nu_0 = 0$ for all other insulators. For metals, ν_0 is not a useful index, but other topological indices apply, that are not discussed in this book.[20]

---

19 Topology is a mathematical classification scheme of geometrical objects. It allows, e.g., to distinguish three-dimensional objects by their number of holes. Therefore, one calculates a specific surface integral, that returns the integer n, the number of holes (Section 5.4.1).

20 A key word is, e.g., Weyl semimetals.

Many other new classes of materials (described by other Hamiltonians) have been predicted based on the approach. They provide a number of exotic properties as, e.g., they contain quasiparticles called Majorana fermions, that are only half a fermion, i.e., the probability of finding an electron in a certain area is exactly 1/2. These particles, of course, can only exist in multiples of two. Interestingly, they are maximally entangled in terms of the constituent electrons, which renders them a favorite candidate for fault tolerant quantum computation (Chapter 3).

Another intriguing aspect of the topological approach is its transferability to other periodic systems. For example, one can build a lattice of optical waveguides, such that the geometry guarantees that the light propagates exclusively along the edges of the geometry. This might be exploited to concentrate the light as required for lasers (Section 2.3.2.4). The approach of topology also works for phonons or for magnons. Since the topological approach is rather new, it is far from being studied exhaustively. The game is partly even completely open, e.g., with respect to possible applications of the novel properties resulting from this approach.

In order to get an insight into the approach, we will follow the historical path of its discovery. Thus, we firstly sketch the reasoning for the quantum Hall effect, which was one of the first solid state systems described in terms of topology.[21] Afterwards, we will describe the distillation of topological numbers for 2D systems and 3D systems without a magnetic field and without electron-electron interaction. A more rigorous treatment of the topological approach would require a field theoretical description, that, however, is too mathematical for an experimental course. Moreover, it would require a long-term experience with this type of description to deliver any intuitive insight. Thus, the arguments will be partly heuristic, only requiring a solid quantum mechanical understanding. However, following the quotations, the interested reader is guided to the papers settling the field theoretical approach.

5.4.1 What is topology?

In topology, as far as it concerns solid state physics, one deals with geometrical properties of objects in mathematically defined spaces. Such a space is, e.g., the $\vec{k}$ space for

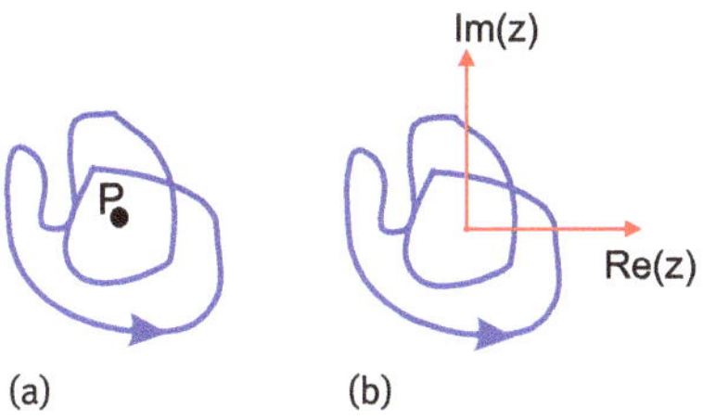

Fig. 5.12: (a) Path (blue line with arrow), which winds twice around the point P counterclockwise. (b) Same curve as in (a) with a coordinate system of the complex plane centered in point P.

21 Nobel prize in Physics 2016, www.nobelprize.org.

electrons, that describes a complete set of quantum numbers, i.e., the relevant Hilbert space. The aim is to classify objects in such spaces according to its geometrical properties, such that different objects of the same class can be transformed into each other by stretching and bending. In contrast, tearing and gluing is needed to transfer an object to another class. Obviously, this classification implies a robustness to stay in the same class with respect to weak perturbations such as stretching or bending.

The art is to firstly show that such a classification exists and, subsequently, to define a calculation, whose output unambiguously puts the object to a certain class.

An example is illustrated in Fig. 5.12. Consider a closed line in space, that avoids the point of observation P. Then, a clearly posed question is how often does the line wind around the point P, before it closes? The answer for Fig. 5.12(a) is obviously twice.

Next, we have to put this answer on solid ground. Experimentally, one would position an observer at point P. The person would follow an object along the closed path visually once and one would count, how often he has turned around its axis.

Mathematically, one defines the path as the function $z(t)$ in the complex plane as depicted in Fig. 5.12(b), where $t \in [0, 1)$, $t \in \mathbb{R}$ runs around the whole curve once and $z \in \mathbb{C}$ is the corresponding point in the complex plane. As usual, one can write:

$$z(t) = |z(t)| \cdot \mathrm{e}^{\mathrm{i}\varphi(t)} . \tag{5.20}$$

Now one can define the integral:

$$Q_{\mathrm{I}}(z) = \frac{1}{2\pi\mathrm{i}} \int_0^1 \frac{\mathrm{d}z(t)/\mathrm{d}t}{z(t)} \mathrm{d}t . \tag{5.21}$$

Q_{I} is exactly the answer to our question for each path $z(t)$ as one sees as follows:

$$Q_{\mathrm{I}}(z) = \frac{1}{2\pi\mathrm{i}} \int_0^1 \frac{\mathrm{d}}{\mathrm{d}t} \ln(z(t))\mathrm{d}t \tag{5.22}$$

$$= \frac{1}{2\pi\mathrm{i}} [\ln(z(t))]_0^1 = \frac{1}{2\pi\mathrm{i}} \ln\left(\frac{|z(1)|\mathrm{e}^{\mathrm{i}\varphi(1)}}{|z(0)|\mathrm{e}^{\mathrm{i}\varphi(0)}}\right) \tag{5.23}$$

$$= \frac{1}{2\pi\mathrm{i}} \cdot (\ln(\mathrm{e}^{\mathrm{i}\varphi(1)}) - \ln(\mathrm{e}^{\mathrm{i}\varphi(0)})) = \frac{\varphi(1) - \varphi(0)}{2\pi} . \tag{5.24}$$

If we define the function $\varphi(t)$ as continuous, such that it does not jump from 2π to 0 after one turn, $|Q_{\mathrm{I}}(z)|$ is the number of turns, we were asking for. Moreover, if the turns are counterclockwise, $Q_{\mathrm{I}}(z)$ is positive and, if the turns are clockwise, $Q_{\mathrm{I}}(z)$ is negative. Mathematicians can show now that a path turning n times clockwise can be transformed into a path turning $n + m$ times clockwise and m times counterclockwise without tearing and gluing. But we do not go into this discussion.

Importantly, we have classified all closed paths in 2D space avoiding the point P (the origin) into classes with identical numbers of counterclockwise turns around the origin. Obviously $Q_{\mathrm{I}}(z) \in \mathbb{Z}$ and, thus, one calls $Q_{\mathrm{I}}(z)$ a $\mathbb{Z}$-type topological invariant: the winding number.

5.4.2 Quantum Hall effect in terms of topology

Now, we apply this concept to the quantum Hall effect, where the geometrical space that we consider, is the Hilbert space of wave functions in magnetic field.

To do so, we start with deriving the Hall conductivity quantum-mechanically. Therefore, we firstly calculate the Hall current for a single electron being within a Bloch state, i.e., within a crystalline solid (eq. (2.90))

$$\psi_{n\vec{k}}(\vec{x}) = u_{n\vec{k}}(\vec{x}) \cdot e^{i\vec{k}\vec{x}} \tag{5.25}$$

with n being the band index and $\vec{k}$ being a wave vector. We will argue later that this description is also relevant in magnetic field. We use first order perturbation theory in the weak electric field $\vec{E} = E_x \cdot \vec{e}_x$ applied in the x direction to the sample. This corresponds to an electric potential

$$\Phi_{\text{el}}(x) = eE_x \cdot x = -i \cdot eE_x \cdot \frac{d}{dk_x}\,, \tag{5.26}$$

where we used the operator replacement $x = -i \cdot d/dk_x$, that can be verified by representing the wave function as a superposition of plane waves.

The perturbed wave function $|n\rangle$ than reads:

$$|n\rangle = |n_0\rangle - \sum_{m_0 \neq n_0} \frac{|m_0\rangle \left\langle m_0 \middle| ieE_x \cdot \frac{d}{dk_x} \middle| n_0 \right\rangle}{E_{n0} - E_{m0}}\,, \tag{5.27}$$

where $|n_0\rangle$, $|m_0\rangle$ are solutions of the unperturbed Hamiltonian, i.e., Bloch states with energies E_{n0} and E_{m0}, respectively. The bracket $\langle || \rangle$ indicates spatial integration.

Next, we determine the velocity of the resulting state in y direction v_y, that is transversal to $\vec{E}$ as the Hall current:

$$\hat{v}_y = \langle n|v_y|n\rangle = \langle n_0|v_y|n_0\rangle - ieE_x \sum_{m_0 \neq n_0} \frac{\langle n_0|v_y|m_0\rangle \left\langle m_0 \middle| \frac{d}{dk_x} \middle| n_0 \right\rangle}{E_{n0} - E_{m0}} + \text{h.c.}\,, \tag{5.28}$$

where h.c. is the complex conjugate and the quadratic term in E_x is neglected. We use

$$\hat{v}_y = \frac{dy}{dt} = -\frac{i}{\hbar}[\hat{H}, \hat{y}]\,, \tag{5.29}$$

where $[.,.]$ is the commutator and $\hat{H}$ is the Hamiltonian of the unperturbed system. The relation (5.29) can be verified straightforwardly by employing the Schrödinger equation $i\hbar d\psi/dt = \hat{H}\psi$. With eq. (5.29), one gets:

$$\langle n_0|v_y|m_0\rangle = -\frac{i}{\hbar}(\langle n_0|\hat{H}y|m_0\rangle - \langle n_0|y\hat{H}|m_0\rangle) = -\frac{i}{\hbar}\langle n_0|y|m_0\rangle \cdot (E_{n0} - E_{m0})\,. \tag{5.30}$$

Using again the operator replacement $y = -i \cdot d/dk_y$, we obtain:

$$\langle n_0|v_y|m_0\rangle = \frac{1}{\hbar}\left\langle n_0 \middle| \frac{d}{dk_y} \middle| m_0 \right\rangle \cdot (E_{n0} - E_{m0}) = \frac{-1}{\hbar}\left\langle \frac{dn_0}{dk_y} \middle| m_0 \right\rangle \cdot (E_{n0} - E_{m0}) \tag{5.31}$$

for all $m_0 \neq n_0$. After insertion into eq. (5.28), we find:

$$\hat{v}_y = \langle n_0|v_y|n_0\rangle + \frac{\mathrm{i}eE_x}{\hbar}\sum_{m_0\neq n_0}\left\langle\frac{\mathrm{d}n_0}{\mathrm{d}k_y}\middle|m_0\right\rangle\left\langle m_0\middle|\frac{\mathrm{d}n_0}{\mathrm{d}k_x}\right\rangle + h.c\,. \tag{5.32}$$

Using the same argument as in eq. (5.29)–(5.31), one can verify that the first term $\langle n_0|v_y|n_0\rangle$ vanishes. After exploiting the completeness relation $|m_0\rangle\langle m_0| = \mathbf{1}$, one gets

$$\hat{v}_y = \frac{\mathrm{i}eE_x}{\hbar}\left(\left\langle\frac{\mathrm{d}n_0}{\mathrm{d}k_y}\middle|\frac{\mathrm{d}n_0}{\mathrm{d}k_x}\right\rangle - \left\langle\frac{\mathrm{d}n_0}{\mathrm{d}k_x}\middle|\frac{\mathrm{d}n_0}{\mathrm{d}k_y}\right\rangle\right). \tag{5.33}$$

One can finally verify straightforwardly (by applying the derivation with respect to k_x and k_y), that the plane wave part of the Bloch state $|n_0\rangle = u_{n\vec{k}}(\vec{x})\cdot \mathrm{e}^{\mathrm{i}\vec{k}\vec{x}}$ does not contribute, such that we end up with

$$\hat{v}_y = \frac{\mathrm{i}eE_x}{\hbar}\left(\left\langle\frac{\mathrm{d}u_{n\vec{k}}(\vec{x})}{\mathrm{d}k_y}\middle|\frac{\mathrm{d}u_{n\vec{k}}(\vec{x})}{\mathrm{d}k_x}\right\rangle - \left\langle\frac{\mathrm{d}u_{n\vec{k}}(\vec{x})}{\mathrm{d}k_x}\middle|\frac{\mathrm{d}u_{n\vec{k}}(\vec{x})}{\mathrm{d}k_y}\right\rangle\right). \tag{5.34}$$

Thus, in this so-called linear response theory, based on the Kubo formula, the transversal velocity in an electric field is a property of the periodic part of a Bloch state. To get the transversal current I_y in electric field E_x, one has to add up the contribution from all occupied states $u_{n\vec{k}}(\vec{x})$ with different $\vec{k}$ and n. The transversal current is, hence, only non-zero, if, for some $u_{n\vec{k}}(\vec{x})$, the derivative with respect to a $\vec{k}$ component (k_x or k_y) is different, if applied to the function or to the complex conjugate of the function. Moreover, this difference must be different for the k_x component and the k_y component. Finally, contributions from different $\vec{k}$ should not cancel. Naively, one might think that will never be the case, but we will see that this is not correct.

Before looking into this, we verify the correctness of the Bloch wave approach in a magnetic field. An elegant verification of the Bloch waves as stationary single particle states without a magnetic field is given by the translational operator $\hat{T}$:

$$\hat{T}(\vec{R}_n) = \exp(\vec{R}_n\cdot\nabla) \tag{5.35}$$

with the lattice vector $\vec{R}_n$. It shifts an arbitrary function by $\vec{R}_n$, i.e.,

$$\hat{T}(\vec{R}_n)\cdot f(\vec{x}) = f(\vec{x}+\vec{R}_n) \tag{5.36}$$

as one can verify, e.g., by Taylor expansion for the three coordinates.[22] Thus, $\hat{T}(\vec{R}_n)$ commutes with the lattice periodic potential $\hat{\Phi}_{\mathrm{el}}(\vec{x}) = \hat{\Phi}_{\mathrm{el}}(\vec{x}+\vec{R}_n)$. It moreover commutes with ∇, since ∇ commutes with itself and, thus, each arbitrary function of ∇ commutes with ∇. Consequently, $\hat{T}(\vec{R}_n)$ commutes with the Schrödinger equation in periodic potential $\hat{H} = -\frac{\hbar^2}{2m}\nabla^2 + e\hat{\Phi}_{\mathrm{el}}(\vec{x})$. This implies that eigenfunctions of the translational operator can be found such that they are eigenfunctions of the Schrödinger equation. These are exactly the Bloch states.[23]

22 This is a general mathematical result by Lagrange known as the shift operator.

23 We do not verify this. A verification can be found, e.g., in Ashcroft/Mermin Solid State Physics: First Proof of Bloch's theorem.

If a homogeneous magnetic field $\vec{B}$ is applied, the Schrödinger equation becomes $\widehat{H}_{\vec{B}} = \frac{1}{2m}(i\hbar\nabla + e\widehat{\vec{A}}(\vec{x}))^2 + e\widehat{\Phi}_{el}(\vec{x})$ with the magnetic vector potential $\vec{A}(\vec{x}) = -\frac{1}{2}(\widehat{\vec{x}} \times \widehat{\vec{B}})$ in symmetric gauge, where $\vec{B}$ is the magnetic induction measured in Tesla. Since $\vec{A}$ depends continuously on $\widehat{\vec{x}}$, $\widehat{T}(\vec{R}_n)$ does not commute with $\widehat{H}_{\vec{B}}$. However, a different operator

$$\widehat{T}_B(\vec{R}_n) = \exp\left(\vec{R}_n \cdot \left(\nabla - \frac{e}{i\hbar}\widehat{\vec{A}}(\vec{x})\right)\right) \tag{5.37}$$

commutes with $i\hbar\nabla + e\widehat{\vec{A}}(\vec{x})$ as one can verify by showing that $\nabla - \frac{e}{i\hbar}\widehat{\vec{A}}(\vec{x})$ commutes with $i\hbar\nabla + e\widehat{\vec{A}}(\vec{x})$ and, thus, $i\hbar\nabla + e\widehat{\vec{A}}(\vec{x})$ commutes with any function of $\nabla - \frac{e}{i\hbar}\widehat{\vec{A}}(\vec{x})$. However, applying $\widehat{T}_B$ to $\widehat{\Phi}_{el}(\vec{x})$ shifts $\widehat{\Phi}_{el}(\vec{x})$ by $\vec{R}_n$ and gives an additional phase factor, i.e.,

$$\widehat{T}_B(\vec{R}_n)\widehat{\Phi}_{el}(\vec{x}) = \exp\left(\vec{R}_n \cdot \frac{e}{i2\hbar}\left(\widehat{\vec{x}} \times \vec{B}\right)\right) \cdot \widehat{\Phi}_{el}(\vec{x} + \vec{R}_n)\,. \tag{5.38}$$

In the next step, one can show that the phase that one adds up by a series of $\widehat{T}_B(\vec{R}_n)$, where the corresponding $\vec{R}_n$ form a closed loop around an area $\widetilde{A}$ is simply $i \cdot \frac{e|\vec{B}|}{\hbar} \cdot \widetilde{A} = i\frac{2\pi eB\widetilde{A}}{h}$. The derivation is similar as for the Aharonov–Bohm effect (Section 1.4.3.5). One employs

$$\oint \vec{A}(\vec{x})\mathrm{d}\vec{x} = \iint \nabla \times \vec{A}(\vec{x})\mathrm{d}\vec{\widetilde{A}} = \iint \vec{B}\mathrm{d}\vec{\widetilde{A}} = |\vec{B}| \cdot A \cdot \mathrm{sgn}(\vec{B}\vec{\widetilde{A}})\,, \tag{5.39}$$

where the last step uses that $\vec{B}$ is constant and perpendicular to the plane of the quantum Hall sample ($\vec{\widetilde{A}}$: area vector). A more rigorous treatment can be found elsewhere.[24] Since $e^{i\cdot 2\pi} = 1$, the phase vector vanishes, if the area $\widetilde{A}$ contains an even number of magnetic flux quanta, i.e., $\Phi = n \cdot h/e = 2n\Phi_0$.[25] This implies a regular lattice of points $\vec{R}_{n,\vec{B}}$, where $\widehat{T}_B(\vec{R}_{n,\vec{B}})$ commutes with $\widehat{H}_B$. The $\vec{R}_{n,\vec{B}}$ are related to each other by the line integral $\int \frac{e}{\hbar}\vec{A}\mathrm{d}\vec{x} = 1$ (eq. (5.38)).

One can hence define a new unit cell, that contains an even number of flux quanta and an integer number of the unit cells without magnetic field. This new unit cell is called the magnetic unit cell with lattice vectors $\vec{R}_{n,\vec{B}}$. The Schrödinger equation commutes with $\widehat{T}_B(\vec{R}_{n,\vec{B}})$. Due to this quasi-periodicity, one gets Bloch waves consisting of a plane wave $e^{i\vec{k}\vec{x}}$ and a periodic function in the magnetic unit cell: $u_{n\vec{k}}(\vec{x})$. Thus, the derivation of v_y leading to eq. (5.34) is also valid for $\vec{B} \neq \vec{0}\,\mathrm{T}$.

However, as a result of the term $\frac{e}{i\hbar}\widehat{\vec{A}}(\vec{x})$ in $\widehat{T}_B(\vec{R}_{n,\vec{B}})$, one obtains a more complex $u_{n\vec{k}}(\vec{x})$, that is the origin of $v_y \neq 0\,\mathrm{m/s}$. One can show that $u_{n\vec{k}}(\vec{x}) = |u_{n\vec{k}}(\vec{x})| \cdot \exp(i \cdot \Theta_{n\vec{k}}(\vec{x}))$ has a phase factor that requires:[26]

$$\oint_{\mathrm{m.u.c.}} \frac{\mathrm{d}\Theta_{n\vec{k}}(\vec{x})}{\mathrm{d}\vec{s}}\mathrm{d}\vec{s} = -2\pi p\,, \tag{5.40}$$

24 See, e.g., [114]: eqs. (7)–(10), [115]: eqs. (2.6)–(2.8).
25 Notice that the flux is a quantity independent on the gauge of $\vec{A}(\vec{x})$.
26 Similar argument as in eqs. (5.37)–(5.39), more rigorously in [115]: eqs. (2.9)–(2.13).

if one integrates counterclockwise around the magnetic unit cell (m.u.c.), where p is the number of pairs of flux quanta within the magnetic unit cell. More intuitively, p is simply the number of times the phase $\Theta_{n\vec{k}}(\vec{x})$ changes by 2π, if one moves around the magnetic unit cell. This is called the vorticity of the wave function. It is easy to imagine that the continuous change of the phase on a closed path by 2π surrounds a zero of the spatially continuous function $u_{n\vec{k}}(\vec{x})$, since it implies positive and negative values at $\Theta_{n\vec{k}}(\vec{x}) = 0$ and $\Theta_{n\vec{k}}(\vec{x}) = \pi$, respectively, on the surrounding path. Shrinking the closed path to the central point $\vec{z}_0$ would lead to a diverging kinetic energy (phase change per path length) of the wave function, except if $\psi_i(\vec{z}_0) = 0$. Hence, the only meaningful solution is $\psi_i(\vec{z}_0) = 0$. As a consequence, the applied magnetic field is imprinted in the geometrical properties of the wave function by zeroes, independent from the functional form of the electrostatic potential $\Phi_{\mathrm{el}}(\vec{x})$. This vorticity of the wave function, that is related to the number of zeroes, is called a topological property, since it is given by the geometric construction of the magnetic unit cell.

By transforming to $\vec{k}$ space, we also get a magnetic Brillouin zone (MBZ) that is given by the reciprocal lattice of the magnetic unit cell (Fig. 5.13).

Finally, we write down the transversal current density j_y carried by a fully occupied, single band n within the MBZ, without spin degeneracy, by integrating eq. (5.34) across the MBZ:

$$\begin{aligned}\hat{j}_y &= -e \iint_{\mathrm{MBZ}} \frac{1}{(2\pi)^2} \hat{v}_y(\vec{k}) \mathrm{d}\vec{k} \\ &= -e \iint_{\mathrm{MBZ}} \frac{1}{(2\pi)^2} \frac{\mathrm{i}eE_x}{\hbar} \left(\left\langle \frac{\mathrm{d}u_{n\vec{k}}(\vec{x})}{\mathrm{d}k_y} \middle| \frac{\mathrm{d}u_{n\vec{k}}(\vec{x})}{\mathrm{d}k_x} \right\rangle - \left\langle \frac{\mathrm{d}u_{n\vec{k}}(\vec{x})}{\mathrm{d}k_x} \middle| \frac{\mathrm{d}u_{n\vec{k}}(\vec{x})}{\mathrm{d}k_y} \right\rangle \right) \mathrm{d}^2\vec{k}\,,\end{aligned} \tag{5.41}$$

where the factor $1/(2\pi)^2$ comes from the density of states in $\vec{k}$-space.

By factorization aiming for $G_0/2 = e^2/h$ as a prefactor, i.e., the conductance carried by a ballistic spin polarized channel (eq. (1.23)), we get:

$$\hat{j}_y = \frac{e^2}{h} E_x \iint_{\mathrm{MBZ}} \frac{1}{2\pi\mathrm{i}} \left(\left\langle \frac{\mathrm{d}u_{n\vec{k}}(\vec{x})}{\mathrm{d}k_y} \middle| \frac{\mathrm{d}u_{n\vec{k}}(\vec{x})}{\mathrm{d}k_x} \right\rangle - \left\langle \frac{\mathrm{d}u_{n\vec{k}}(\vec{x})}{\mathrm{d}k_x} \middle| \frac{\mathrm{d}u_{n\vec{k}}(\vec{x})}{\mathrm{d}k_y} \right\rangle \right) \mathrm{d}^2\vec{k}\,. \tag{5.42}$$

Thus, according to the experimental result (Fig. 5.10(b)), the integral must be an integer at the plateaus of the transversal conductivity $\sigma_{xy} = j_y/E_x = G_{\mathrm{Hall}}$ (Section 5.3.2). This integer is the Chern number n_{Ch}.

To show that it is indeed an integer, we firstly use Stokes theorem, i.e., we write:

$$\begin{aligned}&\left\langle \frac{\mathrm{d}u_{n\vec{k}}(\vec{x})}{\mathrm{d}k_y} \middle| \frac{\mathrm{d}u_{n\vec{k}}(\vec{x})}{\mathrm{d}k_x} \right\rangle - \left\langle \frac{\mathrm{d}u_{n\vec{k}}(\vec{x})}{\mathrm{d}k_x} \middle| \frac{\mathrm{d}u_{n\vec{k}}(\vec{x})}{\mathrm{d}k_y} \right\rangle \\ &\qquad = [\nabla_{\vec{k}} \times \langle u_{n\vec{k}}(\vec{x}) | \nabla_{\vec{k}} | u_{n\vec{k}}(\vec{x}) \rangle]_z := [\nabla_{\vec{k}} \times \vec{A}_{\mathrm{Berry},n}(\vec{k})]_z\,,\end{aligned} \tag{5.43}$$

where $\nabla_{\vec{k}}$ is the gradient operator in $\vec{k}$-space and the subscript z refers to the third component of the resulting vector. The vector

$$\vec{A}_{\mathrm{Berry},n}(\vec{k}) = \langle u_{n\vec{k}}(\vec{x})|\nabla_{\vec{k}}|u_{n\vec{k}}(\vec{x})\rangle \tag{5.44}$$

is called the Berry connection of band n at wave vector $\vec{k}$.

By Stokes theorem, that can be applied, if the integrand is continuous, we would get:

$$\sigma_{xy} = \frac{j_y}{E_x} = \frac{e^2}{h} \cdot \frac{1}{2\pi\mathrm{i}} \oint_{\mathrm{dMBZ}} \vec{A}_{\mathrm{Berry},n}(\vec{k}) \cdot \mathrm{d}\vec{k} := \frac{e^2}{h} \cdot \frac{1}{2\pi\mathrm{i}} \cdot \varphi_{\mathrm{Berry},n} \tag{5.45}$$

with the boundary of the magnetic Brillouin zone dMBZ and the so-called Berry phase $\varphi_{\mathrm{Berry},n}$. Since the MBZ is periodic, i.e., the left edge and the right edge as well as the upper and the lower edge are identical, the line integral along dMBZ is necessarily zero, i.e., $\varphi_{\mathrm{Berry},n} = 0$. This is indeed an integer, but a rather boring one.

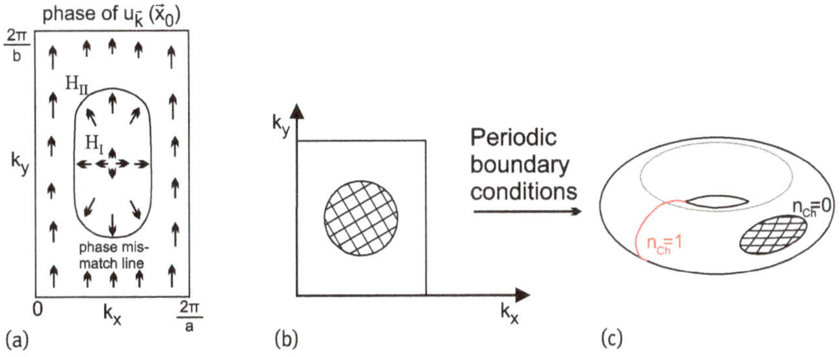

Fig. 5.13: (a) Magnetic Brillouin zone (MBZ) with phase $\Theta_{n\vec{k}}(\vec{x} = \vec{x}_0)$ of the function $u_{n\vec{k}}(\vec{x} = \vec{x}_0)$ at a particular real space position $\vec{x}_0$. $\Theta(\vec{k}) := \Theta_{n\vec{k}}(\vec{x} = \vec{x}_0)$ is symbolized as a vector with a corresponding angle Θ to the vertical axis. Periodic boundary conditions require that the phase is the same on the left and on the right as well as on the top and on the bottom of the MBZ. A particular $\vec{x}_0$ with a single zero in the MBZ is chosen and the required phase mismatch line separating area I and II is drawn. (b) Same as (a) with reduced complexity showing area I as a hatched circle. (c) Transformation of the MBZ to a torus which correctly displays the periodic boundary conditions (PBC). An additional phase mismatch line encircling the torus along one circumference is added in red. The Chern number n_{Ch} is marked for both phase mismatch lines. (a) After [115].

However, $\vec{A}_{\mathrm{Berry},n}(\vec{k})$ cannot be continuous in $\vec{k}$ space, since the zeros of $u_{n\vec{k}}(\vec{x})$ require a circulating phase around them also in $\vec{k}$-space. This does not continuously fit to the periodic boundary conditions as sketched in Fig. 5.13(a). Figure 5.13(a) shows a $\vec{k}$ dependent representation of the phase $\Theta_{n,\vec{k}}(\vec{x})$ (eq. (5.40)) for a selected $\vec{x} = \vec{x}_0$ as a vector with a corresponding angle to the vertical direction. One recognizes the requirement of a phase mismatch line by the clash of phases. One phase is encircling the zero and the other phase has to be the same on both sides due to the periodic boundaries. The topology of closed lines of the winding vector and of closed lines of the periodic vector do not match. Consequently, the Bloch functions of the two areas I and II are different at the phase mismatch line according to:

$$u^{\mathrm{I}}_{n\vec{k}}(\vec{x}) = u^{\mathrm{II}}_{n\vec{k}}(\vec{x}) \cdot \mathrm{e}^{\mathrm{i}\Delta\Theta_{n\vec{x}}(\vec{k})} . \tag{5.46}$$

Here, $\Delta\Theta_{n\vec{x}}(\vec{k})$ describes the phase difference of wave functions at the same $\vec{k}$, but belonging to the two different areas, for a particular $\vec{x}$. To apply Stokes theorem, one has to integrate for area *I* and *II* separately. Taking into account that the line integral at the MBZ-boundary vanishes, using the definitions in eq. (5.43) and (5.46), and calculating the difference in Berry connection $\vec{A}_{\mathrm{Berry},n}(\vec{k}) = \langle u_{n\vec{k}}(\vec{x})|\nabla_{\vec{k}}|u_{n\vec{k}}(\vec{x})\rangle$ between the $u^{\mathrm{I}}_{n\vec{k}}$ and the $u^{\mathrm{II}}_{n\vec{k}}$ explicitly, one gets:

$$\begin{aligned}\sigma_{xy} &= \frac{e^2}{h} \cdot \frac{1}{2\pi\mathrm{i}} \oint\limits_{\mathrm{dPMM}} (\vec{A}^{\mathrm{I}}_{\mathrm{Berry},n}(\vec{k}) - \vec{A}^{\mathrm{II}}_{\mathrm{Berry},n}(\vec{k}))\, \mathrm{d}\vec{k} \\ &= \frac{e^2}{h} \cdot \frac{1}{2\pi\mathrm{i}} \oint\limits_{\mathrm{dPMM}} \int\!\!\int \mathrm{i}\nabla_{\vec{k}}\Delta\Theta_{n\vec{x}}(\vec{k})\, \mathrm{d}^2\vec{x}\mathrm{d}\vec{k}\end{aligned} \tag{5.47}$$

with dPMM being the phase mismatch line. Since $u_{n\vec{k}}$ has to be single-valued in area I and II for each $\vec{x}$, the path integral of the gradient has to be an integer multiple of 2π such that, we finally get

$$\sigma_{xy} = \frac{e^2}{h} \cdot n_{\mathrm{Ch}} \tag{5.48}$$

with $n_{\mathrm{Ch}} \in \mathbb{Z}$.

Thus, we have found a possibility to get a non-zero Chern number by the complex phase vector of the Bloch function according to eq. (5.40) and we have verified that it must be an integer. The requirement is that the integral in eq. (5.42) is continuous in both areas up to the phase mismatch line, such that Stokes theorem can be applied. This is the case, if the MBZ is fully occupied, which is equivalent to a fully occupied, spin-polarized Landau level. Hence, we describe the situation where E_{F} is in between two spin-polarized Landau levels. Determining the value of n_{Ch} gets too complicated for this introductory textbook, such that we restrict ourselves to describe the result. It turns out that the value of n_{Ch} depends on the topology of the mismatch line after transforming the MBZ to a torus and, thus, representing the periodic boundary conditions correctly (Fig. 5.13(c)). Only if the mismatch line is not contractible on the torus, is n_{Ch} different from zero (without argument). Thus, we are back at the winding number from the previous section, asking how many times does the mismatch line surrounds the torus, respectively how many mismatch lines surround it. Most of the mismatch lines surrounding zeros of $u_{n\vec{k}}$ are irrelevant for the Hall conductance, since they do not surround the $\vec{k}$ space torus. Only the ones that surround the torus contribute to the Hall conductivity. In other words, the integer value of σ_{xy} is related to the properties of particular zeros of the wave functions in $\vec{k}$ space, which require a non-contractible phase mismatch line.

Unfortunately, the above result is intuitive, but experimentally useless as described below. Nevertheless, it is a useful introduction into the general approach of topological considerations in solids. Using geometrical properties of the space, where the wave functions are defined (the MBZ), we have found a number n_{Ch} related to a measurable property ($\sigma_{xy} \cdot \frac{h}{e^2} = n_{Ch}$), that is necessarily an integer. The definition of such numbers is at the heart of the topological approach to solids.

Notice, that n_{Ch} is not related to the edge of the sample, but is a property of the wave functions of an infinite 2D system in a magnetic field. Hence, in a real sample with edges, the properties of the edge are not important for the measured value of V_{Hall}.

Finally, we have indeed found a robust prediction of the quantum Hall effect, that does not depend on the details of the edges. Counter-intuitively, it does not contradict our previous description that the current flows at the edge of the sample, but only implies that the origin of this edge current is given by the complex wave functions of the bulk of the sample. This surprising relation between edge properties and bulk properties is dubbed the bulk-boundary correspondence.

5.4.3 General approach to topology

Before proceeding with the quantum Hall effect, we dwell briefly on the generality of the topological approach. Figure 5.14 shows a so-called periodic table of topology. Hamiltonians which describe solids are classified according to fundamental symmetries as particle-hole symmetry (PHS), time-reversal symmetry (TRS) and sublattice symmetry (SLS), also called chiral symmetry. PHS is conserved, e.g., in superconductors, TRS is broken by application of an external or an internal magnetic field and SLS requires two identical atoms in the unit cell of the crystal as present, e.g., in graphene. The Hamiltonians are analyzed for different dimensions d, i.e., for situations where the electrons are only free to move in d linearly independent directions (Section 1.2.3). It is then analyzed by field theoretical means, if a non-trivial integer number for the conductivity is well defined. This eventually reveals results similar to eq. (5.48). The table shows that many of the Hamiltonians exhibit such a number. The quantum Hall case discussed above corresponds to the $d = 2$ unitary class and is $\mathbb{Z}$ topological, thus, exhibits a Chern number $n_{Ch} \in \mathbb{Z}$.

The analysis leading to topological tables is a task for theoretical physicists, but it shows that the approach is applicable to many other types of electron systems. The remaining experimental problem is then mainly to find real materials, that exhibit the robust transport properties predicted by the topological numbers, if these numbers are not zero.

		TRS	PHS	SLS	d=1	d=2	d=3
Standard	A (unitary)	0	0	0	-	Z	-
(Wigner-Dyson)	AI (orthogonal)	+1	0	0	-	-	-
	AII (symplectic)	−1	0	0	-	Z_2	Z_2
Chiral	AIII (chiral unitary)	0	0	1	Z	-	Z
(sublattice)	BDI (chiral orthogonal)	+1	+1	1	Z	-	-
	CII (chiral symplectic)	−1	−1	1	Z	-	Z_2
Bogliubov de Gennes	D	0	+1	0	Z_2	Z	-
	C	0	−1	0	-	Z	-
	DIII	−1	+1	1	Z_2	Z_2	Z
	CI	+1	−1	1	-	-	Z

Fig. 5.14: Periodic table of topology: Ten symmetry classes of single-particle Hamiltonians are classified in terms of the presence or absence of time-reversal symmetry (TRS), particle-hole symmetry (PHS), and sublattice symmetry (SLS), which is relevant, if the unit cell of the crystal contains more than one atom. The absence of symmetries is denoted by 0, while the presence of these symmetries is denoted by +1 or -1, indicating the sign of the result of the squared symmetry operation. The first two columns contain terms used to characterize the Hamiltonian of a certain symmetry by different standards (2nd column) or by traditional terms including names of physicists (1st column). The last three columns show if a topological number is found within the three spatial dimensions symbolized by d. $\mathbb{Z}$ and $\mathbb{Z}_2$ indicate two different types of topological integer numbers, where $\mathbb{Z}$ is related to the Chern number. The dashes indicate the absence of a topological number. The classification ignores electron-electron interaction [116].

5.4.4 Quantum Hall effect and disorder

After this short excursion, we proceed by analyzing the quantum Hall effect.

The derivation of $\sigma_{xy} = \frac{e^2}{h} \cdot n_{\text{Ch}}$ required that the MBZ is densely filled. Otherwise, Stokes theorem is not applicable and we have no means to predict integer values of the transversal conductivity.

If the area of two magnetic flux quanta is commensurable with the lattice unit cell, the bands covering the whole MBZ are simply the Landau levels. Since the area, that contains a magnetic flux of h/e, is typically huge with respect to the unit cell (10 T corresponds to ~ $(20\,\text{nm})^2$ per flux quantum), this condition is approximately fulfilled. Alternatively, one can deal with infinite solids, which will provide some commensurability for each possible B. Then, the commensurable cell contains $2n$ flux quanta and m atomic unit cells with n and m being both integers larger than one. Consequently, the magnetic unit cell gets larger and the MBZ gets smaller. This leads to a backfolding of the Landau levels into the smaller MBZ, i.e., each Landau level consists of n bands. At low B, this does not matter since the bands remain nearly degenerate. But as soon as the cyclotron radius gets similar to the atomic lattice constant, the different bands are split, basically due to a different spatial distribution of the Landau level wave function with respect to the atomic lattice. This leads to a band structure development as shown in Fig. 5.15(d) and called the Hofstadter butterfly. At low field (flux) the Landau fan as in Fig. 5.10(c) is apparent, but at higher B, the Landau levels split in multiple bands with a detailed substructure, that is similar to the texture of a butterfly. The fourfold symmetry is a consequence of the tight binding Hamiltonian, that is the one of eq. (4.29) with an additional B field and, hence, not universal.

Experiments are typically performed at ~ 0.0001 flux units per atomic unit cell and are, hence, not sensitive to this structure. But artificial lattices can be tuned into the interesting region of order 0.1–1 flux units per unit cell and fingerprints of the Hofstadter butterfly have indeed been observed in the transversal conductivity $\sigma_{xy}(B)$ [117–119].

Staying with the simple Landau level picture, Fig. 5.15(a) shows the development of the Landau and spin levels at low $\vec{B}$ (same as Fig. 5.10(c)). Due to the increasing degeneracy of each level, the Fermi level discontinuously jumps down to a lower Landau level. The jump appears, if enough density of states is available in all levels below the actual one, provided that the electron density n_e is constant (Section 5.3.2). Exactly at the positions of the jumps (blue dots), all Landau levels are either completely filled or completely empty, such that Stoke's theorem can be applied and the transversal conductivity is $\sigma_{xy} = \frac{e^2}{h} \cdot n_{Ch}$. Unfortunately, these points are marginal. They appear, if n_e divided by the degeneracy eB/h is an integer. The classical Hall resistivity ρ_{xy} at these $\vec{B}$-fields is:

$$\frac{hn_e}{eB} = n_{Ch} \Longrightarrow B = \frac{hn_e}{en_{Ch}} \Longrightarrow \rho_{xy} = \frac{B}{en_e} = \frac{h}{e^2 n_{Ch}} . \tag{5.49}$$

Thus, the quantized Hall resistivity predicted by topology is identical to the classical Hall resistivity $\rho_{xy} = B/en_e$ at these marginal points, where the Fermi level lies between bands (Fig. 5.15(b)). This supports the calculation, but is sincerely a disappointing result for an intricate derivation.

In other words, the quantization of transversal conductivity deduced by topology exists, but the plateaus observed experimentally can not be explained by the argument.

The riddle can be solved by including disorder. Naturally, one looses periodicity by random disorder and, thus, the MBZ disappears. In order to get a novel description of ρ_{xy}, we firstly recall that the perturbation approach leading to eqs. (5.28) and (5.32) did not rely on Bloch states. Using these two equations, we can write.

$$\hat{v}_y = \langle n|v_y|n\rangle = \langle n_0|v_y|n_0\rangle + ie\hbar E_x \sum_{n\neq m} \frac{\langle n_0|v_y|m_0\rangle\langle m_0|v_x|n_0\rangle}{(E_{n0} - E_{m0})^2} + h.c. \tag{5.50}$$

where $|n_0\rangle$ and $|m_0\rangle$ are now states of the disordered potential. They could even be many-particle ground states and excited states of an interacting electron system, which, however, will not be discussed any further.

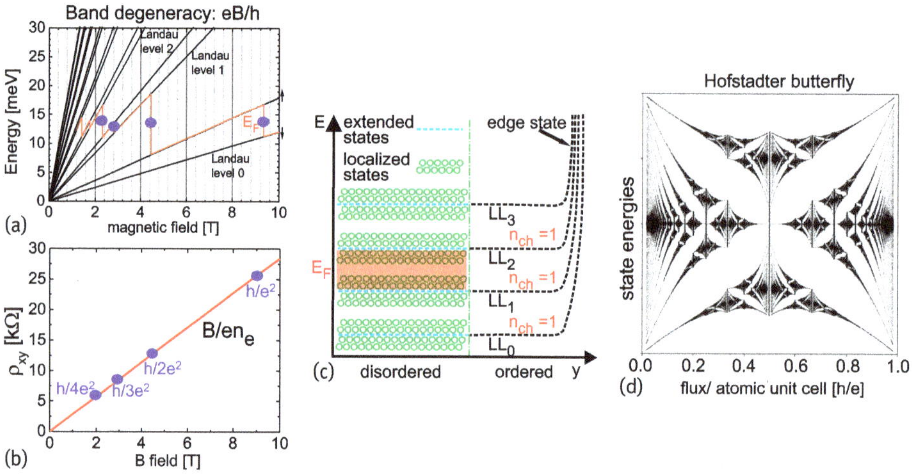

Fig. 5.15: (a) Landau and spin levels (black lines) as a function of $\vec{B}$-field. Spin directions are marked on the right for the lowest Landau level 0. The red line is the Fermi level, if $n_e(\vec{B}) = \text{const}$. The degeneracy of each level increases with $\vec{B}$ as marked on the top and, thus, E_F exhibits jumps. The blue dots mark the positions, where all Landau levels are either completely filled or completely empty. (b) The classical transversal resistivity ρ_{xy} as a function of $\vec{B}$. The blue dots mark the values $\rho_{xy} = 1/\sigma_{xy} = \frac{h}{e^2 n_{\text{Ch}}}$ where the Chern number is necessarily an integer. (c) Right hand side: Landau levels without disorder bending upwards at the rim of the sample, where they cross E_F giving rise to edge states (see also Fig. 5.11). Left hand side: disorder is added leading to extended states and localized states as explained in Section 5.3.2. The Chern number n_{Ch} can be identified with the extended state, such that σ_{xy} is quantized as long as E_F is located within the localized states (brown transparent areas). (d) Hofstadter butterfly: Energy values of electron bands in a cosine periodic potential of the atoms on a square lattice as calculated by a tight binding model. A homogeneous $\vec{B}$ field is added and quantified as flux per lattice unit cell in units of h/e. (d) [120, 121].

The first term vanishes again due to eq. (5.30). The velocity operators depend on the gauge of the $\vec{B}$ field. For the Landau gauge ($\vec{A}(\vec{x}) = |\vec{B}|x \cdot \vec{e}_y$), e.g., we have $\hat{v}_x = \frac{-i\hbar}{m}\frac{d}{dx}$ and $\hat{v}_y = \frac{-i\hbar}{m}\frac{d}{dy} - \frac{eB}{m}x$.

If one takes a finite system of length L_1 and width L_2, the presence of the magnetic field requires a gauge dependent phase shift of all wave functions across the system again being the Aharonov–Bohm phase. The general idea is now firstly to express the velocity operators in terms of these phase factors. They can be described by υL_1 for a shift of L_1 along the x direction and by βL_2 for a shift of L_2 along the y direction. Now one can perform a unitary transformation of the original wave functions $|n\rangle$ to

$$|\tilde{n}\rangle = e^{-i(\upsilon x + \beta y)} \cdot |n\rangle \tag{5.51}$$

This transformation uses the gauge freedom and introduces the prefactors υ and β into the velocity operators of the transformed Hamiltonian $\tilde{\tilde{H}}$ such that

$$\langle n|v_x|m\rangle = \frac{1}{\hbar}\left\langle \tilde{n}\left|\frac{d\tilde{\tilde{H}}}{d\upsilon}\right|\tilde{m}\right\rangle, \quad \langle n|v_y|m\rangle = \frac{1}{\hbar}\left\langle \tilde{n}\left|\frac{d\tilde{\tilde{H}}}{d\beta}\right|\tilde{m}\right\rangle \tag{5.52}$$

This can be verified straightforwardly by transforming the Hamiltonian starting with the Landau gauge and then applying the derivation to the new Hamiltonian $\widetilde{\widetilde{H}}$. It naturally implies a description of $\hat{v}_x$ and $\hat{v}_y$ in terms of the phase factors at the boundaries of the sample by changing eq. (5.50) accordingly:

$$\hat{v}_y = \frac{-ieE_x}{\hbar}\left(\sum_{\tilde{n}_0 \neq \tilde{m}_0} \frac{\left\langle \tilde{n}_0 \middle| \frac{d\widetilde{\widetilde{H}}}{d\beta} \middle| \tilde{m}_0 \right\rangle \left\langle \tilde{m}_0 \middle| \frac{d\widetilde{\widetilde{H}}}{dv} \middle| \tilde{n}_0 \right\rangle}{(E_{n0} - E_{m0})^2} - \frac{\left\langle \tilde{n}_0 \middle| \frac{d\widetilde{\widetilde{H}}}{dv} \middle| \tilde{m}_0 \right\rangle \left\langle \tilde{m}_0 \middle| \frac{d\widetilde{\widetilde{H}}}{d\beta} \middle| \tilde{n}_0 \right\rangle}{(E_{n0} - E_{m0})^2}\right) \tag{5.53}$$

Finally, we use the product rule for $\tilde{m}_0 \neq \tilde{n}_0$, i.e.:

$$\left\langle \tilde{n}_0 \middle| \frac{d\widetilde{\widetilde{H}}}{d\beta} \middle| \tilde{m}_0 \right\rangle = \frac{d}{d\beta}\left(\left\langle \tilde{n}_0 \middle| \widetilde{\widetilde{H}} \middle| \tilde{m}_0 \right\rangle\right) - \left\langle \frac{d\tilde{n}_0}{d\beta} \middle| \widetilde{\widetilde{H}} \middle| \tilde{m}_0 \right\rangle - \left\langle \tilde{n}_0 \middle| \widetilde{\widetilde{H}} \middle| \frac{d\tilde{m}_0}{d\beta} \right\rangle \tag{5.54}$$

$$= 0 - E_{m0}\left\langle \frac{d\tilde{n}_0}{d\beta} \middle| \tilde{m}_0 \right\rangle - E_{n0}\left\langle \tilde{n}_0 \middle| \frac{d\tilde{m}_0}{d\beta} \right\rangle \tag{5.55}$$

$$= -E_{m0}\left\langle \frac{d\tilde{n}_0}{d\beta} \middle| \tilde{m}_0 \right\rangle + E_{n0}\left\langle \frac{d\tilde{n}_0}{d\beta} \middle| \tilde{m}_0 \right\rangle \tag{5.56}$$

$$= \left\langle \frac{d\tilde{n}_0}{d\beta} \middle| \tilde{m}_0 \right\rangle \cdot (E_{n0} - E_{m0}) \tag{5.57}$$

The same trick can be applied to the second term in the product of eq. (5.53) revealing

$$\left\langle \tilde{m}_0 \middle| \frac{d\widetilde{\widetilde{H}}}{dv} \middle| \tilde{n}_0 \right\rangle = \left\langle \tilde{m}_0 \middle| \frac{d\tilde{n}_0}{dv} \right\rangle \cdot (E_{n0} - E_{m0}) \tag{5.58}$$

as well as to the two factors of the product behind the minus sign. With the completeness relation $\sum_{\tilde{m}_0} |\tilde{m}_0\rangle\langle\tilde{m}_0| = \mathbf{1}$, we finally arrive at

$$\hat{v}_y = \frac{-ieE_x}{\hbar}\left[\left\langle \frac{d\tilde{n}_0}{d\beta} \middle| \frac{d\tilde{n}_0}{dv} \right\rangle - \left\langle \frac{d\tilde{n}_0}{dv} \middle| \frac{d\tilde{n}_0}{d\beta} \right\rangle\right] \tag{5.59}$$

or equivalently

$$\hat{v}_y = \frac{-ieE_xL_1L_2}{\hbar}\left[\left\langle \frac{d\tilde{n}_0}{d\phi} \middle| \frac{d\tilde{n}_0}{d\varphi} \right\rangle - \left\langle \frac{d\tilde{n}_0}{d\varphi} \middle| \frac{d\tilde{n}_0}{d\phi} \right\rangle\right] \tag{5.60}$$

with ϕ and φ being the boundary phases at $x = L_1$ and $y = L_2$. This gives by adding up all occupied states $|\tilde{n}_0\rangle$[27]:

$$\sigma_{xy} \sim \frac{ie^2}{\hbar}\sum_{\tilde{n}_0}\left[\left\langle \frac{d\tilde{n}_0}{d\phi} \middle| \frac{d\tilde{n}_0}{d\varphi} \right\rangle - \left\langle \frac{d\tilde{n}_0}{d\varphi} \middle| \frac{d\tilde{n}_0}{d\phi} \right\rangle\right]. \tag{5.61}$$

The decisive argument is now that the particular choice of the boundary phases can not influence the conductivity, if the Fermi level is within localized states with vanishing amplitude, at least, at one boundary. Thus, one can write σ_{xy} also as an

27 We omit the discussion of correct units in the following for the sake of simplicity.

average over the results for different boundary phases.

$$\sigma_{xy} \sim \frac{e^2}{h} \cdot \sum_{\tilde{n}_0} \int_0^{2\pi} \int_0^{2\pi} \frac{1}{2\pi \mathrm{i}} \left[\left\langle \frac{\mathrm{d}\tilde{n}_0}{\mathrm{d}\varphi} \middle| \frac{\mathrm{d}\tilde{n}_0}{\mathrm{d}\phi} \right\rangle - \left\langle \frac{\mathrm{d}\tilde{n}_0}{\mathrm{d}\phi} \middle| \frac{\mathrm{d}\tilde{n}_0}{\mathrm{d}\varphi} \right\rangle \right] \mathrm{d}\phi \mathrm{d}\varphi \,. \tag{5.62}$$

The integral is formally identical to eq. (5.42). Thus, the result of the integral leads to an integer, if a continuous change of any wave function in the space of the boundary phase factors along a closed path does not change the phase factor of the wave function modulo 2π. Thus, as long as the ground state wave function is unique in boundary phase space, the transversal conductance will be

$$\sigma_{xy} = \frac{e^2}{h} \cdot n \tag{5.63}$$

with n being an integer. This is important progress, since the conductance quantization is now predicted for the typical case of a localized state at E_F. The absence of quantization is marginal and appears only when the wave function at E_F is not localized and, thus, an extended state. This is visualized in Fig. 5.15(c), where the transparent brown area marks the typical case for disordered systems with quantized Hall conductance. In the limit of an infinite sample (at $T = 0$ K), there will be only a single state in each spin-polarized Landau level, the extended state, that, if at E_F, gives rise to non-quantized σ_{xy}.

The description in eq. (5.62) makes clear that the integer n_{Ch} (Chern number) is still a geometrical property of individual wave functions, however, in the unusual space of boundary phases of the finite system. Hence, one can sum up n_{Ch} by integers of individual wave functions. It is intuitive that the localized states will not contribute to the Chern number, since they will not depend on the boundary phases. Indeed, one can show that each occupied extended state contributes a unit of one to the Chern number [122]. Thus, the Chern number is encoded in the change of the extended state to changes of the boundary conditions. One can make this even more explicit (without argument): There is exactly one node (zero point) of each extended state that crosses the whole system, if the boundary phase is changed by 2π. Thus, it is again due to the zeros of the wave functions imprinted by the B field, that the Hall conductance is quantized, but now it is the path of the zeros of the extended states in real space, which defines n_{Ch}. In Fig. 5.15(c) this is marked by the states labeled $n_{Ch} = 1$.

An important aspect of the derivation above is again the so-called bulk-boundary correspondence.[28] Since we deduced a finite Hall conductivity for localized, i.e., insulating states of the system at E_F, a natural question is: What are the states that carry the current? Since the bulk states are not able to carry current, the only alternatives are states at the edge of the sample. Thus, the presence of the edge states of Fig. 5.11(d) can

28 Exact mathematical derivations of the bulk-boundary correspondence are lengthy, see e.g. [123]. We will discuss a more rigorous argument in Section 5.4.6.1 on 3D topological insulators.

also be derived on more fundamental, topological grounds. They are a consequence of the occupied, extended bulk states which contain a nontrivial zero due to B field. This bulk property, that is independent of the details of the disorder potential $\Phi_{\mathrm{el}}(\vec{x})$, implies the robustness of the quantum Hall effect.

The argument can even be generalized to any topological number. An insulating material with non-trivial topological number describing the Hall conductance requires conducting states at E_{F}, which are located at the boundaries of the material.

5.4.5 2D topological insulators

After the topological description of the quantum Hall effect, scientists started to look for systems that exhibit a topological number describing σ_{xy}, which is different from zero, without an external magnetic field. This does not work straightforwardly, since a finite σ_{xy} would break the time-reversal symmetry (TRS) in contrast to the Hamiltonian without $\vec{B}$.[29] However, one can separate the Schrödinger equation into two parts for the two opposite spins of the electrons. The two spins exhibit a different spin orbit interaction, that breaks the time reversal symmetry for each spin by acting as an effective magnetic field on the electron with the corresponding spin. Thus, the two spin channels could have opposite σ_{xy}. This would still respect global time reversal symmetry, since the time-inversion inverts spin and current direction and, thus gives the same Hall voltage. The task is, thus, to find a system with a finite σ_{xy} for each spin channel.

It is natural to look for a strong spin-orbit term, e.g., proportional to $\vec{k} \times \vec{\sigma}$ with $\vec{\sigma}$ being the vector of the three Pauli matrices (page 204) representing the electron spin. Such a term naturally leads to a transversal movement of the electrons, which is opposite for the two spins. Within a band structure, the spin-orbit interaction acts much more strongly on bands, where the periodic part of the Bloch function $u_{n\vec{k}}(\vec{x})$ represents atomic states with an orbital quantum number, e.g., p-type, d-type or f-type states, than for bands without, i.e., s-type states. Thus, the different spin components of the p-type valence band of a typical semiconductor shift much more strongly than the s-type conduction band. This results, e.g., in the spin-orbit split band of the valence band of most semiconductors (Si, Ge, GaAs, CdTe, ...). The spin orbit splitting increases, moreover, with the charge number of the atoms constituting the system (Section 4.5.4), i.e., one has to look for semiconductors made of heavy atoms. If the spin-orbit interaction is strong enough, it can even lead to a crossing of the p-type valence band and the s-type conduction band as displayed in Fig. 5.16(a). Mostly, there are additional $\vec{k}$ dependent terms that mix the valence and the conduction band. Formally, one can derive such terms by perturbation theory starting from the atomic

29 If one inverts the time, i.e., the direction of current, the Hall voltage would invert for constant σ_{xy}.

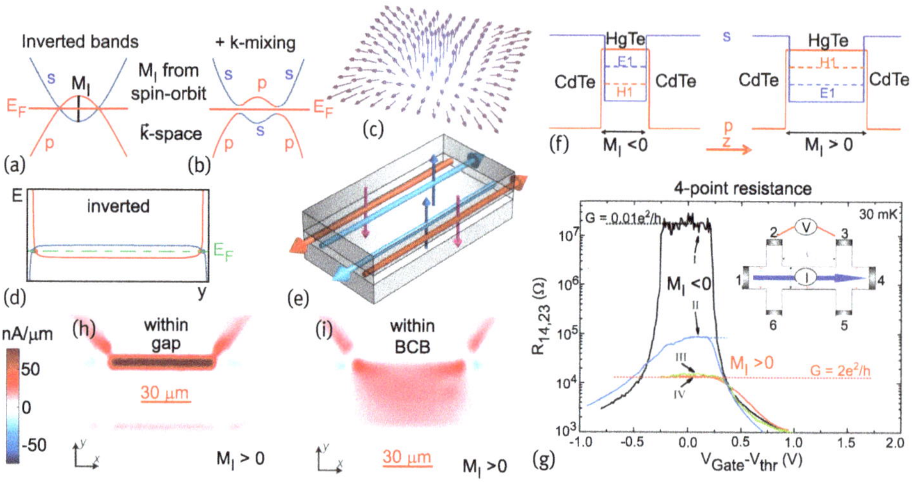

Fig. 5.16: (a), (b) Model band structure of a 2D topological insulator: (a) The spin-orbit interaction, that shifts the p-type valence band (red) upwards by M_I, leads to a band crossing with the s-type conduction band (blue). (b) An additional interaction term between the two bands (parameter A_I in eq. (5.66)) leads to a symmetric and an antisymmetric combination of the two bands, that induces an anticrossing, thus, a band gap and a changing s/p character (colors) of the resulting two bands. The Fermi level E_F is chosen to be in that band gap (red line). (c) Distribution of $\hat{\vec{d}}(\vec{k}) = \vec{d}(\vec{k})/|\vec{d}(\vec{k})|$ (eq. (5.66)) in $\vec{k}$ space close to $\vec{k} = \vec{0}$ /m. (d) Intuitive illustration of the edge states. While the p-type band (blue) bends downwards, the s-type band (red) bends upwards, thus, reestablishing the normal order of bands and implying a conducting state at the edge of the sample. (e) 2D illustration of the edge states (thick arrows) running in opposite directions for the two spin contributions (slim arrows). Note that two edge channels of opposite spin run in the same direction implying $\sigma_{xx} = 2e^2/h$. (f) Vertical band structure for a realization of a 2D topological insulator, i.e., a stacking of CdTe/HgTe/CdTe. The heavier HgTe exhibits a partially inverted band for large thickness ($M_\mathrm{I} > 0$, right), i.e., the hole band H1 is above the electron band E1. For an additional confinement in the z direction, the confinement energy acts oppositely on H1 and E1, such that the normal order (E1 is above H1) can be reestablished ($M_\mathrm{I} < 0$, left). (g) 4-point resistance $R_{14,23}$ of the stacks shown in (f) as a function of the applied gate voltage $V_\mathrm{Gate} - V_\mathrm{thr}$: The resistance for $M_\mathrm{I} < 0$ (curve I) levels at $\sim 1 \times 10^7\ \Omega$, i.e., at the input resistance of the measurement device. This gate voltage region marks the normal band gap. In contrast, the resistance of the $M_\mathrm{I} > 0$ samples (curve III, IV) levels at $R_{14,23} \approx h/2e^2$ as expected for two conducting edge channels. Inset: Sketch to measure $R_{14,23} = V/I$ in (g). (h) Scanning SQUID result, that visualizes the current flow within a $M_\mathrm{I} > 0$ sample with E_F being within the inverted band gap. (i) Same experiment, but with E_F being in the bulk conduction band (BCB). (a), (f) [124], (d), (e), (g) after [125], (h), (i) [126].

orbitals, which is known as $\vec{k} \cdot \vec{p}$ theory. Thus, the states form symmetric and antisymmetric combinations of p-type states and s-type states with different energy and, thus, prohibit a crossing of the two bands. The resulting band structure is sketched in Fig. 5.16(b). If we put the Fermi level into this anticrossing region, it is still within a band gap, albeit the character of the states in the inner part between the anticrossings is inverted with respect to the outer part of the band structure. Since p-states get a negative sign after spatial inversion and s-states are unchanged by spatial inversion, the total inversion symmetry of the product of occupied states for a certain $\vec{k}$ is different for the inner part and the outer part. We will see that this change of symmetry (sign of parity) is enough to predict quantized transversal conductance for each spin.

Next, this situation is analyzed more formally. As described for the quantum Hall effect, the exact derivations can be lengthy and will be partly skipped as marked by (without derivation) or (without argument). The details can be calculated straightforwardly, but this is not required for a general understanding.

Firstly, we separate the two spin channels according to[30]

$$\widehat{H} = \begin{pmatrix} \widehat{H}_\uparrow(\vec{k}) & 0 \\ 0 & \widehat{H}_\downarrow(\vec{k}) \end{pmatrix}, \tag{5.64}$$

where $\widehat{H}_\uparrow$ and $\widehat{H}_\downarrow$ act on the two different spins and $\widehat{H}_\uparrow(\vec{k}) = \widehat{H}^*_\downarrow(-\vec{k})$ in order to respect time-reversal symmetry (TRS).[31] Now we use a two band model consisting of the s-type and the p-type band for both spin channels and write the corresponding Hamiltonian in terms of Pauli matrices $\boldsymbol{\sigma}_i$ and the unity matrix $\mathbf{1}$.

$$\boldsymbol{\sigma}_x = \begin{pmatrix} 0 & 1 \\ 1 & 0 \end{pmatrix} \quad \boldsymbol{\sigma}_y = \begin{pmatrix} 0 & -i \\ i & 0 \end{pmatrix} \quad \boldsymbol{\sigma}_z = \begin{pmatrix} 1 & 0 \\ 0 & -1 \end{pmatrix} \quad \mathbf{1} = \begin{pmatrix} 1 & 0 \\ 0 & 1 \end{pmatrix} \tag{5.65}$$

as

$$\widehat{H}_\uparrow(\vec{k}) = \epsilon(\vec{k}) \cdot \mathbf{1} + \sum_{i=x,y,z} d_i(\vec{k}) \cdot \boldsymbol{\sigma}_i\,, \tag{5.66}$$

where the second term describes the perturbations with prefactor functions $d_i(\vec{k}) : \mathbb{R}^3 \to \mathbb{R}$. The first two Pauli matrices cover the interactions between s-band and p-band. For the lowest order, we can choose its prefactor functions to be linear in $\vec{k}$, i.e., e.g., $d_x(\vec{k}) = A_\mathrm{I} \cdot k_x$ and $d_y(\vec{k}) = -A_\mathrm{I} k_y$ with $A_\mathrm{I} \in \mathbb{R}$ being a constant. These terms are required to guarantee the final band gap opening (transition from Fig. 5.16(a) to (b)). They can be justified by $\vec{k} \cdot \vec{p}$ theory. The final term $d_z \cdot \boldsymbol{\sigma}_z$ describes the energy difference between the two bands and has to be even with respect to inversion, if the crystal is inversion symmetric. We choose it to be $d_z(\vec{k}) = M_\mathrm{I} - B_\mathrm{I}(k_x^2 + k_y^2)$. The term $M_\mathrm{I} \in \mathbb{R}$ describes the strength of the band inversion at $\vec{k} = \vec{0}/\mathrm{m}$ and $B_\mathrm{I} \in \mathbb{R}$ describes

30 Formally, the Hamiltonian describes four bands of different total angular momentum, described by the corresponding magnetic quantum number m_j, where the $m_j = 3/2$ and $m_j = 1/2$ bands form one sector of the block-diagonal shape of the Hamiltonian and the $m_j = -3/2$ and $m_j = -1/2$ states form the other one.

31 Formally, the zeros in the off-diagonal terms require time reversal symmetry and inversion symmetry of the crystal, i.e we do not apply a Rashba term, but only an atomic spin-orbit interaction as in eq. (4.44). Then the two states $|\vec{k},\uparrow\rangle$ and $|\vec{k},\downarrow\rangle$ have to be identical in energy, since time reversal requires $|\vec{k},\uparrow\rangle \to |-\vec{k},\downarrow\rangle$ and inversion requires $|\vec{k}\rangle \to |-\vec{k}\rangle$.

the curvature of the two bands with opposite sign for each band due to $\boldsymbol{\sigma}_z$. Any other details of the bands are covered by $\epsilon(\vec{k})$ and are irrelevant for σ_{xy}.[32]

In order to calculate σ_{xy} by the Kubo formula (eq. (5.34)), similar to the case of the magnetic Brillouin zone (Section 5.4.2), we have to respect the periodicity of the Brillouin zone, i.e., linear terms in (k_x, k_y) are replaced by sine functions and quadratic terms in (k_x, k_y) by cosine functions (eq. (6) in [124]). Then one can show that the vector $\vec{d}(\vec{k}) := (d_x(\vec{k}), d_y(\vec{k}), d_z(\vec{k}))$, respectively, its corresponding unit vector $\hat{\vec{d}}(\vec{k}) = \vec{d}(\vec{k})/|\vec{d}(\vec{k})|$, completely determines the Hall conductivity σ_{xy}, according to (without derivation):[33]

$$\sigma_{xy} = -\frac{e^2}{8\pi^2 h} \int\int_{\mathrm{BZ}} \hat{\vec{d}}(\vec{k}) \cdot \left(\frac{\mathrm{d}\hat{\vec{d}}(\vec{k})}{\mathrm{d}k_x} \times \frac{\mathrm{d}\hat{\vec{d}}(\vec{k})}{\mathrm{d}k_y} \right) \mathrm{d}k_x \mathrm{d}k_y \tag{5.67}$$

with BZ indicating that the integral covers the whole Brillouin zone. Figure 5.16(c) shows the vector $\hat{\vec{d}}(\vec{k})$ close to $\vec{k} = \vec{0}/\mathrm{m}$. The vector $\hat{\vec{d}}(\vec{k})$ is basically the vector within the Bloch sphere of the two-state space spanned by the conduction and the valence band (Section 3.4) with $\vec{k}$ being the parameter controlling the direction of $\hat{\vec{d}}$. Within the center, it points upwards, but depending on the relation between M_I and B_I it can point downwards at the rim of the BZ. The periodicity requires that $\hat{\vec{d}}(\vec{k})$ can only point upwards or downwards at the rim, as one immediately recognizes if one writes down the correct cosine and sine functions. Thus, the number of half windings of the $\hat{\vec{d}}(\vec{k})$ vector from the center to the rim of the BZ is necessarily an integer number.

The question if $\hat{\vec{d}}(\vec{k})$ points in the opposite or in the same direction at the rim and at the center of the BZ (related to the number of half windings on its way) is the decisive topological property. This integer is called the $\mathbb{Z}_2$ index, since it can have only two values. It turns out that, if $0 < M_\mathrm{I} < 4B_\mathrm{I}$, $\hat{\vec{d}}(\vec{k})$ points in opposite directions at the rim and at the center of the BZ. This is equivalent to the existence of a crossing point of the two bands in Fig. 5.16(a), before including the anticrossing terms $d_x(\vec{k})$ and $d_y(\vec{k})$. One can show that the integral in eq. (5.67), that is also called the Skyrmion number of a vector field,[34] probes exactly that property, i.e., the number of half windings of a vector-field. Thus, if the conduction and valence band anticross and, therefore, have a different character in the inner part than in the outer part of the BZ, the Hall conductivity σ_{xy} of a single spin channel is $\sigma_{xy,\uparrow} = 1 \cdot e^2/h$, while σ_{xy} of the other spin channel

32 Due to inversion symmetry, the diagonal terms within $\widehat{H}_\uparrow(\vec{k})$ can only be even in $\vec{k}$ since $\langle \psi_i | \vec{k} | \psi_i \rangle = \langle \psi_i | -\vec{k} | \psi_i \rangle$ is required. One can show that terms different in k_x and k_y do only appear in higher orders of $\vec{k}$. The off-diagonal terms have to be odd in $\vec{k}$, since they mix an odd state (p-state) and an even state (s-state). The sign of the off-diagonal term can be rationalized by writing the orbital moments of the p-states as $|p_x + ip_y\rangle$ and $|-(p_x - ip_y)\rangle$.

33 The calculation uses again the Kubo formula, but without calculating the Bloch functions explicitely. Thus, the integral multiplied by the leading term $\frac{1}{8\pi^2}$ is identical to the Chern number of the Hamiltonian $\widehat{H}_\uparrow(\vec{k})$ [127].

34 if multiplied by $\frac{1}{8\pi^2}$.

is $\sigma_{xy,\downarrow} = -1 \cdot e^2/h$. The effect is dubbed the spin quantum Hall effect (SQHE) and materials that exhibit the SQHE are called 2D topological insulators (2DTI) exhibiting partially inverted bands around E_F.

Next, we use again the bulk-boundary correspondence, which by the requirement that the bulk is insulating, i.e., the Fermi level is located within a bulk band gap, implies conducting edge states of a 2DTI in order to realize a finite σ_{xy}. This is sketched in Fig. 5.16(d), where the inverted bands are inverted back towards the edge in order to match the topological vacuum condition. The edge state is helical, i.e., it runs clockwise for one spin channel and counterclockwise for the other spin channel. Otherwise, it could not realize the different sign of σ_{xy} of the two spin channels (see Fig. 5.11(d) in Section 5.3.2). Thus, the opposite helicity of the two spin edge channels is protected by the different sign of $\sigma_{xy,\uparrow}$ and $\sigma_{xy,\downarrow}$, and, hence, by time-reversal symmetry (eq. (5.64)). One has one ballistic edge channel for each spin on each side running in opposite directions (Fig. 5.16(e)), as required for $\sigma_{xy,\uparrow} = e^2/h$ for both current directions for one spin and $\sigma_{xy,\downarrow} = -e^2/h$ for both current directions for the other spin.[35]

Each spin has one ballistic channel for each direction of current flow, but these channels are on opposite edges of the sample (Fig. 5.16(e)). Each channel contributes e^2/h to the longitudinal conductivity σ_{xx} (Section 5.3.2). Hence, one can verify the presence of the edge states by the quantized longitudinal conductance of $\sigma_{xx} = \sigma_{xx\uparrow} + \sigma_{xx\downarrow} = 2e^2/h$.[36]

This fact has been used as the first hallmark for the existence of 2DTIs. Figure 5.16(f)–(g) shows the experiment. In Fig. 5.16(f), the used samples are sketched. They consist of a heterostructure (Section 1.3.1), where two CdTe layers enclose a HgTe layer, the latter being the 2DTI with a partially inverted band gap ($M_\mathrm{I} \simeq 0.2$ eV). The band inversion is induced by the spin-orbit interaction of the heavy elements Hg and Te, i.e., $M_\mathrm{I} > 0$. So far, there is no material known where $M_\mathrm{I} > 4B_\mathrm{I}$. The band gap of CdTe is much larger (1.56 eV) and not inverted. Thus, 2D layers of HgTe should exhibit $\sigma_{xx} = 2e^2/h$. Indeed, this result has been found, if the HgTe layer is thick enough and an external gate voltage V_Gate is applied to tune E_F into the band gap of HgTe. The inset in Fig. 5.16(g) shows a measurement setup. The measured resistance curves,

35 An explicit derivation of the edge state using an open boundary and the Hamiltonian in eq. (5.66) is given in [128], there Chapter II b. For $A_\mathrm{I}/B_\mathrm{I} > 0$, it results in a function $\psi_i \propto (\mathrm{e}^{-\lambda_1 x} - \mathrm{e}^{-\lambda_2 x})$ with exponential decay lengths $\lambda_{1,2} = a_x^2 \cdot \frac{A_\mathrm{I}}{2B_\mathrm{I}} \pm \sqrt{A_\mathrm{I}^2/(4B_\mathrm{I}^2) - M_\mathrm{I}/B_\mathrm{I}}$ into the bulk of the 2DTI, if $E(\vec{k}) = 0$ eV and $k_y = 0$/m. The variable a_x is the lattice constant in x direction.

36 One can additionally justify that backscattering between the two channels at the same edge is prohibited, such that the channel is indeed ballistic. This is the case for impurities that respect TRS. In order to backscatter, the spin has to be rotated by 180° to fit to the spin of the counter-propagating channel. If TRS is respected, the clockwise and the counterclockwise rotation of the spin must have identical probabilities, such that these two scattering paths interfere. In spin space the rotation by 180° changes the phase by 90°. Hence, the two possible spin rotations differ by a phase of 180° and, consequently, interfere destructively, i.e., the probability for backscattering is exactly zero ([128]: Chapter II.C.1).

labeled III and IV in Fig. 5.16(g), exhibit a gate voltage region where one finds rather exactly $R_{14,23} = h/(2e^2)$.[37] As expected the resistance decreases monotonically, if one moves E_F into the conduction band of HgTe ($(V_{Gate} - V_{thr}) > 0.3$ V). The counter experiment is done for a smaller thickness of the HgTe layer. This induces an additional confinement energy to the bands that is positive for the s-type band (E1) and negative for the p-type band (H1) as expected from the curvature of the bands (different sign of the effective mass). Consequently, the bands can invert back to a normal insulator ($M_I < 0$) with increasing confinement energy, i.e., reduced thickness. Reducing the thickness to 5.5 nm (curve I), the resistance within the band gap indeed dramatically increases to 20 MΩ, which is the maximum resistance that could be detected in this particular experiment [125].

The image in Fig. 5.16(e), moreover, implies that the different spin components are separated at the entrance of the 2DTI. If the current runs from the lower left to the upper right, the spin-up electrons go exclusively to the upper edge, while the spin-down electrons go exclusively to the lower edge. Filtering out, e.g., only the current from the upper edge, reveals a perfect and, thus, very efficient spin filter. This effect has indeed been observed experimentally [129] and raises hopes for a reliable spin filter not requiring any magnetic fields.

Figure 5.16(h) and (i) show the edge channels directly for the $M_I > 0$ HgTe layer. A scanning SQUID[38] has been used to measure the local distribution of the current within the HgTe and, indeed, the current runs only along the edges, if E_F is within the band gap (h), while it runs across the whole layer, if E_F is within the conduction band (i).

Another topological transport effect in 2D is dubbed the quantum anomalous Hall effect (QAHE). The effect is very similar to the QSHE and is found in systems that are additionally magnetic, i.e., they break TRS. One can rationalize that the exchange coupling within a ferromagnet distorts the symmetry between the two spin channels such that $\widehat{H}_\uparrow(\vec{k}) \neq \widehat{H}^*_\downarrow(-\vec{k})$. Hence, the Hamiltonian $\widehat{H}_\uparrow(\vec{k})$ can have a partially inverted band gap implying $\sigma_{xy\uparrow} = e^2/h$, while $\widehat{H}_\downarrow(\vec{k})$ has a trivial band order with $\sigma_{xy\downarrow} = 0$.[39] Then, the edge states of only one spin channel appear at E_F. Thus, one gets $\sigma_{xy} = e^2/h$ that is not compensated by the other spin channel, i.e., a real transversal conductivity. The first successful measurement of the QAHE, i.e., a detection of $\sigma_{xy} = e^2/h$, if E_F is within a band gap, was realized in the 2D system $Cr_{0.15}Bi_{0.18}Sb_{1.67}Te_3$ [131]. The BiSbTe al-

37 The origin of the visible deviation from $R_{14,23} = h/2e^2$ is still not clear.

38 A SQUID (superconducting quantum interference device) described, e.g., in detail in [130], is one of the most sensitive magnetometers detecting $\vec{B}$-fields down to 10^{-18} T (see also Section 3.3.4ff). For this particular experiment, a SQUID is moved with piezomotors across the sample with nm precision. It detects the magnetic stray field of the current. One can reconvert the resulting $\vec{B}$-field map into a current map, that is displayed in Fig. 5.16(h) and (i).

39 Formally, the $\vec{k}$-independent exchange interaction introduces a different M_I into $\widehat{H}_\uparrow$ and $\widehat{H}_\downarrow$ of eq. (5.64) such that, e.g., $M_I < 4B_I$ for one spin channel and $M_I > 4B_I$ for the other spin channel. The two spin channels, thus, exhibit a topological and a normal insulator, respectively.

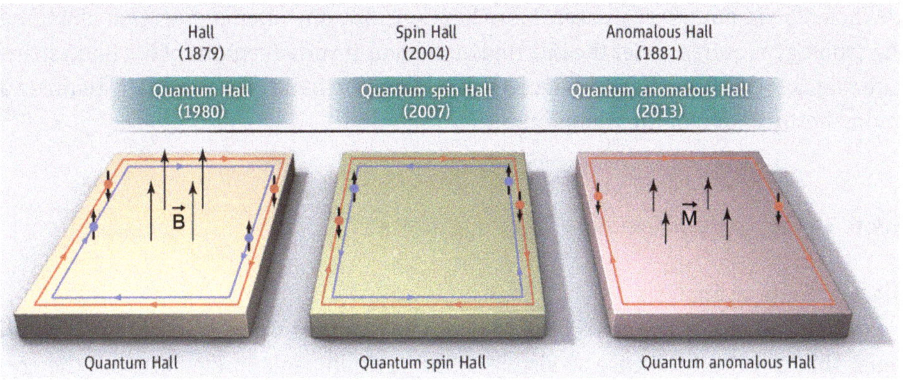

Fig. 5.17: Schematic of the three topological states corresponding to QHE, QSHE and QAHE. The edge states are marked in red and blue with corresponding propagation helicities indicated by in-plane arrows and spin directions of the moving electrons as out-of-plane arrows at the blue and red dots. $\vec{B}$ labels an external magnetic field and $\vec{M}$ an internal magnetization of the sample. The numbers in brackets above the sketches are years of experimental discovery of the corresponding effect. The upper row of terms names the classical and, thus, non-quantized effects with the year of their experimental discovery in brackets. After [132].

loy is a semiconductor again made of heavy elements. It provides a partially inverted band gap ($M_\mathrm{I} > 0$) and the Cr is doped into the system in order to make it ferromagnetic with a magnetization $\vec{M}$ pointing out-of-plane. The additional spin dependent exchange term influences $\widehat{H}_\downarrow(\vec{k})$ oppositely to $\widehat{H}_\uparrow(\vec{k})$.

Figure 5.17 shows a sketch of the three different topological effects discussed so far with the corresponding edge channels, its propagation helicity and its electron spin. These samples cover only two entrances in the topological table of Fig. 5.14, namely the $d = 2$ entrances of the classes A and AII. The QHE and the QAHE have both, neither time-reversal-symmetry, nor particle-hole-symmetry, nor chiral sublattice symmetry belonging to A, i.e., the conduction electrons do not care, if their spin splitting results from an external $\vec{B}$ field or from an exchange coupling.[40] Thus, theoretically, they belong to the same symmetry class, but experimentally it is a big advantage that one can use an intrinsic sample property instead of an external $\vec{B}$ field in order to get a quantized conductance. Some researchers indeed believe that the QAHE might replace the QHE as a resistance standard in the near future.

The robustness of the QSHE and the QAHE with respect to disorder can be shown by using similar arguments as for the quantum Hall effect via the perturbative introduction of disorder. For example, one could again use a description of the transport

---

40 There are other effects related to electron-electron interaction, that are different for exchange interaction and external $\vec{B}$ field, but the periodic table in Fig. 5.14 ignores the electron-electron interaction for the sake of simplicity. One can show that the QSHE and the QAHE are robust even with respect to electron-electron interaction as long as the interaction is not too strong with respect to the kinetic energy of the electrons ([128]): Chapter II.C.2).

coefficients via boundary phases. More heuristically, one often argues that changing the topology requires either the inverted or the non-inverted regions of the band structure (Fig. 5.16(b)) to exchange valence and conduction band, which usually requires a strong perturbation of the Hamiltonian.

5.4.6 3D topological insulators

5.4.6.1 General analysis

Next, one can ask, if the concept of topological numbers is transferable to 3D materials. This is indeed the case as visible in the periodic table in Fig. 5.14, but the corresponding derivation is more difficult than in the former cases. At the end, however, it again boils down to the question if band orders are inverted in some parts of the Brillouin zone?

Firstly, it is intuitively clear, that we have, at least, three different $\mathbb{Z}_2$ indices for the three linearly independent 2D planes in $\vec{k}$ space, e.g., the three planes perpendicular to the k_z, k_x, and k_y direction, that are independent by their linearly independent normal vectors. We can take, e.g, the planes with $k_z = \pi/a$, $k_x = \pi/a$, or $k_y = \pi/a$ for a cubic lattice with lattice constant a, i.e., the edge planes of the Brillouin zone. Each of these planes can be described by a $\mathbb{Z}_2$ index (Section 5.4.5). The $\mathbb{Z}_2$ index of $k_z = \pi/a$ and $k_z = -\pi/a$ are identical, since these are identical planes due to the lattice periodicity. This also applies for the planes at $k_x = \pi/a$ and $k_x = -\pi/a$ as well as for the planes at $k_y = \pi/a$ and $k_y = -\pi/a$. If the bands, additionally, do not touch on the way from $k_z = \pi/a$ to $k_z = -\pi/a$, all planes perpendicular to k_z have the identical $\mathbb{Z}_2$ index, since a change of $\mathbb{Z}_2$ requires a reinversion of the bands, i.e., a closing of the band gap. In this case, the 3D crystal can be regarded as stacks of 2DTIs, where their periodic distance induces the additional Brillouin zone direction k_z. The corresponding $\mathbb{Z}_2$ index marks, as in 2D, the SQHE. Hence, $\sigma_{xy\uparrow}$ and $\sigma_{xy\downarrow}$ are still quantized in each of the layers. The spin polarized 2D $\sigma_{xy\uparrow}$ is a multiple of e^2/h, which depends on the number of layers. For the 3D Hall conductivity of each spin channel, one gets $\sigma_{xy\uparrow} = j_x/E_y = \frac{e^2}{h\cdot a_z}$ with the lattice constant a_z in the z direction by multiplying e^2/h with the density of layers along the z direction. This provides the required different unit S/m reflecting that the current density in 3D is given in A/m^2, while in 2D, it is given in A/m. Thus, in principle, a 3D material consisting of stacked 2DTIs can also be used as an effective spin filter. The same analysis can be provided for the planes perpendicular to the x and y directions provided the bands around E_F do not (anti-)cross anywhere within the Brillouin zone.

The topological classification of 3D materials, is, however more complex, since, firstly, band crossings between parallel $\vec{k}$ planes in the Brillouin zone are not prohibited and, secondly, since there is an additional topological index combining the linearly independent planes, that turns out to be the most robust one. Thus, 3D materials exhibit four topological indices describing transport properties: the three indices describing the $\mathbb{Z}_2$ topology of the linearly independent planes and a general one describing the 3D crystal as a whole. To tackle this in more detail, a sketch of the general derivation of these indices will be firstly given. Afterwards, we will discuss an example Hamiltonian related to a real material that exhibits robust topological properties.

The sketch starts with a 2D material, before generalizing to 3D. Instead of calculating σ_{xy} in 2D as in the previous section, we ask directly, if edge states at E_F are necessarily present? Therefore, we regard a ribbon of an arbitrary crystalline material with finite width in x direction and infinite length in y direction. The question is, if one can transfer a spin from one edge to the other without adding energy? If this is possible by changing the Hamiltonian adiabatically, i.e., by avoiding excitations, on a loop, i.e., by ending with the same Hamiltonian as at the beginning, there is only one solution if the bulk is insulating: there must be edge states at E_F. These edge states can be differently occupied after the process without energy change. More concretely, a spin state at the Fermi level from the left edge of the ribbon has to be emptied in favor of the same spin state at the right edge again being at the Fermi level. If spin is moved without charge movement, the opposite spin has to be moved from the right edge to the left edge of the ribbon.

This means, we can prove the presence of edge states at E_F, if we find an adiabatic loop of the Hamiltonian, that pumps spins from left to right.

The following derivation shows that the possible spin transfer between the edges is a topological property of the bulk band structure of the 2D material. Thus, there is an index that tells us, if such a spin transfer is happening or not, and, thus, if edge states at E_F are necessarily present. *The derivation is quite intricate and does not immediately lead to a transparent understanding. Thus, it is not required to be read in order to understand the subsequent parts. It is, however, provided for more interested students with the clear recommendation to read it only after reading the summary of the calculation given after the bold text following eq.* (5.100). *Then, you can return to this point in order to follow the details of the calculation.*

The parameter of the Hamiltonian, that is tuned adiabatically in a loop, is k_y. It is tuned across the whole Brillouin zone, such that the starting point is identical to the end point.[41] We concentrate on rectangular lattices for the sake of simplicity, but the arguments can be transferred to other type of lattices, too.

The 1D Brillouin zone along k_y has two time reversal invariant points called TRIMs (time reversal invariant momenta), where $-k = k$. They are at the center of the Brillouin zone, where $k = 0/\text{m}$, and at the boundary, where the periodicity requires $k = -k$. In 1D, the states at these points in k space have to be twofold degenerate according to Kramers theorem.[42] In between, there is no requirement for such a spin degeneracy. In 2D, the degenerate states at the two lines in $\vec{k}$ space, where $k_y = -k_y$ can have differ-

41 We later show how that this can be realized within a physical system (Fig. 5.18(c) and (d)).
42 Time reversal inverts the spin and k. Hence, if $k = -k$, we have $E(k,\uparrow) \underset{\text{TRS}}{=} E(-k,\downarrow) \underset{k=-k}{=} E(k,\downarrow)$.

ent k_x values. But they are still forming pairs of states with the same k_y and the same energy.

We are interested in whether a spin is transported from one edge to the other in x direction, if we move k_y from the center ($k_y = 0/\mathrm{m}$) to the boundary ($k_y = \pi/a_y$) of the Brillouin zone.

For the calculation, we firstly return to an infinite 2D system. We start with calculating the movement of charge instead of spin, keeping in mind that spin movement is equivalent to movement of charge with different spins in opposite directions.[43]

Therefore, we define Wannier functions for each occupied band with band index n and each lattice position R_x in the x direction, such that we can track the position of occupied states,[44] i.e.,

$$|R_x, n\rangle_{k_y} = \frac{1}{2\pi} \int_{-\pi/a_x}^{\pi/a_x} \mathrm{e}^{\mathrm{i}k_x(R_x - x)} |u_{\vec{k}n}(\vec{x})\rangle \mathrm{d}k_x \tag{5.68}$$

with lattice constant a_x in the x direction and $\vec{k} = (k_x, k_y)$. By definition, the Wannier functions can look different for each k_y. One can define a charge polarization in the x direction for each k_y referring to the state at $R_x = 0\,\mathrm{m}$:

$$P_x(k_y) = \sum_n \langle 0, n|x|0, n\rangle_{k_y} . \tag{5.69}$$

This describes how far the center of mass of the charge at this k_y value is shifted with respect to $x = 0\,\mathrm{m}$. This charge polarization can be rewritten using $\int \mathrm{e}^{-\mathrm{i}k_1 x} \cdot \mathrm{e}^{\mathrm{i}k_2 x} \mathrm{d}x \propto \delta_{k_1 k_2}$ with $\delta_{k_1 k_2}$ being the Kronecker delta and the operator replacement $\hat{x} \to -\mathrm{i} \cdot \frac{\mathrm{d}}{\mathrm{d}k_x}$:[45]

$$P_x(k_y) = \frac{-1}{2\pi} \oint \mathrm{i} \sum_n \left\langle u_{\vec{k}n} \middle| \frac{\mathrm{d}}{\mathrm{d}k_x} \middle| u_{\vec{k}n} \right\rangle \mathrm{d}k_x = \frac{-1}{2\pi} \int_{-\pi/a_x}^{\pi/a_x} \mathrm{i} \sum_n \left\langle u_{\vec{k}n} \middle| \frac{\mathrm{d}}{\mathrm{d}k_x} \middle| u_{\vec{k}n} \right\rangle \mathrm{d}k_x . \tag{5.70}$$

Here $\oint$ is a closed integral due the periodic boundary conditions along k_x, i.e., $\pi/a_x = -\pi/a_x$. Next, we write the difference of the charge polarization at $k_y = 0/\mathrm{m}$ and $k_y = \pi/a_y$ with a_y being the lattice vector in the y direction.

$$\begin{aligned} &P_x(k_y = \pi/a_y) - P_x(k_y = 0) \\ &= \frac{-1}{2\pi}\left(\oint_{\pi/a_y} \mathrm{i} \sum_n \left\langle u_{\vec{k}n} \middle| \frac{\mathrm{d}}{\mathrm{d}k_x} \middle| u_{\vec{k}n} \right\rangle \mathrm{d}k_x - \oint_0 \mathrm{i} \sum_n \left\langle u_{\vec{k}n} \middle| \frac{\mathrm{d}}{\mathrm{d}k_x} \middle| u_{\vec{k}n} \right\rangle \mathrm{d}k_x \right). \end{aligned} \tag{5.71}$$

43 The derivation can be found in [133].

44 Wannier functions are used, e.g., as the starting point of tight binding calculations of periodic lattices (Section 4.5). They are centered at the lattice position R_x by the constructive overlap of waves with different wavelength at this position.

45 Small warning: Calculation of all steps in detail can be tedious.

This describes if charge is moved for the Wannier state at $R_x = 0\,\text{m}$ on the way from $k_y = 0/\text{m}$ to $k_y = \pi/a_y$. Since R_x is chosen arbitrarily, the result is identical for every lattice position, which are connected to each other by translational symmetry.[46]

Having the expression for the charge polarization, we can now turn to the so-called time reversal polarization. Time reversal symmetry implies that each wave function has a partner of opposite $\vec{k}$ and opposite spin with the same energy, the so called Kramers partner. We track the development of the charge polarization of such a degenerate Kramers pair relative to each other on the way from $k_y = 0/\text{m}$ to $k_y = \pi/a_y$. Thus, we divide the Bloch states at $k_y = 0/\text{m}$ and at $k_y = \pi/a_y$ in states I and II building Kramers pairs each, i.e., having opposite k_x and opposite spin.[47] If the two sectors I and II move in opposite x directions on its way from $k_y = 0/\text{m}$ to $k_y = \pi/a_y$, the charge polarization (effective position of the charge) does not change, but the spin polarization (effective position of the spin) does change, eventually implying movement of a spin from one edge to the other.

For completeness, we write the time reversal operator $\widehat{\widehat{\Theta}}$, connecting the partners of a Kramers pair, explicitly:

$$\widehat{\widehat{\Theta}} = \mathrm{e}^{\mathrm{i}\pi \widehat{S}_y/\hbar}\widehat{K} \tag{5.72}$$

with $\widehat{K}$ being the complex conjugation and $\widehat{S}_y$ being the spin operator. Importantly rotating the spin from $1/2$ to $-1/2$ changes the phase factor only by $\pi/2$. Consequently, rotating the spin from $1/2$ to $-1/2$ and back to $1/2$ rotates by π, such that $\widehat{\widehat{\Theta}}^2 = -1$.[48] The time reversal symmetry requires as a minimal condition for the pairs at $k_y = 0/\text{m}$ or $k_y = \pi/a_y$, i.e., on the time reversal invariant lines, where $\widehat{\widehat{\Theta}}$ does not change k_y:

$$\begin{aligned} \left|u^{\mathrm{I}}_{-k_x n}\right\rangle &= \mathrm{e}^{\mathrm{i}\Theta_{k_x n}}\widehat{\widehat{\Theta}}\left|u^{\mathrm{II}}_{k_x n}\right\rangle \\ \left|u^{\mathrm{II}}_{-k_x n}\right\rangle &= -\mathrm{e}^{\mathrm{i}\Theta_{-k_x n}}\widehat{\widehat{\Theta}}\left|u^{\mathrm{I}}_{k_x n}\right\rangle . \end{aligned} \tag{5.73}$$

The consistency of these equations can be verified by the replacing $-k_x \leftrightarrow k_x$ in the second line and inserting the result into the first line. Importantly, the description includes a possible arbitrary phase $\Theta_{k_x n}$, that does not influence the Kramers degeneracy, and will become decisive in the following.

With these relations, we can calculate the partial charge polarization for the sectors I and II. Starting with sector I, one defines:

$$A^{\mathrm{I}}_{k_y}(k_x) := \frac{1}{\mathrm{i}}\sum_n \left\langle u^{\mathrm{I}}_{\vec{k}n}\left|\frac{\mathrm{d}}{\mathrm{d}k_x}\right|u^{\mathrm{I}}_{\vec{k}n}\right\rangle \tag{5.74}$$

46 Formally, eq. (5.71) describes a boundary integral of the k space torus being closed in the periodic k_x direction (Fig. 5.13(c)). The integrations along k_y at $k_x = \pi/a_x$ and $k_x = -\pi/a_x$, that are required to form a closed path, follow the identical path in inverse directions due to periodicity and, thus, the two integrals cancel each other. The integral along a closed line could be the starting point to apply Stokes theorem, but we will not follow this path.

47 At these points $k_y = -k_y$, trivially for $k_y = 0/\text{m}$ and due to the lattice periodicity for $k_y = \pi/a_y$.

48 This observation is at the heart of the fermionic description of electrons.

and analogously for $A^{\mathrm{II}}_{k_y}(k_x)$. Notice, that the definition is very similar to the sum of the Berry connections $A_{\mathrm{Berry},n}(\vec{k})$ (eq. (5.44)) at $\vec{k} = (k_x, k_y)$ of all occupied bands n, but it is calculated separately for each partner sector of the Kramers pairs. With this definition, we get from eq. (5.70):

$$\begin{aligned} P^{\mathrm{I}}_x(k_y) &= \frac{1}{2\pi}\int_{-\pi/a_x}^{\pi/a_x} A^{\mathrm{I}}_{k_y}(k_x)\mathrm{d}k_x = \frac{1}{2\pi}\int_0^{\pi/a_x} A^{\mathrm{I}}_{k_y}(k_x) + A^{\mathrm{I}}_{k_y}(-k_x)\mathrm{d}k_x \\ P^{\mathrm{II}}_x(k_y) &= \frac{1}{2\pi}\int_{-\pi/a_x}^{\pi/a_x} A^{\mathrm{II}}_{k_y}(k_x)\mathrm{d}k_x = \frac{1}{2\pi}\int_0^{\pi/a_x} A^{\mathrm{II}}_{k_y}(k_x) + A^{\mathrm{II}}_{k_y}(-k_x)\mathrm{d}k_x\,. \end{aligned} \tag{5.75}$$

Next, we use $A^{\mathrm{I}}_{k_y}(-k_x) = A^{\mathrm{II}}_{k_y}(k_x) + \sum_n \frac{\mathrm{d}\Theta_{k_x n}}{\mathrm{d}k_x}$, that can be shown by inserting eq. (5.73) into eq. (5.74).[49] Defining $A_{k_y}(k_x) := A^{\mathrm{I}}_{k_y}(k_x) + A^{\mathrm{II}}_{k_y}(k_x)$, we get:

$$\begin{aligned} P^{\mathrm{I}}_x(k_y) &= \frac{1}{2\pi}\left(\int_0^{\pi/a_x} A_{k_y}(k_x)\mathrm{d}k_x + \sum_n \int_0^{\pi/a_x} \frac{\mathrm{d}\Theta_{k_x n}}{\mathrm{d}k_x}\mathrm{d}k_x\right) \\ &= \frac{1}{2\pi}\left(\int_0^{\pi/a_x} A_{k_y}(k_x)\mathrm{d}k_x + \sum_n (\Theta_{k_x=\pi/a_x,n} - \Theta_{k_x=0,n})\right). \end{aligned} \tag{5.76}$$

We have chosen $\Theta_{k_x n}$ such, that it changes continuously across the Brillouin zone. Importantly, the two remaining phases $\Theta_{k_x n}$ can be different, since the integral covers only half the Brillouin zone in x direction. Moreover, the two phases appear with opposite sign in $P^{\mathrm{II}}_x(k_y)$, that can be shown by repeating the calculation for $P^{\mathrm{II}}_x(k_y)$. The difference between the two partial polarizations $P^{\Delta}_x(k_y) = P^{\mathrm{I}}_x(k_y) - P^{\mathrm{II}}_x(k_y)$ can be calculated more elegantly by realizing that $P_x(k_y) = P^{\mathrm{I}}_x(k_y) + P^{\mathrm{II}}_x(k_y)$ (eq. (5.70)) leading to:

$$\begin{aligned} P^{\Delta}_x(k_y) &= 2\cdot P^{\mathrm{I}}_x(k_y) - P_x(k_y) \\ &= \frac{2}{2\pi}\left(\int_0^{\pi/a_x} A_{k_y}(k_x)\mathrm{d}k_x + \sum_n (\Theta_{k_x=\pi/a_x,n} - \Theta_{k_x=0,n})\right) \\ &\quad - \frac{1}{2\pi}\int_{-\pi/a_x}^{\pi/a_x} A_{k_y}(k_x)\mathrm{d}k_x \\ &= \frac{1}{2\pi}\left(\int_0^{\pi/a_x} A_{k_y}(k_x)\mathrm{d}k_x - \int_{-\pi/a_x}^{0} A_{k_y}(k_x)\mathrm{d}k_x + 2\sum_n (\Theta_{k_x=\pi/a_x,n} - \Theta_{k_x=0,n})\right). \end{aligned} \tag{5.77}$$

Obviously, the phase factors survive the final subtraction $P^{\mathrm{I}}_x(k_y) - P^{\mathrm{II}}_x(k_y)$.

49 We employ the sum rule for the derivative $d/\mathrm{d}k_x$, applied once to the phase factor $\mathrm{e}^{\mathrm{i}\Theta_{k_x n}}$ and once to $\widetilde{\Theta} u^{\mathrm{II}}_{k_x n}$, as well as $\langle \widetilde{\Theta} u^{\mathrm{II}}_{k_x n} | \frac{\mathrm{d}}{\mathrm{d}k_x} | \widetilde{\Theta} u^{\mathrm{II}}_{k_x n} \rangle = -\langle \widetilde{\Theta} u^{\mathrm{II}}_{k_x n} | \widetilde{\Theta} \frac{\mathrm{d}}{\mathrm{d}k_x} | u^{\mathrm{II}}_{k_x n} \rangle = \langle u^{\mathrm{II}}_{k_x n} | \frac{\mathrm{d}}{\mathrm{d}k_x} | u^{\mathrm{II}}_{k_x n} \rangle$.

In the following, we clarify that $P_x^{\Delta}(k_y)$, i.e., the difference in charge polarization between the two time-reversal partners, is either zero or one. Thus, it can topologically classify the time reversal invariant lines at $k_y = 0/\mathrm{m}$ and at $k_y = \pi/a_y$.

Therefore, we define the matrix $\mathbf{w}(k_x)$ with elements

$$w_{lm}(k_x) := \langle u_{-k_x l}|\widehat{\widetilde{\Theta}}|u_{k_x m}\rangle \ , \tag{5.78}$$

where l and m cover the occupied states from both sectors I and II reading $l, m \in (1\mathrm{I}, 1\mathrm{II}, 2\mathrm{I}, 2\mathrm{II}, 3\mathrm{I}, 3\mathrm{II}, \ldots)$ with $1, 2, 3, \ldots$ describing the different bands. Such a matrix exists for each k_y, which we drop as an index for the sake of simplicity. The matrix consists of antisymmetric blocks of the form:

$$\begin{pmatrix} 0 & \mathrm{e}^{\mathrm{i}\Theta_{k_x n}} \\ -\mathrm{e}^{\mathrm{i}\Theta_{-k_x n}} & 0 \end{pmatrix} \tag{5.79}$$

for the elements belonging to the indices

$$((l, m)) = \begin{pmatrix} (2n-1, 2n-1) & (2n-1, 2n) \\ (2n, 2n-1) & (2n, 2n) \end{pmatrix} . \tag{5.80}$$

This can be verified by inserting eq. (5.73) into eq. (5.78) and using $\widehat{\widetilde{\Theta}}^2 = -1$ as well as the orthogonality of the wave functions. The matrix does only contain the phase factors of Kramers pairs. At $k_x = 0/\mathrm{m}$ and $k_x = \pi/a_x$, where $k_x = -k_x$, the matrix is antisymmetric. At these points, one can define the so-called Pfaffian of the Matrix $\mathrm{Pf}(\mathbf{w})$, that exists for any antisymmetric $2N \times 2N$ matrix and is defined as

$$\mathrm{Pf}(\mathbf{w}) := \frac{1}{2^N N!} \sum_{\tilde{\sigma} \in S_{2N}} \mathrm{sgn}(\tilde{\sigma}) \prod_{n=1}^{N} w_{\tilde{\sigma}(2n-1)\tilde{\sigma}(2n)} \tag{5.81}$$

with $\tilde{\sigma}$ marking all possible permutations between the $2N$ elements and $\mathrm{sgn}(\tilde{\sigma})$ being -1 $(+1)$ for an odd (even) number of permutations. Thus, the product without permutation, i.e., with elements $w_{(2n-1)(2n)}$, is the line above the diagonal of the matrix tracing any second element (eq. (5.79)). One can show that this is the only summand of the Pfaffian, if the only non-zero elements of the matrix are in the lines above and below the diagonal.[50] Thus, we get

$$\mathrm{Pf}(\mathbf{w}(k_x = 0)) = \prod_{n=1}^{N} \exp\left(\mathrm{i}\Theta_{0n}\right) \tag{5.82}$$

$$\mathrm{Pf}(\mathbf{w}(k_x = \pi/a_x)) = \prod_{n=1}^{N} \exp\left(\mathrm{i}\Theta_{\pi/a_x n}\right) . \tag{5.83}$$

50 See Wikipedia: Pfaffian.

Notice that both Pfaffians are complex numbers with absolute value $|\mathrm{Pf}(\boldsymbol{w})| = 1$. Inserting the Pfaffians into eq. (5.76) results in

$$P_x^{\mathrm{I}}(k_y) = \frac{1}{2\pi}\left(\int_0^{\pi/a_x} A(k_x)\mathrm{d}k_x + \mathrm{i}\cdot\ln\left(\frac{\mathrm{Pf}(\boldsymbol{w}(k_x = \pi/a_x))}{\mathrm{Pf}(\boldsymbol{w}(k_x = 0))}\right)\right) \tag{5.84}$$

and, hence, we find (eq. (5.77)):

$$P_x^{\Delta}(k_y) = \frac{1}{2\pi}\left(\int_0^{\pi/a_x} A(k_x)\mathrm{d}k_x - \int_{-\pi/a_x}^{0} A(k_x)\mathrm{d}k_x + 2\mathrm{i}\cdot\ln\left(\frac{\mathrm{Pf}(\boldsymbol{w}(k_x = \pi/a_x))}{\mathrm{Pf}(\boldsymbol{w}(k_x = 0))}\right)\right). \tag{5.85}$$

One can show that $\mathrm{Pf}(\boldsymbol{w})^2 = \mathrm{Det}(\boldsymbol{w})$ for any matrix, where the Pfaffian is defined.[51] This leaves two possible signs of the Pfaffian with respect to the determinant. These different signs turn out to be the decisive distinction between the two topological classes in 2D: trivial insulator and topological insulator.

The remaining task is now to show that the first two terms are related to the determinant of the matrix $\boldsymbol{w}$. Therefore, we firstly rewrite the first two terms of eq. (5.85) in terms of the matrix $\boldsymbol{w}$ (eq. (5.78)). It is obvious from the definition of $A(k_x)$ (eq. (5.74)) that we need the derivative of w:

$$\frac{\mathrm{d}w_{ml}}{\mathrm{d}k_x} = \left\langle \frac{\mathrm{d}u_{-k_x m}}{\mathrm{d}k_x}\middle|\widehat{\widehat{\Theta}}\middle|u_{k_x l}\right\rangle - \left\langle u_{-k_x m}\middle|\widehat{\widehat{\Theta}}\middle|\frac{\mathrm{d}u_{k_x l}}{\mathrm{d}k_x}\right\rangle . \tag{5.86}$$

The minus sign comes from the fact that $\widehat{\widehat{\Theta}}$ and $\frac{\mathrm{d}}{\mathrm{d}k_x}$ anti-commute, since $\frac{\mathrm{d}}{\mathrm{d}k_x}$ corresponds to $\mathrm{i}\cdot\hat{x}$ in real space and $\widehat{\widehat{\Theta}}$ includes complex conjugation K or, since time reversal means that the momentum k must point in the opposite direction. For the second term on the right, one can get rid of the $\langle u_{-k_x m}|$ by multiplying with $\boldsymbol{w}^\dagger$ according to:

$$\begin{aligned}\left(\boldsymbol{w}^\dagger\cdot\frac{\mathrm{d}\boldsymbol{w}}{\mathrm{d}k_x}\right)_{ol,2} &= \sum_m \langle u_{k_x o}|\widehat{\widehat{\Theta}}^\dagger|u_{-k_x m}\rangle\left\langle u_{-k_x m}\middle|\widehat{\widehat{\Theta}}\middle|\frac{\mathrm{d}u_{k_x l}}{\mathrm{d}k_x}\right\rangle \\ &= \left\langle u_{k_x o}\middle|\widehat{\widehat{\Theta}}^\dagger\widehat{\widehat{\Theta}}\middle|\frac{\mathrm{d}u_{k_x l}}{\mathrm{d}k_x}\right\rangle = \left\langle u_{k_x o}\middle|\frac{\mathrm{d}u_{k_x l}}{\mathrm{d}k_x}\right\rangle ,\end{aligned} \tag{5.87}$$

where the 2 in the index on the left marks that only the second summand from eq. (5.86) is displayed. We used again the completeness relation $\sum_m |u_{-k_x m}\rangle\langle u_{-k_x m}| = \mathbf{1}$ as well as $\widehat{\widehat{\Theta}}^\dagger = -\widehat{\widehat{\Theta}}$ and $\widehat{\widehat{\Theta}}^2 = -1$.[52] This already resembles the definition of $A(k_x)$ in eq. (5.74). In more detail, we have to take the trace of the matrix $\boldsymbol{w}^\dagger\cdot\frac{\mathrm{d}\boldsymbol{w}}{\mathrm{d}k_x}$ multiplied with the imaginary $1/\mathrm{i}$.

$$A(k_x) = \frac{1}{\mathrm{i}}\mathrm{Tr}\left(\boldsymbol{w}^\dagger\cdot\frac{\mathrm{d}\boldsymbol{w}}{\mathrm{d}k_x}\right)_2 . \tag{5.88}$$

51 see Wikipedia: Pfaffian. Recall that the determinant of a $N\times N$ matrix is given by $\mathrm{Det}(\boldsymbol{w}) = \sum_{\tilde{\sigma}}\mathrm{sgn}(\tilde{\sigma})\prod_{n=1}^{N} w_{n\tilde{\sigma}(n)}$.

52 The first equation becomes obvious considering that complex conjugation gets reversed by † and also the spin operator gets reversed (see eq. (5.72)).

Using the trace directly, we can get also the first term on the right hand side of eq. (5.86) into this form as shown in the following. Therefore, we use $\frac{\mathrm{d}}{\mathrm{d}k_x}\langle u_{k_x m}|\widetilde{\widetilde{\Theta}}|u_{k_x m}\rangle = 0$, that is an obvious consequence of orthogonality, and the fact that $\widetilde{\widetilde{\Theta}}|u_{k_x m}\rangle$ is also an eigenfunction due to time reversal symmetry. Moreover, we apply the product rule of derivatives, the clever inclusion of a **1**, and the same properties of $\widetilde{\widetilde{\Theta}}$ as above. We obtain

$$\begin{aligned}
\operatorname{Tr}\left(\boldsymbol{w}^\dagger \cdot \frac{\mathrm{d}\boldsymbol{w}}{\mathrm{d}k_x}\right)_1 &= \sum_{n,m} \left\langle u_{k_x n}\middle|\widetilde{\widetilde{\Theta}}^\dagger\middle|u_{-k_x m}\right\rangle \left\langle \frac{\mathrm{d}u_{-k_x m}}{\mathrm{d}k_x}\middle|\widetilde{\widetilde{\Theta}}\middle|u_{k_x n}\right\rangle \\
&= \sum_{l,n,m} \left\langle u_{k_x n}\middle|\widetilde{\widetilde{\Theta}}^\dagger\middle|u_{-k_x m}\right\rangle \left\langle \frac{\mathrm{d}u_{-k_x m}}{\mathrm{d}k_x}\middle|u_{-k_x l}\right\rangle \left\langle u_{-k_x l}\middle|\widetilde{\widetilde{\Theta}}\middle|u_{k_x n}\right\rangle \\
&= \sum_{l,n,m} \left\langle u_{-k_x l}\middle|\widetilde{\widetilde{\Theta}}\middle|u_{k_x n}\right\rangle \left\langle u_{k_x n}\middle|\widetilde{\widetilde{\Theta}}^\dagger\middle|u_{-k_x m}\right\rangle \left\langle \frac{\mathrm{d}u_{-k_x m}}{\mathrm{d}k_x}\middle|u_{-k_x l}\right\rangle \\
&= \sum_{l,m} \left\langle u_{-k_x l}\middle|\widetilde{\widetilde{\Theta}}\widetilde{\widetilde{\Theta}}^\dagger\middle|u_{-k_x m}\right\rangle \left\langle \frac{\mathrm{d}u_{-k_x m}}{\mathrm{d}k_x}\middle|u_{-k_x l}\right\rangle \\
&= \sum_{l,m} \langle u_{-k_x l}|u_{-k_x m}\rangle \left\langle \frac{\mathrm{d}u_{-k_x m}}{\mathrm{d}k_x}\middle|u_{-k_x l}\right\rangle \\
&= -\sum_{l,m} \langle u_{-k_x l}|u_{-k_x m}\rangle \left\langle u_{-k_x m}\middle|\frac{\mathrm{d}u_{-k_x l}}{\mathrm{d}k_x}\right\rangle \\
&= -\sum_{l} \left\langle u_{-k_x l}\middle|\frac{\mathrm{d}u_{-k_x l}}{\mathrm{d}k_x}\right\rangle = -\mathrm{i}\cdot A(-k_x)\,.
\end{aligned}$$

The index 1 means that we use only the first term of eq. (5.86) for the derivative. Hence, we have, using both terms of eq. (5.86):

$$\operatorname{Tr}\left(\boldsymbol{w}^\dagger \cdot \frac{\mathrm{d}\boldsymbol{w}}{\mathrm{d}k_x}\right) = \mathrm{i}(A(k_x) - A(-k_x))\,. \tag{5.89}$$

Thus, we reached our goal to replace the first two terms of eq. (5.85) by an expression of $\boldsymbol{w}$, that now reads

$$P_x^\Delta = \frac{-\mathrm{i}}{2\pi}\left(\int_0^{\pi/a_x} \operatorname{Tr}\left(\boldsymbol{w}^\dagger \cdot \frac{\mathrm{d}\boldsymbol{w}}{\mathrm{d}k_x}\right)\mathrm{d}k_x - 2\cdot\ln\left(\frac{\operatorname{Pf}(\boldsymbol{w}(k_x=\pi/a_x))}{\operatorname{Pf}(\boldsymbol{w}(k_x=0))}\right)\right). \tag{5.90}$$

The next task is to rewrite the new term as an expression of determinants. Therefore, we need a few rules for determinants and for the derivative of determinants. The most simple ones are

$$\begin{aligned}
\operatorname{Det}(\boldsymbol{A}\cdot\boldsymbol{B}) &= \operatorname{Det}(\boldsymbol{A})\cdot\operatorname{Det}(\boldsymbol{B}) \\
\operatorname{Det}(\mathbf{1}+\tilde{\varepsilon}\cdot\boldsymbol{B}) &= 1+\tilde{\varepsilon}\cdot\operatorname{Tr}(\boldsymbol{B})+\tilde{\varepsilon}^2\ldots
\end{aligned} \tag{5.91}$$

The latter is valid for small $\tilde{\varepsilon}$. Now we derive the derivative of a determinant explicitly:

$$\frac{\mathrm{d}\operatorname{Det}(\boldsymbol{w}(k_x))}{\mathrm{d}k_x} = \lim_{\tilde{\varepsilon}\to 0}\frac{\operatorname{Det}(\boldsymbol{w}(k_x+\tilde{\varepsilon})) - \operatorname{Det}(\boldsymbol{w}(k_x))}{\tilde{\varepsilon}}\,. \tag{5.92}$$

With $\boldsymbol{w}(k_x+\tilde{\varepsilon}) = \boldsymbol{w}(k_x) + \tilde{\varepsilon}\cdot\frac{\mathrm{d}\boldsymbol{w}(k_x)}{\mathrm{d}k_x}$ and $\boldsymbol{B} := \frac{\mathrm{d}\boldsymbol{w}(k_x)}{\mathrm{d}k_x}$ one gets

$$\begin{aligned}\mathrm{Det}(\boldsymbol{w}(k_x)+\tilde{\varepsilon}\cdot\boldsymbol{B}) &= \mathrm{Det}(\boldsymbol{w}(k_x)\boldsymbol{w}(k_x)^{-1}(\boldsymbol{w}(k_x)+\tilde{\varepsilon}\cdot\boldsymbol{B}))\\ &= \mathrm{Det}(\boldsymbol{w}(k_x))\,\mathrm{Det}(\boldsymbol{w}(k_x)^{-1}(\boldsymbol{w}(k_x)+\tilde{\varepsilon}\cdot\boldsymbol{B}))\\ &= \mathrm{Det}(\boldsymbol{w}(k_x))\,\mathrm{Det}(\boldsymbol{1}+\tilde{\varepsilon}\cdot\boldsymbol{w}(k_x)^{-1}\cdot\boldsymbol{B})\\ &= \mathrm{Det}(\boldsymbol{w}(k_x))\cdot(1+\tilde{\varepsilon}\cdot\mathrm{Tr}(\boldsymbol{w}(k_x)^{-1}\cdot\boldsymbol{B}))\end{aligned}$$

with $\boldsymbol{1}$ being the identity matrix. This leaves us with

$$\frac{\mathrm{d}\,\mathrm{Det}(\boldsymbol{w}(k_x))}{\mathrm{d}k_x} = \mathrm{Det}(\boldsymbol{w}(k_x))\cdot\mathrm{Tr}\left(\boldsymbol{w}(k_x)^{-1}\cdot\frac{\mathrm{d}\boldsymbol{w}(k_x)}{\mathrm{d}k_x}\right). \tag{5.93}$$

Finally, we get rid of the first determinant by the derivative of the natural logarithm:

$$\begin{aligned}\frac{\mathrm{d}\ln(\mathrm{Det}(\boldsymbol{w}(k_x)))}{\mathrm{d}k_x} &= \frac{1}{\mathrm{Det}(\boldsymbol{w}(k_x))}\cdot\mathrm{Det}(\boldsymbol{w}(k_x))\cdot\mathrm{Tr}\left(\boldsymbol{w}(k_x)^{-1}\cdot\frac{\mathrm{d}\boldsymbol{w}(k_x)}{\mathrm{d}k_x}\right)\\ &= \mathrm{Tr}\left(\boldsymbol{w}(k_x)^{-1}\cdot\frac{\mathrm{d}\boldsymbol{w}(k_x)}{\mathrm{d}k_x}\right).\end{aligned} \tag{5.94}$$

Using $\boldsymbol{w}(k_x)^{-1} = \boldsymbol{w}(k_x)^{\dagger}$, that can be verified by inserting eq. (5.79), we see that we can replace the trace term in eq. (5.90) by a determinant and can perform the integration such that

$$P_x^{\Delta}(k_y) = \frac{-\mathrm{i}}{2\pi}\left[\int_0^{\pi/a_x}\frac{\mathrm{d}\ln(\mathrm{Det}(\boldsymbol{w}(k_x)))}{\mathrm{d}k_x}\mathrm{d}k_x - 2\cdot\ln\left(\frac{\mathrm{Pf}(\boldsymbol{w}(k_x=\pi/a_x))}{\mathrm{Pf}(\boldsymbol{w}(k_x=0))}\right)\right] \tag{5.95}$$

$$= \frac{-\mathrm{i}}{2\pi}\left[\ln\left(\frac{\mathrm{Det}(\boldsymbol{w}(k_x=\pi/a_x))}{\mathrm{Det}(\boldsymbol{w}(k_x=0))}\right) - 2\cdot\ln\left(\frac{\mathrm{Pf}(\boldsymbol{w}(k_x=\pi/a_x))}{\mathrm{Pf}(\boldsymbol{w}(k_x=0))}\right)\right] \tag{5.96}$$

$$= \frac{-\mathrm{i}}{\pi}\ln\left(\frac{\pm\sqrt{\mathrm{Det}(\boldsymbol{w}(k_x=\pi/a_x))}}{\pm\sqrt{\mathrm{Det}(\boldsymbol{w}(k_x=0))}}\,\frac{\mathrm{Pf}(\boldsymbol{w}(k_x=0))}{\mathrm{Pf}(\boldsymbol{w}(k_x=\pi/a_x))}\right). \tag{5.97}$$

Thus, we are at the desired stage, where the so-called time reversal polarization[53] P_x^{Δ} at either $k_y = 0/\mathrm{m}$ or at $k_y = \pi/a_y$ depends only on Pfaffians and determinants. Moreover, these factors include only the phase factors $\Theta_{k_x n}$ of the Kramers pairs at the two time-reversal invariant momenta (TRIM) in k_x direction ($k_x = 0/\mathrm{m}$, $k_x = \pi/a_x$). Recall that our calculation was performed for $k_y = 0/\mathrm{m}$ or $k_y = \pi/a_y$ only, since otherwise the degeneracy leading to eq. (5.73) would not hold. Thus, only the phase vectors of the Kramers partners at the four 2D TRIMS matter.

Recall, moreover, that the Pfaffians are complex numbers with absolute value 1 (eq. (5.82)) and, by $\mathrm{Det}(\boldsymbol{w}) = \mathrm{Pf}(\boldsymbol{w})^2$, the determinants are complex numbers with absolute value 1, too. This implies that the natural logarithm (ln) only measures the

53 It is not called spin polarization, since spin is not a good quantum number in the presence of spin-orbit interaction. The two partners are related by time reversal, hence the name.

phase of its argument (multiplied by i), that itself is only defined modulo 2π, such that the possible distinct values are $P_x^\Delta \in [0, 2)$.

Moreover, we find from the relations between Pfaffians and determinants

$$\frac{\pm\sqrt{\mathrm{Det}(\mathbf{w}(k_x = \pi/a_x))}}{\mathrm{Pf}(\mathbf{w}(k_x = \pi/a_x))} = \pm 1$$

and

$$\frac{\pm\sqrt{\mathrm{Det}(\mathbf{w}(k_x = 0))}}{\mathrm{Pf}(\mathbf{w}(k_x = 0))} = \pm 1 .$$

In complex notation, we have $+1 = e^{i0}$ and $-1 = e^{i\pi}$. Consequently, there are only two distinct results for P_x^Δ: zero or one. The question, if there is a charge polarization in opposite directions for the two partners of a Kramers pair can be distinctly answered by yes (one) or no (zero).

Of course, the branches of the square root (+ or −) have to be chosen. But if the branches are chosen to be continuous on the way from $k_x = 0/\mathrm{m}$ to $k_x = \pi/a_x$, the question is only, if, for a distinct k_y, the two fractions between determinant and Pfaffian at the two relevant k_x points are equal or if they are different.

Thus, we can rewrite the distinction between the two possible polarizations as:

$$(-1)^{P_x^\Delta(k_y)} = \frac{\sqrt{\mathrm{Det}(\mathbf{w}(k_x = \pi/a_x))} \cdot \mathrm{Pf}(\mathbf{w}(k_x = 0))}{\mathrm{Pf}(\mathbf{w}(k_x = \pi/a_x)) \cdot \sqrt{\mathrm{Det}(\mathbf{w}(k_x = 0))}} \tag{5.98}$$

$$= \frac{\sqrt{\mathrm{Det}(\mathbf{w}(k_x = \pi/a_x))} \cdot \sqrt{\mathrm{Det}(\mathbf{w}(k_x = 0))}}{\mathrm{Pf}(\mathbf{w}(k_x = \pi/a_x)) \cdot \mathrm{Pf}(\mathbf{w}(k_x = 0))} . \tag{5.99}$$

It turns out that only the distinction between the two numbers for P_x^Δ is unique, but not its absolute value, which depends on the chosen gauge of the wave functions (without argument)[54]. However, the distinction is enough to distinguish, if the time reversal polarization $P_x^\Delta(k_y)$ is changed on the way from $k_y = 0/\mathrm{m}$ to $k_y = \pi/a_y$ or not. If it is changed, one partner of each Kramers pair is moved to the right as much as the other partner is moved to the left. This is displayed in Fig. 5.19(b).[55] Thus, a gauge-invariant topological index ν_0 reads:

$$(-1)^{\nu_0} = \prod_{i=1}^{4} \frac{\sqrt{\mathrm{Det}(\mathbf{w}(\Gamma_i))}}{\mathrm{Pf}(\mathbf{w}(\Gamma_i))} \tag{5.100}$$

using all four non-equivalent TRIMs Γ_i as marked in Fig. 5.19(a). Notice, that the whole topological information, if polarization is changed or not on the way from $k_y = 0/\mathrm{m}$

54 Heuristically, one can choose the original partners either at the $(k_y = 0)$ line or at the $(k_y = \pi/a)$ line before asking if they change their partner towards the other TRIM line.

55 Recall that Kramers pairs are required at $k_y = 0/\mathrm{m}$ and $k_y = \pi/a_y$ albeit with opposite k_x values. By construction of the Wannier functions (eq. (5.68)), this implies identical lattice positions for two states with opposite spin.

to $k_y = \pi/a_y$, is encoded in the occupied wave functions at the TRIMs of the Brillouin zone Γ_i, where $-\vec{k} = \vec{k}$. Formally, one can attribute an index δ_i to each TRIM i being $\delta_i = \frac{\sqrt{\mathrm{Det}(\mathbf{w}(\Gamma_i))}}{\mathrm{Pf}(\mathbf{w}(\Gamma_i))}$ and multiply these indices to get the topological invariant ν_0.

This is extremely convenient and gets even more convenient, if the system is inversion symmetric. Then, the sign of $\delta_i = \frac{\sqrt{\mathrm{Det}(\mathbf{w}(\Gamma_i))}}{\mathrm{Pf}(\mathbf{w}(\Gamma_i))}$ is simply the product of the parity of all occupied wave functions at the TRIM[56] (see below).

To make our argument complete in terms of an adiabatic change of the Hamiltonian along a closed loop, we have to move on from $k_y = \pi/a_y$ to $k_y = 2\pi/a_y$ such that the complete movement ends up with identical wave functions. This second movement is identical to a movement from $k_y = -\pi/a_y$ to $k_y = 0/\mathrm{m}$. Due to time reversal symmetry, the states running from $k_y = 0/\mathrm{m}$ to $k_y = \pi/a_y$ must run exactly the same way as states with opposite spin from $k_y = 0/\mathrm{m}$ to $k_y = -\pi/a_y$, respectively from $k_y = 2\pi/a_y$ to $k_y = \pi/a_y$. As sketched in Fig.5.19(b), this leads to an identical movement from $k_y = \pi/a_y$ to $k_y = 2\pi/a_y$ as from $k_y = 0$ to $k_y = \pi/a_y$. Thus, the complete movement moves all spins by two lattice constants in opposite directions, i.e., it returns the system to its original time reversal polarization defined modulo 2. But, if we cut a finite ribbon out of the material, spin is effectively moved from left to right in the x direction, since the most left solid lines and the most right dashed lines in Fig. 5.19(b) cannot join with a partner from the right or the left, respectively. Simultaneously, the wave functions of the ribbon must be exactly the same by changing k_y adiabatically in one closed cycle.

Thus, we have the situation, which was needed for spin polarized edge states at the Fermi level at both sides of the insulating ribbon. We repeat that spin movement from the left to the right, without energy penalty, requires states at E_F, which are not available in the insulating bulk. Since the calculation is based on occupied wave functions of the bulk only (The matrix $\mathbf{w}(k_x)$ in eq. (5.79) is defined in terms of occupied bulk states), the edges states are a bulk property.

Since this property is not changed by moving E_F within the bulk band gap, there have to be edge states at each energy within the bulk band gap. Consequently, the edge states have to connect conduction band and valence band as shown in Fig. 5.18(b).

Here, we readopt the majority of students with a short summary of what has been derived. Firstly, we have defined pairs of states that are time reversal partners, i.e., we get the second state of the partner from the first state by applying the time reversal operator, that changes the wave vector $\vec{k}$ to $-\vec{k}$ and rotates the spin from $\uparrow$ to $\downarrow$ or vice versa. These states are called Kramers pairs. We have considered the Kramers pairs at $k_y = 0/\mathrm{m}$ and $k_y = \pi/a_y$, since at these points one gets $k_y = -k_y$. This implies that the states are spin degenerate via

$$E(k_y, \uparrow) \underset{\mathrm{TRS}}{=} E(-k_y, \downarrow) \underset{k_y=-k_y}{=} E(k_y, \downarrow) \,.$$

56 The parity is +, if the wave function $\psi_i(\vec{x})$ does not change sign after the inversion operation and –, if the wave function changes sign.

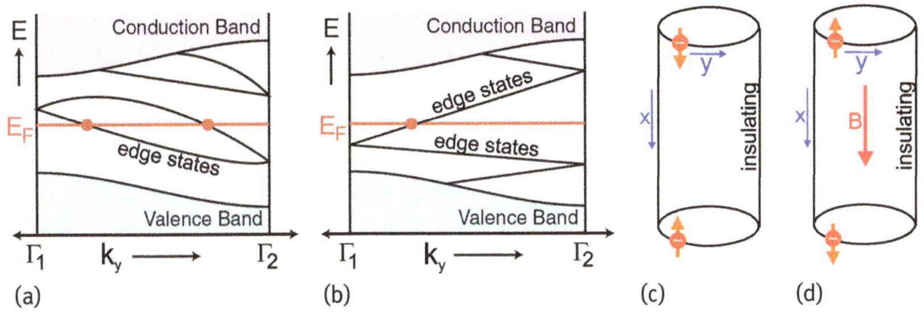

Fig. 5.18: (a), (b) One dimensional $E(k_y)$ band structures along the edge direction of a 2D ribbon of finite width in x direction and infinite length in y direction. The two topologically distinct cases of edge states (black lines) are displayed for $\nu_0 = 0$ (a) and $\nu_0 = 1$ (b). The blue and pink areas mark the projected band structure of the 2D bulk to the edge direction. Γ_1 and Γ_2 mark the time-reversal invariant points (TRIMs) where states have to be, at least, twofold degenerate. The edge states in (b) cover the whole bulk band gap as required for topologically protected edge states, while the edge states in (a) only cover part of the band gap and can not be topological. Moreover, the edge state bands in (a) exhibit two crossings at E_F and only one crossing in (b) (red points). (c), (d) Sketch of the hollow cylinder geometry without and with pierced $\vec{B}$ field in order to physically realize the spin pump. The spin exchange at the edges is realized by adding two flux quanta to the insulating cylinder with $\nu_0 = 1$. (a)–(b) [134].

Secondly, we have calculated, via construction of Wannier states in x direction (Section 4.5), if the charge of the two partners of the pair is moved in opposite directions along x, if we proceed our Hamiltonian from $k_y = 0/\mathrm{m}$ to $k_y = \pi/a_y$. Such a movement is displayed in Fig. 5.19(b). It turns out that there are only two mathematically distinct values for this movement, which are either zero or one. These numbers are the topological $\mathbb{Z}_2$ index ν_0. If $\nu_0 = 0$, the charges of the two partners reunite at $k_y = \pi/a_y$. If $\nu_0 = 1$, the charges of the partners move in opposite directions and they join with other partners at $k_y = \pi/a_y$ as sketched in Fig. 5.19(b). This implies, for a finite ribbon in x direction, that spin is moved effectively from the left to the right.[57]

Thirdly, we have shown that, if the Kramers pair is moved in opposite directions from $k_y = 0/\mathrm{m}$ to $k_y = \pi/a_y$, then it keeps moving in opposite directions from $k_y = \pi/a_y$ to $k_y = 2\pi/a_y$. This results in no spin movement from $k_y = 0/\mathrm{m}$ to $k_y = 2\pi/a_y$ for an extended 2D system, since a movement by two lattice constants is mathematically identical to a movement by zero lattice constants. This sounds physically reasonable, since we move between nominally identical points in $\vec{k}$ space. Mathematically, it is due to the fact that one has a gauge freedom, that makes the movement by two lattice constants identical to the one without movement. However, if we cut out a ribbon with finite width in x direction, the moved charges at the edge of the ribbon can not return to a partner, such that the movement by two lattice constants is not identical to the movement by zero lattice constants. Within a ribbon, spin is moved from left

57 All pairs at different atomic sites have to move identically due to lattice periodicity.

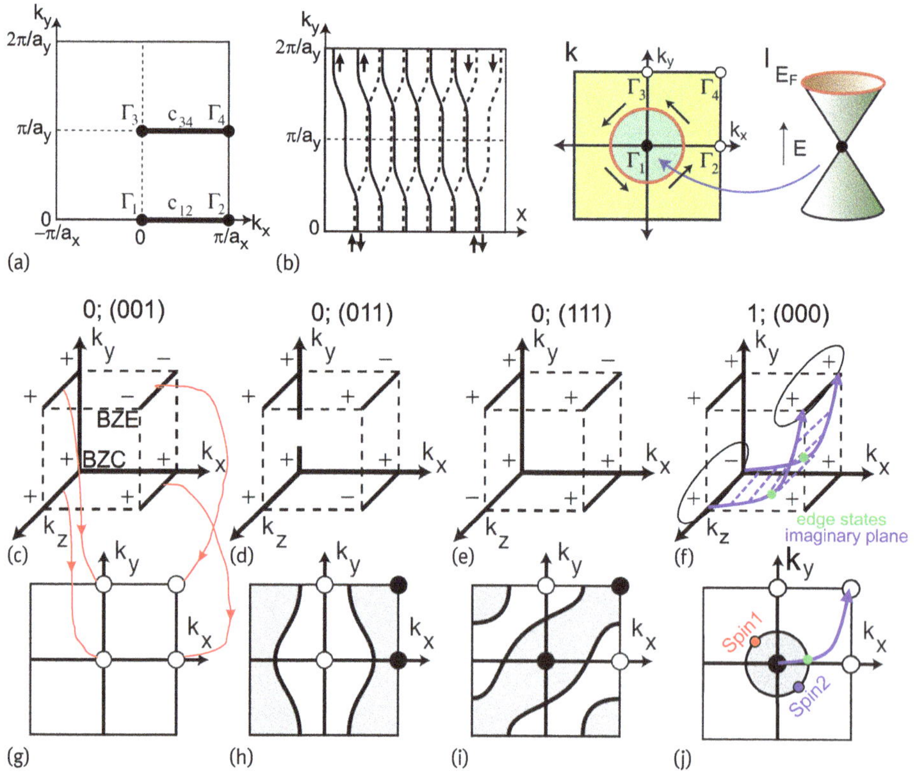

Fig. 5.19: (a) Sketch of a 2D Brillouin zone with the four time reversal invariant momenta (TRIMs) Γ_i ($i = 1, 2, 3, 4$) marked as used in eq. (5.100). The thick lines c_{12} and c_{34} are used to calculate the time reversal polarization $P_x^{\Delta}(k_y)$ for the two time reversal invariant k_y values: $k_y = 0/\mathrm{m}$, $k_y = \pi/a_y$. (b) Sketch of the changing polarization of Wannier states after separating them into two Kramers partners (dashed and solid lines marked by opposite spin arrows). The states switch partners on the way from $k_y = 0/\mathrm{m}$ to $k_y = \pi/a_y$ and again on the way from $k_y = \pi/a_y$ to $k_y = 2\pi/a_y$. (c)–(f) Sketches of a 3D Brillouin zone (Brillouin zone center (BZC) and Brillouin zone corner (BZE) are marked in (c)) with signs of δ_i at the TRIMs marked for a particular gauge choice. Pairs of TRIMs used for the projection in (g)–(j) are connected by thick lines. The topological indices $\nu_0; (\nu_x \nu_y \nu_z)$ are given above the images. In (f), a hypothetical 2D plane (blue), with edges in the front and back surface of the cube, is additionally drawn with the relevant time-reversal invariant lines encircled. The fact that the product of the TRIM states of these lines changes sign between the two lines requires spin polarized edge states at E_F as marked by green dots at an arbitrary (k_x, k_y) value. (g)–(j) Projected surface Brillouin zone in the k_z direction (front plane in (c)–(f)) with white (black) circles marking a + (-) for the product of the δ_i connected by the thick lines in (c)–(f). The connection is additionally marked by the red, curved arrows between (c) and (g). Thick black lines separating grey and white areas mark the necessarily present Fermi lines of surface states. These surface states at E_F have to exist at each path from black to white dots. The projected blue plane of (f) is shown in (j) (blue line) with resulting edge state (green dot) marked. Grey (white) areas indicate the resulting areas of occupied (empty) surface states. The inverted spin directions required by time reversal symmetry are marked in (j). (k) Surface Brillouin zone identical to (j) with red Fermi line and occupied (empty) states marked in dark (bright) green. The black arrows mark a possible spin direction of the states at the Fermi line. (l) Corresponding $E(\vec{k})$ of the surface state called the Dirac cone with the marked Dirac point (black dot) and Fermi circle (red). (a)–(b) [133], (c)–(j) [135], (k)–(l) [134].

to right by going from $k_y = 0/\mathrm{m}$ to the nominal identical $k_y = 2\pi/a_y$. This is exactly the situation discussed above, that requires states of both spins at both edges of the ribbon at E_F, if the Fermi level is within a bulk band gap. The edge states at E_F are required to enable that spin is moved from left to right without an energy penalty.[58] Hence, $\nu_0 = 1$ requires edge states for both spins at both edges.

One can show that ν_0 is identical to the spin dependent Chern index discussed with respect to 2DTIs via eq. (5.67). But since we would have to change from the description in eq. (5.67) to a more abstract one, we do not show this explicitly.[59]

A more intuitive understanding can be gained via the Boltzmann relaxation model of electric current. An electric field E_y implies a change of the k_y value of all occupied states according to $eE_y = \hbar\dot{k}_y$. For a full band, this does not lead to a current in the y direction, since the global occupation of the band does not change, if all states are moved forward by Δk_y due to the periodic boundary conditions. However, if spin is moved simultaneously in the perpendicular direction to k_y as required by $\nu_0 = 1$, a transversal spin current results, hence $\sigma_{xy\uparrow} \neq 0\,\mathrm{S}$. Quantitatively, one can use $E_y = \hbar\dot{k}_y/e = \hbar\Delta k_y/e\Delta t$ with $\Delta k_y = 2\pi/a_y$ for the full adiabatic cycle described above. This transports one spin $\uparrow$ from left to right and one spin $\downarrow$ from right to left per unit cell in y direction,[60] such that two extra spins appear at each side as required by Fig. 5.19(b). Hence,

$$\begin{aligned} I_\uparrow &= -I_\downarrow = \frac{e}{\Delta t} \\ \Rightarrow \quad j_{x\uparrow} &= -j_{x\downarrow} = \frac{e}{a_y \Delta t} \\ \Rightarrow \quad \sigma_{xy\uparrow} &= \frac{j_{x\uparrow}}{E_y} = \frac{e}{a_y\Delta t}\cdot\frac{e\Delta t}{\hbar\Delta k_y} = \frac{e}{a_y\Delta t}\cdot\frac{e\Delta t a_y}{\hbar 2\pi} = \frac{e^2}{h} = -\sigma_{xy\downarrow}\,. \end{aligned} \tag{5.101}$$

Thus, the topological index ν_0 is equal to the spin Chern index as introduced on page 298. The difference between $\sigma_{xy\uparrow}$ and $\sigma_{xy\downarrow}$ implies that the dispersion of the states cannot be spin degenerate, since this would lead to identical Hall conductivities for both spins (Fig. 5.16(e)). Thus, the edge states are also spin polarized having a Kramers partner of opposite spin at opposite k_y.

Fourthly, we have shown that ν_0 can be calculated from the Bloch wave functions of an infinite 2D crystal. Basically, we have calculated the spin movement by the adiabatic change of the Hamiltonian within the bulk, and, finally, we have cut out a ribbon with edges. Thus, we get the properties of the edge without having an edge in the initial calculation, but just by the information, that spin is moved on the way between

---

58 The adiabatic charge pump, i.e., the movement of charges within an adiabatic cycle without energy penalty, has been observed experimentally within a lattice of atoms confined in the potential of standing light waves [136].

59 Some details of the complete derivation are found in [133], but they are based on a number of references quoted there.

60 The reduction to one unit cell gets obvious, while discussing Fig. 5.18 below.

$k_y = 0/\text{m}$ to $k_y = \pi/a_y$. The question, if edge states are present, is, thus, a bulk property. This implies automatically that such an edge state has to cross the whole bulk band gap of the 2D system, since the properties of the occupied bulk states do not change throughout the band gap. Such an edge state is sketched in Fig. 5.18(b) in the $E(k)$ diagram.

Fifthly, we have found that only the occupied wave functions at the time reversal invariant momenta (TRIMs), i.e., at the points in 2D $\vec{k}$ space, where $\vec{k} = -\vec{k}$, contribute to ν_0. These TRIMs are sketched in Fig.5.19(a) as Γ_1, Γ_2, Γ_3, and Γ_4. The reduction of relevant $\vec{k}$ points is very convenient for a numerical determination of ν_0.

Sixthly, we have shown that we can attribute an index to each TRIM Γ_i in the Brillouin zone being either $+1$ or -1. The index results from a phase factor matrix of all occupied states called $\boldsymbol{w}$ (eq. (5.78)–(5.80)). The resulting integer value $\delta_i = \frac{\sqrt{\text{Det}(\boldsymbol{w}(\Gamma_i))}}{\text{Pf}(\boldsymbol{w}(\Gamma_i))}$ is calculated resulting in $\delta_i = -1$ or $\delta_i = 1$. However, the value of the index δ_i is not well defined, i.e., it is gauge dependent. Only the product of the indices at the four TRIMs in the Brillouin zone is gauge independent and, thus, meaningful. If the product is -1 or $+1$, we have $\nu_0 = 1$ or $\nu_0 = 0$, respectively, meaning that we necessarily have spin polarized edge states with opposite chirality through the whole bulk band gap or we have not, respectively.

Finally, we describe a system with edges in the x direction, that can indeed be moved from $k_y = 0/\text{m}$ to $k_y = 2\pi/a_y$ by a physical operation. This can be done formally, e.g., by considering a hollow cylinder of the 2D material with the circumference of a single unit cell in the y direction and a finite length in the x direction (Fig. 5.18(c)–(d)). The cylinder will be pierced by an adjustable $\vec{B}$ field in the x direction. Changing the according magnetic flux by h/e introduces a phase shift of 2π per unit cell in the y direction and, thus, an increase of the wave vector by $2\pi/a_y$ as required. Thus, firstly, the electron wave functions are identical after the operation, since the wave vectors are only defined modulo $2\pi/a_y$ and, secondly, we can perform the process adiabatically without inducing energy. The latter can be seen by realizing that the field along the cylinder axis is parallel to the 2D plane of the system and, thus, cannot influence the energies of the 2D states. Consequently, we have transported spin from one edge of the cylinder to the other without changing the band structure and without energy supply exactly as required for the arguments above. The procedure is performed for one unit cell in y direction and transports one electron with spin $\uparrow$ from left to right and one electron with spin $\downarrow$ from right to left. This justifies that we have chosen $j_{x\uparrow} = \frac{e}{a_y \Delta t}$ in eq. (5.101).

Thus, if the product of phase factors of Kramers pairs (as represented by $\boldsymbol{w}$ in eq. (5.79)) is different at one TRIM with respect to the other TRIMs, spin is moved from left to right by an adiabatic cycle in $\vec{k}$ space. This implies the spin quantum Hall effect (SQHE) via $\sigma_{xy\uparrow} = -\sigma_{xy\downarrow} = e^2/h$, that is equivalent to two oppositely chiral, spin polarized edge state, one for for each spin, at E_F (Fig. 5.16(e)).

Notice, that $\nu_0 = 0$ does not imply the absence of edge states. For example, edge states as shown in Fig. 5.18(a) can still be present, e.g., due to a particular edge chemistry. However, these kind of edge states do not necessarily fill the whole bulk band gap and, thus, they are distinct from the topologically protected edge states. Moreover, these edge states can be removed by modifying the edge, e.g., by H adsorption, while the topological edge states cannot be removed by a procedure modifying the edge only. The removement requires a change of the bulk properties, i.e., a change of the phase factors of occupied states at the TRIMs.

Now, we turn to the 3D case referring to the results of the intricate calculation of eqs. (5.68)–(5.100) for the 2D case. Recall that the calculation of the toplogical 2D index ν_0 requires knowing the signs of the multiplied phase factors of the occupied wave functions at the 4 TRIMs, i.e., at the non-equivalent points of the Brillouin zone, where $\vec{k} = -\vec{k}$. The product of the four signs is either – or +, implying edge states via $\nu_0 = 1$ (– sign) or no edge states for a topologically trivial index $\nu_0 = 0$ (+ sign), respectively.

Next, we transfer this approach of topological indices to 3D, again for a rectangular lattice for the sake of simplicity. We have now eight non-equivalent TRIMs as sketched in Fig. 5.19(c). They are at the center of the Brillouin zone (1) as marked by BZC, at a corner of the Brillouin zone (1) as marked by BZE, in the middle of Brillouin zone boundary lines (3), and in the middle of Brillouin zone boundary planes (3). They are marked with different signs of $\frac{\sqrt{\mathrm{Det}(\mathbf{w}(\Gamma_i))}}{\mathrm{Pf}(\mathbf{w}(\Gamma_i))} := \delta_i$ (for a particular gauge choice) in Fig. 5.19(c)–(f). As already mentioned, the sign is related to the inversion symmetry of all occupied wave functions at Γ_i, if the crystal is inversion symmetric. Each occupied wave function $\psi_i(\vec{x})$ gets either a factor of -1 or a factor of $+1$ after the inversion. The sign is called the parity of the wave function. The product of the parity numbers of all occupied states at Γ_i is δ_i.

In order to determine the topological index, that tells us, if a surface state is present at a particular surface, we again, firstly combine two TRIMs. This is visualized by the thick lines in Fig. 5.19(c)–(f), which are equivalent to the thick lines for the 2D case in Fig. 5.19(a). We ask the question, if the product of δ_i of the pair changes sign on the way between two pairs of TRIMs, as we did in 2D. One can imagine a 2D plane, that connects the two pairs as drawn in Fig. 5.19(f) by full and dashed blue lines. This plane must have an edge state since it connects TRIMs in the k direction along the ribbon with different pair products of δ_i across the ribbon (pairs are encirceled by ellipses). The resulting edge states are marked as green points in Fig. 5.19(f). Due to TRS and $\sigma_{xy\uparrow} = -\sigma_{xy\downarrow} = e^2/h$, they have to obey $E(k_x, k_y, \uparrow) = E(-k_x, -k_y, \downarrow) = E_F$. The same applies for each possible plane between the two pairs of TRIMs. Thus, we can deform the blue plane, with edges within the two surfaces perpendicular to k_z, and always get edge states there. Since adjacent planes are similar, they will have similar edge states, implying a continuous line of edge states on the chosen $\vec{k}$-space surface. This naturally leads to a closed contour, which is a closed line of surface states at E_F (thick black lines in Fig. 5.19(h)–(j). Alternatively, one can argue that an interrupted line of surface states would enable the construction of a virtual plane without edge states that connects the two pairs of TRIMs with opposite product of δ_i. As mentioned, the surface states along the closed line must have opposite spin for opposite (k_x, k_y) due to TRS, as sketched in Fig. 5.19(j).

We see that if one projects pairs of δ_i to the chosen surface, the sign of the δ_i product of the pair is decisive. This product sign is sketched by the different colors of the dots at the TRIMs in Fig. 5.19(g)–(j)[61]. Moreover, we learn that points with different color have to be separated by a spin polarized surface state at E_F as drawn by a thick line in Fig. 5.19(h)–(j). The fact that δ_i is a bulk property requires that a closed line of surface states exists at any energy within the bulk band gap. Hence, an $E(\vec{k})$ band of surface states connects the valence band and the conduction band. Time reversal symmetry implies that the surface states of this band have to be spin degenerate at the surface TRIMs, where $(k_x, k_y) = -(k_x, k_y)$, i.e., the spin polarized states have to merge, e.g., at $(k_x, k_y) = \vec{0}$/m. Finally, since (k_x, k_y) is a good quantum number of the surface, we expect that the band has only one surface state per (k_x, k_y) point implying an $E(k_x, k_y)$ dispersion.

A typical band of surface states obeying these requirements is sketched in Fig. 5.19 (l). It exhibits a chiral spin in 2D $\vec{k}$ space. One possibility is that the spin direction is always perpendicular to $\vec{k}$, as favored by the spin-orbit interaction. Such a state, if isotropic, is called the topological Dirac cone, since it is the solution of the massless 2D Dirac equation with a reduced velocity of light.

The procedure to connect pairs of TRIMs in order to get the decisive signs of the δ_i products on a particular surface can be repeated for any surface of the crystal. It is obvious, that the case of Fig. 5.19(f), where only one of the 3D TRIMs has an inverted sign of δ_i with respect to the other TRIMs, implies surface states on any surface, since one never gets rid of the inverted sign by the pairwise products.[62] These systems are called strong 3D topological insulators (3DTIs). However, if a pair of 3D TRIMs has an inverted sign with respect to the other TRIMs, one can find a surface, where the two inverted signs compensate, such as in Fig. 5.19(c) and (g). These surfaces, thus, do not have a topological surface state. For other surfaces, the two inverted signs do not compensate, but imply two points of different colors and, hence, at least, two connected lines of surface states as shown in Fig. 5.19(d)–(e) and (h)–(i). These systems are called weak 3D topological insulators.[63] Only, if all TRIMs have the same sign of δ_i, we have a trivial insulator, not implying surface states on any surface. As in 2D, however, there can still be so-called trivial surface states, that are not caused by the bulk topology, i.e., they can be removed by a change of the surface chemistry.

---

61 We use colors instead of + and – signs, since the product of two δ_i is not gauge invariant, while only the distinction between the products is gauge invariant.

62 The same applies for three inverted signs, i.e., e.g., three times $\delta_i = 1$ and five times $\delta_i = -1$.

63 Since two bands of surface states are necessarily present on such a surface, one can anticipate a danger that disorder couples them, i.e., it produces symmetric and antisymmetric combinations of the wave functions. This could result in a gap and, thus, would destroy the robust surface conductivity. However, it turns out that also the conductivity of these so-called weak 3DTIs is rather robust.

In turn, the Fermi lines of a topological surface state can only be destroyed either by adding terms that break time-reversal symmetry, such that our whole analysis does not apply, or by exchanging the parity (for the case of inversion symmetric crystals) of, at least, one occupied wave function at, at least, one TRIM. Thus, one has to exchange a conduction and a valence band of different parity. In that sense, the Dirac cone is robust with respect to parameter changes of the Hamiltonian (including disorder), which do not invert the bands back into normal/trivial order. Thus, we can adapt a very similar scenario as in 2D (Fig. 5.16(a) and (b)). Inverting the p- and s-bands at one TRIM and leaving the normal order at all other TRIMs will lead to topological surface states on all surfaces. We will discuss a corresponding example in Section 5.4.6.2.

With this intuitive understanding in mind, we shortly describe the calculation of topological indices of 3DTIs. Firstly, we attribute three subscripts n_x, n_y, n_z to each δ_i. These subscripts are 0 (1), if the projection of the vector from $\vec{k} = 0/\mathrm{m}$ to the TRIM is 0 (π/a_i) in the respective direction x, y, or z. The indices ν_j ($j = 0, x, y, z$) are then:

$$\begin{aligned}(-1)^{\nu_0} &= \prod_{n_x,n_y,n_z=0,1} \delta_{n_x n_y n_z}\\(-1)^{\nu_{k=x,y,z}} &= \prod_{n_{j\neq k}=0,1,\ n_k=1} \delta_{n_x n_y n_z}\end{aligned} \tag{5.102}$$

Thus, for ν_0, we multiply all eight δ_i. This implies a strong topological insulator, if $\nu_0 = 1 \mod 2$, as expected from our projection argument. If $\nu_0 = 0$, we do not necessarily have surface states on all surfaces and the other indices matter. For the three ν_k, we have to multiply four selected δ_i. If all three indices ν_k are zero, we have a trivial insulator. Otherwise, one has a weak 3DTI. The triple (ν_x, ν_y, ν_z) then marks the surface normal of the only surface without robust conductivity, i.e., without required surface states (without argument). Thus, we are back at our original imagination of 2DTIs stacked into the direction (ν_x, ν_y, ν_z) and having robust conductivity at any perpendicular surface.

5.4.6.2 Materials and experiments

To find a candidate material for a strong 3DTI, one could look for a material with band inversion at the center of the Brillouin zone, called the Γ point, which inverts back towards the edges of the Brillouin zone. This trick works straightforwardly for inversion symmetric crystals. Since we already know that spin-orbit interaction helps to invert bands, we look for heavy atoms. The combination of elements often results in band gaps due to anticrossings within the bulk, which is the second requirement. The first material being a strong 3DTI was a BiSb alloy [137], but the real breakthrough came with the materials depicted in Fig. 5.20: Bi_2Se_3, Sb_2Te_3, and Bi_2Te_3. They are inversion symmetric, consist of heavy elements, and are known to exhibit a bulk band gap. The structure is layered distinguishing the bonding in the z direction and the (x, y)-direction. Figure 5.20(c) shows a model, how the states at Γ evolve by adding interaction terms between the atoms using Bi_2Se_3 as an example (see also Fig. 4.18–4.20). The

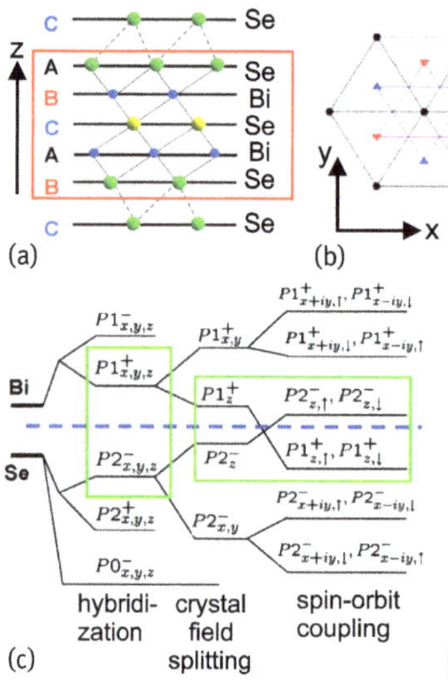

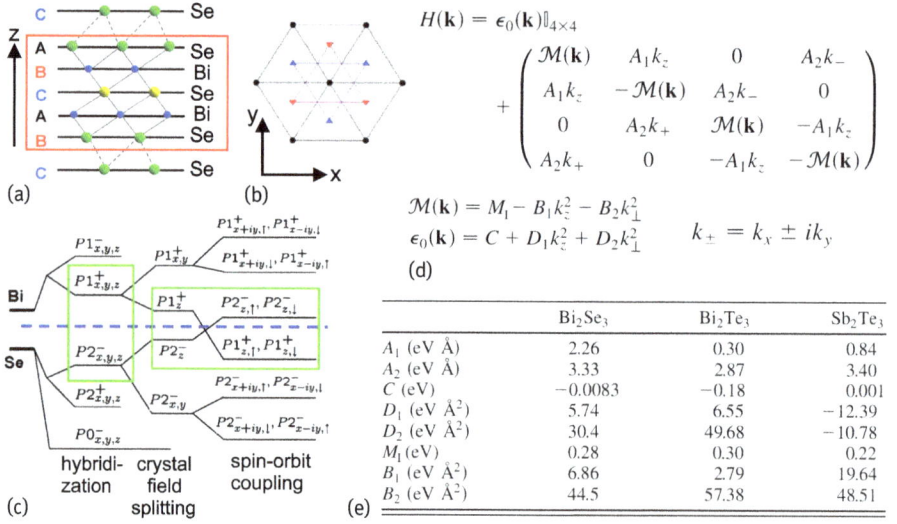

$$H(\mathbf{k}) = \epsilon_0(\mathbf{k})\mathbb{I}_{4\times4} + \begin{pmatrix} \mathcal{M}(\mathbf{k}) & A_1 k_z & 0 & A_2 k_- \\ A_1 k_z & -\mathcal{M}(\mathbf{k}) & A_2 k_- & 0 \\ 0 & A_2 k_+ & \mathcal{M}(\mathbf{k}) & -A_1 k_z \\ A_2 k_+ & 0 & -A_1 k_z & -\mathcal{M}(\mathbf{k}) \end{pmatrix}$$

$$\mathcal{M}(\mathbf{k}) = M_1 - B_1 k_z^2 - B_2 k_\perp^2$$
$$\epsilon_0(\mathbf{k}) = C + D_1 k_z^2 + D_2 k_\perp^2 \qquad k_\pm = k_x \pm i k_y$$

(d)

(e)

	Bi_2Se_3	Bi_2Te_3	Sb_2Te_3
A_1 (eV Å)	2.26	0.30	0.84
A_2 (eV Å)	3.33	2.87	3.40
C (eV)	−0.0083	−0.18	0.001
D_1 (eV Å^2)	5.74	6.55	−12.39
D_2 (eV Å^2)	30.4	49.68	−10.78
M_1 (eV)	0.28	0.30	0.22
B_1 (eV Å^2)	6.86	2.79	19.64
B_2 (eV Å^2)	44.5	57.38	48.51

Fig. 5.20: (a) Side view of the atomic structure of the layered material Bi_2Se_3 consisting of quintuples of stacked layers (red box) in stacking sequence Se-Bi-Se-Bi-Se. The forces between the quintuples, and, thus, between adjacent Se layers, are weak van der Waals forces such that the crystal can be cleaved between the outer Se layers of the quintuple. (b) Top view of the crystal showing the ABC stacking of the hexagonal layers marked each by different symbols and colors. Sb_2Te_3 and Bi_2Te_3 as well as mixtures of these three crystals exhibit the same structure. (c) Model of the energy levels at the center of the Brillouin zone (Γ point). The p-levels of Bi and Se (first column) get first split by chemical hybridization into bonding and antibonding states marked by their parity (+, −) and their preferential orientation (x, y, z). Five p_x-,p_y-, and p_z-states result due to the five atoms of the unit cell corresponding to the five layers in the quintuple (second column). Secondly, the crystal field lifts degeneracies between p_{xy} and p_z in favor of the stronger bonds in the (x, y) plane (third column). Finally, the spin orbit coupling splits the p_{xy} states and additionally mixes all three p-states, since there are also electric fields in the z-direction, which induce the spin-orbit terms. This leads to level repulsion, e.g., between $|P1_z^+ \uparrow\rangle$ and $|P1_{x+iy}^+ \downarrow\rangle$ and, thus, to a level crossing between the states closest to E_F (fourth column). (d) Matrix description of a minimal model Hamiltonian for the structure shown in (a)–(c) describing only the central states $P2_{z\uparrow}^-$, $P2_{z\downarrow}^-$, $P1_{z\uparrow}^+$, and $P1_{z\downarrow}^+$ of (c). (e) Parameters for the model Hamiltonian deduced from first-principle calculations for three different materials. (a)–(e) [128].

main players close to E_F are p-orbitals (Bi 6p and Se 4p), since the corresponding s-orbitals and d-orbitals are completely filled. The lattice period of five hexagonal layers in the z direction gives five hybridized levels for p_x, p_y and p_z each (in sum 15), that get split due to the different strength of the bonding in the z- and in the (x, y)-direction (crystal field splitting, Section 4.6.1). The decisive step is then provided by the spin-orbit interaction (Section 4.5.4), that exchanges a state with positive parity and a state with negative parity at E_F and, thus, inverts the sign of δ_i at Γ. The dispersion of these two bands inverts these two states back on the way to the Brillouin zone boundaries, such that only the Γ point has a different sign of δ_i. Consequently, we get a strong

3DTI. The (4×4) model Hamiltonian, that includes only the bands $P2^-_{z\uparrow}$, $P2^-_{z\downarrow}$, $P2^+_{z\uparrow}$, and $P2^+_{z\downarrow}$, is given in Fig. 5.20(d). The signs + and – define the parity of the wave function. The real numbers $A_1, A_2, B_1, B_2, C, D_1, D_2$, and M_I define the terms in the matrix similarly to the ones in the 2D case (A_I, B_I, M_I in eq. (5.66)ff.).

As long as M_I is positive and smaller than B_1 and B_2 multiplied by the corresponding k^2 at the Brillouin zone boundary and, in addition, A_1 and A_2 are not both zero, we get the topological band inversion exclusively at Γ and a gap at the position of the back-inversion, thus, a strong 3DTI. Parameters for three binary materials are given in Fig. 5.20(e). They are deduced from density functional theory (DFT) calculations and all fulfill these requirements for a strong 3DTI.[64]

Experimentally, the most direct proof of a strong 3DTI is ARPES,[65] that probes the band structure $E(k_x, k_y)$ of a particular surface, where the angle of the emitted photoelectrons is directly related to the $\vec{k}$ vector within the surface.[66] Modern instruments can map the $E(\vec{k})$ dispersion in one particular $\vec{k}$ direction within one shot such that mapping of a surface band structure becomes a routine task. Since the 3DTIs Bi_2Se_3, Sb_2Te_3 and Bi_2Te_3 can be cleaved in ultra high vacuum, the preparation of the clean crystal surfaces is simple. An ARPES analyzer can, moreover, be combined with a Mott detector, which can analyze the spin direction of particular photoelectrons belonging to a selected E and $\vec{k}$. An experimental setup is sketched in Fig. 5.21(a). The photoelectrons are selected in energy by the spherical capacitor and in angle by the entrance slit position into this capacitor with respect to the light point on the sample. After the final $(E, \vec{k})$ selection by the exit slit, the electrons are accelerated and directed to a heavy metal foil (Au foil in the image), where they get scattered. Due to the strong spin-orbit interaction within the foil, the scattering direction depends on spin. For example, the spin-up (spin-down) electrons get preferentially scattered to the left (right), which after calibration allows determination of the spin polarization in up-down direction from the difference of the photoelectron current in the left and the right detector. The same can be done for the left-right spin polarization using the differences of the upper and the lower detector. The detectors are drawn in Fig. 5.21(a) as blue planes with arrows symbolizing the spin detection direction.

Figure 5.21(b) shows a calculated surface band structure by DFT with the indicated Dirac cone (surface states). The crossing point of the Dirac cone, the so-called Dirac point, is in the center of the surface Brillouin zone, which is called the $\overline{\Gamma}$ point. It is, moreover, exactly at E_F. ARPES results of the same surface are shown in the inset. One recognizes reasonable agreement between experiment and theory, except that the Fermi level in the experiment is not at the Dirac point, but within the conduction band. Hence, the material is an n-doped semiconductor. This is caused by va-

64 Using the model Hamiltonian, one can explicitly show the presence of a single Dirac cone on each surface ([128]: Chapter III.B).

65 Angularly resolved photoelectron spectroscopy.

66 The method is described in detail, e.g., in [138].

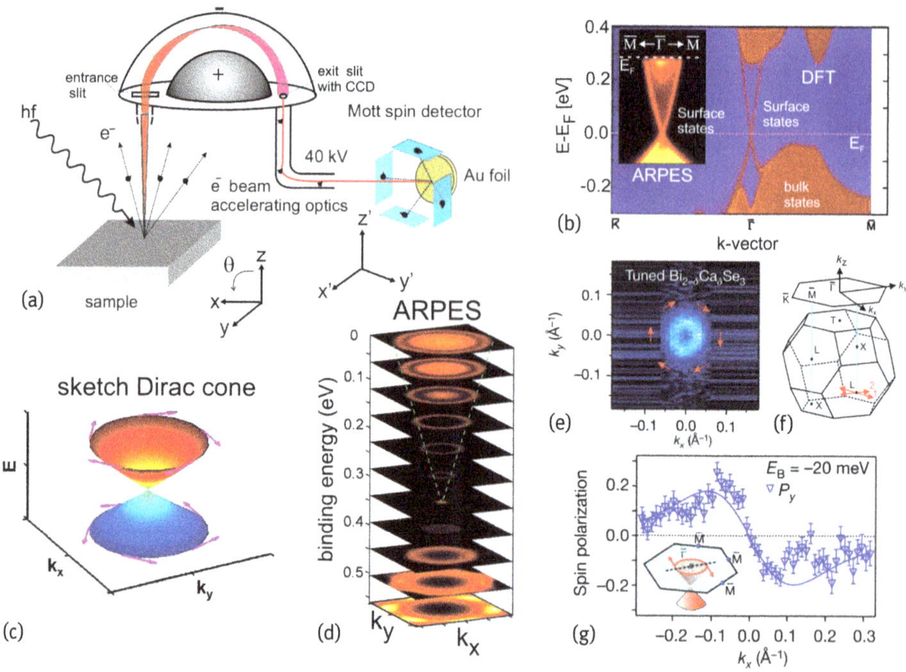

Fig. 5.21: (a) Sketch of the experimental setup for spin-polarized ARPES: hf labels the incoming photons and the red-orange areas mark the electron paths contributing to the detector signal without spin polarization. The electrons are detected by a CCD chip selecting different energies in radial direction and different angles perpendicular to the radial direction of the spherical capacitor. Polarities at the plates of the analyzing capacitor are marked. (b) DFT calculation of the surface band structure of Bi_2Se_3 at the cleavage plane. Filled, red areas are projected bulk states, while red lines are surface states as marked. Inset: ARPES data of the same surface along the $\overline{\Gamma M}$ direction in $\vec{k}$ space. Note the different positions of E_F in DFT and ARPES results. (c) 3D representation of a surface Dirac cone. The arrows indicate the expected spin directions of the states. (d) 3D representation of ARPES results of the cleavage surface of Bi_2Se_3. (e) Fermi line (at $E = E_F$) of Ca doped Bi_2Se_3 as measured by ARPES. (f) Brillouin zone of Bi_2Se_3 with labeled high symmetry points and surface Brillouin zone of the cleavage plane drawn above. (g) Measured spin polarization of a cut at constant energy ($E - E_F = -20$ meV) through the $\vec{k}$ space showing that the spin polarization inverts with the sign of $\vec{k}$ as required by time-reversal symmetry. The spin polarization is strongest in the direction perpendicular to $\vec{k}$ (polarization measured in the y direction for $\vec{k}$ in the x direction) as sketched in the inset and in (e). The bad resolution in the k_x direction is due to the Mott detector which reduces the photoelectron intensity dramatically. (a), (f) [134], (b), (c) [128], inset (b) [139], (d) [140], (e), (g) [139].

cancies and antisite defects unintentionally produced during the preparation of the crystal. These defects lead to strong n-doping of the material. Thus, the predicted transport properties of 3DTIs are not expected for Bi_2Se_3. The same is true for Bi_2Te_3 and oppositely for Sb_2Te_3, which is strongly p-doped. However, for ARPES, this does not matter and one can map the complete Dirac cone (Fig. 5.21(d)), the Fermi line of the surfaces (Fig. 5.21(e)) and the spin polarization including its preferential direction (Fig. 5.21(g)).[67]

Using spin polarized ARPES, it has also been shown that magnetic doping, e.g., of Bi_2Se_3 with Mn atoms, that breaks time-reversal symmetry, indeed opens a gap within a Dirac cone [142], that counter-doping, e.g., with Ca, can be used to shift the Fermi level of Bi_2Se_3 to the Dirac point [139], that also thin films exhibit a gap in the Dirac cone because of the interaction of the two Dirac cones of opposite surfaces [143], such that a minimum thickness of about five quintuple layers is required to get the topological properties, and that the phase transition from a trivial insulator to a topological insulator can be tuned by material composition using, e.g., $BiTl(S_{1-x}Se_x)Se_2$ with different x [144]. Also another topologically protected state has been found by ARPES, where the pairs to be considered for the polarization difference $P_x^\Delta(k_y)$ (eq. (5.95)) are not pairs related by time reversal symmetry, but pairs related by a mirror symmetry of the crystal [145]. These materials are called topological crystalline insulators, however, with Dirac cones being only robust to disorder types or perturbations, which do not break the mirror symmetry of the crystal.

The second technique used to verify topological properties is scanning tunneling spectroscopy (STS), that measures the local density of states (LDOS), i.e., the sum of squared wave functions of the system at a particular energy E selected by the applied voltage V according to $E - E_F = eV$ [146]. One major contribution to the LDOS are standing waves, that are the overlap of incoming and scattered waves from a barrier, e.g., a step edge or a defect (Fig. 5.1(f)). Figure 5.22(a) shows a LDOS map of the cleavage surface of the 3DTI $Bi_{1-x}Sb_x$ [147]. The changing intensity across the surface (red, blue, yellow) is given by the standing electron waves originating from the scattering at potential fluctuations.

The Fourier transformation in Fig. 5.22(b) shows the $\vec{k}$-space distribution of the standing waves. It should be compared with the $\vec{k}$-space distribution of the states of the same surface determined by ARPES (Fig. 5.22(c)). Therefore, the possible scattering vectors $\Delta\vec{k} = \vec{k}_1 - \vec{k}_2$ between two states at $\vec{k}_1$ and $\vec{k}_2$ have to be calculated[68], and subsequently to be weighted with the density of states at the contributing $\vec{k}$ points as sketched in Fig. 5.22(d). This results in Fig. 5.22(e), that obviously deviates from the experimental data in Fig. 5.22(b). This deviation directly shows the topological character of the contributing states. As described above (footnote 36 on page 299), a direct backscattering between Kramers pairs leads to a Berry phase of π between the required possible spin rotations, being clockwise or counterclockwise, and, thus, to destructive interference. Thus, all scattering vectors combining states with $\vec{k}_1 = -\vec{k}_2$ are eliminated from the standing wave pattern by destructive interference, i.e.,

67 The absolute values of spin polarization have to be taken with care since the polarization of the light also contributes to it via selection rules [141].

68 To determine the standing wave, one has to calculate $|\psi_1(\vec{x}) + \psi_2(\vec{x})|^2 \propto |e^{i\vec{k}_1\cdot\vec{x}} + e^{i\vec{k}_2\cdot\vec{x}}|^2 \propto 1 + \cos((\vec{k}_1 - \vec{k}_2)\cdot\vec{x})$.

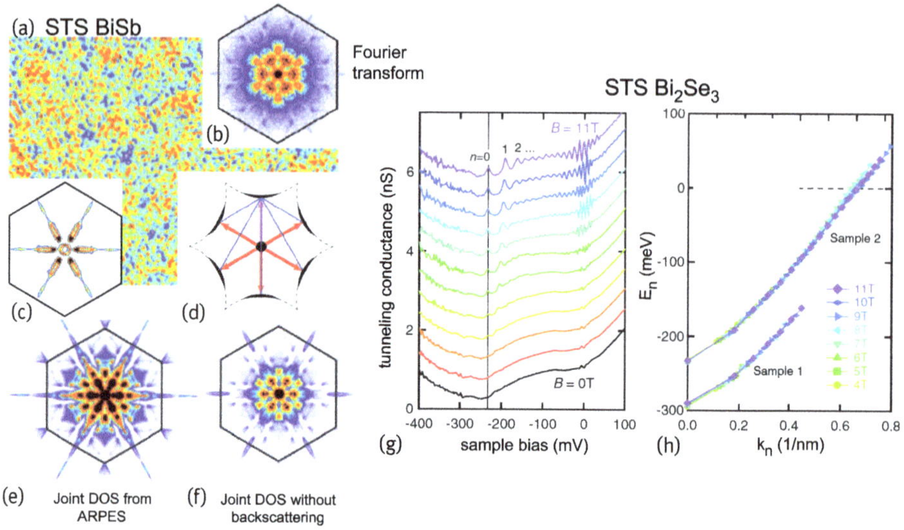

Fig. 5.22: (a) STS image of $Bi_{1-x}Sb_x$ recorded in the bulk band gap. The contrast is caused by overlapping standing electron waves scattered at multiple impurities at the surface. (b) Fourier transformation of the data in (a) representing the $\vec{k}$ space distribution of standing waves. (c) ARPES data of the same surface at the same energy representing the $\vec{k}$ space distribution of states. (d) Sketch showing how to transform ARPES data (black areas) into the Fourier transform of the STS data, namely, by defining scattering vectors in $\vec{k}$ space (red and blue arrows), which correspond to the expected wave vectors of the standing waves. Scattering vectors combining areas with large state density from ARPES contribute more than the ones combining areas with low state density. (e) Resulting expected Fourier transformation of STS images deduced from the ARPES data in (c) and the construction in (d). (f) Same as (e), but neglecting all scattering vectors combining states at $+\vec{k}$ with states at $-\vec{k}$, i.e., states that are scattered by 180°. Note that (f) fits much better to (b) than (e). (g) STS curves on the cleavage surface of Bi_2Se_3 representing the LDOS as a function of energy at different $\vec{B}$ field perpendicular to the surface as marked at the top and at the bottom curve. The numbers n mark consecutive Landau levels. (h) $E(|\vec{k}|)$ dispersion deduced from the Landau levels for two different samples. The wave number k_n corresponds to the quantized circular motion of the electron at constant velocity $|\vec{v}|$. (a)–(c), (e), (f) [147], (d) [148], (g), (h) [149].

by the helical spin character of the Dirac cone. Indeed, erasing these standing wave contributions as in Fig. 5.22(f), the experimental Fourier transformation in Fig. 5.22(b) is reproduced. Thus, STS can show directly, that the surface states of a strong 3DTI are spin helical.

Another application of STS was the indirect measurement of the $E(|\vec{k}|)$ dispersion of the surface states using Landau level spectroscopy (Fig. 5.22(g) and (h)). The experiments firstly show that the distance between Landau levels decreases with distance from the Dirac point, which is at $E_D - E_F = e \cdot V = -230$ meV. This is expected for a linear dispersion $E \propto |\vec{k}|$ that results in energies for the nth Landau level being

$E_n^{\mathrm{LL}} \propto \sqrt{nB}$.[69] Albeit the application of $\vec{B}$ breaks time-reversal symmetry, this effect is close to negligible, since the spin splitting by the $\vec{B}$-field is about 1 meV only, while the spin-orbit interaction is about several 100 meV. Thus, the Dirac cone might be gapped in an external field, but to a negligible amount.

Transport measurements, so far, often suffer from the intrinsic doping and the strong disorder potential of the materials. Moreover, the surface of the topological insulator is typically exposed to air and several chemicals for making Hall bar devices and contacting the surface, such that the surface chemistry is strongly changed. As pointed out above, this should not lead to a destruction of the topological surface states, but can lead to strong surface doping shifting the Fermi level at the surface in an uncontrolled manner. Thus, for reliable transport measurements, the surface of the 3DTI has to become an interface to a normal insulator. Nevertheless, seamless ambipolar transport, i.e., the change from an electron current to a hole current by gating has been observed [150] as well as the quantum Hall effect [151]. Experiments exploiting the spin helicity of the surface state by transport are under way, albeit still rather indirect [152].

The progress with weak 3DTIs is much slower, but, at least, one system has been identified and produced, where weak 3DTI properties are found. It consists of layers in a honeycomb structure made of the heavy elements Bi, I and Rh. This structure forms a 2DTI in DFT calculations with an inverted gap at the Γ point of 0.3 eV. The layers are stacked in the 3D material and are additionally separated by spacer layers of the normal insulator BiI, such that they form a weak 3DTI [153]. The material exhibits, e.g., topologically protected edge states on all step edges of the surface perpendicular to the stacking direction [154].

5.4.7 Other topological properties

There are many other fascinating effects arising from the topological considerations as described above. Most of them still await their experimental verification.

A few of them are sketched in Fig. 5.23.[70] For example, there is a so called magnetoelectric response, i.e., a static electric field can induce a robust, static magnetic field, via the Hall current circulating around the electric field, if the degeneracy of the two counter-circulating edge states is broken, e.g., by ferromagnetic layers (FM) (Fig. 5.23(a)). The same effect induces a magnetic monopole as the image of an electric monopole on top of a gapped 3DTI surface (Fig. 5.23(b)). Putting, additionally, superconductors on top of a 3DTI induces another type of gap, the superconducting gap into the Dirac cone. This gap of a so-called topological superconductor is unusual

69 This can be verified, e.g., by using the 2D DOS of a linear $E(\vec{k})$ dispersion (eq. 1.20) and the degeneracy of Landau levels $2eB/h$ exploiting that the integrated 2D DOS must fit into the Landau levels.
70 Arguments can be found in [128]: Chapter III.D and V.C.

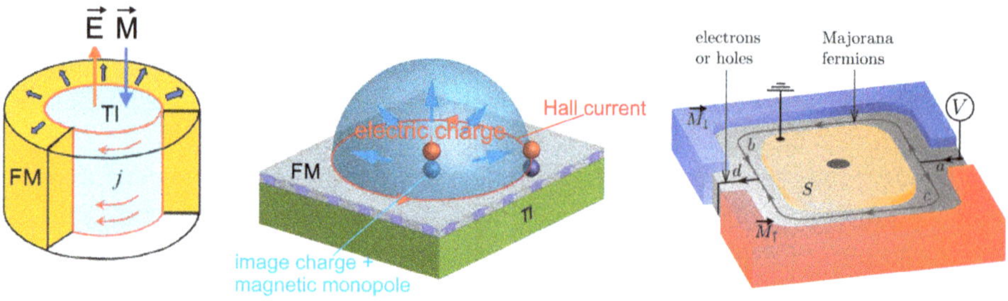

Fig. 5.23: (a) Sketch explaining the magnetoelectric effect: An electric field $\vec{E}$ in a weak 3DTI or a strong 3DTI with a gapped surface and a ferromagnetic layer (yellow, FM) on the remaining topological surfaces induces a Hall current perpendicular to the electric field. This Hall current is circular and, thus induces a magnetic field antiparallel to the electric field applied. (b) An electric monopole (red) above a gapped surface of a 3DTI, e.g., by a ferromagnetic layer (grey, FM), induces an image charge (blue) and an image magnetic monopole (blue) due to the Hall current (red line) at the interface between gapped areas of different sign. (c) Surface of a strong 3DTI (grey) covered with two magnets (red, blue) having the opposite sign of the magnetization $\vec{M}$, thus, leaving a state at the Fermi level at the interface between the two domains (black line with arrows). A superconductor (S, orange) induces Majorana channels (grey lines with arrows) at the interface between the superconductor and the ferromagnet. Applied voltage V, ground at S, and the grey dot marking a magnetic vortex in the superconductor can be used to probe unusual interference effects. (a) [155], (b) [156], (c) [157].

in the sense that the superconductor is made of only one spin species for each $\vec{k}$ rendering it a so-called spinless superconductor. One can show that states at the Fermi level appear at the edges of such superconductors. These states are equally distributed on opposite sides of the superconductor [158]. They are called Majorana modes and can be found, e.g., at the end of a 1D topological superconductor, where some experimental confirmation for the Majorana modes has been found [159]. Other hosts for Majorana modes are vortices within topological superconductors or edges between areas gapped by magnets and gapped by superconductors. The experimental proposal shown in Fig. 5.23(c) aims at the separation of an injected electron into two Majorana modes that encircle the central superconducting island, which leads to unusual interference effects.

We close this section by commenting on the robustness of the topological properties. This robustness originates from the integer values of the topological indices, which cannot be changed continuously. Hence, a qualitative change of the system is required to change its topology. For example, an occupied band has to be exchanged at one TRIM by a former empty band and the two bands must have a different character, e.g., exhibiting a different parity. However, such an exchange of bands can be realized by a minimal change of the total energy of the system. Multiple examples show that slight changes of the strain or of the stoichiometry of a specific sample can change its topological index, mostly if the system is already energetically close to a topological phase transition. Consequently, the celebrated robustness of topological properties is, a priori, not a robustness with respect to the total energy of the system, but is related to the robustness of boundary effects, that can only be changed if the character of the material within the boundaries is changed. In this sense, all topological systems are similar to the Quantum Hall system, where the presence of the edge states is not related to any details of the edge, but can only be removed if the energetics of the bulk of the 2D system is changed.

5.5 Consequences of electron-electron interaction

5.5.1 Zero-dimensional electron systems

Next, we discuss the more complex electron-electron interaction for itinerant electrons. Corresponding electron phases can also be partly classified by topological indices, but we do not follow this path, since it gets too abstract for an intuition. The most transparent introduction into the electron-electron interaction can be given at complete quantization of the kinetic energy in quantum dots (Sections 1.3 and 3.3.1.2), i.e., within 0D systems. Such systems are also called artificial atoms, since they provide discrete energy levels similar to atoms, but are confined within much larger volumes ($10\,\text{nm}^3$–$1\,\mu\text{m}^3$). A similar approach has been followed for the electron-electron interaction of localized electrons in Section 4.4, where relevant effects were first described for atoms and molecules, before turning to the extended solids. We will start our approach by describing the single particle spectra of quantum dots, before we move to the consequences of the electron-electron interaction.[71]

5.5.1.1 Single particle spectrum

The setup of a typical quantum dot used for the study of interaction effects is shown in Fig. 5.24(a). Firstly, one grows a layered system consisting of n-doped GaAs, AlGaAs and InGaAs (Section 1.3.1). The system contains a 2D electron system confined within the InGaAs layer, that has the smallest band gap of the three materials. The thickness of the InGaAs layer is typically about 10 nm. The 2D electron system (2DES) in the InGaAs layer is filled with electrons due to adequate doping close to the surface. In order to confine the 2DES in the remaining lateral directions, lithography and etching methods can be used.[72] One can, e.g., produce a cylinder with diameter $d_{\text{QD}} \simeq 500$ nm (Fig. 5.24(a)). A scanning electron microscopy image of such a cylinder is shown in Fig. 5.24(b). This cylinder will then be surrounded by a metal, the so called side gate. Afterwards metallic contacts to the upper and lower n-GaAs are established, which

71 We largely follow [160].

72 The other approach using gates is described in Sections 1.3.1.1 and 3.3.1.2.

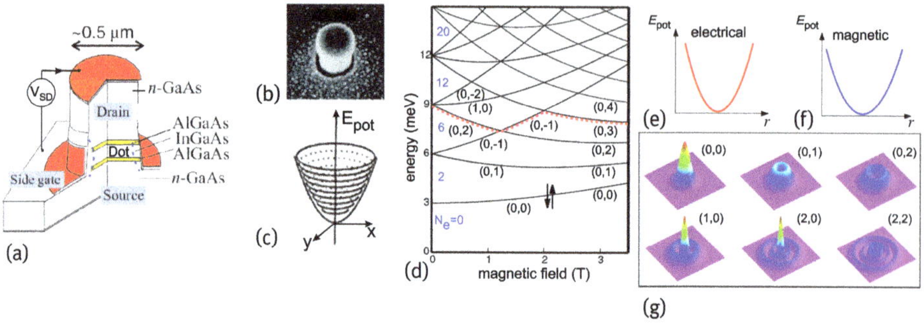

Fig. 5.24: (a) Schematic of a quantum dot device. The side gate can be used to vary the charge within the quantum dot. (b) Scanning electron micrograph of the quantum dot sketched in (a) before adding contacts. (c) Electrostatic potential within the plane of the quantum dot. (d) Single particle states of the quantum dot as a function of $\vec{B}$ field perpendicular to the plane of the quantum dot. Numbers in brackets mark radial and azimuthal quantum numbers of the states (n, m). Blue numbers at the left mark the number of electrons N_e within the dot, if E_F is in the corresponding gap. Red dotted line marks the highest occupied state for seven electrons within the quantum dot. (e), (f) Schematic of the electrostatic and the magnetostatic potential, respectively. (g) $|\psi_i(\vec{x})|^2$ of some single particle states with labeled quantum numbers (n, m) at $B = 0$ T. (a), (b), (d), (g) [160].

act as source and drain electrodes allowing current transport across the AlGaAs tunnel barriers and, thus, through the InGaAs quantum dot. At the interface between side gate and semiconducting cylinder, negative charges appear due to the difference in work function of the two materials. These charges cause a parabolic, electrostatic potential for the electrons within the quantum dot, which is mostly rotationally symmetric within the plane (Fig. 5.24(c),(e)). The single particle states of this 2D harmonic oscillator can be calculated analytically in a $\vec{B}$ field. The central idea of the calculation is to map the Schrödinger equation of a free electron in a $\vec{B}$ field to the harmonic oscillator potential, as is typically shown explicitly within standard textbooks of quantum mechanics. The effective, harmonic potential due to $\vec{B}$, as shown in Fig. 5.24(f), leads to the equidistant Landau levels also discussed in Section 5.3.2. The squared wave functions $|\psi_i|^2$ of the states of the quantum dot at $B = 0$ T are shown in Fig. 5.24(g). Since the symmetry of the potential implies separation in radial and azimuthal coordinates, one gets states classified by radial and azimuthal quantum numbers (n, m) $(n \in \mathbb{N}^0, m \in \mathbb{Z})$. Thereby, n counts the number of antinodes in radial direction, while m belongs to the function $e^{im\varphi}$ with φ being the azimuthal angle, i.e., it is related to the orbital momentum of the quantum dot.

At $B = 0$ T, equidistant energies $E_{n,m} = (2n + |m| + 1) \cdot \hbar\omega_0$ result.[73] It is easy to see that the degeneracy of the energy levels increases by two, if one jumps from a

73 The eigenfrequency of the quantum dot ω_0 depends on the curvature of the parabolic potential $E_{pot}(\vec{x})$ according to $\omega_0 = \sqrt{d^2 E_{pot}(\vec{x})/d|\vec{x}|^2/m^*}$.

given energy level to the next one higher in energy and considers the additional spin degree of freedom. Hence, the lowest level has a degeneracy of two, the next level a degeneracy of four, then a degeneracy of six, and so on (Fig. 5.24(d)).

Within a $\vec{B}$ field, three things are changing:

1. The additional magnetic confinement (Fig. 5.24(f)) increases the energy of all states. Formally, ω_0 is replaced by $\sqrt{\omega_0^2 + \omega_c^2/4}$ with $\omega_c = eB/m^*$. This enhancement in energy is called the diamagnetic shift.
2. The orbital momentum of the electrons is accompanied by a magnetic dipole oriented oppositely to the orbital momentum. This implies an additional energy proportional to $\vec{B}$: $E_{\text{dipole}} = -m \cdot \hbar\omega_c/2$. This is called the paramagnetic energy shift.
3. Finally, the degeneracy of the spin degree of freedom $s = \pm 1/2$ is lifted in a $\vec{B}$ field by $E_Z = s \cdot g\mu_B|\vec{B}|$ called the Zeeman energy.

Consequently, one finds:

$$E_{n,m,s} = (2n + |m| + 1)\hbar \cdot \sqrt{\omega_0^2 + \frac{\omega_c^2}{4}} - \frac{m}{2} \cdot \hbar\omega_c + s \cdot g\mu_B|\vec{B}| . \tag{5.103}$$

This is called the Fock–Darwin spectrum of a quantum dot. Assuming $g = 0$, the resulting state energies are displayed in Fig. 5.24(d). Importantly, one observes crossing points of energy levels, i.e., the seventh electron added to the quantum dot (dotted, red line in Fig. 5.24(d)) would firstly be placed into the sixfold degenerate state at $E \simeq 9$ meV, before it will occupy, consecutively with increasing $\vec{B}$, the states $(0, 2)$, $(0, -1)$, and $(0, 3)$. The crossing points are decisive for the determination of the strength of particular electron-electron interaction effects (Section 5.5.1.2).

5.5.1.2 Interaction effects in quantum dots

The single particle picture as described above neglects the Coulomb interaction between the electrons. This is massively wrong, if more than one electron reside within the quantum dot. We consider a quantum dot with a typical $\hbar\omega_0 = 1$ meV for $m^* = m_e$, i.e., a potential $E_{\text{pot}}(\vec{x}) = 7.25\ \mu\text{eV/nm}^2 \cdot \vec{x}^2$ and, thus, a lateral extension $2r_{00}$ of the ground state of $2r_{00} \simeq 2|\vec{x}(E_{\text{pot}} = 1\,\text{meV})| \simeq 11$ nm. The Coulomb interaction between the two possible electrons within this state at a dielectric constant of $\varepsilon = 1$ can be estimated classically to be $E_{\text{ee}}^{\text{cl}} \simeq 1/(4\pi\varepsilon\varepsilon_0) \cdot e^2/2r_{00} = 130$ meV, i.e., two orders of magnitude larger than the confinement energy $\hbar\omega_0$. Only at much smaller confinement areas with radii of 0.1 nm (atoms), the confinement energy dominates the Coulomb repulsion. Within quantum dot experiments, the employed materials mostly exhibit $m^* < m_e$ and $\varepsilon > 1$, such that the relation $\hbar\omega_0/E_{\text{ee}}^{\text{cl}} \propto \varepsilon/m^*$ is larger (InGaAs: $m^* \simeq 0.04 m_e$ and $\varepsilon \simeq 13$). Nevertheless, the Coulomb energy between the electrons dominates for quantum dots with diameters above ~ 100 nm.

The primary consequence of the Coulomb interaction is that it costs charging energy E_c to bring an additional electron to the quantum dot. As discussed in Sec-

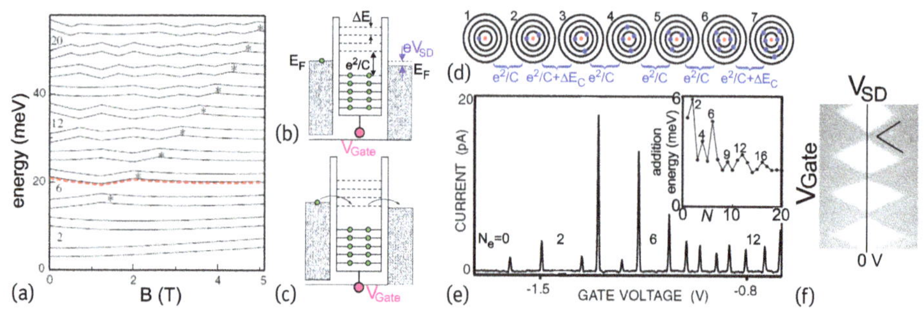

Fig. 5.25: (a) Single particle energies of a quantum dot, where the energy of the nth level is shifted by $n \cdot e^2/C$ with respect to Fig. 5.24(d). The kinks within the curves are identical to the crossing points of Fig. 5.24(d). The stars mark the last crossing point of a particular state. (b) Representation of the states within a quantum dot. The occupied single particle states are represented by lines with filled green circles. They differ by the confinement energy. The unoccupied states are additionally shifted by the charging energy e^2/C, since its occupation would cost this energy in addition to the confinement energy. Grey areas are filled electronic states up to the Fermi levels of the contacts. If no state of the dot is between the Fermi levels of the contacts (as drawn), current can not flow as long as $k_B T \ll e^2/2C$. The pink lever V_{Gate} indicates that the ensemble of states can be shifted by an external electrostatic potential. (c) Same as (b), but at a gate voltage V_{Gate}, where unoccupied states of the dot are found between the Fermi levels of source and drain electrode such that current can flow. (d) Schematic of quasi-atomic orbitals of a quantum dot, that correspond to the degenerate single particle states of the dot. Electrons (filled blue circles) are distributed consecutively onto the orbits (rings). The curved brackets below the orbits are marked with the energy differences between the different numbers of electron occupation of the dot at $B = 0$ T, where ΔE_C is the difference between adjacent single particle energies $E_{n,m}$. (e) Current between source and drain electrode of a quantum dot, such as in Fig. 5.24(a), as a function of gate voltage V_{Gate} at fixed voltage between source and drain $V_{SD} \leq 1$ mV. Inset: Energy distance between neighboring peaks. (f) Coulomb diamonds: the current is displayed in greyscale as a function of V_{Gate} and V_{SD}, i.e., the darker the area, the larger the current. The two black lines mark so-called excited states. (a), (d), (e) [160].

tion 3.3.1.2, one can estimate $E_c \simeq e^2/C$ with C being the capacitance of the quantum dot with respect to its surrounding. Within the capacitance model, E_c is the same for all electrons. In order to get an additional electron onto the quantum dot, confinement energy and charging energy are required. They are provided by a positively charged side gate (Fig. 5.24(a)), that attracts electrons to the dot. The single particle states are, thus, energetically shifted by e^2/C with respect to each other as shown in Fig. 5.25(a). Figure 5.25(b) and (c) display the energies of different states for two different side gate voltages V_{Gate} (see also Fig. 3.3(d)–(e)). The source and drain electrodes on the left and right are filled up to its corresponding E_F (grey area). Within the quantum dot (middle), the occupied states (green circles on full lines) are separated by the single particle energies $E_{n,m}$. In principle, the lowest level would be filled by two electrons, the second lowest level by four electrons, and so on, but this is reduced within the sketch to two electrons per level for the sake of simplicity. The electron-electron interaction between the occupied states is an offset in energy being the same for all occupied states. In order to put another electron onto the quantum dot, one has to pay the energy difference to an unoccupied single particle level and the charging energy e^2/C. The corresponding energy levels are displayed in Fig. 5.25(b) as broken lines, located above the Fermi levels of source and drain electrode, i.e., the current through the quantum dot amounts to $I \simeq 0$ A at low temperature. The energy of the quantum dot states with respect to source and drain can be changed by the side gate (eq. (3.4)). In a first approximation, all the states move parallel as shown in Fig. 5.25(c). As soon as an unoccupied state is located between the Fermi levels of source and drain, an electron can tunnel onto the quantum dot, thereby shifting all other states upwards by e^2/C. Afterwards, the electron can tunnel into the drain, such that $I > 0$ A. If the originally unoccupied state is moved below E_F of the drain, the level gets permanently occupied having a distance of e^2/C to all unoccupied states again, such that the current disappears again. If an additional charging energy of e^2/C is contributed by the side gate, the next unoccupied level gets shifted between the Fermi levels of source and drain. This leads to the rather regular peaks in current as a function of side gate voltage as displayed in Fig. 5.25(e) (see also Fig. 3.3(f)). Recall that these current peaks are called Coulomb peaks. These Coulomb peaks have been exploited as charge detectors for a neighboring qubit (Section 3.3.1.3).

Increasing the voltage between source and drain V_{SD}, increases the area of side gate voltages where a state is located between the E_F's of source and drain, such that the area of side gate voltages with $I > 0$ A continuously increases, until the source-drain voltage V_{SD} is larger than e/C and current can flow independently of V_{Gate}. Displaying the current in grey scale as a function of V_{Gate} and V_{SD} reveals the characteristic Coulomb diamonds of suppressed current as shown in Fig. 5.25(f) (see also Fig. 3.3(g)).

A closer inspection of Fig. 5.25(e) reveals an irregularity of the peak distances, which is additionally displayed in the inset. The irregularity is firstly a consequence of the confinement energies, which are often called single particle energies. If a confinement energy level, respectively an orbit of the artificial atom, is completely filled, the distances between the Coulomb peaks are larger. This is the case after two electrons for the first confinement energy, $2 + 4 = 6$ electrons for the second confinement energy, $2 + 4 + 6 = 12$ electrons for the third confinement energy, and so on. After these fillings, the next electron requires the energy penalty e^2/C for charging and the confinement energy ΔE_C, being the energy difference between the last and the next orbit. Such a successive filling of orbits is sketched in Fig. 5.25(d) including the required energies to add the next electron. The expected larger gate voltage distances of peaks after 2, 6, and 12 electrons (filled orbit) are indeed observed as most clearly shown in the inset of Fig. 5.25(e).

The inset, moreover, exhibits larger energy distances after filling 4, 9 and 16 electrons into the dot. This indicates a second consequence of the electron-electron interaction, the exchange interaction (Section 4.4). As in real atoms, it leads to Hund's rules. The first rule is that filling an electron orbit with electrons of the same spin is more favorable than mixing the spins due to the antisymmetry of the many-particle wave function. The symmetric spin function for parallel spins implies an antisymmetric spatial wave function with larger electron-electron distances and, hence, reduced Coulomb repulsion. Thus, parallel spins are energetically favorable for the same orbit. They are also favorable for different orbits, if the gain in exchange energy is larger than ΔE_C. Because of Pauli's principle, parallel spins in the same orbit are only possible up to half filling of the orbit.

For example, within the third orbit of the quantum dot (degeneracy: 6), we can put electron 7, 8 and 9 with the same spin and, thus, with reduced electron-electron repulsion energy, respectively reduced charging energy E_C. After the 9th electron, the spin of the 10th electron must be antiparallel to 7, 8, and 9, such that the spatial wave function gets symmetric for interchanging electron 10 with electron 7, 8, or 9, leading to a maximum in E_C. This causes the peaks at 4, 9 and 16 within the inset of Fig. 5.25(e). More formally, the two-particle exchange energy of a pure two-electron system E_{Ex} is given by the first term of $\tilde{J}$ of eq (4.13) (Section 4.4) reading $E_{Ex} \propto \pm \int\int \psi_i^*(\vec{x}_1)\psi_j^*(\vec{x}_2) \cdot 1/|\vec{x}_1 - \vec{x}_2| \cdot \psi_i(\vec{x}_2)\psi_j(\vec{x}_1)\, d^3\vec{x}_1\, d^3\vec{x}_2$. It has to be added to e^2/C with − (+) sign for parallel (antiparallel) spins.

In turn, the exchange energies as well as the single particle energies of the states in the quantum dot can be estimated from the distance variations of the Coulomb peaks. Another possibility to probe the single particle states is called spectroscopy of the excited states and uses the Coulomb diamonds. Increasing the source-drain bias V_{SD} allows having two (or more) energy levels within the energy window between the two E_F's of source and drain. This increases the tunneling current, since the probability to tunnel into or out of the dot scales with the number of available states for tunneling. Thus, if an additional state contributes by increasing V_{SD}, the current increases. This is visible as a step in the current. The step runs parallel to the onset of current within the Coulomb diamonds as displayed for one example as a black line in Fig. 5.25(f). The reason is, that the energy distance between the lowest unoccupied energy state, to be shifted into the V_{SD} window, and the next unoccupied state remains roughly the same, independent of V_{Gate}. In order to increase the visibility of these steps, one mostly displays dI/dV_{SD}. This transforms the steps into peaks, respectively into lines within a (V_{SD}, V_{Gate}) plot. Hence, one finds additional lines outside the Coulomb diamonds (Fig. 3.3(g)). Notice, that it does not matter, if two occupied states within the V_{SD} window increase the probability to tunnel out of the quantum dot or if two unoccupied states increase the probability to tunnel into the quantum dot, at least, as long as the two tunnel barriers are not too different. Notice further, that the tunneling probability can also be increased by other additional processes, e.g., tunneling and exciting an optical phonon at the same time within a second order process. This also leads to additional steps in the Coulomb diamond pattern, if getting energetically possible within the V_{SD} window, e.g., if eV_{SD} gets larger than the required phonon energy. Consequently, the more general term of excitation spectroscopy is mostly used.

The role of the exchange interaction can be probed more directly by applying a $\vec{B}$ field. Figure 5.26(a) shows the principle for four electrons within the quantum dot. The degenerate states at $B = 0\,\mathrm{T}$ belonging to the second energy level, (0, 1) and (0, −1), increasingly separate with increasing $\vec{B}$. At small $\vec{B}$, the exchange interaction dominates, such that the two electrons, besides the two electrons within the state (0, 0), exhibit parallel spin, thus, occupy both levels, (0, 1) and (0, −1). With increasing $\vec{B}$, this gets increasingly unfavorable, due to the different single particle energies of (0, 1) and (0, −1), until one spin flips in order to occupy the lower level (0, 1), too. This happens as soon as the exchange energy E_{ex} gets smaller than the energy distance between the single particle states (0, 1) and (0, −1). Consequently and counter-intuitively, the magnetization (= magnetic dipolar moment/volume) of the artificial atom dramatically decreases above a certain $\vec{B}$ field.

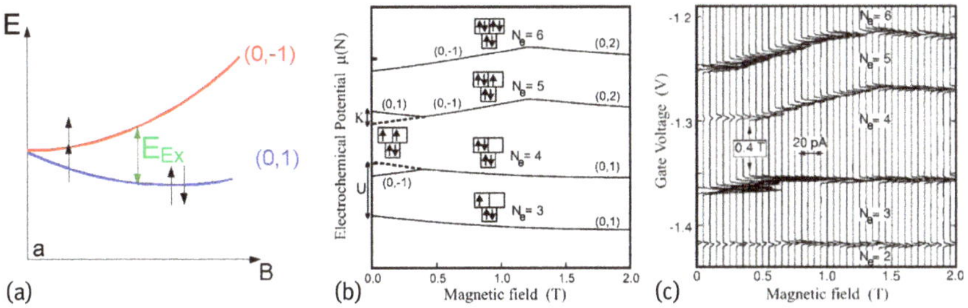

Fig. 5.26: (a) Principle of a triplet-singlet transition within a quantum dot. As soon as the energy distance between the single particle levels (blue and red line) is larger than the exchange energy E_{Ex}, the two electrons occupy the same energy level with opposite spin. (b) Course of the single particle energies marked by quantum numbers (n, m), if the exchange interaction is included in addition to the energies displayed in Fig. 5.25(a). The electron configurations for the individual states are marked using arrows to represent the spin direction within a particular single particle level (n, m) displayed as a box. (c) $I(V_{Gate})$ curves recorded for different $\vec{B}$ as marked on the x axis. The state configuration can be deduced from the course of the steps. N_e marks the number of electrons within the quantum dot. The spin flip at $B = 0.4\,\mathrm{T}$ is marked by a double arrow. (b), (c) [161].

The resulting course of Coulomb peaks including charging energies, single particle energies, and exchange energies is shown schematically in Fig. 5.26(b) with the number of electrons N_e present between the charging energies marked. Figure 5.26(c) shows corresponding experimental data. Both reveal a kink in slope for the charging lines (or steps) surrounding $N_e = 4$ at $B \approx 0.4\,\mathrm{T}$. This kink indicates the transition from a triplet state (two spins parallel) to a singlet state (two spins antiparallel).

The quantum dot also exhibits reverse singlet-triplet transitions, e.g., for $N_e = 2$ electrons within the dot. If the two levels (0, 0) and (0, 1) in Fig. 5.24(d) get closer to each other than E_{Ex}, one electron will change from (0, 0) to (0, 1) while switching the spin in order to gain by the reduced electron-electron repulsion (exchange energy) for parallel spins.

Many other effects of electron-electron interaction can be studied within quantum dots. This includes such effects that are driven by correlation terms of the Hamiltonian, such that more than two states are involved in the corresponding interaction term (scattering process) according to $E_{\mathrm{XC}} = u_{ijkl}\hat{c}_i^+\hat{c}_j^+\hat{c}_k\hat{c}_l$. The most prominent example is the Kondo effect. Generally, it describes the interaction of two electrons via two extended and two localized states. Both electrons exhibit a fermionic spin, e.g., $S = 1/2$. It turns out that a scattering process of the second order, i.e., the incoming extended state electron switches its spin twice ($\uparrow \rightarrow \downarrow$ and $\downarrow \rightarrow \uparrow$), while the localized state does the spin flip twice in opposite order ($\downarrow \rightarrow \uparrow$ and $\uparrow \rightarrow \downarrow$) reveals a diverging energy gain for the system at $T = 0$ K, similarly to the divergence found for a charge density wave in 1D (eq. (5.13), Section 5.2.1). This drives an increased density of extended states to E_{F}, such that the states can gain from this attractive scattering process.[74] The result is a peak in the density of states at E_{F} called the Kondo peak. For quantum dots, the dot itself represents the localized spin, if filled, e.g., with an odd number of electrons, and the extended spin is given by electrons from the electrodes. The process of double spin-flip (second order scattering process) can transfer an electron from the source to the drain, even if no QD state is within the V_{SD} window. Formally, a virtual excitation of a quantum dot electron into a level above E_{F} of the source and the drain takes place, while the intinerant electron is scattered from a source state to a drain state. Hence, the Kondo effect becomes directly visible, as a conductance, that increases with decreasing temperature in areas of side gate voltages away from the Coulomb peaks. Since the process is most favorable for identical energies for initial and final state (the localized state cannot change its energy by flipping the spin twice), the Kondo conductance is restricted, moreover, to $V_{\mathrm{SD}} \simeq 0$ mV. Finally, the Kondo effect depends exponentially on the exchange interaction between localized and extended electrons, such that it only appears if the tunnel barriers are relatively weak [162].

Since, within the quantum dots, all parameters such as V_{SD}, strength of tunnel barrier, temperature, and spin occupation of the dot are tunable, the quantum dots reveal an excellent test bed for studying the Kondo effect.

5.5.2 Electron-electron interactions in two dimensions

After this introduction into electron-electron interaction processes using artificial atoms, we will discuss electron phases of itinerant electrons employing two dimensions, that are driven by electron-electron interactions. The first two phases describe a particular arrangement driven by the Coulomb repulsion and the exchange interaction, respectively. The final phase, described in Section 5.5.2.3, exhibits completely

74 Recall that an interaction, as described by a perturbation term, requires a scattering of the electrons between occupied and empty states. The most energy efficient way for such a scattering takes place at E_{F}.

novel types of quasi-particles with surprising properties as, e.g., a measurable fractional charge $Q = e/n$ ($n \in \mathbb{N}$).

Before starting, it is important to recall that a many-particle wave function $\Psi_{\rm el}(\vec{x}_1, \vec{x}_2, \ldots, \vec{x}_{N_{\rm e}}, t)$ can consist of more than $N_{\rm e}$ single particle wave functions, such that the relation between electrons and single particle wave functions is not biunique. Transitions between different many-particle wave functions, e.g., after absorbing a photon (Section 2.3.1), can, thus, not be exactly mapped to single particle transitions, which eventually leads to astonishing results in measurements as, e.g., the exchange of fractional charge e/n between two electron systems.

5.5.2.1 Wigner crystal

A classically understandable example of a phase driven by electron-electron interaction is the Wigner crystal. The charge density $n_{\rm e}(\vec{x})$, that evolves at low temperature, is shown in Fig. 5.27(a). A hexagonal pattern appears with one electron per maximum similar to patterns of close-packed atomic crystals. One can show that such a regular arrangement is the optimal pattern for classical electrons at $T = 0\,{\rm K}$.

For the sake of simplicity, we will restrict ourselves to 1D showing that the favorable position of a charged object with charge Q_1 between two identically charged objects with charge Q_2, which are in mutual distance x_0, is the central position at $x = x_0/2$. Therefore, we determine the minimum of the potential repulsion energy $E_{\rm pot}$ for Q_1:

$$\frac{{\rm d}E_{\rm pot}(x)}{{\rm d}x} = \frac{{\rm d}\frac{Q_1 Q_2}{4\pi\varepsilon_0}\cdot\left(\frac{1}{|x|}+\frac{1}{|x-x_0|}\right)}{{\rm d}x} = 0 \Longrightarrow x = x_0/2 \qquad \frac{{\rm d}^2 E_{\rm pot}(x_0/2)}{{\rm d}x^2} > 0 \tag{5.104}$$

The repulsion between the electrons, thus, keeps electrons at identical distances with respect to each other, similarly as the interaction between atoms in a solid keeps the atoms at equal distances. At $T > 0\,{\rm K}$ and constant electron density $n_{\rm e}$, the minimum of free energy $F = U_{\rm int} - TS$ ($U_{\rm int}$: internal energy) can favor the disordered phase because of the corresponding gain in entropy S, that eventually leads to a melting temperature $T_{\rm melt}$ of the Wigner crystal. Since the electrons still strongly interact above $T_{\rm melt}$, a short-range order remains and one calls the phase above $T_{\rm melt}$ a Landau liquid or Fermi liquid (Section 4.2).

Quantitatively, the melting temperature depends on the ratio of the ordering electron-electron repulsion energy being on average $E_{\rm ee}^{\rm cl} \propto 1/\bar{r}_{\rm ee} \propto n_{\rm e}^{1/d}$ ($\bar{r}_{\rm ee}$: average distance between two electrons, d: dimension of the system) and the thermal energy $E_{\rm therm} \propto k_{\rm B}T$ preferring a disordered system, such that one gets in first approximation $T_{\rm melt} \propto n_{\rm e}^{1/d}$.

Quantum mechanically, the periodic arrangement requires additional kinetic energy, since the periodic charge arrangement is realized by the superposition of plane waves with the same phase. Consequently, one needs multiples of the wave vector $k_0 = 2\pi/(2\bar{r}_{\rm ee})$, respectively a kinetic energy of $E_{\rm kin} \propto \hbar^2 k_0^2/2m^* \propto n_{\rm e}^{2/d}$. Since

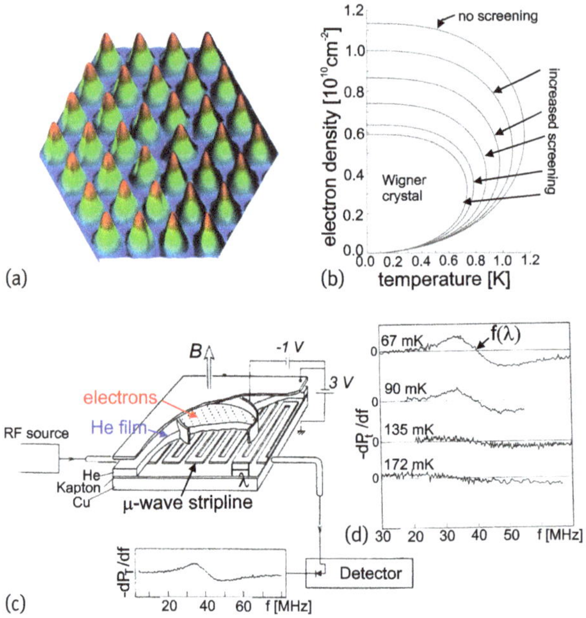

Fig. 5.27: (a) Electron density $n_e(\vec{x})$ of a 2D Wigner crystal. (b) Phase diagram of a 2D electron system. The inner curves include the screening of electron-electron interaction by the substrate. (c) Experimental setup used to probe a Wigner crystal of electrons deposited on suprafluid helium with exemplary absorption spectrum below (P_t: transmitted power, f: frequency). (d) Microwave absorption spectra as a function of frequency measured by the setup in (c) at different T as marked. The frequency $f(\lambda)$ marks the probed phonon of the Wigner crystal at $n_e = 6 \cdot 10^7\,\text{cm}^{-2}$. (b) After [163]. (c)–(d) [164].

this energy cost increases more strongly with n_e than the gain in Coulomb energy ($E_{ee}^{cl} \propto n_e^{1/d}$), a critical density n_c appears, above which the Wigner crystal is not the ground state even at $T = 0\,\text{K}$.

The resulting phase diagram for $d = 2$ is shown in Fig. 5.27(b). The different curves result from the fact that the probed 2D electron system has to be supported by a 3D crystal which screens the electron-electron repulsion ($\varepsilon > 1$, $\lambda_{\text{screen}} < \infty$). Consequently, the area of the Wigner crystal within the phase diagram shrinks.

The calculated phase diagram reveals a required temperature of $T < 1\,\text{K}$ and a required electron density of $n_e < 10^{10}/\text{cm}^2$, i.e., $\bar{r}_{ee} > 100\,\text{nm}$. This differs from the above approximation $E_{ee}^{cl} > E_{\text{kin}} \Longrightarrow \bar{r}_{ee} > 1\,\text{nm}$. Such a discrepancy is also known for atomic crystals. The reason is that melting does not happen by pushing individual atoms or electrons out of its equilibrium position, but starts already, if dislocations (collective shifts of electrons out of its lattice positions) within the ordered lattice become favorable in terms of free energy. This melting by dislocations results in a T_{melt} much smaller than E_{ee}^{cl}/k_B. Nevertheless, the ratio $E_{ee}^{cl}/E_{\text{kin}}$, respectively $E_{ee}^{cl}/(k_B T)$, remains the decisive ratio, albeit with a prefactor of ~ 0.01, for the determination of the phase area where the electrons crystalize.

Experimentally, Wigner crystals were primarily observed on thin films of liquid He. An experimental setup is shown in Fig. 5.27(c). Helium is firstly deposited on a flat substrate at $T < 1$ K. Helium remains a liquid down to lowest temperatures and forms a flat film (thickness ~ 100 nm) on the substrate. Onto this film, electrons are deposited that are created, e.g., by thermal emission from a hot filament. The He film is cut by a circular, negatively charged electrode which keeps the electrons within its area and can be used to change n_e. Cooling the whole film below $T < T_{melt}$ can now create a 2D Wigner crystal of electrons on the He film. To probe the existence of this Wigner crystal, an elegant method measures the lattice vibrations of the crystal, which disappear above T_{melt}. Therefore, a meandering metallic stripe (μ-wave stripline) with distance between the stripes of $L \simeq 20\,\mu$m is positioned below the He film. On the stripe, one applies an ac voltage with tunable frequency f. This ac voltage periodically attracts and repels the electrons being about 3 µm above the stripe, such that the electron lattice gets periodically compressed and expanded. If this excitation is in resonance with a vibration of the lattice, i.e., if f matches the frequency of the phonon of the Wigner crystal with wave length $\lambda = L$, the applied ac voltage loses energy and, thus, the ac resistance of the metallic stripe increases. One measures the transmitted power $P_t(f)$ along the stripe as a function of f. More precisely, one determines the ratio of $P_t(f)$ with and without the electrons on top of the He film and differentiates this ratio with respect to f by lock-in technique in order to improve the signal-to-noise ratio. The result for different T but fixed n_e is shown in Fig. 5.27(d). A resonance appears below $T = 0.1$ K at $f \simeq 35$ MHz, that indicates the phonon frequency of the Wigner crystal at $\lambda = 20\,\mu$m. The resulting $f(\lambda)$ nicely agrees with calculations of the phonon dispersion of a 2D Wigner crystal.

A schematic of a Wigner crystal on top of a He film deposited on a metal is shown in Fig. 5.28(a). Because of the attraction of the electrons by the image charge within the metal, the He is repelled around each electron such that a little valley appears. If the electron lattice gets moved by an electric field, the valley structure will be moved with it, such that the viscosity of the He film contributes to the electrical resistance of the Wigner crystal. The corresponding resistance is shown in Fig. 5.28(b) as measured with the experimental setup sketched in the inset. One can apply a voltage pulse to electrode B1 and measure the time delay until it appears at electrode B3. In practice, the phase difference $\Delta\varphi$ between an ac voltage coupled to B1 and the resulting ac voltage at B3 is determined. The additional grounded electrode in between reduces the direct capacitive coupling between B1 and B3. The measured time delay $\Delta t = \Delta\varphi/(2\pi f)$ can be used to determine the average drift velocity $v_d = L_{13}/\Delta t$ of the Wigner crystal with L_{13} being the distance between B1 and B3. Using the applied voltage amplitude to numerically calculate the electric field $|\vec{E}|$ within the Wigner crystal, and knowing n_e, the specific conductivity reads $\sigma = n_e e v_d / |\vec{E}|$. This determines the resistance of the Wigner crystal after considering additional geometrical factors. Figure 5.28(b) shows

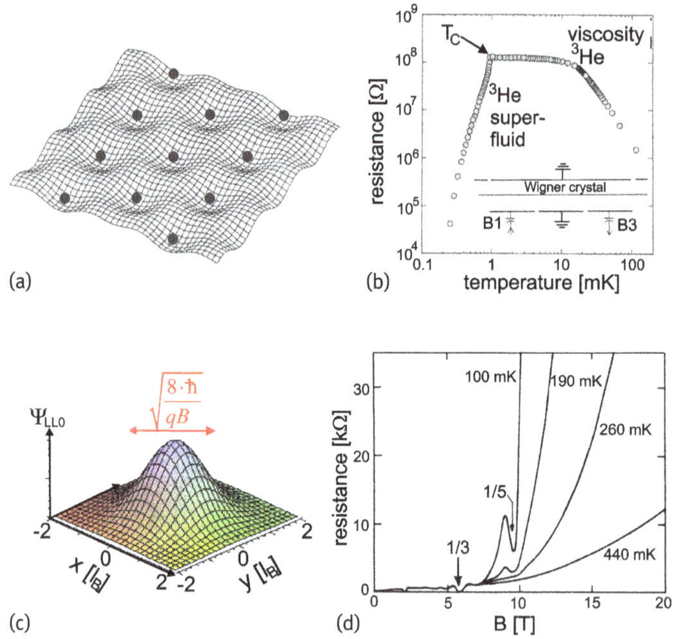

Fig. 5.28: (a) Schematic of a Wigner crystal of electrons (grey balls) on a liquid He film, deposited on a metallic substrate. (b) Resistance of a Wigner crystal on top of liquid He as a function of T, $n_e = 2 \cdot 10^8/\text{cm}^{-2} \Longrightarrow T_{\text{melt}} = 0.27$ K. Inset: Schematic of the experimental setup. (c) $\psi_i(\vec{x})$ of a single particle state within the Landau level with lowest energy. (d) Resistance of a 2D electron system within GaAs as a function of $\vec{B}$ field perpendicular to the plane of the electron system at different T, $n_e = 4 \cdot 10^{10}$ cm^{-2}, electron mobility $\mu = 3 \cdot 10^5$ cm^2/(Vs). Fractional numbers with arrows indicate the filling of the lowest spin-polarized Landau level (filling factor ν). (b) [165], (d) [166].

that the resistance at high temperature decreases with increasing T, since the viscosity η of the ^{3}He decreases with T ($\eta \propto T^{-2}$). Moreover, it drops dramatically below $T \simeq 1$ mK that marks the transition of ^{3}He into the suprafluid phase that exhibits $\eta = 0$ Ns/m^2. Thus, obviously, the Wigner crystal can also be used to probe properties of the underlying He film.

Within solids housing 2D electron systems, the Wigner crystal has not been unambiguously observed at $B = 0$ T.[75] This is primarily due to the electrostatic disorder potential induced by defects, that competes with $E_{\text{ee}}^{\text{cl}}$. In addition, the 2D electron systems with lowest disorder are housed in GaAs/AlGaAs heterostructures (Section 1.3.1), that exhibit a large dielectric constant $\varepsilon = 13$ decreasing $E_{\text{ee}}^{\text{cl}}$ and a low effective mass $m^* = 0.07$ increasing E_{kin}. Consequently, the Wigner crystal is expected at

75 Some evidence for Wigner crystal formation has been found in 1D electron systems in solids housed, e.g., in carbon nanotubes [167].

$n_e < 10^8/\text{cm}^2$, respectively $\bar{r}_{ee} > 1\,\mu\text{m}$, $E_{ee}^{cl} < 0.1$ meV, and $T_{melt} < 10$ mK. In turn, one can roughly estimate that a potential disorder with amplitude 0.1 meV and length scales of 1 μm would destroy the Wigner crystal in favor of an arrangement profiting by placing the electrons into the minima of the electrostatic disorder potential (Section 5.3.1).

In order to create a Wigner crystal in solids at larger density, one has to reduce E_{kin}. The most simple way is applying a $\vec{B}$ field perpendicular to the 2D plane. If $\vec{B}$ is large enough to put all electrons in the lowest spin polarized Landau level ($\vec{B}[T] > n_e/(2.4 \cdot 10^{10}/\text{cm}^2)$)[76], the kinetic energy of all electrons is identical. It is simply the zero point energy $\hbar\omega_c/2$. The corresponding probability density of a single particle state in the lowest Landau level is shown in Fig. 5.28(c). It is a 2D Gaussian function with a full width at half maximum of about the cyclotron diameter $d_c = 50\,\text{nm}/\sqrt{B[T]}$. As soon as $d_c \ll \bar{r}_{ee}$, the Wigner crystal does not require any kinetic energy anymore. Additionally, the antisymmetrization of the spatial many-particle wave function for parallel spins favors a disjunct distribution of the corresponding Gaussian single particle states. Thus, using $B = 10$ T, a Wigner crystal should be stable at about $n_e < 2 \cdot 10^{11}$ cm^{-2}. In reality, it turns out that the Wigner crystal in a $\vec{B}$ field competes at the smallest potential disorders with particular liquid phases that reduce the total electron repulsion energy E_{ee} even better than the Wigner crystal (Section 5.5.2.3). Consequently, the Wigner crystal appears only at fillings of the lowest spin polarized Landau level of less than 1/5. Since this Wigner crystal is typically pinned at the smallest potential disorders, very similar to the charge density waves of Section 5.2.1, it exhibits an insulating conductivity $\sigma(T) \propto \exp(-E_{pin}/k_B T)$ (E_{pin}: pinning energy at a defect). This is indeed found, e.g., in the data of Fig. 5.28(d) for $n_e = 4 \cdot 10^{10}$ cm^{-2}. At $B > 10$ T, corresponding to a filling of the lowest spin polarized Landau level by less than 1/5, the resistance strongly increases with decreasing temperature. The additional observation that the resistance gets strongly reduced at sufficiently large voltage (not shown), corresponding to a depinning of the Wigner crystal from the defects as well as the observed frequency dependence of the resistance corresponding to a phonon excitation of the Wigner crystal (as in Fig. 5.27(d)) confirms the existence of the Wigner cyrstal [165].

5.5.2.2 Stripe phase

A second electron phase in 2D electron systems in a $\vec{B}$ field driven by electron-electron interaction is the so-called stripe phase. It is strongly influenced by the exchange interaction and appears, if an energetically higher, spin-polarized Landau level is half filled. Surprisingly, electrons within the stripe phase prefer to be closer to each other than given by their average distance in order to reduce their overall repulsion. To understand this counterintuitive behavior, we firstly recall the behavior of electrons in

76 The degeneracy of a spin polarized Landau level is $n_{LL}/2 = eB/h$ (Section 5.3.2).

a $\vec{B}$ field without electron-electron interactions as described in Section 5.3.1 and 5.3.2. The $\vec{B}$ field implies Landau and spin levels with energies $E_{n,s} = (n+1/2)\cdot\hbar eB/m^* + s\cdot g\mu B$ with $n = 0, 1, 2, \ldots$ and $s = \pm 1/2$. The levels are broadened by the disorder potential (Fig. 5.10(g)). The corresponding density of states is shown again in the lower part of Fig. 5.29(a). The ratio between electron density n_e and the degeneracy density of a spin-polarized Landau level $n_\mathrm{LL}/2 = e\cdot B/h$ is called the filling factor

$$\nu := \frac{n_\mathrm{e}}{n_\mathrm{LL}/2} = \frac{n_\mathrm{e}}{eB/h} \,. \tag{5.105}$$

Recall that $n_\mathrm{LL}/2$ is identical to the areal density of pairs of magnetic flux quanta penetrating the 2D system. For $\nu = 9/2$, the lowest two Landau levels are completely filled and, additionally, half of the lower spin level of the third Landau level ($n = 3$, $s = -1/2$) is filled with electrons. One could expect that the electrons within the ($n = 3$, $s = -1/2$) level distribute themselves in regular lateral distances as in a Wigner crystal (Section 5.5.2.1) in order to reduce their electron-electron repulsion energy. However, the charge densities of individual electronic states in Landau levels with $n > 0$ are ring structures (Fig. 5.29(b)) and not Gaussian functions making their ordering less straightforward.

The radius of the ring corresponds to the quantized cyclotron radius $r_\mathrm{c} = \sqrt{2n+1}\cdot l_B$ and its width is about the magnetic length $l_B = \sqrt{\hbar/eB}$. The resulting Wigner crystal at half filling (e.g., $\nu = 9/2$) is sketched in Fig. 5.29(c). The boxes mark areas containing a flux of h/e with density $n_\mathrm{LL}/2$ and the red rings mark the occupied states of the half-filled Landau level. Obviously, the states overlap strongly, i.e., the resulting direct Coulomb repulsion, called Hartree energy E_Hartree (eq. (4.12)),

$$E_\mathrm{Hartree} = \int\int |\psi_i(\vec{x}_i)|^2 \frac{e^2}{4\pi\varepsilon\varepsilon_0\cdot|\vec{x}_i - \vec{x}_j|} |\psi_j(\vec{x}_j)|^2 \; \mathrm{d}^2\vec{x}_i \mathrm{d}^2\vec{x}_j \tag{5.106}$$

deviates strongly from the Coulomb repulsion of point charges in the center of the boxes. A more detailed calculation reveals that the Coulomb repulsion energy barely changes if two overlapping rings are pushed closer together. The screening of the electron-electron repulsion by electrons from lower Landau levels, moreover, implies that the Coulomb repulsion energy drops dramatically, if the distance between two electron ring centers gets larger than $2r_\mathrm{c}$. Consequently, the arrangement of the rings in stripes, as shown in Fig. 5.29(d), does increase the repulsion with neighbors in the stripe only slightly, but, at the same time, reduces the interactions with electrons in neighboring stripes more strongly. It, thus, appears favorable to separate the stripes as much that rings of neighboring stripes do not overlap anymore.

Notice that the total charge density $n_\mathrm{e}(\vec{x})$ is only slightly changed by this rearrangement. The amplitude of the resulting charge density wave perpendicular to the stripes (Fig. 5.29(a), top), is only 10–20% of the average charge density within the half-filled Landau level. One can calculate more precisely, that the optimal wave length of the charge density wave is $\lambda \simeq 2.7\cdot r_\mathrm{c}$ as marked in Fig. 5.29(e), where only the centers of the electron rings are marked as dots. For electrons in the third Landau level at

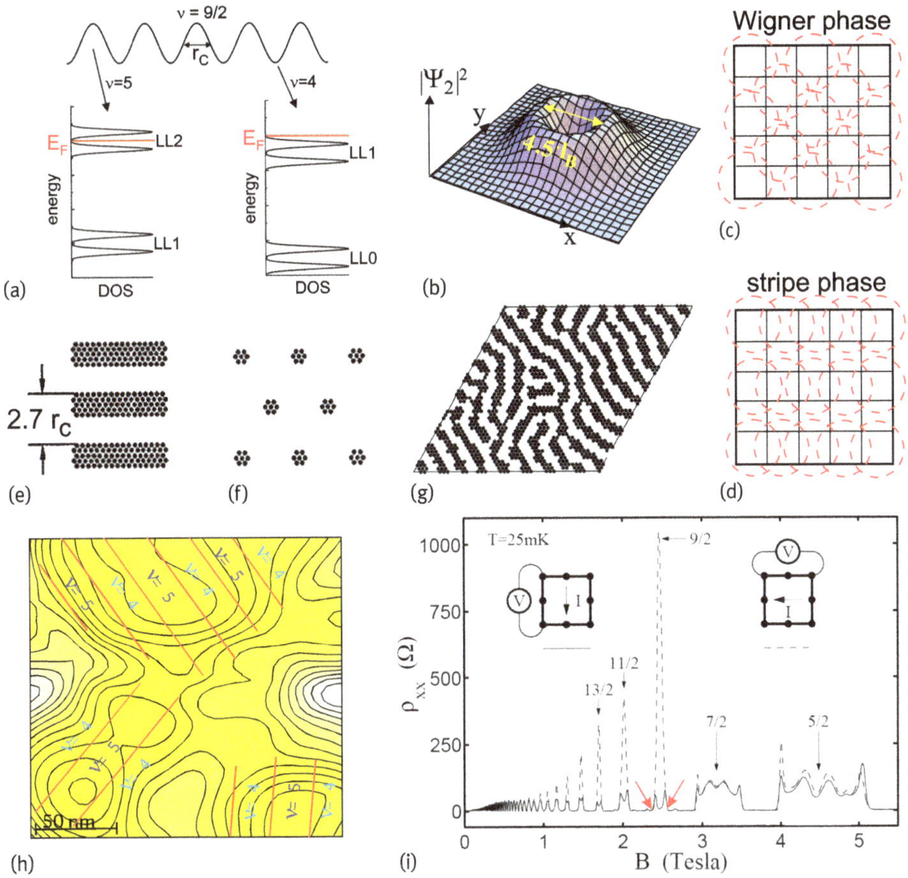

Fig. 5.29: (a) Schematic of the stripe phase. Top: Charge density $n_e(x)$ perpendicular to the stripes at $\nu = 9/2$ with marked cyclotron radius r_c. Bottom: Local density of states corresponding to the areas marked by arrows with indicated local filling factor. (b) $\psi_i(\vec{x})$ of a state in Landau level LL2. (c), (d) Schematic of the ordered electron rings in a Wigner crystal (c) and in the stripe phase (d) for half filling of a spin polarized Landau level LL2, i.e., at $\nu = 9/2$. (e) Sketch of the stripe phase obtained at half filling of higher Landau levels, where the dots symbolize the center of the rings in (c), (d). (f) Bubble phase observed at a quarter filling of a spin polarized Landau level. (g) Numerically simulated stripe phase on a triangular lattice. The stripes meander which indicates that the system is not in the energetic ground state but only close to it. (h) Sketch of a disordered stripe phase (red lines with indicated local filling factor ν in between the lines) in the presence of an electrostatic disorder potential as displayed in color scale with black equipotential lines. (i) Longitudinal resistance of a 2D electron system in GaAs as a function of $\vec{B}$. The resistance is measured for current along two perpendicular directions (full and dashed lines) as marked in the insets, $n_e = 2.3 \cdot 10^{11}\ \text{cm}^{-2}$, $\mu = 9 \cdot 10^7\ \text{cm}^2/(\text{Vs})$. Fractional numbers with arrows mark the filling factor ν. Red arrows indicate likely bubble phases. (e)–(g) [168], (i) [169].

$B = 10\,\mathrm{T}$ one finds $\lambda \simeq 60$ nm. Quantitatively, the repulsion term E_{Hartree} of the stripe phase at this λ does not differ from E_{Hartree} of a Wigner crystal, that would imply that a mixture of both should appear in experiment. The stripe phase is, however, favored by the exchange interaction[77]

$$E_{\mathrm{Ex}} = \int\int \psi_i^*(\vec{x}_m)\psi_j^*(\vec{x}_n)\frac{e^2}{4\pi\varepsilon\varepsilon_0 \cdot |\vec{x}_n - \vec{x}_m|}\psi_j(\vec{x}_m)\psi_i(\vec{x}_n)\ \mathrm{d}^2\vec{x}_m\mathrm{d}^2\vec{x}_n\ . \tag{5.107}$$

For parallel spins of electrons, as present in a spin polarized Landau level, E_{Ex} contributes with a negative sign decreasing the electron-electron repulsion. Consequently, the optimal arrangement favors small E_{Hartree} and large E_{Ex}. A large E_{Ex}, however, requires a direct overlap of the contributing wave functions (eq. (5.107)), i.e., $\int\int \psi_i^*(\vec{x}_n)\cdot\psi_j(\vec{x}_n)\mathrm{d}^2\vec{x}_n \gg 0$, contrary to E_{Hartree}, that barely depends on the overlap at small distance. Thus, E_{Ex} favors the strong overlap of the wave functions within the stripe phase. The stripe phase is, thus, a consequence of both, the strong reduction of the direct Coulomb repulsion E_{Hartree} for large distances between the electrons ($> 2r_{\mathrm{c}}$) and the increasing exchange interaction for increasing overlap of the occupied wave functions, that is favorable for electrons with parallel spin.

The quantitative comparison between the different phases requires complex, partly numerical calculations [168, 170] and will not be described in detail. In short, firstly one transforms the Schrödinger equation of all electrons in a way, such that the completely filled Landau levels are approximated by a wave number ($q = |\vec{q}|$) dependent dielectric constant $\varepsilon(q)$. For the electron-electron interaction within the half-filled Landau level, one obtains $\varepsilon(q) = \varepsilon\varepsilon_0\cdot(1+2/(qa_{\mathrm{B}}^*)\cdot(1-J_0^2(qr_{\mathrm{c}}))$ with effective Bohr radius $a_{\mathrm{B}}^* = \varepsilon m_{\mathrm{e}}/m^* \cdot 0.5$ Å and zeroth Bessel function J_0. The completely filled levels are, thus, described as a polarizable background. Then, one calculates wave vector dependent interaction energies E_{Hartree} and E_{Ex} for the half filled Landau level using the Fourier transform of the interaction potential $\Phi_{\mathrm{el}}(q) = 2\pi e^2/\varepsilon(q)q$. This results in an oscillating Hartree term being approximately (besides a q-independent term): $E_{\mathrm{Hartree}}(q) \propto 1/(2 + q\cdot a_{\mathrm{B}}^*)\cdot J_0^2(qr_{\mathrm{c}})$. The exchange term is positive for all $q > 0$, i.e., it always prefers the arrangement in stripes, while E_{Hartree} determines the distance between the stripes.

The approximate ground state derived from a numerical calculation for the arrangement of the electron rings on a triangular lattice for a half filled spin-polarized tenth Landau level is shown in Fig. 5.29(g). One recognizes that the stripes slightly meander. This might be due to the finite size effects in the simulation or due to the fact that the simulation did not find the absolute ground state. It also shows that a bending of the stripes is easily possible, implying that the stripes can follow a moderate

77 Recall that the electron-electron interaction energy contains terms of the form $E_{\mathrm{ee},ijkl} = \int\int \psi_i^*(\vec{x}_m)\psi_j^*(\vec{x}_n)\frac{e^2}{4\pi\varepsilon\varepsilon_0\cdot|\vec{x}_n-\vec{x}_m|}\psi_k(\vec{x}_m)\psi_l(\vec{x}_n)\ \mathrm{d}^2\vec{x}_m\mathrm{d}^2\vec{x}_n$ including four different single particle wave functions. But mostly, the two terms with only two single particle wave functions E_{Hartree} ($i = k, j = l$) and E_{Ex} ($i = l, j = k$) are relevant. All other terms are called correlation terms (Section 4.4).

disorder potential as sketched in Fig. 5.29(h), where areas of large filling are placed in potential minima. In addition, the stripe order costs entropy leading to a melting transition at higher T. For typical electron densities $n_e = 10^{11} - 10^{12}/\text{cm}^2$, the melting temperature can be estimated to be $T_{\text{melt}} \simeq 1$–$10\,\text{K}$.

Experimental evidence for the stripe phase is shown in Fig. 5.29(i). The resistance of a 2D electron system with high mobility μ appears to be largely anisotropic at filling factors 9/2, 11/2, etc. Temperature dependent measurements at these ν show metallic conductivity in one direction and insulating conductivity in the perpendicular direction. This can be easily rationalized by the stripe phase. Within the transition regions between different integer filling factors of the stripes (Fig. 5.29(a)), edge type states appear at E_F following the stripe directions (Section 5.3.2). They can transport current along the stripes, but not perpendicular to it. Thus, one gets metallic behavior along the stripes and insulating behavior perpendicular to the stripes.

In order to measure this, one needs very low T (25 mK) and high electron mobility μ such that the stripes are oriented in the same direction over macroscopic distances. For example, the electrodes for the experiments shown in Fig. 5.29(i) are 5 mm apart and the mean free path of the electrons within the 2D system is 0.5 mm at 25 mK. Most likely, some effects related to the crystal structure prefer one of the two directions for the orientation of the stripes by small differences in potential energy, but this is not finally clarified.

The combination of large overlap between neighboring wave functions (large E_{Ex}) and large distances between other neighboring wave functions (small E_{Hartree}) can, of course, also be realized differently as shown, e.g., in Fig. 5.29(f). In fact, calculations predict that such phases, called bubble phases, can appear for Landau levels at $\nu \in [n + 1/n, n + 0.4]$ with $n > 2$. These bubble phases are similar to the Wigner crystal, but with many electrons within a density maximum (bubble). The reason, that they are more stable than the stripe phase, is that the electrons at the edges of the electron islands can avoid the Hartree repulsion from neighboring islands more effectively than electrons can avoid each other within the stripe phase. With increasing filling factor, the islands start to overlap. This then favors the stripe phase requiring less free space in order to avoid overlap of electrons from different entities. So far, there are some indications of the existence of these bubble phases close to integer filling factors $\nu \approx n$, that, moreover, seem to appear and to disappear consecutively as a function of ν, but an unambiguous proof is missing.[78] At very small differences to integer fillings $\nu - n < 1/n$, the bubble phase with one electron per bubble appears in theory and is identical to the Wigner crystal discussed in Section 5.5.2.1.

Notice that the stripe phase appears for a half filled band due to the ring like wave functions within the band. This should be contrasted with the Mott–Hubbard transi-

78 Micro-wave experiments and non-linear $I(V)$ characteristics indicate, moreover, that the peaks marked by red arrows in Fig. 5.29(i) are related to insulating bubble phases.

tion, which also appears for a half filled band, but for a band, that consists of nearly point-like electron wave functions (Fig. 4.14 in Section 4.5.3). Hence, details of the wave functions can be decisive for the electronic phase and can lead to very different transport behaviors (anisotropic metal versus insulator).

5.5.2.3 Fractional quantum Hall effect

Within the lowest Landau level ($\nu < 1$), we already found the Wigner crystal at $\nu < 1/5$ as an insulating state (Section 5.5.2.1). Experimentally, a number of minima of the longitudinal resistance accompanied by plateaus of the Hall resistance are found at larger ν as shown in Fig. 5.30(a). The plateaus exhibit $R_{\mathrm{Hall}} = h/e^2 \cdot 1/\nu$ with ν being the filling factor at the corresponding minimum of the longitudinal resistance. The minima appear at rational filling factors $\nu \in \mathbb{Q}$ with odd denominator.[79] They are also

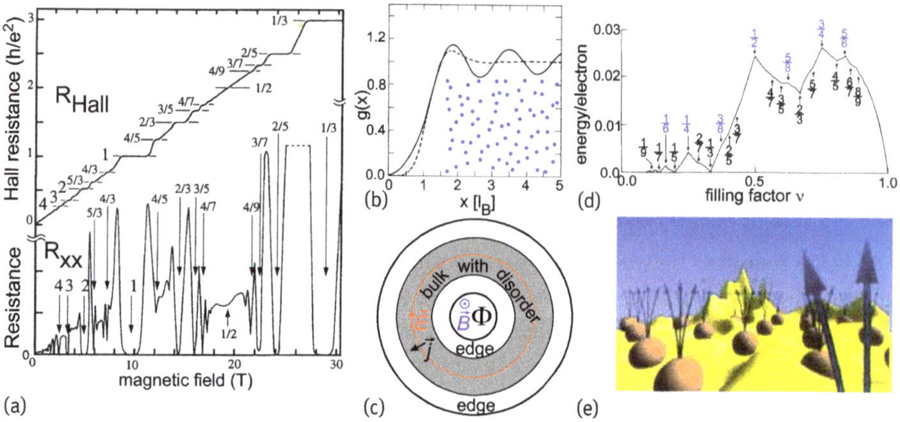

Fig. 5.30: (a) Longitudinal resistance R_{xx} and Hall resistance R_{Hall} of a 2D electron system at high mobility $\mu \simeq 10^8\,\mathrm{cm^2/Vs}$, $n_e = 2 \cdot 10^{11}/\mathrm{cm}^2$. Fractional numbers at the plateaus of R_{Hall} and at the arrows marking the minima of R_{xx} are filling factors ν. (b) Spatial correlation function $g(x)$ of a Wigner crystal (full line) and of the Laughlin liquid at $\nu = 1/3$ (dashed line) with x being the distance between two electrons. Inset: Snap shot of the electrons (blue points) in a Laughlin liquid according to numerical simulations. (c) Corbino geometry (ring area between thick lines) with edge (white areas) and insulating bulk (grey area) due to disorder. The ring is pierced by a magnetic flux Φ, respectively a $\vec{B}$ field perpendicular to the plane. The red line marked with electric field $\vec{E}$ and crossed by current density $\vec{j}$ is used for the integration in eq. (5.113). It explains the appearance of fractional charge (see text). (d) Numerically calculated energy per electron of the ground state as a function of ν. The energy is given in units of $e^2/(4\pi\varepsilon\varepsilon_0 l_B)$. Filling factors with odd (even) denominator are given in black (blue). (e) Sketch of the stable quasiparticles (electron + three fluxes of h/e) at $\nu = 1/3$. The yellow background is a sketch of the electrostatic disorder potential. (a) [171], (b)–(d) [172], (e) [173].

79 More exotic electron phases with minima at even denominators exist, e.g., at $\nu = 5/2$, but are beyond the scope of this book.

somehow symmetric around $\nu = 1/2$. The effect is called the fractional quantum Hall effect (FQHE). It can be explained by a new type of quasiparticles arising from the many-particle wave function of the 2D electron system.

The explanation of the FQHE is based on the preference of certain electron-electron distances in $\vec{B}$ field, that can be understood semi-classically. The Coulomb repulsion is rotationally symmetric and implies that two electrons circle around each other within the $\vec{B}$ field. Since the phase of their single-particle wave functions ψ_i is varied by encircling a magnetic flux and the phase of ψ_i has to be unambiguous, i.e., it must be a multiple of 2π per circle, one obtains preferential distances $\overline{r}_{\text{ee}} = |\vec{x}_n - \vec{x}_m|$ between two electrons n and m encircling each other: $\overline{r}_{\text{ee}} \simeq l_B \cdot \sqrt{2n}$ $(n \in \mathbb{N})$. This argument can be transferred to a system of many electrons encircling each other in pairs at the preferential distance. Consequently, one finds electron densities n_{e} that are favorable, if the number of fluxes h/e is identical to a multiple of the number of electrons. The corresponding electron phases are called incompressible phases, since they exhibit a band gap for excitations implying that a change of electron density by a minimal amount costs a finite energy penalty. Such a system with band gap is an insulator that explains the minima in R_{xx} straightforwardly (Section 5.3.2). This semi-classical idea would predict minima at fractional numbers $\nu = 1/2, 1/3, 1/4, \ldots$, but it does not explain all minima observed in experiment and wrongly predicts minima at fractional numbers with even denominators. The latter is due to the fact that it neglects the fermion character of the electrons implying an antisymmetric exchange statistics.

A more detailed argument has been given by Bob Laughlin. He used three requirements from general physical principles in order to guess a many-particle wave function, that allowed a reasonable quantitative description of the most prominent minimum at $\nu = 1/3$.

The requirements are:

1. Since electron-electron interaction as well as $\vec{B}$ field are rotationally symmetric in the 2D plane, the many-particle wave function $\Psi_{\text{el}}(\vec{x}_1, \ldots \vec{x}_{N_{\text{e}}})$ must be an eigenfunction of the total orbital momentum.
2. Since electrons are fermions, the many-particle wave function must be odd with respect to an exchange of two coordinates. Since the parallel spins in the lowest spin polarized Landau level ($\nu < 1$) imply an even spin function, the spatial wave function has to be odd.
3. If the electron-electron interaction at large enough $\vec{B}$ is significantly smaller than $\hbar\omega_{\text{c}}$, the many-particle wave function consists of single-particle wave functions $\psi_i(\vec{x})$ of the lowest Landau level[80]

$$\psi_i(\vec{x}) = \exp\left(\frac{x^2 + y^2}{4}\right) \cdot \left(\frac{\text{d}}{\text{d}x} + \text{i} \cdot \frac{\text{d}}{\text{d}y}\right) \cdot \exp\left(-\frac{x^2 + y^2}{2}\right). \tag{5.108}$$

80 For the sake of simplicity, we neglect normalization factors of the wave functions, that are irrelevant for the following arguments.

In order to make the representation simple, we use the coordinates $z := (x + iy)/l_B = |z| \cdot \exp(i\varphi)$. Then, the arguably most simple function obeying all three requirements for N_e electrons is:

$$\Psi_{m,\mathrm{el}}(z_1, z_2, z_3, \ldots, z_{N_e}) = \prod_{k=1}^{N_e} \prod_{j<k} (z_j - z_k)^m \cdot \exp\left(-1/4 \cdot \sum_{l=1}^{N_e} |z_l|^2 \right). \tag{5.109}$$

This function exhibits a quantized orbital momentum of $L_{\mathrm{tot}} = \hbar \cdot \frac{N_e \cdot (N_e - 1)}{2} \cdot m$ (without derivation) and is antisymmetric with respect to exchange of two coordinates, if m is odd. Implicitly, we use the idea that all electrons within the many-particle wave function are identical (m is the same for each pair (j, k)) and that the electron-electron repulsion, which has to be minimized, interacts between pairs of electrons only, i.e., the polynomial factors consist of differences of only two coordinates. An analysis of this function reveals that it exhibits a lower total energy than the Wigner crystal for most ν. This can be rationalized by regarding the correlation function[81] $g(x)$ shown for $m = 3$ in Fig. 5.30(b) (dashed line). While the exchange-correlation hole around the selected electron can be approximated by a x^2 function for the Wigner crystal (solid line), it is a x^{2m} function for the Laughlin wave function (dashed line). Consequently, electrons in the Laughlin function are less close to each other. This results in a smaller Coulomb repulsion energy. Additionally, the Laughlin function shows barely a minimum at the position of a neighboring electron, while the Wigner crystal shows long range oscillations of $g(x)$ indicating that the Laughlin wave function represents an electron liquid, while the Wigner crystal is an electron solid. The inset of Fig. 5.30(b) is a typical snapshot of the electrons within the liquid as obtained by numerical calculations [172]. It reveals the long range disorder in the electron distribution and the preferentially large distance between neighbors by naked eye.

Interpreting the square of $\Psi_{m,\mathrm{el}}$ of the Laughlin liquid as a classical probability distribution of particles at finite temperature T within a potential energy E_{pot}, one can understand the preference of filling factor $\nu = 1/3$. Therefore, one writes:

$$|\Psi_{m,\mathrm{el}}|^2 = e^{-E_{\mathrm{pot}}(z_1,\ldots,z_{N_e})/k_B T} \tag{5.110}$$

with

$$E_{\mathrm{pot}}(z_1, \ldots, z_{N_e}) := -2m^2 \sum_{k=1}^{N_e} \sum_{j<k} \ln |z_j - z_k| + m/2 \cdot \sum_{l=1}^{N_e} |z_l|^2 \quad \text{and} \quad k_B T := 2m. \tag{5.111}$$

Here, $k_B T = 2m$ is chosen such that E_{pot} becomes identical to the potential of a classical 2D plasma made out of N_e charges with charge m. The charges pairwise repel

81 A correlation function $g(x)$ describes the probability to find one of the other electrons in a distance x from a selected electron.

each other via a logarithmic function (first term on the right of $E_{pot} \ldots = \ldots$).[82] Moreover, they are confined in a parabolic background potential (second term on the right of $E_{pot} \ldots = \ldots$). The plasma is, hence, accompanied by a homogeneous circle of charges with opposite sign. The charge density of this background is given by $1/(2\pi l_B^2)$, if one includes the constants correctly within eq. (5.109) and calculates the potential of a homogeneously charged disk.[83] These background charges keep the charges m within the area. It is energetically favorable to keep the plasma electrically neutral on small length scales, i.e., each area of a flux of h/e gets $1/m$ from the charge carriers with charge m. Consequently, the ideal charge density is $N_e = N_\Phi/m$ with N_Φ being the number of magnetic flux portions of strength h/e. Thus, we found the preferential filling factor to be $\nu = 1/m$ (Fig. 5.31(b)). Each additional m-type charge or each additional flux h/e (background charge) would locally charge the equivalent plasma. This is energetically unfavorable.

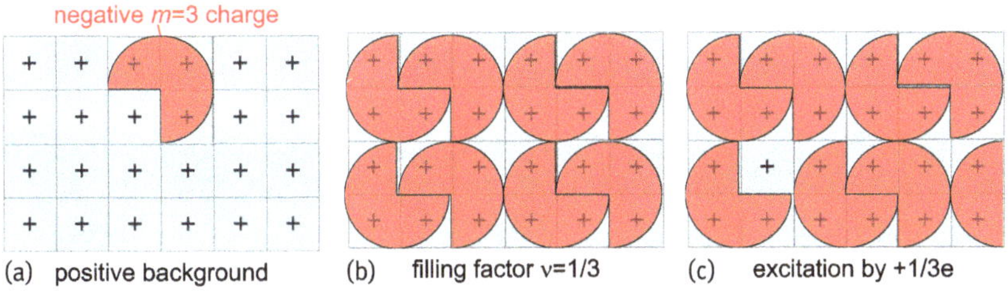

Fig. 5.31: Visualization of the plasma analogue of the fractional quantum Hall effect close to $\nu = 1/3$. (a) Positive background (blue) with individual +1 charges and a single $-m$ charge (red object) representing an electron for the case $m = 3$. (b) Favorable filling of the −3 charges at $\nu = 1/3$. (c) Charge excitation by a third of a −3 charge, that represents the excitation by charge of $e/3$, slightly away from $\nu = 1/3$.

Consequently, the Laughlin liquid is most stable at filling factors $1/m$ with an odd integer m. This phase is separated by a band gap from the phase with an additional charge. This has also been found numerically, i.e., the 2D electron system at $\nu = 1/3, 1/5 \ldots$ is an insulator with a band gap. Hence, it requires a minimum energy to add an additional quasiparticle. Typical excitation energies at $m = 3$ are $\Delta E \simeq 0.027e^2/(4\pi\varepsilon\varepsilon_0 l_B)$.[84] In fact, the strongest minimum of $R_{xx}(B)$ (Fig. 5.30(a)), that implies the largest excitation gap, is observed at $\nu = 1/3$. At lower $\nu \leq 1/5$, the

82 The logarithmic term becomes plausible by using Maxwell's equation in 2D reading $\oint \vec{E} d\vec{s} = Q/\epsilon\epsilon_0$ leading to $|\vec{E}| \propto (r - r_0)^{-1}$ and $E_{pot} \propto \ln(r - r_0)$, where $r - r_0$ is the distance to a particular charge.
83 One can show that a homogeneous circle of charge leads to a parabolic confinement potential (see also Section 5.5.1.1, Fig. 5.24(c)).
84 Using $\varepsilon = 13$ as for GaAs and $B = 10$ T implying $l_B = 9$ nm, we obtain $\Delta E = 0.3$ meV.

Wigner crystal often gets energetically favorable, such that our theory only applies for $\nu = 1/3$ and partly, depending on the details of the system, for $\nu = 1/5$.

Interestingly, the charged excitations of these ground states carry a charge of $Q = e/m$ ($e = 1.6 \cdot 10^{-19}$ C). Intuitively, this is the charge, that can be moved between the flux boxes (Fig. 5.31(c)). More formally, this gets evident, if one analyzes the ring geometry in Fig. 5.30(c), that is pierced by a current carrying coil in the center marked by the magnetic flux Φ and the corresponding $\vec{B}$ field pointing out of the plane. If the interior of the 2D sample (inner ring) is insulating as for $\nu = 1/m$ or due to disorder forming localized states (Section 5.3.2), the tensor of resistivity reads:

$$\boldsymbol{\rho} = \begin{pmatrix} 0 & B/(n_\mathrm{e} e) \\ B/(n_\mathrm{e} e) & 0 \end{pmatrix} . \tag{5.112}$$

Changing Φ adiabatically, i.e., slowly, with the help of the coil implies by Maxwell's equations (eqs. (2.41)–(2.44), eq. (1.1) (Ohm's law)):

$$\begin{aligned} &\frac{\mathrm{d}\Phi}{\mathrm{d}t} = \iint \frac{\mathrm{d}\vec{B}}{\mathrm{d}t}\, \mathrm{d}\vec{A} = -\oint \vec{E}\, \mathrm{d}\vec{s} = -\oint \boldsymbol{\rho} \cdot \vec{j}\, \mathrm{d}\vec{s} \\ \Longrightarrow \quad &\frac{\mathrm{d}\Phi}{\mathrm{d}t} = -\rho_{xy} \oint \frac{\mathrm{d}\rho_{\mathrm{c}\perp}(\vec{x})}{\mathrm{d}t} \mathrm{d}|\vec{s}| \quad \Longrightarrow \quad \Phi(t) - \Phi(0) = -\rho_{xy} \cdot \Delta Q_\perp . \end{aligned} \tag{5.113}$$

The corresponding contour of the closed integration path (red line) and the vectors $\vec{E}$ and $\vec{j}$ are sketched in Fig. 5.30(c). Naturally, $\vec{E}$ is perpendicular to $\vec{B}$ by Maxwell's equation and $\vec{j}$ is perpendicular to $\vec{E}$ by the exclusively non-diagonal elements of $\boldsymbol{\rho}$. Finally, we used the definition of the current density $|\vec{j}| = \mathrm{d}\rho_\mathrm{c}/\mathrm{d}t$ with charge density $\rho_\mathrm{c}(\vec{x})$ (Section 1.2.1) and considered that the charge flow is perpendicular to the integration contour ($\vec{E}$) via the non-diagonal matrix $\boldsymbol{\rho}$. This is symbolized by $\rho_{\mathrm{c}\perp}$ and the total charge $\Delta Q_\perp$ traversing this contour (red line). Thus, the addition of a magnetic flux causes a charge transport $\Delta Q_\perp$ through a closed ring with arbitrary radius within the insulator. Adding a flux of $\Phi(t) - \Phi(0) = h/e$ implies a change of the single particle wave functions by the irrelevant factor $\mathrm{e}^{\mathrm{i}2\pi}$. Thus, the transport of charge across the ring, i.e., from the inner edge to the outer edge or vice versa, is equivalent to an excitation of the system without changing the contributing $\psi_i(\vec{x})$ and without changing the energy, if the process is performed adiabatically, i.e., slowly. The Laughlin wave function describing the interior of the ring, consequently remains unchanged. With $n_\mathrm{e} = 1/m \cdot eB/h$ at $\nu = 1/m$ using the degeneracy of a spin polarized Landau level eB/h (Section 5.3.2, page 278), we find

$$\Delta Q_\perp \underset{\text{eq. (5.113)}}{=} \frac{h}{e\rho_{xy}} \underset{\text{eq. (5.112)}}{=} \frac{hen_\mathrm{e}}{eB} = \frac{he^2B}{heBm} = \frac{e}{m} . \tag{5.114}$$

The edge channels, thus, exchange fractional charges e/m by redistribution of the electron wave functions during the adiabatic cycle and, consequently, can also transport fractional charges e/m. This fits with the intuition that, e.g., the positive charge in

Fig. 5.31(c) can be moved through the lattice by adequate rearrangement of the $m = 3$ negative charges.

A more formal quantum mechanical way to explain the fractional charges is to add the flux h/e to Laughlin's wave function (eq. (5.109)). Such an additional flux must be represented by an additional zero for each single particle wave function, since surrounding the flux gives a phase factor of 2π. This would lead to a diverging kinetic energy after being contracted to a point, except if this point has a weight of zero (Section 5.4.2, Fig. 5.13(a)). The additional flux can, thus, be represented by a factor $\prod_{j=1}^{N_e}(z_0 - z_j)$ in front of Laughlin's wave function (eq. (5.109)). However, this factor differs from the usual factor of an additional electron with index k, $\prod_{j=1}^{N_e}(z_j - z_k)^m$, in a way that adding an additional electron to the system would be equivalent to adding three such fluxes. Thus, adding a flux of h/e to the 2D electron system, which is an allowed operation, one formally adds a fractional charge of $Q = e/m$.

The fractional charge of the quasi-particles can be directly measured by the square of the amplitude of the current noise S_I with units A^2/Hz. Therefore, one uses a setup as sketched in Fig. 5.32(a). One drives a current from source to drain, i.e., along the upper edge channel. Within the constriction, that can be changed in width by the two marked gate voltages, charge carriers can be scattered from the upper edge channel to the lower one and, thus, can be detected at the electrode labeled "Noise." The strength of the noise of this current is given by the arbitrary back scattering of individ-

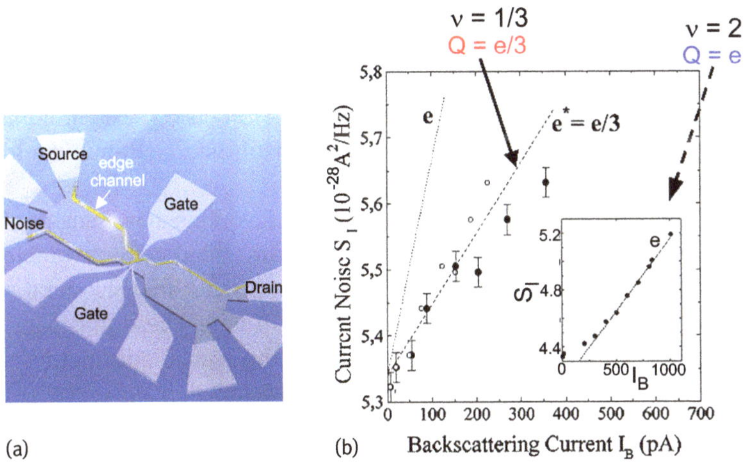

Fig. 5.32: (a) Sketch of the sample used to measure the fractional charge within the fractional quantum Hall regime. The voltage is applied between source and drain and drives a current through the edge channels of the 2DES, while the backscattered current from the constriction at the gates is analyzed via the contact "Noise." (b) Square of the amplitude of the current noise per frequency interval S_I as a function of the current I_B detected at the electrode "Noise" at filling factor $\nu = 1/3$ (main image) and filling factor $\nu = 2$ (inset). Lines mark expected noise levels for different charges of the quasiparticles as marked. (b) [174].

ual charges at the constriction. The noise at given backscattered current gets reduced, if the individual charges get smaller, since then the current flows in smaller pieces and, hence more continuously. Figure 5.32(b) shows indeed that the current noise at $\nu = 1/3$ is significantly smaller than expected for an individual charge of e and can be fitted assuming a charge of $Q = e/3$. This result is not contradicting the quantization of the elementary charge in units of e, but appears due to the correlated motion of all electrons within the system. This correlated motion makes $Q = e/m$ a measurable quantity.

Introducing disorder, as sketched in Fig. 5.30(e), leads to insulating behavior, even if ν slightly deviates from $\nu = 1/m$. The argument is similar as in the integer quantum Hall regime (Section 5.3.2). The excited quasiparticles with charge e/m (electrons or holes) are localized in the disorder potential (Fig. 5.30(e)), such that they cannot move through the system and, hence, they cannot transport charge. Only the edge channels contribute to the conductivity. As for the integer QHE (Section 5.3.2), they form a ballistic, one-dimensional conductor, that, however, transports particles with fractional charge $Q = e/m$. Since each ballistic channel contributes a conductivity of $Q \cdot e/h$ (derivation at the end of Section 5.3.2 or in Section 1.4.1), we get a Hall resistance: $R_{\text{Hall}} = h/eQ = h/e \cdot m/e = hm/e^2$. This explains the plateau value in R_{xy} around $\nu = 1/3$ (Fig. 5.30(a)). With the help of the symmetry of electrons and holes within the lowest Landau level, it also explains the plateau around $\nu = 2/3$, that represents the lowest Landau level occupied by $1/3$ with holes.

But what about the other minima in R_{xx} and plateaus in R_{Hall}? Figure 5.30(d) shows the energy of the two-dimensional electron system in a $\vec{B}$ field as resulting from a numerical calculation for a few electrons. It displays the ground state energy per electron as a function of ν. Besides the minima at $\nu = 1/3$ and $\nu = 2/3$, minima appear at other ν, that, at least, partly correspond to the ν of the $R_{xx}(B)$-minima in Fig. 5.30(a).

A possible explanation for the additional minima is a favorable arrangement of the quasi particle excitations (charge: $e/3$ = flux h/e) that are added to the favorable state at $\nu = 1/3$. Formally, one repeats the argument of the plasma analogue for the remaining excited quasiparticles not belonging to the $\nu = 1/3$ state. These quasiparticles move in an equivalent background of electrons. Since the excited quasiparticles are formally similar to bosons[85], one obtains a Laughlin wave function (eq. (5.109)) of them with even exponent, that is called k to distinguish it from the odd exponent m. Hence, even integers k of the inverse filling factor N_e/N_{qp} (N_{qp}: quasi particle number) are preferred within the plasma analogue. A favorable filling factor is, thus, defined

85 Using the wave functions of the excited quasiparticles of the form $\prod_{j=1}^{N}(z_0 - z_j)$ does not lead to a negative sign after exchanging two quasiparticle terms z_j, i.e., they are bosons. As a side remark, we mention that the real exchange statistics of the $\nu = 1/3$ quasiparticles with charge $e/3$ is fractional, i.e., an exchange of two quasiparticles reveals a phase factor $e^{i\pi/q} \neq \pm 1$. This is called anyon statistics [175].

by (N_Φ: number of fluxes h/e):

$$N_e = k \cdot N_{qp}\,, \quad N_\Phi = mN_e + \alpha_\pm \cdot N_{qp}\,, \tag{5.115}$$

with $\alpha_\pm = \pm 1$ in order to describe electron and hole excitations of the $\nu = 1/3$ state on the same footing. Solving for the filling factor $\nu = N_e/N_\Phi$ reveals

$$\nu = \frac{kN_{qp}}{mN_e + \alpha_\pm N_{qp}} = \frac{k}{mk + \alpha_\pm} = \frac{1}{m + \alpha_\pm/k}\,. \tag{5.116}$$

This novel stable state can again be excited by fractionally charged quasiparticles with symmetric exchange statistics, such that one gets a hierarchy of favorable filling factors:

$$\nu = \cfrac{1}{m + \cfrac{\alpha_{1,\pm}}{k_1 - \cfrac{\alpha_{2,\pm}}{k_2 - \cfrac{\alpha_{3,\pm}}{k_3 - \dots}}}} \tag{5.117}$$

with k_i (m) being even (odd) integers and $\alpha_{i,\pm} = \pm 1$.

This indeed explains many of the states (plateaus) observed in the FQHE regime, but fails to explain its hierarchy, i.e., states of higher hierarchy in this model appear at larger temperature than states of lower hierarchy.

The correct hierarchy can instead be obtained within the so-called composite fermion description. Therefore, one realizes that Laughlin's wave function for $\nu = 1/3$ can be rewritten as $\prod_j \prod_{i<j}(z_i - z_j)^{m-1} \cdot \Psi_1^{LL}(z_1, z_2, \dots, z_{Ne})$ with Ψ_1^{LL} being the favorable wave function for $\nu = 1$ (Laughlin's wave function for $m = 1$ or Slater determinant of the single particle wave functions of the lowest Landau level). Formally, one takes the many-particle wave function of the lowest Landau level, that is insulating (Section 5.3.2), and adds two fluxes of h/e to each electron for $m = 3$. The system at $\nu = 1/3$ is ergo a system of electrons with two attached fluxes h/e, that fill exactly one Landau level. This corresponds to an effective $\vec{B}$ field B_{eff} for the new quasiparticles, consisting of one electron and two fluxes of h/e, that is exactly 1/3 of the real $\vec{B}$ field or $B_{eff} = B - B_{1/2}$ with $B_{1/2}$ being the B field at $\nu = 1/2$. Thus, the two fluxes of h/e, caused by the concerted motion of the other electrons around a particular electron, compensate an external magnetic field of $B_{1/2}$. One can easily show that the attachment of a flux of $2 \cdot n \cdot h/e$ ($n \in \mathbb{N}$) to an electron does not change the fermionic statistics [176]. Intuitively, it is still a fermion since the exchange of two coordinates in the additional factor $\prod_j \prod_{i<j}(z_i - z_j)^{2n}$ reveals $+1$, such that the exchange statistics is not changed. A system with electron density of one electron per flux $2h/e$ is, thus, a system of novel fermions, consisting of two fluxes h/e and one electron, with $B_{eff} = 0\,\mathrm{T}$. These new fermions are called composite fermions. At a real external $\vec{B}$ field of x fluxes of h/e ($x \in \mathbb{R}$) per electron the composite fermions are prone to $x - 2n$ fluxes of h/e per composite fermion. For $\nu = 1/3$ and $n = 1$, one gets one such flux per composite fermion and, thus, a completely filled lowest Landau level.

Formally, the addition of flux to the electrons is performed via the so called Chern–Simons gauge transformation. More intuitively, electrons at $\nu = 1/2$ encircle each other in a way that the external field is perfectly compensated for each electron. Applying an electric field then moves the composite fermions similar to free electrons, i.e., they behave like Bloch waves. However, the whole encircling environment has to be moved with the electron in order to keep the effective $\vec{B}$ field at zero. This leads to a larger inertial mass for the composite fermions than for the bare electrons. Indeed, the mass m^*_{CF} depends on the density of the electrons at $\nu = 1/2$ according to $m^*_{\mathrm{CF}} \propto 1/l_B \propto \sqrt{n_e}$, respectively, the Fermi circle of the composite fermions exhibits a Fermi wave vector of $k_{\mathrm{F}}^{\mathrm{CF}} \propto 1/l_B$. These composite fermions have been found by experiments at $\nu = 1/2$. For example, applying a very small magnetic field to the composite fermions leads to a movement of the composite fermions in relatively large circles (diameter: $\sim 1\,\mu$m) as expected from the Lorentz force [177].

The number of composite fermions N_{CF} is fixed by the electron number at the field $B_{1/2}$ for filling factor $\nu = 1/2$ according to $N_{\mathrm{CF}} = 1/2 \cdot \frac{eB_{1/2}}{h} A$. Here, A is the sample area and eB/h is the degeneracy of a spin polarized Landau level. By deviating the B field from $B_{1/2}$, the composite fermions effectively feel the field $B_{\mathrm{eff}} = |B - B_{1/2}|$ implying an effective number $N_{\widetilde{\Phi}}(B_{\mathrm{eff}})$ of flux quanta $\widetilde{\Phi}$. Consequently, the composite fermions form Landau levels and show quantum Hall plateaus around effective integer filling factors $\widetilde{\nu} = N_{\mathrm{CF}}/N_{\widetilde{\Phi}}(B_{\mathrm{eff}}) \in \mathbb{N}$ (eq. (5.105)).

Next, we want to deduce the corresponding real filling factors $\nu = \frac{n_e}{eB/h}$ (eq. (5.105)). We employ $N_{\widetilde{\Phi}} = \frac{B_{\mathrm{eff}}A}{h/e}$ and $N_{\mathrm{CF}} = 1/2 \cdot \frac{eB_{1/2}}{h} A$. This leads to

$$\widetilde{\nu} = \frac{N_{\mathrm{CF}}}{N_{\widetilde{\Phi}}(B_{\mathrm{eff}})} = \frac{eB_{1/2}}{2h} \cdot \frac{h}{eB_{\mathrm{eff}}} = \frac{B_{1/2}}{2 \cdot |B - B_{1/2}|} \tag{5.118}$$

$$\Rightarrow \quad |B - B_{1/2}| = \frac{B_{1/2}}{2\widetilde{\nu}} \tag{5.119}$$

$$\Rightarrow \quad B = B_{1/2} \cdot \frac{2\widetilde{\nu} \pm 1}{2\widetilde{\nu}} \,. \tag{5.120}$$

Since $n_e = \frac{1}{2}\frac{B_{1/2}}{h/e} = \nu \frac{B}{h/e}$ (Section 5.3.2), one gets $\frac{B}{B_{1/2}} = \frac{1}{2\nu}$ leading to

$$\nu = \frac{\widetilde{\nu}}{2\widetilde{\nu} \pm 1} \,. \tag{5.121}$$

Obviously, the state at $\nu = 1/3$ corresponds to the filling factor $\widetilde{\nu} = 1$ with + sign for the composite fermions, such that the initial idea of the derivation naturally reappears. Moreover, the apparent symmetry of the resistance minima around $\nu = 1/2$ (Fig. 5.30(a)) and the fact that the minima get deeper at increasing B distance from $\nu = 1/2$ can be naturally explained. The former is due to the symmetry between negative and positive B_{eff} for the composite fermions and the latter is due to the increased absolute value of B_{eff} increasing the energy distance between adjacent Landau levels of the composite fermions.

Such fractional states, explained by Landau levels of composite fermions, can also be observed in higher Landau levels (see, e.g., the minima at $\nu = 4/3$ and $\nu = 5/3$ in Fig. 5.30(a)), where the completely filled spin polarized Landau level is just a polarizable background (Section 5.5.2.2). The argument also applies for other even numbers of fluxes h/e attached to the electrons as, e.g., four such fluxes per electron, revealing quasi-particles without effective $\vec{B}$ field at $\nu = 1/4$.

The system of composite fermions shows, in addition, collective behaviors known from normal electrons, such as ferromagnetism with domain patterns or superconductivity driven by Cooper pairs of the composite fermions. This provides the advantage that the collective effects can be switched on and off by either changing n_e via a gate voltage or by changing $\vec{B}$.[86]

The predictive power of the composite fermion model, thus, appears to be enormous, albeit it is based only on a well-guessed wave function (eq. (5.109)). The FQHE is another excellent example showing that the electron-electron interaction can provide completely unexpected behavior as, e.g., the fractional charge. Since there is no way to solve the corresponding Schrödinger equation by ab-initio methods, the results have not been predicted prior to their experimental observation. Moreover, they needed to be rationalized by a mixture of electrodynamic and quantum mechanical arguments prior to the development of adequate models, that support the arguments quantitatively.

Another prominent example of this kind is the description of interacting electrons in 1D without $\vec{B}$, that is called the Luttinger liquid description. Here, the low-energy excitations are either collective spin density waves or collective charge density waves. These two types of excitations exhibit different velocities, such that inducing an electron into the system at one side leads to a different velocity of its charge degree of freedom and its spin degree of freedom. Hence, the charge degree of freedom arrives earlier on the other side of the system than its spin counterpart. However, describing this behavior more formally by so-called bosonization of a model Hamiltonian is beyond the scope of this book and can be found, e.g., in [178–180].

5.6 Summary

The last two chapters of the book have described several electronic properties, that would not exist within the well established description of the electrons by Bloch waves. They arise because of additional terms in the many-particle Schrödinger equation describing electron-phonon, electron-disorder, spin-orbit, or electron-electron interaction. The resulting electron phases such as charge density waves, integer quantum Hall phases, Mott–Hubbard insulators, or fractional quantum Hall phases do not drop immediately from the Schrödinger equation, which is too complex to be solved rigidly. Often, such phases have been discovered by chance and are described theoretically much later. The most prominent example is superconductivity, discovered in 1911 and explained in 1957, albeit the required Schrödinger equation has been known

86 Conventional superconductivity can also be switched off by a $\vec{B}$ field.

since the late 1920s. The high temperature superconductivity, discovered in 1986, remains unexplained up to now, i.e., even the question if electron-phonon interaction, electron-electron interaction or both dominate the pairing of the Cooper pairs is not clarified unambiguously. The Kondo effect, briefly described in Section 5.5.1.2, was discovered in 1934 and required 30 years before being explained by a second order spin dependent electron-electron interaction. Of course, also the opposite way of a theoretical prediction, which was difficult to be found in experiments, appears. For example, the concept of a Luttinger liquid was introduced in 1963, while first experimental hints of its existence were found only in the 1990s, that are still not unambiguous.

Importantly, the rigorous understanding of individual effects can have far reaching consequences. Most strikingly, the mathematical understanding of the exact quantization of the quantum Hall effect in terms of topological indices opened the whole new field of topology in solid state physics. This field continues to predict novel physical properties of well studied materials by theoretical analysis. Interestingly, these properties have been mostly overlooked in previous studies or partly have been found without any reasonable explanation. This tells us that adequate analysis of the Schrödinger equation might still bear novel possibilities for the exploitation of electrons, albeit it has been tackled for about a century.

In other words, albeit the describing equation for electron systems in solids is known exactly since nearly 100 years, the investigation of such systems still provides groundbreaking results due to the impossibility of a rigid solution of the Schrödinger equation, neither analytically nor numerically. It is easy to imagine that combinations of the different interaction terms might lead to additional insights in the near future.

Bibliography

[1] T. Ihn. *Semiconductor Nanostructures*. Oxford University Press, Oxford, 2010.

[2] H. Ibach and H. Lüth. *Solid-State Physics: An Introduction to Principles of Materials Science, 4th edition*. Springer, Berlin, 2009.

[3] H. Lüth. *Surfaces and Interfaces of Solid Materials*. Springer, Berlin, 1995.

[4] M. Katsnelson. *Graphene - Carbon in Two Dimensions*. Cambridge Univ. Press, Cambridge, 2012.

[5] B. J. van Wees, H. van Houten, C. W. J. Beenakker, J. G. Williamson, L. P. Kouwenhoven, D. van der Marel, and C. T. Foxon. Quantized conductance of point contacts in a two-dimensional electron gas. *Phys. Rev. Lett.*, 60(9):848–850, 1988.

[6] C. W Groth, M. Wimmer, A. R Akhmerov, and X. Waintal. Kwant: a software package for quantum transport. *New J. Phys.*, 16:063065, 2014.

[7] J. Dauber, M. Oellers, F. Venn, A. Epping, K. Watanabe, T. Taniguchi, F. Hassler, and C. Stampfer. Aharonov-Bohm oscillations and magnetic focusing in ballistic graphene rings. *Phys. Rev. B*, 96(20):205407, 2017.

[8] M. Huefner, F. Molitor, A. Jacobsen, A. Pioda, C. Stampfer, K. Ensslin, and T. Ihn. The Aharonov–Bohm effect in a side-gated graphene ring. *New J. Phys.*, 12(4):043054, 2010.

[9] R. A. Webb, S. Washburn, C. P. Umbach, and R. B. Laibowitz. Observation of the Aharonov-Bohm Oscillations in Normal-Metal Rings. *Phys. Rev. Lett.*, 54(25):2696–2699, 1985.

[10] D. Y. Sharvin and Y. V. Sharvin. Magnetic-flux quantization in a cylindrical film of a normal metal . *JETP Lett.*, 34(5):272–275, 1981.

[11] G. Bergmann. Weak localization in thin films. *Phys. Rep.*, 107(1):1–58, 1984.

[12] G. D. Mahan. *Many-Particle Physics, 3rd edition*. Kluver Acadmic, New York, 2000.

[13] R. Mansfield, S. Abboudy, and P. Fozooni. Hopping conduction in n-type indium phosphide. *Phil. Mag. B*, 57:777–789, 1988.

[14] E. Abrahams, P. W. Anderson, D. C. Licciardello, and T. V. Ramakrishnan. Scaling Theory of Localization: Absence of Quantum Diffusion in Two Dimensions. *Phys. Rev. Lett.*, 42:673–676, 1979.

[15] M. Fox. *Optical Properties of Solids, 2nd edition*. Oxford Master Series in Physics. Oxford University Press, Oxford, 2013 (Reprint).

[16] J. R. Chelikowsky and M. L. Cohen. Nonlocal pseudopotential calculations for the electronic structure of eleven diamond and zinc-blende semiconductors. *Phys. Rev. B*, 14(2):556–582, 1976.

[17] M. L. Cohen and J. R. Chelikowsky. *Electronic Structure and Optical Properties of Semiconductors*. Springer, Berlin Heidelberg, 1988.

[18] E.D. Palik. *Handbook of the optical constants of solids*. Academic Press, San Diego, 1985.

[19] M. Fox. *Optical Properties of Solids, 1st edition*. Oxford Master Series in Physics. Oxford University Press, Oxford, 2007 (Reprint).

[20] G. W. Fehrenbach, W. Schäfer, and R. G. Ulbrich. Excitonic versus plasma screening in highly excited gallium arsenide. *J. Lumin.*, 30(1-4):154–161, 1985.

[21] M. D. Sturge. Optical Absorption of Gallium Arsenide between 0.6 and 2.75 eV. *Phys. Rev.*, 127(3):768–773, 1962.

[22] H. Ehrenreich, H. R. Philipp, and B. Segall. Optical Properties of Aluminum. *Phys. Rev.*, 132(5):1918–1928, 1963.

[23] B. Segall. Energy Bands of Aluminum. *Phys. Rev.*, 124(6):1797–1806, 1961.

[24] C. K. N. Patel and R. E. Slusher. Light Scattering by Plasmons and Landau Levels of Electron Gas in InAs. *Phys. Rev.*, 167:413–415, 1968.

https://doi.org/10.1515/9783110438321-006

[25] C. Sönnichsen, T. Franzl, T. Wilk, G. von Plessen, and J. Feldmann. Plasmon resonances in large noble-metal clusters. *New J. Phys.*, 4:93, 2002.

[26] John D. Trimmer. The Present Situation in Quantum Mechanics: A Translation of Schrödinger's "Cat Paradox" Paper. *Proceedings of the American Philosphical Society*, 124:323–338, 1980. [E. Schrödinger, Die gegenwärtige Situation in der Quantenmechanik. Naturwissenschaften 23(48) (1935): 807–812].

[27] M. A. Nielsen and I. L. Chuang. *Quantum Computation and Quantum Information*. Cambridge University Press, Cambridge, 2000.

[28] R. P. Feynman. Simulating physics with computers. *Int. J. Theor. Phys.*, 21(6-7):467–488, 1982.

[29] S. Wildermuth, S. Hofferberth, I. Lesanovsky, E. Haller, L. M. Andersson, S. Groth, I. Bar-Joseph, P. Krüger, and J. Schmiedmayer. Microscopic magnetic-field imaging. *Nature*, 435:440, 2005.

[30] D. P. DiVincenzo. The Physical Implementation of Quantum Computation. *Fortschr. Phys.*, 48(9-11):771–783, 2000.

[31] W. K. Wootters and W. H. Zurek. A single quantum cannot be cloned. *Nature*, 299:802–803, 1982.

[32] R. Blatt and D. Wineland. Entangled states of trapped atomic ions. *Nature*, 453(7198):1008–1015, 2008.

[33] L. DiCarlo, J. M. Chow, J. M. Gambetta, L. S. Bishop, B. R. Johnson, D. I. Schuster, J. Majer, A. Blais, L. Frunzio, S. M. Girvin, and R. J. Schoelkopf. Demonstration of two-qubit algorithms with a superconducting quantum processor. *Nature*, 460(7252):240–244, 2009.

[34] I. Chiorescu, Y. Nakamura, C. J. P. M. Harmans, and J. E. Mooij. Coherent Quantum Dynamics of a Superconducting Flux Qubit. *Science*, 299(5614):1869–1871, 2003.

[35] J. Zhang, G. Pagano, P. W. Hess, A. Kyprianidis, P. B. Becker, H. Kaplan, A. V. Gorshkov, Z. X. Gong, and C. Monroe. Observation of a many-body dynamical phase transition with a 53-qubit quantum simulator. *Nature*, 551(7682):601–604, 2017.

[36] B. Lekitsch, S. Weidt, A. G. Fowler, K. Molmer, S. J. Devitt, C. Wunderlich, and W. K. Hensinger. Blueprint for a microwave trapped ion quantum computer. *Sci. Adv.*, 3(2):e1601540, 2017.

[37] L. P. Kouwenhoven, T. H. Oosterkamp, M. W. S. Danoesastro, M. Eto, D. G. Austing, T. Honda, and S. Tarucha. Excitation Spectra of Circular, Few-Electron Quantum Dots. *Science*, 278(5344):1788–1792, 1997.

[38] J. M. Elzerman, R. Hanson, L. H. W. van Beveren, B. Witkamp, L. M. K. Vandersypen, and L. P. Kouwenhoven. Single-shot read-out of an individual electron spin in a quantum dot. *Nature*, 430:431–435, 2004.

[39] S. Amasha, K. MacLean, I. P. Radu, D. M. Zumbühl, M. A. Kastner, M. P. Hanson, and A. C. Gossard. Electrical control of spin relaxation in a quantum dot. *Phys. Rev. Lett.*, 100(4):046803, 2008.

[40] C. B. Simmons, J. R. Prance, B. J. Van Bael, T. S. Koh, Z. Shi, D. E. Savage, M. G. Lagally, R. Joynt, M. Friesen, S. N. Coppersmith, and M. A. Eriksson. Tunable Spin Loading and T_1 of a Silicon Spin Qubit Measured by Single-Shot Readout. *Phys. Rev. Lett.*, 106(15):156804, 2011.

[41] A. J. Leggett. Testing the limits of quantum mechanics: motivation, state of play, prospects. *J. Phys.: Condens. Mat.*, 14(15):R415–R451, 2002.

[42] B. D. Josephson. Supercurrents through barriers. *Adv. Phys.*, 14(56):419–451, 1965.

[43] J. M. Martinis. Superconducting phase qubits. *Quantum Inf. Process.*, 8(2–3):81–103, 2009.

[44] J. E. Mooij. Josephson Persistent-Current Qubit. *Science*, 285(5430):1036–1039, 1999.

[45] B. L. T. Plourde, T. L. Robertson, P. A. Reichardt, T. Hime, S. Linzen, C.-E. Wu, and J. Clarke. Flux qubits and readout device with two independent flux lines. *Phys. Rev. B*, 72(6):060506, 2005.

[46] K. W. Lehnert, K. Bladh, L. F. Spietz, D. Gunnarsson, D. I. Schuster, P. Delsing, and R. J. Schoelkopf. Measurement of the Excited-State Lifetime of a Microelectronic Circuit. *Phys. Rev. Lett.*, 90(2):027002, 2003.
[47] F. H. L. Koppens, C. Buizert, K. J. Tielrooij, I. T. Vink, K. C. Nowack, T. Meunier, L. P. Kouwenhoven, and L. M. K. Vandersypen. Driven coherent oscillations of a single electron spin in a quantum dot. *Nature*, 442:766–771, 2006.
[48] M. Veldhorst, J. C. C. Hwang, C. H. Yang, A. W. Leenstra, B. de Ronde, J. P. Dehollain, J. T. Muhonen, F. E. Hudson, K. M. Itoh, A. Morello, and A. S. Dzurak. An addressable quantum dot qubit with fault-tolerant control-fidelity. *Nat. Nanotechnol.*, 9(12):981–985, 2014.
[49] A. A. Clerk, M. H. Devoret, S. M. Girvin, F. Marquardt, and R. J. Schoelkopf. Introduction to quantum noise, measurement, and amplification. *Rev. Mod. Phys.*, 82(2):1155–1208, 2010.
[50] E. A. Chekhovich, M. N. Makhonin, A. I. Tartakovskii, A. Yacoby, H. Bluhm, K. C. Nowack, and L. M. K. Vandersypen. Nuclear spin effects in semiconductor quantum dots. *Nat. Mater.*, 12(6):494–504, 2013.
[51] M. Devoret, B. Huard, R. Schoelkopf, and L. F. Cugliandolo. *Quantum Machines: Measurement and Control of Engineered Quantum Systems*. Ecole De Physique Des Houches. Oxford University Press, Oxford, 2011.
[52] A. Abragam. *The principles of Nuclear Magnetism*. Oxford University Press, Oxford, London, 1961.
[53] A. V. Khaetskii, D. Loss, and L. Glazman. Electron Spin Decoherence in Quantum Dots due to Interaction with Nuclei. *Phys. Rev. Lett.*, 88(18):186802, 2002.
[54] I. A. Merkulov, A. L. Efros, and M. Rosen. Electron spin relaxation by nuclei in semiconductor quantum dots. *Phys. Rev. B*, 65:205309, Apr 2002.
[55] J. R. Petta, A. C. Johnson, J. M. Taylor, E. A. Laird, A. Yacoby, M. D. Lukin, C. M. Marcus, M. P. Hanson, and A. C. Gossard. Coherent manipulation of coupled electron spins in semiconductor quantum dots. *Science*, 309(5744):2180–2184, 2005.
[56] F. H. L. Koppens, K. C. Nowack, and L. M. K. Vandersypen. Spin echo of a single spin in a quantum dot. *Phys. Rev. Lett.*, 100:236802, 2008.
[57] H. Bluhm, S. Foletti, I. Neder, M. Rudner, D. Mahalu, V. Umansky, and A. Yacoby. Dephasing time of GaAs electron-spin qubits coupled to a nuclear bath exceeding 200 μs. *Nat. Phys.*, 7(2):109–113, 2010.
[58] W. M. Witzel and S. Das Sarma. Concatenated dynamical decoupling in a solid-state spin bath. *Phys. Rev. B*, 76(24):241303, 2007.
[59] W. Yao, R.-B. Liu, and L. J. Sham. Restoring coherence lost to a slow interacting mesoscopic spin bath. *Phys. Rev. Lett.*, 98:077602, 2007.
[60] H. Bluhm, S. Foletti, D. Mahalu, V. Umansky, and A. Yacoby. Enhancing the coherence of a spin qubit by operating it as a feedback loop that controls its nuclear spin bath. *Phys. Rev. Lett.*, 105(21):216803, 2010.
[61] I. T. Vink, K. C. Nowack, F. H. L. Koppens, J. Danon, Y. V. Nazarov, and L. M. K. Vandersypen. Locking electron spins into magnetic resonance by electron–nuclear feedback. *Nat. Phys.*, 5(10):764–768, 2009.
[62] K. C. Nowack, F. H. L. Koppens, Y. V. Nazarov, and L. M. K. Vandersypen. Coherent Control of a Single Electron Spin with Electric Fields. *Science*, 318:1430–1433, 2007.
[63] E. A. Laird, C. Barthel, E. I. Rashba, C. M. Marcus, M. P. Hanson, and A. C. Gossard. Hyperfine-Mediated Gate-Driven Electron Spin Resonance. *Phys. Rev. Lett.*, 99(24):246601, 2007.
[64] D. J. Reilly, C. M. Marcus, M .P. Hanson, and A. C. Gossard. Fast single-charge sensing with a rf quantum point contact. *App. Phys. Lett.*, 91:162101, 2007.
[65] C. Barthel, D. J. Reilly, C. M. Marcus, M. P. Hanson, and A. C. Gossard. Rapid single-shot measurement of a singlet-triplet qubit. *Phys. Rev. Lett.*, 103(16):160503, 2009.

[66] M. D. Shulman, S. P. Harvey, J. M. Nichol, S. D. Bartlett, A. C. Doherty, V. Umansky, and A. Yacoby. Suppressing qubit dephasing using real-time Hamiltonian estimation. *Nat. Commun.*, 5:5156, 2014.

[67] J. R. Petta, J. M. Taylor, A. C. Johnson, A. Yacoby, M. D. Lukin, C. M. Marcus, M. P. Hanson, and A. C. Gossard. Dynamic Nuclear Polarization with Single Electron Spins. *Phys. Rev. Lett.*, 100:067601, Feb 2008.

[68] D. J. Reilly, J. M. Taylor, E. A. Laird, J. R. Petta, C. M. Marcus, M. P. Hanson, and A. C. Gossard. Measurement of Temporal Correlations of the Overhauser Field in a Double Quantum Dot. *Phys. Rev. Lett.*, 101(23):236803, 2008.

[69] D. J. Reilly, J. M. Taylor, J. R. Petta, C. M. Marcus, M. P. Hanson, and A. C. Gossard. Exchange control of nuclear spin diffusion in a double quantum dot. *Phys. Rev. Lett.*, 104(23):236802, 2010.

[70] I. Neder, M. Rudner, H. Bluhm, S. Foletti, B. Halperin, and A. Yacoby. Semi classical model for dephasing of a two-electron spin qubit coupled to a coherently evolving nuclear spin bath. *Phys. Rev. B*, 84:035441, 2011.

[71] W. M. Witzel and S. Das Sarma. Quantum theory for electron spin decoherence induced by nuclear spin dynamics in semiconductor quantum computer architecture. *Phys. Rev. B.*, 74:035322, 2006.

[72] W. Yao, R.-B. Liu, and L. J. Sham. Theory of electron spin decoherence by interacting nuclear spins in a quantum dot. *Phys. Rev. B*, 74:195301, 2006.

[73] Ł. Cywiński, W. M. Witzel, and S. Das Sarma. Pure quantum dephasing of a solid-state electron spin qubit in a large nuclear spin bath coupled by long-range hyperfine-mediated interactions. *Phys. Rev. B*, 79:245314, Jun 2009.

[74] Ł. Cywiński, W. M. Witzel, and S. Das Sarma. Electron Spin Dephasing due to Hyperfine Interactions with a Nuclear Spin Bath. *Phys. Rev. Lett.*, 102:057601, Feb 2009.

[75] R. Schirhagl, K. Chang, M. Loretz, and C. L. Degen. Nitrogen-Vacancy Centers in Diamond: Nanoscale Sensors for Physics and Biology. *Annu. Rev. Phys. Chem.*, 65:83–105, 2014.

[76] R. B. Laughlin and D. Pines. The Theory of Everything. *Proc. Natl. Acad. Sci. U.S.A.*, 97(1):28–31, 2000.

[77] T. Vicsek. Complexity: The bigger picture. *Nature*, 418(6894):131, 2002.

[78] J. G. Bednorz and K. A. Müller. Possible high T_c superconductivity in the Ba–La-Cu–O system. *Z. Phys. B Con. Mat.*, 64(2):189–193, 1986.

[79] A. M. Haghiri-Gosnet and J. P. Renard. CMR manganites: physics, thin films and devices. *J. Phys. D: Appl. Phys.*, 36(8):R127–R150, 2003.

[80] G. Binasch, P. Grünberg, F. Saurenbach, and W. Zinn. Enhanced magnetoresistance in layered magnetic structures with antiferromagnetic interlayer exchange. *Phys. Rev. B*, 39(7):4828–4830, 1989.

[81] M.-H. Phan and S.-C. Yu. Review of the magnetocaloric effect in manganite materials. *J. Magn. Magn. Mater.*, 308(2):325–340, 2007.

[82] M. Fiebig. Revival of the magnetoelectric effect. *J. Phys. D: Appl. Phys.*, 38(8):R123–R152, 2005.

[83] E. J. W. Verwey. Electronic Conduction of Magnetite (Fe_3O_4) and its Transition Point at Low Temperatures. *Nature*, 144(3642):327–328, 1939.

[84] G. D. Barrera, J. A. O. Bruno, T. H. K. Barron, and N. L. Allan. Negative thermal expansion. *J. Phys. Condens. Mat.*, 17(4):R217–R252, 2005.

[85] E. Dagotto. Complexity in Strongly Correlated Electronic Systems. *Science*, 309(5732):257–262, 2005.

[86] N.W. Ashcroft and N.D. Mermin. *Solid State Physics*. College ed. Fort Worth: Hartcourt College Publ., 2000.

[87] K. Held, O. K. Andersen, M. Feldbacher, A. Yamasaki, and Y.-F. Yang. Bandstructure meets many-body theory: the LDA +DMFT method. *J. Phys. Condens. Mat.*, 20(6):064202, 2008.
[88] N. F. Mott. Metal-Insulator Transition. *Rev. Mod. Phys.*, 40(4):677–683, 1968.
[89] S. Blügel, T. Brückel, and C. M. Schneider, editors. *Magnetism goes nano : electron correlations, spin transport, molecular magnetism; Lecture manuscripts of the 36th IFF Spring School.* Schriften des Forschungszentrums. Forschungszentrum Jülich, Zentralbibliothek, Jülich, 2005.
[90] P. W. Anderson. Theory of Magnetic Exchange Interactions: Exchange in Insulators and Semiconductors. *Solid State Phys.*, 14:99–214, 1963.
[91] P. Fazekas. *Lecture Notes on Electron Correlation and Magnetism.* Series in Modern Condensed Matter Physics 5. World Scientific, Singapore, New Jersey, London, Hong Kong, 1999.
[92] K. I. Kugel and D. I. Khomskii. The Jahn-Teller effect and magnetism: transition metal compounds. *Sov. Phys. Usp.*, 25(4):231–256, 1982.
[93] T. Chatterji, editor. *Colossal Magnetoresistive Manganites.* Kluwer Academic Publ., Dordrecht, 2004.
[94] A. Urushibara, Y. Moritomo, T. Arima, A. Asamitsu, G. Kido, and Y. Tokura. Insulator-metal transition and giant magnetoresistance in $La_{1-x}Sr_xMnO_3$. *Phys. Rev. B*, 51(20):14103–14109, 1995.
[95] A. J. Millis, P. B. Littlewood, and B. I. Shraiman. Double Exchange Alone Does Not Explain the Resistivity of $La_{1-x}Sr_xMnO_3$. *Phys. Rev. Lett.*, 74(25):5144–5147, 1995.
[96] G. H. Rao, K. Bärner, and I. D. Brown. Bond-valence analysis on the structural effects in magnetoresistive manganese perovskites. *J. Phys. Condens. Mat.*, 10(48):L757–L763, 1998.
[97] D. H. Templeton, L. K. Templeton, J. C. Phillips, and K. O. Hodgson. Anomalous scattering of X-rays by cesium and cobalt measured with synchrotron radiation. *Acta Crystallogr. A*, 36(3):436–442, 1980.
[98] Y. Murakami, J. P. Hill, D. Gibbs, M. Blume, I. Koyama, M. Tanaka, H. Kawata, T. Arima, Y. Tokura, K. Hirota, and Y. Endoh. Resonant X-Ray Scattering from Orbital Ordering in $LaMnO_3$. *Phys. Rev. Lett.*, 81(3):582–585, 1998.
[99] J. Stöhr and H. C. Siegmann. *Magnetism: From Fundamentals to Nanoscale Dynamics.* Springer Series in Solid State Science. Springer, Berlin, Heidelberg, New York, 2006.
[100] T. Brückel, R. Waser, S. Blügel, and C. M. Schneider. *Electronic Oxides: Correlation Phenomena, Exotic Phases and Novel Functionalities; Lecture manuscripts of the 41st IFF Spring School.* Schriften des Forschungszentrums Jülich. Forschungszentrum Jülich, Zentralbibliothek Jülich, 2010.
[101] J. Klijn, L. Sacharow, C. Meyer, S. Blügel, M. Morgenstern, and R. Wiesendanger. STM measurements on the InAs(110) surface directly compared with surface electronic structure calculations. *Phys. Rev. B*, 68(20):205327, 2003.
[102] C. Wittneven, R. Dombrowski, M. Morgenstern, and R. Wiesendanger. Scattering States of Ionized Dopants Probed by Low Temperature Scanning Tunneling Spectroscopy. *Phys. Rev. Lett.*, 81(25):5616–5619, 1998.
[103] T. Schäpers. *Semiconductor Spintronics.* De Gruyter Textbook. De Gruyter, Berlin, 2016.
[104] J. M. Ziman. *Principles of the Theory of Solids.* Cambridge University Press, Cambridge, 1972.
[105] S. Kagoshima, H. Nagasawa, and T. Sambongi. *One-Dimensional Conductors.* Springer, Berlin Heidelberg, 1988.
[106] T. Straub, T. Finteis, R. Claessen, P. Steiner, S. Hüfner, P. Blaha, C. S. Oglesby, and E. Bucher. Charge-Density-Wave Mechanism in 2H-$NbSe_2$: Photoemission Results. *Phys. Rev. Lett.*, 82(22):4504–4507, 1999.
[107] J. A. Silva-Guillén, P. Ordejón, F. Guinea, and E. Canadell. Electronic structure of 2H-$NbSe_2$ single-layers in the CDW state. *2D Mater.*, 3(3):035028, 2016.

[108] G. K. Shenoy, B. D. Dunlap, and F. Y. Fradin, editors. *Proceedings of Conference on Ternary Superconductors, Lake Geneva 1980*. Elsevier, North Holland, Amsterdam, 1981.

[109] J. Billy, V. Josse, Z. Zuo, A. Bernard, B. Hambrecht, P. Lugan, D. Clément, L. Sanchez-Palencia, P. Bouyer, and A. Aspect. Direct observation of Anderson localization of matter waves in a controlled disorder. *Nature*, 453(7197):891–894, 2008.

[110] F. Evers, A. Mildenberger, and A. D. Mirlin. Multifractality of wavefunctions at the quantum Hall transition revisited. *Phys. Rev. B*, 64:241303, 2001.

[111] P. P. Edwards and M. J. Sienko. Universality aspects of the metal-nonmetal transition in condensed media. *Phys. Rev. B*, 17(6):2575–2581, 1978.

[112] K. Hashimoto, C. Sohrmann, J. Wiebe, T. Inaoka, F. Meier, Y. Hirayama, R. A. Römer, R. Wiesendanger, and M. Morgenstern. Quantum Hall Transition in Real Space: From Localized to Extended States. *Phys. Rev. Lett.*, 101(25):256802, 2008.

[113] U. Klass, W. Dietsche, K. von Klitzing, and K. Ploog. Image of the dissipation in gated quantum Hall effect samples. *Surf. Sci.*, 263(1-3):97–99, 1992.

[114] E. Brown. Bloch Electrons in a Uniform Magnetic Field. *Phys. Rev.*, 133(4A):A1038–A1044, 1964.

[115] M. Kohmoto. Topological invariant and the quantization of the Hall conductance. *Ann. Phys.*, 160(2):343–354, 1985.

[116] A. P. Schnyder, S. Ryu, A. Furusaki, and A. W. W. Ludwig. Classification of topological insulators and superconductors in three spatial dimensions. *Phys. Rev. B*, 78(19):195125, 2008.

[117] M. C. Geisler, J. H. Smet, V. Umansky, K. von Klitzing, B. Naundorf, R. Ketzmerick, and H. Schweizer. Detection of a Landau Band-Coupling-Induced Rearrangement of the Hofstadter Butterfly. *Phys. Rev. Lett.*, 92(25):256801, 2004.

[118] C. R. Dean, L. Wang, P. Maher, C. Forsythe, F. Ghahari, Y. Gao, J. Katoch, M. Ishigami, P. Moon, M. Koshino, T. Taniguchi, K. Watanabe, K. L. Shepard, J. Hone, and P. Kim. Hofstadter's butterfly and the fractal quantum Hall effect in moiré superlattices. *Nature*, 497(7451):598–602, 2013.

[119] L. A. Ponomarenko, R. V. Gorbachev, G. L. Yu, D. C. Elias, R. Jalil, A. A. Patel, A. Mishchenko, A. S. Mayorov, C. R. Woods, J. R. Wallbank, M. Mucha-Kruczynski, B. A. Piot, M. Potemski, I. V. Grigorieva, K. S. Novoselov, F. Guinea, V. I. Fal'ko, and A. K. Geim. Cloning of Dirac fermions in graphene superlattices. *Nature*, 497(7451):594–597, 2013.

[120] D. R. Hofstadter. Energy levels and wave functions of Bloch electrons in rational and irrational magnetic fields. *Phys. Rev. B*, 14(6):2239–2249, 1976.

[121] D. Pfannkuche and R. R. Gerhardts. Theory of magnetotransport in two-dimensional electron systems subjected to weak two-dimensional superlattice potentials. *Phys. Rev. B*, 46(19):12606–12626, 1992.

[122] D. P. Arovas, R. N. Bhatt, F. D. M. Haldane, P. B. Littlewood, and R. Rammal. Localization, wave-function topology, and the integer quantized Hall effect. *Phys. Rev. Lett.*, 60(7):619–622, 1988.

[123] G. M. Graf and M. Porta. Bulk-Edge Correspondence for Two-Dimensional Topological Insulators. *Com. Math. Phys.*, 324(3):851–895, 2013.

[124] B. A. Bernevig, T. L. Hughes, and S.-C. Zhang. Quantum Spin Hall Effect and Topological Phase Transition in HgTe Quantum Wells. *Science*, 314(5806):1757–1761, 2006.

[125] M. König, S. Wiedmann, C. Brüne, A. Roth, H. Buhmann, L. W. Molenkamp, X.-L. Qi, and S.-C. Zhang. Quantum Spin Hall Insulator State in HgTe Quantum Wells. *Science*, 318(5851):766–770, 2007.

[126] K. C. Nowack, E. M. Spanton, M. Baenninger, M. König, J. R. Kirtley, B. Kalisky, C. Ames, P. Leubner, Chr. Brüne, H. Buhmann, L. W. Molenkamp, D. Goldhaber-Gordon, and K. A.

Moler. Imaging currents in HgTe quantum wells in the quantum spin Hall regime. *Nat. Mater.*, 12(9):787–791, 2013.

[127] X.-L. Qi, Y.-S. Wu, and S.-C. Zhang. Topological quantization of the spin Hall effect in two-dimensional paramagnetic semiconductors. *Phys. Rev. B*, 74(8):085308, 2006.

[128] X.-L. Qi and S. C. Zhang. Topological insulators and superconductors. *Rev. Mod. Phys.*, 83(4):1057–1110, 2011.

[129] C. Brüne, A. Roth, H. Buhmann, E. M. Hankiewicz, L. W. Molenkamp, J. Maciejko, X.-L. Qi, and S. C. Zhang. Spin polarization of the quantum spin Hall edge states. *Nat. Phys.*, 8(6):485–490, 2012.

[130] W. Buckel and R. Kleiner. *Superconductivity: Fundamentals and Applications*. Wiley VCH, Weinheim, 2004.

[131] C.-Z. Chang, J. Zhang, X. Feng, J. Shen, Z. Zhang, M. Guo, K. Li, Y. Ou, P. Wei, L.-L. Wang, Z.-Q. Ji, Y. Feng, S. Ji, X. Chen, J. Jia, X. Dai, Z. Fang, S.-C. Zhang, K. He, Y. Wang, L. Lu, X.-C. Ma, and Q.-K. Xue. Experimental Observation of the Quantum Anomalous Hall Effect in a Magnetic Topological Insulator. *Science*, 340(6129):167–170, 2013.

[132] S. Oh. The Complete Quantum Hall Trio. *Science*, 340(6129):153–154, 2013.

[133] L. Fu and C. L. Kane. Time reversal polarization and a Z_2 adiabatic spin pump. *Phys. Rev. B*, 74(19):195312, 2006.

[134] M. Z. Hasan and C. L. Kane. Colloquium: Topological insulators. *Rev. Mod. Phys.*, 82(4):3045–3067, 2010.

[135] L. Fu, C. L. Kane, and E. J. Mele. Topological Insulators in Three Dimensions. *Phys. Rev. Lett.*, 98(10):106803, 2007.

[136] M. Lohse, C. Schweizer, O. Zilberberg, M. Aidelsburger, and I. Bloch. A Thouless quantum pump with ultracold bosonic atoms in an optical superlattice. *Nat. Phys.*, 12(4):350–354, 2016.

[137] D. Hsieh, D. Qian, L. Wray, Y. Xia, Y. S. Hor, R. J. Cava, and M. Z. Hasan. A topological Dirac insulator in a quantum spin Hall phase. *Nature*, 452(7190):970–974, 2008.

[138] S. Hüfner. *Photoelectron Spectroscopy: Principles and Applications*. Springer, Berlin Heidelberg, 2003.

[139] D. Hsieh, Y. Xia, D. Qian, L. Wray, J. H. Dil, F. Meier, J. Osterwalder, L. Patthey, J. G. Checkelsky, N. P. Ong, A. V. Fedorov, H. Lin, A. Bansil, D. Grauer, Y. S. Hor, R. J. Cava, and M. Z. Hasan. A tunable topological insulator in the spin helical Dirac transport regime. *Nature*, 460(7259):1101–1105, 2009.

[140] Y. L. Chen, J.-H. Chu, J. G. Analytis, Z. K. Liu, K. Igarashi, H.-H. Kuo, X. L. Qi, S. K. Mo, R. G. Moore, D. H. Lu, M. Hashimoto, T. Sasagawa, S. C. Zhang, I. R. Fisher, Z. Hussain, and Z. X. Shen. Massive Dirac Fermion on the Surface of a Magnetically Doped Topological Insulator. *Science*, 329(5992):659–662, 2010.

[141] C. Jozwiak, C.-H. Park, K. Gotlieb, C. Hwang, D.-H. Lee, S. G. Louie, J. D. Denlinger, C. R. Rotundu, R. J. Birgeneau, Z. Hussain, and A. Lanzara. Photoelectron spin-flipping and texture manipulation in a topological insulator. *Nat. Phys.*, 9(5):293–298, 2013.

[142] S.-Y. Xu, M. Neupane, C. Liu, D. Zhang, A. Richardella, L. A. Wray, N. Alidoust, M. Leandersson, T. Balasubramanian, J. Sánchez-Barriga, O. Rader, G. Landolt, B. Slomski, J. H. Dil, J. Osterwalder, T.-R. Chang, H.-T. Jeng, H. Lin, A. Bansil, N. Samarth, and M. Z. Hasan. Hedgehog spin texture and Berry's phase tuning in a magnetic topological insulator. *Nat. Phys.*, 8(8):616–622, 2012.

[143] Y. Zhang, K. He, C.-Z. Chang, C.-L. Song, L.-L. Wang, X. Chen, J.-F. Jia, Z. Fang, X. Dai, W.-Y. Shan, S.-Q. Shen, Q. Niu, X.-L. Qi, S.-C. Zhang, X.-C. Ma, and Q.-K. Xue. Crossover of the three-dimensional topological insulator Bi_2Se_3 to the two-dimensional limit. *Nat. Phys.*, 6(8):584–588, 2010.

[144] S.-Y. Xu, Y. Xia, L. A. Wray, S. Jia, F. Meier, J. H. Dil, J. Osterwalder, B. Slomski, A. Bansil, H. Lin, R. J. Cava, and M. Z. Hasan. Topological Phase Transition and Texture Inversion in a Tunable Topological Insulator. *Science*, 332(6029):560–564, 2011.
[145] P. Dziawa, B. J. Kowalski, K. Dybko, R. Buczko, A. Szczerbakow, M. Szot, E. Łusakowska, T. Balasubramanian, B. M. Wojek, M. H. Berntsen, O. Tjernberg, and T. Story. Topological crystalline insulator states in $Pb_{1-x}Sn_xSe$. *Nat. Mater.*, 11(12):1023–1027, 2012.
[146] R. Wiesendanger. *Scanning Probe Microscopy and Spectroscopy*. Cambridge University Press, Cambridge, 1994.
[147] P. Roushan, J. Seo, C. V. Parker, Y. S. Hor, D. Hsieh, D. Qian, A. Richardella, M. Z. Hasan, R. J. Cava, and A. Yazdani. Topological surface states protected from backscattering by chiral spin texture. *Nature*, 460(7259):1106–1109, 2009.
[148] T. Zhang, P. Cheng, X. Chen, J.-F. Jia, X. Ma, K. He, L. Wang, H. Zhang, X. Dai, Z. Fang, X. Xie, and Q.-K. Xue. Experimental Demonstration of Topological Surface States Protected by Time-Reversal Symmetry. *Phys. Rev. Lett.*, 103(26):266803, 2009.
[149] T. Hanaguri, K. Igarashi, M. Kawamura, H. Takagi, and T. Sasagawa. Momentum-resolved Landau-level spectroscopy of Dirac surface state in Bi_2Se_3. *Phys. Rev. B*, 82(8):081305, 2010.
[150] D. Kim, S. Cho, N. P. Butch, P. Syers, K. Kirshenbaum, S. Adam, J. Paglione, and M. S. Fuhrer. Surface conduction of topological Dirac electrons in bulk insulating Bi_2Se_3. *Nat. Phys.*, 8(6):459–463, 2012.
[151] R. Yoshimi, A. Tsukazaki, Y. Kozuka, J. Falson, K.S. Takahashi, J.G. Checkelsky, N. Nagaosa, M. Kawasaki, and Y. Tokura. Quantum Hall effect on top and bottom surface states of topological insulator $(Bi_{1-x}Sb_x)_2Te_3$ films. *Nat. Commun.*, 6(1):6627, 2015.
[152] Y. Shiomi, K. Nomura, Y. Kajiwara, K. Eto, M. Novak, K. Segawa, Y. Ando, and E. Saitoh. Spin-Electricity Conversion Induced by Spin Injection into Topological Insulators. *Phys. Rev. Lett.*, 113(19):196601, 2014.
[153] B. Rasche, A. Isaeva, M. Ruck, S. Borisenko, V. Zabolotnyy, B. Büchner, K. Koepernik, C. Ortix, M. Richter, and J. van den Brink. Stacked topological insulator built from bismuth-based graphene sheet analogues. *Nat. Mater.*, 12(5):422–425, 2013.
[154] C. Pauly, B. Rasche, K. Koepernik, M. Liebmann, M. Pratzer, M. Richter, J. Kellner, M. Eschbach, B. Kaufmann, L. Plucinski, C. M. Schneider, M. Ruck, J. van den Brink, and M. Morgenstern. Subnanometer-wide electron channels protected by topology. *Nat. Phys.*, 11(4):338–343, 2015.
[155] X.-L. Qi, T. L. Hughes, and S.-C. Zhang. Topological field theory of time-reversal invariant insulators. *Phys. Rev. B*, 78(19):195424, 2008.
[156] X.-L. Qi, R. Li, J. Zang, and S.-C. Zhang. Inducing a Magnetic Monopole with Topological Surface States. *Science*, 323(5918):1184–1187, 2009.
[157] A. R. Akhmerov, J. Nilsson, and C. W. J. Beenakker. Electrically Detected Interferometry of Majorana Fermions in a Topological Insulator. *Phys. Rev. Lett.*, 102(21):216404, 2009.
[158] J. Alicea. New directions in the pursuit of Majorana fermions in solid state systems. *Rep. Prog. Phys.*, 75(7):076501, 2012.
[159] V. Mourik, K. Zuo, S. M. Frolov, S. R. Plissard, E. P. A. M. Bakkers, and L. P. Kouwenhoven. Signatures of Majorana Fermions in Hybrid Superconductor-Semiconductor Nanowire Devices. *Science*, 336(6084):1003–1007, 2012.
[160] L. P. Kouwenhoven, D. G. Austing, and S. Tarucha. Few-electron quantum dots. *Rep. Prog. Phys.*, 64(6):701–736, 2001.
[161] S. Tarucha, D. G. Austing, T. Honda, R. J. van der Hage, and L. P. Kouwenhoven. Shell Filling and Spin Effects in a Few Electron Quantum Dot. *Phys. Rev. Lett.*, 77(17):3613–3616, 1996.
[162] M. Pustilnik and L. Glazman. Kondo effect in quantum dots. *J. Phys.: Condens. Mat.*, 16(16):R513–R537, 2004.

[163] F. M. Peeters. Two-dimensional Wigner crystal of electrons on a helium film: Static and dynamical properties. *Phys. Rev. B*, 30(1):159–165, 1984.
[164] G. Deville, A. Valdes, E. Y. Andrei, and F. I. B. Williams. Propagation of Shear in a Two-Dimensional Electron Solid. *Phys. Rev. Lett.*, 53(6):588–591, 1984.
[165] Y. P. Monarkha and V. E. Syvokon. A two-dimensional Wigner crystal (Review Article). *Low Temp. Phys.*, 38(12):1067–1095, 2012.
[166] R. L. Willett, H. L. Stormer, D. C. Tsui, L. N. Pfeiffer, K. W. West, and K. W. Baldwin. Termination of the series of fractional quantum Hall states at small filling factors. *Phys. Rev. B*, 38(11):7881–7884, 1988.
[167] V. V. Deshpande and M. Bockrath. The one-dimensional Wigner crystal in carbon nanotubes. *Nat. Phys.*, 4(4):314–318, 2008.
[168] M. M. Fogler, A. A. Koulakov, and B. I. Shklovskii. Ground state of a two-dimensional electron liquid in a weak magnetic field. *Phys. Rev. B*, 54(3):1853–1871, 1996.
[169] M. P. Lilly, K. B. Cooper, J. P. Eisenstein, L. N. Pfeiffer, and K. W. West. Evidence for an Anisotropic State of Two-Dimensional Electrons in High Landau Levels. *Phys. Rev. Lett.*, 82(2):394–397, 1999.
[170] I. L. Aleiner and L. I. Glazman. Two-dimensional electron liquid in a weak magnetic field. *Phys. Rev. B*, 52(15):11296–11312, 1995.
[171] J. P. Eisenstein and H. L. Stormer. The Fractional Quantum Hall Effect. *Science*, 248(4962):1510–1516, 1990.
[172] R. E. Prange and S. M. Girvin, editors. *The Quantum Hall Effect*. Springer, New York, 1990.
[173] Wiley VCH, *Phys. Bl.*, 54(12): title page, 1998.
[174] L. Saminadayar, D. C. Glattli, Y. Jin, and B. Etienne. Observation of the e/3 Fractionally Charged Laughlin Quasiparticle. *Phys. Rev. Lett.*, 79(13):2526–2529, 1997.
[175] F. Wilczek. Quantum Mechanics of Fractional-Spin Particles. *Phys. Rev. Lett.*, 49(14):957–959, 1982.
[176] A. Lopez and E. Fradkin. Fractional quantum Hall effect and Chern-Simons gauge theories. *Phys. Rev. B*, 44(10):5246–5262, 1991.
[177] V. J. Goldman, B. Su, and J. K. Jain. Detection of composite fermions by magnetic focusing. *Phys. Rev. Lett.*, 72(13):2065–2068, 1994.
[178] K. Schönhammer and V. Meden. Fermion–boson transmutation and comparison of statistical ensembles in one dimension. *Am. J. Phys.*, 64(9):1168–1176, 1996.
[179] K. Schönhammer. Interacting fermions in one dimension: The Tomonaga-Luttinger model. eprint. *arXiv:cond-mat/9710330*, 1997.
[180] K. Schönhammer. Luttinger Liquids: The Basic Concepts. eprint. *arXiv:cond-mat/0305035*, 2003.

List of Figures

https://doi.org/10.1515/9783110438321-007

List of Symbols

i, j, k, p, n, m, N	Integer numbers
$\vec{e}_x, \vec{e}_y, \vec{e}_z,$	Unit vectors in principal directions
x, y, z	Coordinates in real space
$z = x + iy$	Complex coordinate in 2D
$\vec{a}, \vec{b}, \vec{c},$	Basis vectors of crystal unit cell
$\alpha, \beta, \theta, \varphi\ \phi$	Angles
∇	Gradient operator, Nabla operator
$\nabla_{\vec{k}}$	Gradient operator in $\vec{k}$-space
$\widehat{A}, \widehat{B}, \ldots$	Quantum mechanical operators
a	Atomic lattice constant
$A, \widetilde{A}$	Area
$\vec{A}$	Electromagnetic vector potential
$\vec{\widetilde{A}}$	Area vector
$\|a_1, a_2, \ldots, a_N\rangle$	N-qubit state
A_{21}	Einstein coefficient for spontaneous emission
a_{B}	Bohr radius (0.5 Å)
a_{B}^*	Bohr radius of donor atom
$\vec{A}_{\mathrm{Berry}}$	Berry connection
a_{c}	Ring area encircled by electron paths
A_{I}	Band parameter of two band model
$\widetilde{A}_j$	Hyperfine coupling constant
$A_{k_y}(k_x)$	Function similar to Berry connection used in topological analysis
a_{N}	Nanoparticle radius
A_p	Prefactor of semiclassical Green's function for path p
a_p	Encircled area of path p, sign describes path direction
a_x, a_y, a_z	Lattice constant in x, y, z direction
a_X	Exciton Bohr radius
$\vec{B}$	Magnetic B field, Magnetic induction
$B_{\perp}$	B field perpendicular to the 2D plane of two-dimensional electron system
$B_{1/2}$	B field at filling factor 1/2
B_{12}	Einstein coefficient for absorption
B_{21}	Einstein coefficient for stimulated emission
B_{ac}, B_1	Amplitude of oscillating magnetic field
B_{C}	Critical $\vec{B}$-Field
B_{eff}	Effective $\vec{B}$ field for composite fermions

https://doi.org/10.1515/9783110438321-008

B_{ext}	External $\vec{B}$ field
B_{I}	Band parameter of two band model
B_{nuc}	Overhauser / hyperfine field due to nuclear spins
$B_{\text{nuc}}^{\perp}$	Nuclear field strength perpendicular to external $\vec{B}$ field
$\hat{b}_{\vec{q},j}$	Annihilation operator for a phonon of branch j and wave vector $\vec{q}$
$\hat{b}_{\vec{q},j}^{+}$	Creation operator for a phonon of branch j and wave vector $\vec{q}$
c	Speed of light
C	Capacitance
$\tilde{C}$	Coulomb term of H_2 molecule
c_0	Speed of light in vacuum
$\hat{c}_{i,\sigma}$, $\hat{c}_{n,\vec{k}}$	Annihilation operator of electron (e.g., for site i and spin σ or band n and wave vector $\vec{k}$)
$\hat{c}_{i,\sigma}^{+}$, $\hat{c}_{n,\vec{k}}^{+}$	Creation operator of electron (e.g., for site i and spin σ or band n and wave vector $\vec{k}$)
$c_j(t)$	Time-dependent coefficient of wave function in superposition state
$c(t)$	Sign function for π pulse sequence as function of time
$C(t)$	Return probability density after time t
C_{Σ}	Total capacitance of a quantum dot
d	Dimensionality
D	Diffusion constant
$\vec{D}$	Dielectric displacement field
$\tilde{D}$	Effective spring constant between atoms
d_{c}	Cyclotron diameter
$\mathcal{D}_{\text{c}}(E)$	Density of states of conduction band
$\mathcal{D}_{d}(E)$	Density of states at dimension $d = 1, 2, 3$
$\mathcal{D}(E)$	Density of states
$\vec{d}(\vec{k})$	Vector describing two-band Hamiltonian in Pauli matrix basis
$\hat{\vec{d}}(\vec{k})$	Bloch vector describing pair of electron bands
D_p	Classical deflection factor for trajectories close to path p
$dPMM$	Phase mismatch line path
$\mathcal{D}_{\text{R}}^{(m)}(E)$	Density of states of right moving electrons of subband m
$\mathcal{D}_{\text{v}}(E)$	Density of states of valence band
e	Elementary charge
E	Energy
$\vec{E}$	Electric field
$E_{+/-}$	Energies of the two states of a qubit

$\tilde{\vec{E}}$	Electric field (in complex notation)
$\vec{E}_0$	Amplitude of oscillating electric field indicating polarization direction
E_c	Charging energy
E_C	Critical Energy, Mobility edge
E_{CBM}	Energy of conduction band minimum
E^{center}	Energy of center-of-mass motion of electron-hole pair
$\vec{E}_{critical}$	Critical electric field for field ionization of exciton
E_D	Dirac point energy
E_{dipole}	Paramagnetic energy shift in quantum dot
E_e	Electron energy
E_{ee}	Total quantum-mechanical electron-electron repulsion energy
E_{ee}^{cl}	Classical electron-electron repulsion energy
$E_{e\text{-}ion}$	Attraction energy between electron and ion in charge density wave
$E_{elastic}$	Elastic energy
E_{Ex}	Exchange energy of two electrons
E_F	Fermi energy
$E_{F,e}$	Quasi-Fermi energy of electrons
$E_{F,h}$	Quasi-Fermi energy of holes
e_g	Representation of crystalline field states with e_g symmetry
E_g, E_{Gap}	Band gap energy
E_h	Hole energy
$E_{Hartree}$	Bare Coulomb repulsion energy of two electrons, Hartree energy
$E_{i/f}$	Energy of initial/final state of an optical transition
E_J	Josephson energy
$\vec{e}_j$	Unit vector in vibration direction of phonon
E_{kin}	Kinetic Energy
E_n^{LL}	Energy of nth Landau level
E_{osc}	Oscillation energy of light field
E_{ph}	Phonon energy
E_{pin}	Pinning energy
E_{pot}	Potential energy
E^{rel}	Energy of relative electron-hole motion
$E_{Ryd,X}$	Excitonic binding energy
E_{Ryd}	Rydberg energy (13.6 eV)
E_S	Singlet energy
E_S	Band edge energy
E_T	Triplet energy
E_{Th}	Thouless energy
E_{therm}	Thermal energy ($\sim k_B T$)

$E_{x/y}$	Electric field in x/y direction
E_Z	Zeeman energy
$E_{Z,\,tot}$	Total Zeeman energy including contributions from the nuclear B field
E_γ	Photon energy
f	Frequency
F	Free energy
$\vec{F}$	Force vector
F_+, F_-	Quantum mechanical amplitudes of two time-reversed paths
$\vec{F}_C$	Restoring force acting on electrons in nanoparticle
$f(E, T)$	Fermi-Dirac distribution function at temperature T
$f_{e/h}(E, T)$	Quasi-stationary Fermi-Dirac distribution function of electrons/holes after optical excitation
$\tilde{f}_i$	Form factor of resonant x-ray transition from initial state i
f_{ij}^{Bose}	Function accounting for Bose-Einstein distribution of phonons
f_j	Spectral weight of jth optical transition
f_k	Fermi-Dirac distribution for a particular wave vector k
F_p^{nm}	Integrated weight of a path bundle connecting subbands n and m
f_{Rabi}	Rabi frequency
f_{res}	Resonance frequency
$F(\omega t)$	Filter function corresponding to a pulse sequence
g, g_e	Electron gyromagnetic factor
G	Conductance
$\vec{G}$	Reciprocal lattice vector
$\tilde{g}$	Dimensionless conductance, Thouless number
G_0	Conductance quantum $(2e^2/h)$
G_c	Contact conductance
$\tilde{g}_C$	Critical dimensionless conductance
G_{cl}	Classical conductance
G_{Hall}	Hall conductance
$g(\hbar\omega)$	Joint density of states for optical transitions
G_{qm}	Quantum mechanical conductance
G_S	Intrinsic conductance
g_{th}	Threshold of gain coefficient for lasing
$g(x)$	Spatial correlation function
$G(y_2, y_1, k_F)$	Retarded Green's function for path from y_1 to y_2 at wave vector k_F
$G^{SC}(y_2, y_1, k_F)$	Semiclassical Green's function for path from y_1 to y_2 at wave vector k_F
g_ω	Gain coefficient of a laser

$h, \hbar$	Planck's constant
$\vec{H}$	Magnetic H-field
$\widehat{H}$	Hamilton operator
$\widehat{H}_{\uparrow/\downarrow}$	Hamiltonian for only ↑ electrons/ only ↓ electrons
$\widehat{H}_0$	Hamilton operator without perturbations
$\widehat{H}_\mathrm{B}$	Hamilton operator of a bath
$h.c.$	Complex conjugate
$\widehat{H}_\mathrm{e,ph}$	Hamiltonian of electron-phonon interaction
$\widehat{H}_{jk}$	Perturbation operator for transition $\psi_j \to \psi_k$
$\widehat{H}_\mathrm{Q}$	Hamilton operator of a qubit
$\widehat{H}_\mathrm{QB}$	Interaction Hamilton operator between qubit and bath
I	Current
$I_{\uparrow/\downarrow}$	Spin current for ↑ electrons/ ↓ electrons
$\hat{\vec{I}}, \widehat{I}_z, \widehat{I}_\eta,$	Nuclear spin operators
I_B	Backscattered current
I_c	Critical current of a superconductor
I_e	Electron current ($= -I$)
I_el	Photoelectron intensity
I_in	Injection current of laser diode or LED
I_p	Intensity of light
$\overline{I}_\mathrm{p}$	Time-averaged intensity of light
I_pc	Photocurrent
I_QPC	Current through quantum point contact
I_th	Threshold current of a laser diode
I_y	Transversal current
$J, J(\epsilon)$	Exchange parameter (for Heisenberg model or between singlet and triplet states in qubit)
J, J_eff	Quantum number of total orbital momentum
$\vec{J}$	Total angular momentum
$\tilde{J}$	Exchange term of H_2 molecule
$\vec{j}$	Current density
J_AF	Antiferromagnetic exchange parameter
$J_\mathrm{Dipole,\,n}$	Dipolar coupling strength of nuclear spins
J_dir	Direct exchange parameter
J_H	Hund's exchange parameter
J_kin	Kinetic exchange parameter
$J_m(x)$	mth Bessel function
$j_{x,\uparrow/\downarrow}$	Spin current density for ↑ electrons/ ↓ electrons in x direction
$J_x^{(m)}$	Electron flux at position x
j_y	Transversal current density in 2D

k	Wave number
$\vec{K}$	Wave vector of center of mass motion
$\widehat{K}$	Operator for complex conjugation
$\vec{k}$	Wave vector of electron or electromagnetic wave
$\tilde{k}$	Wave number (in complex notation)
k_0	Wave number of light in vacuum
k_B	Boltzmann constant
$\vec{k}_d$	Drift wave vector of electrons
$\vec{k}_F$	Fermi wavevector with length $2\pi/\lambda_F$
$\vec{k}_{i/f}$	Wave vector of initial/final Bloch state for optical transition
$K_\beta(\Delta t)$	Correlation function of noise
$\vec{k}_\gamma$	Photon wave vector
l	Orbital quantum number
L, l	Lengths
$\vec{L}$	Orbital angular momentum
$\hat{\vec{L}}, \hat{L}_z$	Orbital momentum operators
L^*	Optimal hopping length
$\|L\rangle, \|R\rangle$	Eigenstates of a double well charge qubit
L_0	Characteristic length describing contact resistance
l_B	Magnetic length
l_{el}	Elastic mean free path of electrons
L_I	Inductance
l_{in}	Inelastic mean free path of electrons
L_J	Josephson inductance
l_m	Mean free path of electrons
L_M	Macroscopic system size
l_{OV}	Overlap integral of two wave functions
L_p	Length of path p
l_T	Thermal length describing the limit of phase coherent propagation of a thermally broadened electron wave packet
L_{tot}	Total orbital momentum
l_φ	Phase coherence length
m	Azimuthal angular quantum number, magnetic quantum number
M	Number of open modes for electron transfer
$\vec{M}$	Magnetization
$\widetilde{M}$	Optical matrix element
m_0	Electron rest mass
m_e	Free electron mass
m_e^*, m^*	Effective mass of electron

m_h^*	Effective mass of hole
M_I	Band inversion parameter for toy model with pair of electron bands
M_n	Mass of nucleus
$M^{j,n,n'}_{\vec{q}+\vec{G},\vec{k}}$, $M_{\vec{q},\vec{k}}$	Matrix element for electron-phonon coupling
n	(Real) refractive index
n	Principal quantum number, subband number
$\vec{n}$	Bloch vector
$\tilde{n}$	Complex refractive index
$\hat{n}$ ($\hat{n}_n$)	Occupation operator (at site n)
$\|n\rangle$	Quantum mechanical single-particle state
$\|n_0\rangle$, $\|m_0\rangle$	Unperturbed single-particle states
n_A	Number density of atoms
N_A	Number of atoms
N_CF	Number of composite fermions
n_Ch	Chern number
N_CP	Number of Cooper pairs
n_d	Dopant density
$n_\mathrm{d,c}$	Critical dopant density
n_e	Number density of electrons
N_e	Number of electrons
$n_\mathrm{e,h}$	Carrier density of electrons (e) or holes (h)
$N_\mathrm{e,h}$	Number of charged particles
n_h	Number density of holes
N_I	Number of nuclear spins
$\|n_\mathrm{L}\rangle$	nth logical qubit state
n_LL	Degeneracy density of Landau level
$\|nm\rangle$	Two-particle state
N_qp	Number of quasiparticles
N_uc	Number of unit cells
n_x	Carrier density moving in x-direction
N_γ	Number of photons
N_Φ	Number of magnetic flux quanta
$N_{\tilde{\Phi}}$	Number of effective magnetic flux quanta for composite fermions
p	Path
p	Pressure
P	Power
$\vec{P}$	Dielectric polarization
$P_{+/-}$	Probability of state at north pole/south pole of Bloch sphere

$P_{\uparrow/\downarrow}$	Probability of spin ↑/spin ↓ state
$P_{\leftrightarrow}(\ldots)$	Permutation function
$\widetilde{P}$	Dielectric polarization (in complex notation)
$\vec{p}$	Momentum vector
p_+, p_-	Two time-reversed paths
$P(a)$	Weighted probability distribution of path area a
$P(a, l_\varphi)$	Probability density for encircling the path area a phase coherently at given phase coherence length l_φ
$P_{\mathrm{cl}}^{(\mathrm{ret})}$	Classical return probability of electron path
$\vec{p}_E$	Electric dipole moment
$\overline{P}_{\mathrm{em}}$	Radiated light power integrated over one oscillation period
$\vec{p}_{E,\mathrm{res}}$	Dipole moment of resonantly excited electron
P_{error}	Error probability per gate operation
$\vec{p}_{\mathrm{F}}$	Fermi momentum
$\mathrm{Pf}\,(\boldsymbol{w})$	Pfaffian of matrix $\boldsymbol{w}$
$\overline{P}_i$	Time-averaged incident power
P_{ij}	Hopping probability between sites i and j
$P_{\lvert n\rangle}$	Probability to find an electron in state $\lvert n\rangle$
$\vec{\widetilde{P}}_{\mathrm{nonres}}$	Electric polarization from nonresonant charge oscillations (in complex notation)
P_{out}	Output power
$P_{\mathrm{qm}}^{(\mathrm{ret})}$	Quantum mechanical return probability of electron path
$\overline{P}_r$	Time-averaged reflected light power
$\vec{\widetilde{P}}_{\mathrm{res}}$	Electric polarization from resonant charge oscillations (in complex notation)
$P^{(\mathrm{ret})}$	Return probability of electron path
$p(\vec{r}, t)$	Probability density of electron at position $\vec{r}$ and time t
P_{scat}	Scattered light power
$P_{\mathrm{S/T}}$	Probability of singlet/triplet state
$\overline{P}_{\mathrm{t}}$	Average transmitted power
$P_x(k_y)$	Charge polarization in x direction for states at wave number k_y
$P_x^{\Delta}(k_y)$	Relative polarization of time-reversal partners in x direction for states at wave number k_y
$\vec{p}_\gamma$	Photon momentum
Q	Charge
$\vec{Q}$	Scattering vector
$\vec{q}$	Phonon wave vector
r	Radial coordinate
R	Electrical resistance

$\Re$	Real part of a complex number
$\vec{R}$	Center-of-mass coordinate of electron-hole pair
$\vec{r}$	Position vector, relative coordinate of electron-hole pair
$\tilde{R}$	Radius
$\overline{R}$	Average reflectivity
$\vec{R}_A$, $\vec{R}_B$, $\vec{R}_n$	Atomic coordinates
r_c	Cyclotron radius
$\overline{r}_{ee}$	Average distance between two electrons
R_{Hall}	Hall resistance
r_{ij}	Distance between sites i and j
R_{ij}	Bond length between neighboring atoms i and j
$R_{ij,nm}$	Resistance measured with current running from i to j and voltage probed between n and m
R_l	Reflectivity of light
r_n	Radius of wave function with quantum number n
r_{nm}	Reflection matrix element from mode n to mode m
R_n	Reflection probability at interface n
$\vec{R}_{n,\tilde{B}}$	Translation vector of magnetic unit cell
R_{nm}	Reflection probability from mode n to mode m
r'_{nm}	Reflection amplitude from mode n to mode m
R_{ph}	Responsivity of photodiode
$\overline{R}_{qm}$	Average quantum mechanical reflection probability
R_x	Lattice position in x direction
R_{xx}	Longitudinal resistance
R_y	Reflectivity of light
s	Average number of electrons per atom
S	Entropy
S	Singlet state
$\hat{\vec{S}}$, $\hat{S}_x$, $\hat{S}_y$, $\hat{S}_z$	Spin operators
S_I	Current noise
s_{ij}	Valence of bond between atom i and atom j
S_p	Classical action of path p
$S_\beta(\omega)$, $S_V(\omega)$	Noise spectral density
t	Hopping matrix element
t	Thickness
t	Time
T	Temperature
$\overline{T}$	Average transmission coefficient
T_0, T_+, T_-	Triplet states

T_1	Energy relaxation time
T_2	Dephasing time due to fast dynamic fluctuations with respect to a single measurement run, decoherence time
T_2^*	Dephasing time due to slow dynamic fluctuations with respect to a single measurement run
t_{2g}	Representation of crystalline field states with t_{2g} symmetry
t_c	Tunnel coupling between two quantum dots
T_C	Transition temperature, Critical temperature
$\overline{T}_{cl}$	Average classical transmission probability
T_{Curie}	Curie temperature
T_e	Electron temperature
T_{ion}	Tolerance factor accounting for misfit between ionic radii
T_{JT}	Temperature of Jahn–Teller transition
T_{melt}	Melting temperature
T_n	Transmission probability at interface n
T_{Neel}	Neel temperature
$t_{n,m}$	Matrix element for hopping between sites n and m
t_{nm}	Transmission matrix element from mode n to mode m
T_{nm}	Transmission probability from mode n to mode m
t'_{nm}	Transmission amplitude from mode n to mode m
t_{noise}	Correlation time of noise
T_{noise}	Noise temperature
$\overline{T}_{qm}$	Average quantum mechanical transmission probability
$\hat{T}(\vec{R})$	Operator for translation by a vector $\vec{R}$
$\mathrm{Tr}\,(\boldsymbol{w})$	Trace of matrix $\boldsymbol{w}$
T_{SD}	Time constant of electron spin dephasing due to nuclear spin diffusion
T_V	Verwey transition temperature
T_{VRH}	Temperature scale for variable range hopping
t_w	Waiting time for relaxation measurement
T_γ	Transmissivity of light
T_ω	Time period of oscillation
U (U_{3s})	On-site Coulomb repulsion energy (for two electrons in a 3s state)
U_{AB}	On-site Coulomb repulsion energy for two electrons in different states A and B of the same atom
$u_{i/f}(\vec{r})$	Lattice-periodic factor in initial/final Bloch state of optical transition
u_{ijkl}	Exchange/correlation term
U_{int}	Internal energy

$u_{n\vec{k}}(\vec{x})$, $u_{\vec{k}}(\vec{r})$	Lattice periodic part of Bloch function
u_q	Wave vector dependent displacement amplitude of the nuclei from its equilibrium position
$\widehat{U}(t)$	Time evolution operator
$u(\omega)$	Spectral energy density of light at frequency ω
$\vec{v}$	Velocity
V	Voltage
V	Volume
$\vec{v}_0$	Initial velocity
v_a	Velocity of acoustic phonon
V_{bi}	Built-in voltage of photodiode
$\vec{v}_d$	Drift velocity
V_{ext}	External voltage
$\vec{v}_F$	Fermi velocity
V_{Gate}, V_G	Gate voltage
V_{Hall}	Hall voltage
V_i	Valence of cation i
V_p	Plunger gate voltage
V_{QD}	Electrostatic potential of quantum dot
V_{SD}	Source-drain voltage
v_{th}	Thermal velocity
V_{thr}	Threshold voltage
V_{uc}	Volume of one unit cell
$\vec{v}_x$ ($\vec{v}_y$)	Velocity in x (y) direction
V_{xx}	Longitudinal voltage
W	Geometric width
W	Band width
$\widetilde{W}$	Energetic width of potential disorder
$W_{i\to f}$, $W_{nk\to n'k'}$	Transition rates between the states given in the subscript
$\mathbf{w}(\vec{k})$	Matrix containing phase factors of all occupied states at wave vector $\vec{k}$
$\vec{x}$	Position vector, displacement vector of electron
$\vec{x}_0$	Amplitude of electron oscillation
Z	Atomic number
z_{NN}	Number of nearest neighbors of an atom
$z(t)$	Path function in the complex plane

α	Absorption coefficient
α_{loss}	Attenuation coefficient of laser light
$\beta(\tilde{g})$	Scaling function
β_n	Average value of noise in nth time interval
$\beta(t)$	Time dependent noise function
γ	Damping rate
Γ	Tunnel rate
Γ_i	ith point in Brillouin zone with time reversal invariant momentum (TRIM)
$\Gamma(x)$	Gamma function
δ	Phase of time dependent Rabi field
δE	Energy level spacing
ΔE_{C}	Difference between confinement energy levels in quantum dot
δG_{qm}	Quantum mechanical correction of conductance
δG_{R}	Conductance correction due to reflection
δ_i	Phase factor index at TRIM i
δ_{ij}	Kronecker delta
δ_L	Energy distance between states separated by a geometric distance L
$\delta P_{\mathrm{qm}}^{(\mathrm{ret})}$	Quantum mechanical correction to the return probability of an electron
δ_{skin}	Skin depth of electromagnetic field, Penetration depth
$\delta(x)$	Delta function
Δ	Generalized tunnel coupling between the two states of a qubit
Δ_1	Amplitude of time dependent Rabi field
$\Delta a_{p,p'}$	Difference of encircled area between paths p and p'
Δ_{CF}	Crystal field splitting parameter
ΔE_{ij}	Energy distance between states i and j
$\Delta E_{N_{\mathrm{e}}}$	Addition energy for the N_{e}th electron in a quantum dot
$\Delta L_{p,p'}$	Length difference between paths p and p'
δ_ζ	Energy separation between overlapping states
$\delta\rho_{\mathrm{c},q}$	Amplitude of the charge density wave
$\delta\rho_{\mathrm{c}}(x)$	Electron density in 1D charge density wave
$\delta\rho_{\mathrm{qm}}$	Quantum mechanical correction to the resistivity
$\delta\sigma_{\mathrm{qm}}$	Quantum mechanical correction to conductivity
ε	Dielectric constant of material
$\tilde{\varepsilon}$	Infinitesimal small number

ε_0	Electrical field constant in vacuum
$\varepsilon_\mathrm{m}(\omega)$	Dielectric function of metal
$\varepsilon(q)$	Wave vector dependence of dielectric function
$\varepsilon_\mathrm{u}(\omega)$	Dielectric function of material surrounding a nanoparticle
$\varepsilon(\omega)$	Dielectric function
ϵ	Generalized detuning energy between the two states of a qubit
ϵ^*	Detuning energy, where $S(0,2)$ and $T_+(1,1)$ cross
$\epsilon_{\vec{k}}$, ϵ_at	Single particle energies
ϵ_s	Strain
ζ	Spin-orbit parameter
ζ_c	Correlation length
η	Viscosity
η_L	Fraction of injected electrons which emit laser photons
η_q	Quantum efficiency
η_R	Radiative quantum efficiency
θ	Polar angle
$\widehat{\widehat{\Theta}}$	Time reversal operator
Θ_D	Debye temperature
$\Theta_{k_x n}$	Possible arbitrary phase distinguishing time reversal partner wave functions of band n at wave number k_x
$\Theta_{n\vec{k}}(\vec{x})$	Phase of lattice periodic part of single particle wave function
κ	Extinction coefficient of light
$\kappa(q)$	Polarizability function
λ	Wavelength
λ_0	Wavelength of light in vacuum
λ_F	Fermi wave length
λ_I	Coupling constant for two-qubit gate
λ_n	nth eigenvalue of an observable
λ_screen	Screening length
μ	Carrier mobility
$\widetilde{\mu}$	Reduced mass of pair of particles
μ_0	Vacuum permeability
μ_B	Bohr's magneton
μ_D	Chemical potential of drain electrode
$\vec{\mu}_{jk}$	Dipole matrix element between states ψ_j and ψ_k

μ_{M}	Permeability of a material
μ_{N}	Nuclear magneton
$\mu_{\text{QD},N_{\text{e}}}$	Chemical potential of quantum dot with N_{e} electrons
μ_{S}	Chemical potential of source electrode
μ_X	Electrochemical potential of object X
ν	Filling factor of spin-polarized Landau level
$\widetilde{\nu}$	Effective filling factor for composite fermions
ν_0, ν_n, ν_k	Topological indices
ν_{C}	Critical exponent at phase transition
ν_p	Maslov index of path p
ξ	Localization length
ρ	Resistivity
$\widehat{\rho}$	Density matrix
$\boldsymbol{\rho}$	Resistivity tensor
ρ_{c}	Density of free charges
ρ_{xx}	Longitudinal resistivity
ρ_{xy}	Transversal resistivity
σ	Conductivity
$\boldsymbol{\sigma}$	Conductivity tensor
$\vec{\boldsymbol{\sigma}}$	Vector of Pauli matrices
$\widetilde{\sigma}(n)$	Permutation operator
$\vert\sigma\rangle$	Spin eigenstate $\vert\downarrow\rangle$ or $\vert\uparrow\rangle$
$\sigma_{B_{\text{nuc}}}$	Standard deviation of nuclear field distribution
σ_{cl}	Classical conductivity, Drude conductivity
σ_{qm}	Quantum mechanical conductivity
$\widehat{\sigma}_x$, $\widehat{\sigma}_y$, $\widehat{\sigma}_z$	Pauli or $\widehat{\sigma}$-matrices or operators (see p. 204)
$\boldsymbol{\sigma}_x$, $\boldsymbol{\sigma}_y$, $\boldsymbol{\sigma}_z$	Pauli matrices
σ_{xx}	Longitudinal conductivity
σ_{xy}	Transversal conductivity
$\sigma_{xy\uparrow}$, $\sigma_{xy\downarrow}$	Spin polarized transversal conductivity
τ	Total evolution time in Hahn echo sequence
$\vec{\tau}$	Torque
τ_B	Magnetic time scale
$\tau_{\text{el-def}}$	Relaxation time of electron due to scattering at defects
$\tau_{\text{el-el}}$	Relaxation time of electron due to electron-electron scattering
$\tau_{\text{el-ph}}$	Relaxation time of electron due to scattering with phonons
τ_n	Time of the nth π pulse

τ_{NR}	Nonradiative lifetime
τ_{R}	Radiative lifetime
τ_{S}	Evolution time of singlet-triplet precession in double quantum dot
τ_{sc}	Scattering time
τ_{Th}	Thouless time scale
τ_{φ}	Phase coherence time
φ, ϕ	Phase of electron wave function, superconducting wave function, light field, complex number
φ	Azimuthal angle
φ_{AB}	Aharonov-Bohm phase
φ_{Berry}	Berry phase
Φ_{Coul}	Coulomb potential
$\widetilde{\Phi}^{\mathrm{eI}}_{\vec{q}}$	Fourier component of the ionic potential to wave vector $\vec{q}$
$\Phi^{\mathrm{eI}}(\vec{x})$	Ionic potential
Φ_{el} ($\Phi_{n,\mathrm{el}}$)	Electrostatic potential (at atom site n)
$\Phi^{j}_{\mathrm{electrons}}(\vec{x})$	Effective electrostatic potential of all other electrons for electron j
$\widetilde{\Phi}_q$	Amplitude of the electrostatic potential of the frozen phonon in a charge density wave
$\Phi_{\mathrm{xc}}(n_{\mathrm{e}})$	Exchange correlation functional for local density approximation
Φ	Magnetic flux
Φ_0	Magnetic flux quantum ($h/2e$)
χ, $\chi(\omega)$	Susceptibility
$\chi_{\mathrm{nonres}}(\omega)$	Susceptibility from non-resonant excitations
$\chi_{\mathrm{res}}(\omega)$	Susceptibility from resonant excitations
$\psi(\vec{x})$	Single-particle wave function
$\widetilde{\psi}(\vec{x}, t)$	Time-dependent single particle wave function
$\lvert\psi\rangle$	Quantum state (ket notation)
$\psi_{\mathrm{c}}(\vec{x})$	Confined electron wave function
$\lvert\psi_{\mathrm{E}}\rangle$	Entangled state
$\psi_{\mathrm{i/f}}(\vec{r})$	Wave function of initial/final state of optical transition
$\psi_{\mathrm{p}}(\vec{x})$	Plane-wave type electron wave function
Ψ	Many-particle wave function
Ψ_{el}	Many-particle wave function of electrons
$\Psi_{0,\mathrm{el}}$	Unperturbed many-particle wave function of electrons
Ψ^{LL}_{n}	Many-particle wave function of a completely filled nth Landau level
$\Psi(\vec{r}_e, \vec{r}_h)$	Two-particle wavefunction of electron-hole pair

$\Psi_\Gamma(x)$	Digamma function
ω	Angular frequency
ω_0	Eigenfrequency
ω_c	Cyclotron frequency
ω_p	Plasma frequency
ω_ph	Phonon frequency
ω_Rabi	Rabi frequency

Index

https://doi.org/10.1515/9783110438321-009

Also of Interest

Collective Effects in Condensed Matter Physics Applications in Biology and Medicine
Vladimir V. Kiselev, 2018
ISBN 978-3-11-058509-4, e-ISBN 978-3-11-058618-3

Crystal Growth of Intermetallics
Peter Gille, Yuri Grin (Eds.), 2018
ISBN 978-3-11-049584-3, e-ISBN 978-3-11-049678-9

Space – Time – Matter
Jochen Brüning, Matthias Staudacher (Eds.), 2018
ISBN 978-3-11-045135-1, e-ISBN 978-3-11-045215-0

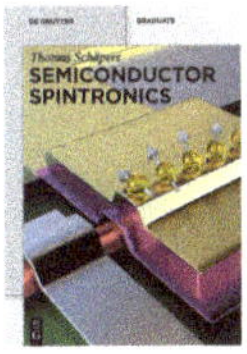

Semiconductor Spintronics
Thomas Schäpers, 2016
ISBN 978-3-11-036167-4, e-ISBN 978-3-11-042544-4

www.ingramcontent.com/pod-product-compliance
Lightning Source LLC
Chambersburg PA
CBHW080658220326
41598CB00033B/5245
* 9 7 8 3 1 1 0 4 3 8 3 1 4 *